Camp Colt
to
Desert Storm

Camp Colt
to
Desert Storm

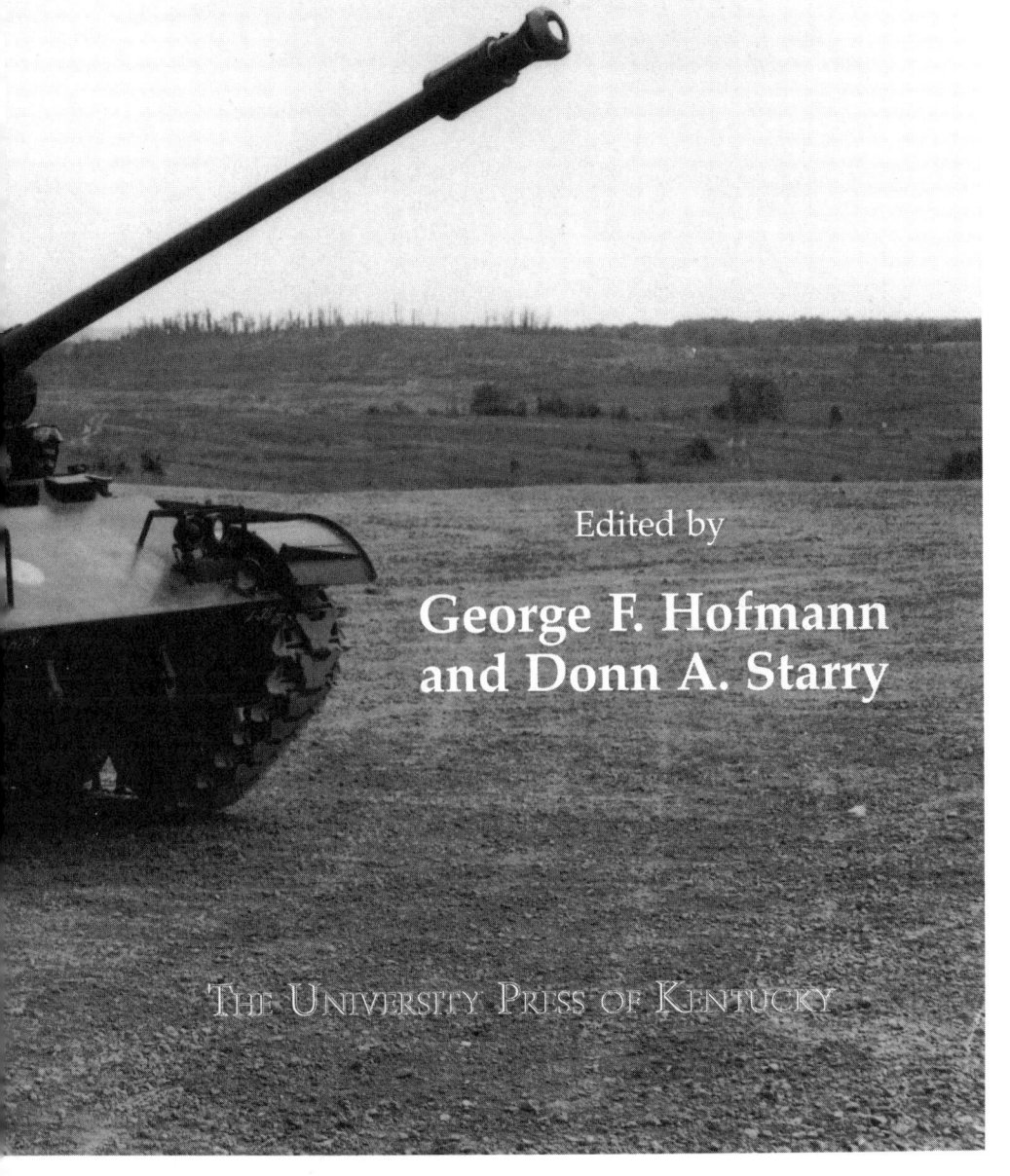

The History of
U.S. Armored Forces

Edited by

George F. Hofmann
and Donn A. Starry

THE UNIVERSITY PRESS OF KENTUCKY

Copyright © 1999 by The University Press of Kentucky

Scholarly publisher for the Commonwealth,
serving Bellarmine College, Berea College, Centre
College of Kentucky, Eastern Kentucky University,
The Filson Club Historical Society, Georgetown College,
Kentucky Historical Society, Kentucky State University,
Morehead State University, Murray State University,
Northern Kentucky University, Transylvania University,
University of Kentucky, University of Louisville,
and Western Kentucky University.
All rights reserved.

Editorial and Sales Offices: The University Press of Kentucky
663 South Limestone Street, Lexington, Kentucky 40508-4008

03 02 01 00 99   5 4 3 2 1

Library of Congress Cataloging-in-Publication Data

Camp Colt to Desert Storm : the history of U.S. armored forces
    / edited by George F. Hofmann and Donn A. Starry.
        p.   cm.
    Includes bibliographical references and index.
    ISBN 0-8131-2130-2 (alk. paper)
    1. Mechanization, Military—United States—History.
    2. United States—Armed Forces—Armored troops—History.
    I. Hofmann, George F. II. Starry, Donn A. (Donn Albert), 1931-
UE160.C36   1999
358'.18'09730904—dc21   99-28924

This book is printed on acid-free recycled paper
meeting the requirements of the American National Standard
for Permanence of Paper for Printed Library Materials.

Manufactured in the United States of America

And the Lord was with Judah;
and he drove out the inhabitants of the mountain;
but could not drive out the inhabitants of the valley,
because they had chariots of iron.
—The Book of Judges 1:19

With great admiration, respect,
and considerable affection,
this book is dedicated to
the Chariots of Iron,
and the iron men
who take them
to battle.

Contents

Preface

The idea for a history of American armored forces began in 1976 at the U.S. Army Armor Conference held at Fort Knox, Kentucky. We briefly discussed the possibility; however, the timing then was inappropriate. Years later in Gettysburg, Pennsylvania, at the 1995 meeting of the Society for Military History we again raised the issue of an armor history, especially considering the remarkable success of U.S. military forces in Operation Desert Storm, the hundred-hour Persian Gulf War. Nowhere could we find a suitable one-volume American armor history covering the twentieth century experience from the first lumbering tanks used by the American Expeditionary Forces Tank Corps in World War I to the very impressive performance of Abrams tanks, Bradley fighting vehicles, and Marine Corps M60A1s that defeated the Iraqis in 1991. We decided then to assemble a talented group of military historians who specialized in armor history. To this group we added armor officers, both Army and Marines, who had made history with their considerable post–World War II contributions to the development of doctrine, equipment, organization, and training that eventually led to the exceptional battle achievements in the Gulf War.

This team was tasked to address various historical aspects in the long, turbulent story of U.S. armored force development. The combined effect of the team's contributions make this book the first to record in a single volume the significant events of American armor history. Our goal was to make this book unique. Each chapter was written to appeal to specialists as well as to the more casual military history reader.

To undertake such an ambitious project it was necessary to seek start-up funds. Almost immediately the 6th Armored Division Association generously provided a grant. It was their desire to perpetuate armor's history for future generations. These World War II 6th Armored Division veterans deserve special recognition for their commitment to the project, especially Edward Reed, the secretary-treasurer, and Forrest Herbert, a past president of the association. Unfortunately, due to the passage of time and its members' advancing age, the 6th Armored Division Association will cease to exist after its farewell banquet in Louisville on 16 September 2000. It is with deep regret that we watch this proud veterans' organization retire its colors. Nevertheless, through our efforts these veterans, and all armor veterans, both Army and Marines, will not be forgotten. They left a legacy of duty to country.

In his farewell to the 6th Armored Division, Maj. Gen. Robert W. Grow said: "You have made history, history that will be recorded and read as long as men cherish gallantry and glory in the record of success in combat. For your story is the story of men who never failed. If and when we meet, you and I, or you and you, let there be a bond of close personal friendship that was cemented by the trials of battle." Those words apply just as well to all Army and Marine armor veterans.

We are equally appreciative to Mr. Gordon England from General Dynamics Corporate Headquarters, and Lt. Gen. Donald Pihl USA (Ret.) and Mr. Peter Keating from General Dynamics Land Systems for their financial support and for making available numerous action photos of the Abrams tanks and the Marine Corps's Advanced Amphibious Assault Vehicle.

Additional financial support came from the University of Cincinnati's College of Evening and Continuing Education, which provided the means to copy thousands of pages required in crafting the manuscripts. We would also like to extend our gratitude to the University of Cincinnati Army ROTC Detachment, especially to Lt. Col. Wade Johnson and Capt. Adam R. Grijalba for providing U.S. Army Training and Doctrine Command documents pertinent to our history.

We are deeply indebted to Lt. Col. Terry A. Blakely and the staff at *ARMOR* magazine for their support. Special thanks go to Managing Editor Jon T. Clemens, who graciously reviewed a number of the manuscripts and provided numerous pictures. His suggestions were very constructive. We also thank Editorial Assistant Vivian Oertle and Connie Bright, secretary of the United States Armor Association, for their assistance. We are grateful as well to the staff at the Patton Museum of Cavalry and Armor for providing pictures from their collections, especially Director John M. Purdy and Librarians Katie Baldwin and Candace L. Fuller. In addition, we sincerely thank Judy Stephenson of the Armor School Library for providing excellent reference services.

A special acknowledgment goes to Charles Lemons, Patton Museum curator, who not only assisted in providing pictures of armored vehicles, but trained coeditor Hofmann to qualify and drive a World War II M3A1 light tank. In addition, this editor-author extends his deepest appreciation to Maj. Jess M. Simpson and the tankers from 1st Battalion, 147th Armor, of the Ohio National Guard for extending an invitation to spend a Fort Knox weekend with M1A1 Abrams tank simulators. This hands-on experience and comparison were very revealing in understanding the striking differences in handling a World War II tank as compared to one of the best-designed main battle tanks in the world today.

Our sincere thanks also go to Lt. Col. Donald F. Bittner, Ph.D., USMC (Ret.) from the Marine Corps Command and Staff College of the Marine Corps University and to University Archivist Janet Kennelly for their support when we were researching USMC armor history, especially for their generosity in always furnishing pictures at the last minute.

To the Marine Corps Association, *Leatherneck, Marine Corps Gazette,* and to *Army* magazine, published by the Association of the U.S. Army, for providing the initial historical framework for a number of contributors to this book, we are most thankful. In addition, we would like especially to recognize the Society for Military History, which publishes the *Journal of Military History*—formerly *Military Affairs*—for providing the expression of military thought and history by a number of the contributing authors.

Our sincere obligations in the project are many. While ideas, insights, and support came from many sources, special thanks are due the following: Col. Joseph Ameel, USA (Ret.); Lt. Gen. Robert J. Baer, USA (Ret.); Lt. Gen. Donald M. Babers, USA (Ret.); Lt. Col. Dennis Beal, USMC; Dr. John T. Broom; Maj. Michael F. Campbell, USMC; Col. Christopher V. Cardine, USA (Ret.); Rep. Steve Chabot; Col. Richard L. Coffman, USA (Ret.); former Sgt. Hubert D. Crotts, USMC; Col. Louis F. Dixon, USA (Ret.); Col. Edward J. Driscoll, USMC (Ret.); Col. Gregory Fontenot, USA; Col. Robert Harrington, USA (Ret.); Mr. Richard P. Hunnicutt; Col. Jerry B. Houston, USA (Ret.); Col. Michael D. Jackson, USA (Ret.); Col. Richard L. Knox, USA (Ret.); Col. James H. Leach, USA (Ret.); Dr. Philip W. Lett; Maj. Gen. John E. Longhouser, USA (Ret.); Maj. Gen. Thomas P. Lynch, USA (Ret.); Maj. Gen. Peter M. McVey, USA (Ret.); Prof. Allan R. Millett; Gen. Glenn K. Otis, USA (Ret.); Col. Robert J. Putnam, USMC (Ret.); Maj. Gen. Joseph Raffiani Jr., USA (Ret.); Ms. Sharon Reynolds; Lt. Gen. George Sammet Jr., USA (Ret.); Maj. Gen. Stan R. Sheridan, USA (Ret.); Lt. Gen. Martin R. Steele, USMC (Ret.); Mrs. JoAnn Sunell; Brig. Gen. Donald P. Whalen, USA (Ret.); and Mr. Justus "Judd" P. White. We would especially like to acknowledge Gen. Frederick M. Franks Jr., USA (Ret.), who provided us with guidance and valuable comments on a study in command during the Gulf War.

Doctrine, and the ensuing development of equipment, organization, and training reflected in the United States force that deployed to Saudi Arabia in 1990-91, especially the U.S. Army force, was the product of seventeen years of intense and dramatic change following termination of U.S. involvement in the Vietnam War. While that force reflected the U.S. experience in both World Wars and the wars in Korea and Vietnam, it also reflected the experience of other armed forces, especially British and German, and most importantly of the Israeli Defense Force in the Arab-Israeli wars, especially in the 1973 Yom Kippur War. While many advised, we owe special thanks to the following: In Israel, the officers and soldiers of the IDF Armoured Corps; Maj. Gen. Moshe "Musa" Peled, IDF (Ret.), who led his division to the rescue of Israeli forces on the Golan Heights in October 1973; Maj. Gen. Israel Tal, IDF (Ret.), who breathed fire into the Armoured Corps and later fathered the Israeli Merkava tank; and Brig. Gen. Avigdor Kahalani, IDF (Ret.), the intrepid commander of "COURAGE 77." In Germany, Gen. Frido "Deide" von Senger und Etterlin, distinguished panzer leader and former commander in chief of NATO's Central European Command; Lt. Gen. Rudolph Reichenberger and Lt. Gen. Eberhard Burandt, both deputy Inspekteurs of the Bundeswehr; Lt. Gen. Heinz-Georg Lemm and Lt. Gen. Horst Wenner, both great combat leaders and later chiefs of the Heeresamt. In the United Kingdom, the officers and soldiers of the Royal Armoured Corps; Gen. Sir Richard Worsley, former commander 1st British Corps and later Quartermaster General of the British Army; and Lt. Gen. Robin Carnegie, the pioneer Director General of Army Training. Those of us who engineered the changes during those seventeen years that resulted in today's U.S. armored force stand ever in the debt of these remarkable soldiers.

We have been fortunate to have the counsel of Emeritus Prof. Edward M. Coffman of the University of Wisconsin, who directed us to The University Press of Kentucky. This was a very appropriate move since Fort Knox is the Home of Armor and Cavalry and the Armor School. Fort Knox has been and is Armor, and as a result, has become an important part of Kentucky history. We are sincerely grateful to the staff of The University Press of Kentucky. We were very fortunate to have one of the foremost copyeditors in military history, Dr. Dale E. Wilson, who also contributed to this anthology.

Introduction

George F. Hofmann
and
Donn A. Starry

Several years ago the distinguished Nigerian novelist Chinua Achebe explained human society's need for what he called drummers, warriors, and storytellers. Drummers to stir up the will of the people and line them up behind causes, warriors to fight for the causes, and storytellers to "make us what we are . . . create history." He then explained that of the three, storytellers are the most important, for they are tellers of important events.

This book is an anthology that seeks to identify milestones in the history of the mechanization of the U.S. Army and, at least in part, of the U.S. Marine Corps. It begins with World War I and ends with Operation Desert Storm—the U.S.-led coalition campaign to drive Iraqi military forces out of Kuwait—in 1991. Each chapter is written by a storyteller describing important historical events. Each chapter illuminates segments of the larger landscape of that three quarters of a century, from mechanization's slender beginnings with tanks tactically supporting infantry in World War I France, to the impressively synchronized combined arms campaign at the operational level of war that defeated a heavily armed and armored enemy in the Arabian deserts nearly seventy-five years later. The strength of this anthology is the combination of research and writing on American armor history both by academics and by former military men—Soldiers and Marines who have had firsthand experience with the subjects about which they have written.

However, this book is not intended to be a definitive history of mechanization and the development, or a lack thereof, of operational art in the U.S. Army and Marine Corps. Nor is it intended to be another reference book covering the technical details of American tanks and armored fighting vehicles, already most effectively described in the excellent works of Richard P. Hunnicutt. The purpose of this book is to set forth a perspective on mechanization and operational-level intellectual development in the U.S. armed forces against the background of the necessarily continual, if persistent, struggle to define concepts. It traces the development of doctrine for operations at the tactical and operational levels of war; the translation of war-fighting doctrine into the development of equipment and organizations; and describes the evolution of the training and education of Soldiers and Marines, noncommissioned and commissioned officers, and units to fight successfully and win the first and succeeding battles of the next, not the last, war.

In this context, doctrine—operational concepts for battle fighting—must drive all else; doctrine is the keystone of military architecture, the engine of change, the "first and great commandment." Doctrine in turn is driven by two imperatives: threat and technology.

Threat definition identifies the most challenging threats to the country's vital national interests, how those threats are represented in military capabilities of potential adversaries, and how best to achieve the nation's political goals, given a decision to employ armed forces for that purpose. This reality is especially demanding in the post–Cold War world, for gone is the once overwhelming Soviet conventional threat to the security of western Europe and the familiar, carefully crafted, and relatively tidy nuclear deterrent framework in which that confrontation was embedded. It is not that threats to vital national interests have gone away, it is just that they have changed; for we yet live in a dangerous world. With this in mind, it is time to consider the need for doctrinal change.

Not only is doctrine driven by political imperatives, doctrine itself must change as technology advances on either side. Doctrine may seek to overcome perceived advantages made possible by new technology developed by the opposing side, or doctrine may seek to embrace a new technology that, if fielded, promises a singular advantage over potential adversaries. This world of measures and countermeasures is made particularly vexing by the rapid pace of technological development in many fields. It is a pace that imposes incredible strain on traditional models of research, development, and acquisition. Operational doctrine must therefore take into account the timely modernization of armed forces either to counter new technology threats or to take advantage of new technology opportunities.

Traditionally, technology zealots proclaim that, if left to its own, technology will inevitably provide new opportunities. Thus, they contend, technology should drive doctrinal development. Those who devise battle-fighting concepts insist that concepts should drive technological development. The truth, of course, is that there must be an ongoing symbiosis between the battle fighters' ideas and technological possibilities. Otherwise, the two will forever talk past one another, each believing his argument to be absolute when in fact neither is totally correct. A historic inability to achieve such a symbiotic relationship in the United States has been, and remains, a persistent problem aggravated by the accelerating pace of technological change. Modern doctrinal thinkers must therefore think faster, more precisely, and argue more persuasively than did their predecessors. The alternative amounts to throwing ever more of a decreasing apportionment of national treasure at technology, hoping against hope that something good may someday come of it all. The nation cannot now, if indeed it ever could, afford that. In that context, this anthology of American arms is badly needed today as military planners again struggle to identify threats to America's national security and, from that definition, determine what must be expected of our armed forces in the first and succeeding battles of the next war.

The U.S. air-ground force that deployed to and so successfully fought Desert Storm in 1991 was equipped, organized, and trained to dominate the battlefield with a combined arms team able to outmaneuver and outshoot its adversaries on any terrain, in any weather, day or night. The tactics and operational-level concepts it employed were the result of seventy-five years of thought, experimentation, and reflections on the lessons learned in two world wars and the wars in Korea and Vietnam. The weapon systems employed reflected the requirements of battle-fighting doctrine. Critical direct-fire systems were the M1-series Abrams tank, the M2 and M3 Bradley infantry and cavalry fighting vehicles, and the AH-64 Apache attack helicopter. In the direct-fire fight, the Abrams tank was capable of closing with and destroying enemy forces using long-range, highly lethal firepower and rapid maneuver— day and night. Indirect fire was provided by the M109 155mm self-propelled howitzer—an upgrade of a system first fielded in 1963—and the Multiple Launch Rocket System (MLRS). The force was organized to take best advantage of its equipment, both at the tactical and operational levels of war. Its Soldiers and leaders were well trained both as individuals and as units.

This story of the mechanization of America's armed forces is set forth as a chronology. In the chapters that follow *mechanized* means armed and armored vehicles, such as the tank, which is an armed tracked fighting vehicle with armor protection. The reader will be able to insert equipment-specific chapters into the overriding accounting of armor doctrine development—the battle-fighting concepts, tactics, techniques, and procedures that underlie the whole.

Dale E. Wilson's chapter on the birth of U.S. armor begins the year after President Woodrow Wilson boasted during his reelection campaign that he kept the United States out of the European war that began in 1914. Author Wilson's armor history sets the framework for the history of the American Expeditionary Forces's Tank Corps. It is a story of combat leaders who strove mightily to best employ new weapons technology to overcome the devastating effects of machine guns and artillery on massed infantry in trench warfare. It is also a story of materiel managers who attempted against great odds to get tanks into battle. Wilson examines how the U.S. Army incorporated the tank and, almost as an afterthought, developed doctrine, organization, and trained and fielded a tank force in France only to see it all evaporate in the rapid postwar demobilization.

In Chapter 2, Timothy K. Nenninger chronicles armor history from the time of accelerated demobilization after World War I to the eve of World War II in the context of equipment and organizational developments that were dominated by obsolete World War I tanks whose primary function was seen only as an aid to advancing infantry. In describing the ensuing post–World War I branch conflict over the control of tanks and the development of mechanization during the 1930s, Nenninger explains how Army Chief of Staff Gen. Douglas MacArthur's branch-oriented policy on mechanization brought a significant change to cavalry, and how mechanization for all arms was severely restrained by what General MacArthur once characterized as "trivial" budgets. He goes on to show how the world military situation in 1940 finally brought change to the traditional branch chief structure of the Army with the creation of a separate Armored Force.

George F. Hofmann in Chapter 3 describes the first assault amphibian tank, its contentious inventor, J. Walter Christie, and the tank's influence on the operational and tactical side of post–World War I Marine Corps amphibious doctrine. In addition, he discusses problems the Marine Corps encountered applying new technology to doctrine when the Corps was affected by retrenchment, budget problems, and an uncertain future. Finally, he addresses the history of the adaptation of the civilian-designed Roebling tractor, which played a key role in the Pacific during World War II.

In Chapter 4, Hofmann focuses on the U.S. Army's debate over how equipment related to doctrine and the failure to develop operational level concepts of warfare during the interwar period. He argues that the Army was short on vision, demonstrating little interest in either an appropriate tank-development strategy or a concept for operational-level employment of a modern mechanized combined arms force with the tank, such as the celebrated Christie, as the dominant maneuver weapon. Although limited budgets were a major constraint, Hofmann maintains that fact did not of itself preclude the Army's leaders from modernizing their minds. U.S. Army attaché reports from Europe during the 1930s cited the development in Germany and in the Soviet Union of ideas for mechanized warfare at the operational level. In the Soviet Union the Red Army positioned the Christie-type tank as the main maneuver element, especially for deep operations. Within this context, Hofmann examines the controversy surrounding one of the most controversial tank designers in U.S. military history, J. Walter Christie, and the effect of the controversy on American tank-development policy.

Christopher R. Gabel in Chapter 5 discusses how the Armored Force organized for World War II, its initial bloody disillusionment in North Africa, its subsequent adjustment, and, finally, its employment primarily in pursuit and exploitation in northern and central Europe. He argues that while armor did not dominate ground combat tactically or operationally, it figured prominently in most operations. Gabel maintains that the 1944-1945 campaign in Europe demonstrated the battle advantage, tactically and operationally, of combined arms, thus demonstrating the wisdom of basic employment concepts for armor as it exists today. He concludes, moreover, that the execution of operational-level blitzkrieg-type operations by American ground forces was the exception and not the rule, reinforcing Hofmann's hypothesis regarding the Army's lack of vision concerning the operational level of war.

Joseph H. Alexander heads a second group of authors, those who made and lived, in part, armor history. In Chapter 6 he discusses Marine Corps armor history during World War II, relating the steady development of doctrine and the acquisition and employment of tracked amphibians and tanks that led the assaults across the Pacific. More specifically, he describes armor tactics and technology as first applied in the bloody battle for Tarawa in 1943 and the subsequent lessons learned.

Philip L. Bolté in Chapter 7 continues where Gabel left off, describing the postwar organization of Army armor units. Bolté argues that tank programs were hampered by post–World War II demobilization and an overreliance on airpower, which together with nuclear weapons would surely eliminate the need for maintaining a large ground force, or so it was perceived. As a result, the U.S. Army again went to war unprepared. Bolté offers a firsthand account of the frustrations encountered by tankers during the Korean War, where armor was predominately used for infantry support. His tactical analysis concludes with a reminder that the tankers who were asked to fight and die in Korea while employing World War II equipment and makeshift organizations were yet another clear signal about the importance of readiness and the need to modernize equipment, organization, and training.

In Chapter 8, Kenneth W. Estes examines the Marine Corps's struggle with post–World War II doctrine. As a result, Estes sees clear evidence in postwar policy that a first-rate main battle tank fleet was desired, but proved difficult to buy, first because of fiscal constraints and later because of Army research and design and production delays. The Korean War affected little in Marine Corps armor doctrine or policy. However, the post-Korea fascination with the tactical promise of the helicopter fostered the enduring debate of "heavy vs. light" that eroded the previous emphasis on armor-protected firepower for Marine Corps divisions. By 1965, when the Corps began to expand and deploy forces to Vietnam, tank tactics and organization remained Korean War vintage.

Oscar C. Decker in Chapter 9 chronicles tank development and acquisition during the Cold War—primarily the story of the Patton tank series, a development that set the stage for the Abrams main battle tank development in the 1970s. From the perspective of the developer, he describes the continuing struggle to strike the right balance between firepower, protection, and mobility as it relates to doctrine. Following World War II, he notes, tank production moved from no new production to rapid production of several interim tank models, all reflecting lessons from the Korean War, ongoing funding constraints, and the newfound concern for weapons commonality amongst North Atlantic Treaty Organization (NATO) allies in Europe. This is a story of how equipment relates to doctrine.

Lewis Sorley, in Chapter 10, recounts the history of the mounted force in the unique combat environment of Vietnam, a war without fronts. He describes a wide range of armor and mechanized units deployed to Vietnam, their equipment, and operations. Sorley describes how at first armored operations were not considered feasible in Vietnam largely due to misperceptions about the country, weather, and terrain. However, he concludes, once committed, armored forces proved to be the most cost-effective combat organization.

In Chapter 11, Richard M. Swain provides a corollary to Hofmann's chapter on the Army's failure to develop concepts for battle at the operational level of war, describing the development of a new operational-level doctrine, AirLand Battle. Swain begins by relating how the Army in 1973 set about recovering from the ravages of the Vietnam War, developing doctrine, equipment, organization, and training to meet the demands of the next war. By 1982, AirLand Battle doctrine appeared in field manual form. It included concepts for deep offensive operations employing mobile forces equipped with new materiel: principally a main battle tank, a new infantry and cavalry fighting vehicle, and an attack helicopter. AirLand Battle dealt with war at the operational level; it foresaw large maneuver forces—combined arms forces integrated in both close and deep battle. AirLand Battle was the doctrine the U.S. Army employed in Desert Storm in 1991.

Two direct-fire weapons systems were key to the tactical battle and the operational level of war. They were the M2 and M3 Bradley infantry and cavalry fighting vehicles and the M1 Abrams main battle tank. As in Decker's chapter, Diane L. Urbina and Robert J. Sunell recount the history of equipment development as it relates to doctrine. Their chapters deal with the attempt to fit machines to tactical ideas that resulted in the introduction of two innovative fighting systems necessary to engage in the operational level of war, a concept that was emerging as a result of AirLand Battle doctrine.

In Chapter 12, Urbina tells the story of the Bradley's rocky progress through a number of design changes. She analyzes the never-ending debate over whether tactics should drive design or technical capabilities should determine doctrine. By 1981, Urbina notes, most serious problems were resolved through compromise, and the Bradley system—now designed for its intended combined arms role—began a new phase in mechanized warfare for Army mechanized infantry and armored cavalry units.

Robert J. Sunell in Chapter 13 provides an insider's history of the Abrams tank system in which armored vehicle design and development were moved from the traditional arsenal-centered process to competitive design and development by industry. Sunell's chapter not only deals with the Abrams main battle tank and the competition between contractor prototypes, but tells as well the difficult story of evolution through innovation. He also relates the political and international controversies surrounding the Abrams, demonstrating how key personnel and their organizations made the tank a reality during its tumultuous developmental years.

In Chapter 14, Kenneth W. Estes discusses Marine Corps tank history from the Vietnam era to the Persian Gulf War. As with other armed services, he notes, Marine Corps armor underwent post-Vietnam adjustments that included an increase in firepower, with the tank, nonetheless, remaining an infantry-support weapon. He summarizes his chapter by discussing the role Marine Corps armor played during Operation Desert Storm, concluding that Marine Corps leaders' limited tactical experience with armor and their continuing fixation on lighter-weight vehicles remain obstacles to creating and operating a modern armored force.

Steven A. Bourque's chapter is the corollary to Hofmann's and Swain's chapters dealing with the issue of the operational level of war. Bourque describes the successful execution of AirLand Battle doctrine and the dominant capabilities of the Abrams main battle tank, the Bradley fighting vehicle, and the Apache attack helicopter during Operation Desert Storm. Noting that the Persian Gulf War saw the first large-scale maneuver of U.S. armor divisions since World War II, Bourque covers armor doctrine, training, and technology, and then demonstrates its brilliant application in ground combat dominated by the Army maneuver force. He concludes by demonstrating that once the short war was over there was no doubt about the effectiveness of the U.S. Army combined arms team centered on the Abrams, the Bradley, and the Apache.

In the final chapter, Donn A. Starry summarizes the history of mechanization in the United States armed forces by setting the preceding chapters against the broader framework of legacies from the past and his own historic role in the Army's transformation to one of the most powerful armored forces in the world.

The Industrial Revolution made the mechanization of warfare inevitable. But many Industrial Revolution concepts, now deeply institutionalized in America's armed forces, are anachronisms that are difficult to change and probably counterproductive. While concepts for warfare at the tactical and operational levels have developed—along with equipment, organization, and training—into what Heidi and Alvin Toffler have styled in their book *The Third Wave* (1980), as Third Wave (Information Age) systems, several critical functions remain Second Wave (Industrial Revolution) systems. Principal among the delinquents are the Army's individual personnel replacement system; service and joint logistics systems; the Defense Department's materiel research, development, and acquisition system; command and control systems now beset by burgeoning information technology; and the continuing struggle to define relevant operational-level war-fighting concepts.

So the search for means, both intellectual and technical, to achieve victory in battle continues. While the notion persists that the United States is the most technically advanced nation on earth, equally persistent is America's historic ineptitude at designing operational-level concepts that make the best use of its perceived technical advantage.

This idea of operational concepts will recur, so it deserves comment here. As already noted, several of our authors refer to an operational level of war or to operational art. Neither term has a liturgically correct definition. Therefore, for the sake of consistency in this anthology, operational art is defined as the study and definition of military operations in a tripartite system that divides military art into strategy as the study of war, operational art as the study of operations, and tactics as the study of battle (Georgi Isserson, *War and Revolution*, 1932). The operational level of war, also frequently referred to herein, is therefore what lies between strategy and tactics (frequently at the theater level of operations). Beginning in the late nineteenth century, with the impact of the Industrial Revolution and later of "modern technology," as demonstrated in World War I with machine guns and rapid-firing artillery, and with nuclear weapons at the end of World War II, there developed a lacuna between strategy and tactics. The attempt to fill that gap was begun by the Soviet military in the post–World War I period, and continues to this day, with more or less success.

Consider again the tank. It appeared in battle in World War I as a technical counter to the appearance and the technical advantage of machine guns and massed artillery—massed firepower over massed infantry. However, it proved impossible to overcome the deeply entrenched biases of a moribund bureaucracy, both military and political, to effectively use the new technology to reduce the risk of slaughter of millions of massed Soldiers afoot. Generals and field marshals who simply threw more infantry into the devastating maw of massed firepower were no more or less guilty than their political masters who failed—or were unwilling—to acknowledge the urgent need to seek political alternatives to the annihilation of generations of their nation's young.

Tanks were at first considered so mechanically unreliable as to be impractical on the battlefield, a notion that was not without substance. As mechanical reliability improved, direct-fire antitank weapons appeared—first guns, then rocket launchers, and later long-range guided missiles. In the hands of individual Soldiers or crews, mounted on ground and then on aerial platforms, many believed long-range missiles would at last render tanks obsolete. Indeed, in the technical euphoria over the appearance of the long-range antitank guided missile in the 1973 Yom Kippur War in the Middle East, technology zealots in the United States proposed arming the citizenry of western Europe with an antitank system of some kind for each household. So armed, it was said, NATO's citizens would simply step from behind their kitchen doors and blast away at and destroy oncoming Soviet tanks. This new capability would of course eliminate the need for many battalions of NATO tanks, thus saving millions, if not billions of precious NATO dollars. A cost-effectiveness analysis was actually drawn up. It was, in truth, so ludicrous as to appear a spoof, until one realized its authors were serious. For a few nervous weeks it was the rage around the Defense Department cocktail circuit in Washington. Fortunately a modicum of reason eventually prevailed.

Nevertheless, the tank remains—and for some time is likely to remain—the centerpiece of a devastating combined arms team of fighting systems: the direct-fire, indirect-fire, ground and air systems whose awesome combat power was so masterfully demonstrated in Desert Storm in 1991.

We are therefore not so much in search of dominating technology as we are in search of the intellectual power to understand the possibilities and limitations of burgeoning technology, and the moral courage to step out in new directions. While there is no lack of new technology, the intellectual power and moral courage to use it properly seem ever wanting.

This anthology is a chronicle of that latter search.

1

World War I

The Birth of
American Armor

Dale E. Wilson

One of history's great ironies is that the nation that spawned the technology from which the tank was created did not play a role in that vehicle's conception. It is equally ironic that the United States, which later became known as the "arsenal of democracy," was unable to produce a single armored vehicle that saw combat with its Tank Corps. Although the Army trained more than twenty thousand tank officers and crewmen in less than a year, and shipped more than half of them to France, it was able to send only three battalions into combat—in vehicles borrowed from its European allies.[1] Finally, in what can only be called one of history's most prescient acts, the U.S. Army chose George S. Patton Jr.—whose name would become synonymous with armored warfare a generation later—to be the first Soldier in its ranks assigned to duty with tanks.

That is the rough framework for the story of the World War I Tank Corps. It is a fascinating tale, fraught with important lessons for combat leaders charged with preparing men to employ new weapons in battle and materiel managers who must work to get those weapons into Soldiers' hands. Sadly, historians have largely ignored it.[2] What follows is a brief look at how the U.S. Army first incorporated the tank into its force structure, developed doctrine for its employment, and trained and fielded a small but competent tank force on the battlefields of France during World War I.

In June 1917, Gen. John J. Pershing, commander of the recently arrived American Expeditionary Forces (AEF), read a report on British and French tank operations submitted to the director of the Army War College by the American military mission in Paris. Pershing, greatly impressed by the report, which included the personal observations of Maj. Frank Parker, a liaison officer who observed French tank operations in the April offensive, immediately appointed several committees to study tank warfare. He also instructed several members of his staff to visit the front lines and look at British and French tank organization and tactics. Despite misgivings expressed by some of these observers, Pershing concluded that a mix of British heavy tanks and French light tanks would be a valuable asset when the AEF went into battle.[3]

At the time, the Allies had only five tanks in production: the British Mark IV and V heavy tanks, and the French Schneider, St.-Chamond, and Renault vehicles. However, neither the Schneider nor the St.-Chamond could truly be classified as a tank. They were actually lightly armored tracked artillery carriers that had to be accompanied by infantry skirmishers who carefully marked the routes the vehicles should follow. All of the American observers agreed that the Schneider and St.-Chamond were unsuited for tank operations.

Colonel Frank Parker (center) and Louis Renault (left) inspect a Renault Char FT light tank at the Renault tank production facility at Billancourt in August 1917. Renault Communications.

Inspired by the observers' reports—and the inability of members of a joint British-French tank board to reconcile differences in the Allies' theories on tactics and equipment—Pershing directed that a board of officers be convened to perform a detailed study of British heavy tanks and the French Renault Char FT (*faible* [light] *tonnage*) light tank. The members of the board (Cols. Fox Conner and Frank Parker, Lt. Col. Clarence C. Williams, and Maj. Nelson E. Margetts) submitted a report of their findings on 1 September. They concluded that the tank would play an important role in the war and that Pershing should create a separate Tank Department with a single chief reporting directly to him. They further recommended that a force of more than two thousand tanks be procured by the AEF, with a 10-to-1 mix of light to heavy tanks, and that production be geared to provide for a 15 percent monthly replacement rate.[4]

Pershing responded by assigning Lt. Col. LeRoy Eltinge, an operations staff officer, the job of drafting specific requirements for a "Combat Tank Service" for the AEF. Working closely with other members of the AEF staff, Eltinge determined that a force of six hundred heavy and twelve hundred light tanks, more than eight hundred trucks and automobiles, 180 motorcycles, and nearly fifteen thousand men would be needed to support an Army consisting of twenty fighting and ten replacement divisions.

Eltinge reported that the French were willing to permit manufacture of the Renault in the United States. They agreed to supply detailed plans and a production copy of the vehicle in exchange for two thousand copies of an American-made version. The British, acting in the same spirit of cooperation, offered to provide complete plans and specifications so the United States could produce their Mark VI (a 27-to 30-ton heavy tank that was never built) design.[5]

Word of plans to create a Tank Corps quickly filtered through the AEF. Among those it reached was Capt. George Patton, post adjutant and commander of the AEF headquarters company at Chaumont. Patton, a cavalry officer who had been Pershing's aide-de-camp during the Punitive Expedition in Mexico, was frustrated in his current position and anxiously sought assignment to a combat unit. After discussing the anticipated role of tanks in the AEF with Eltinge and Col. Frank R. McCoy, the Assistant Chief of Staff, Patton submitted a request for transfer on 3 October.

On 10 November, AEF headquarters issued orders directing Patton to report to the Commandant of the AEF schools at Langres for the purpose of establishing a tank training program. First Lt. Elgin Braine, an artillery officer with a background in mechanical engineering, was detailed to serve as Patton's assistant.[6] The pair had been at Langres for little more than a week when they were ordered to report to the French light tank center at Chamlieu near Paris for two weeks of training, then on to the Renault tank production facility at Billancourt.

During their visit to the training center Patton and Braine observed or participated in all phases of individual and crew training, watched maneuver training, and talked at length with members of the center staff. Patton also had several meetings with Brig. Gen. Jean E. Estienne, commander of all French tank forces. On 1 December Patton joined Col. Frank Parker for a visit to the Royal Tank Corps's (RTC) headquarters at Albert. The two Americans met with Col. J.F.C. Fuller, the RTC operations officer, and the trio discussed the mass employment of tanks in the recent British offensive at Cambrai, tank doctrine, and tactics. Two days later, Patton linked up with Braine and together they toured the Renault tank works at Billancourt.[7] After seeing all that went into construction of the vehicles, they recommended four minor improvements which the French later incorporated: a self-starter, a self-sealing fuel tank, an interchangeable mount so that each tank could carry either a 37mm gun or a machine gun, and a firewall between the crew and engine compartments.[8]

Patton was particularly concerned over the "great difficulty" French tankers had getting their manufacturers to cooperate. Foreseeing the possibility of American builders being equally recalcitrant, he included a veiled warning in his subsequent report on light tanks calling for officers charged with tank procurement to take a hard line when dealing with manufacturers.[9]

The two tank officers returned to GHQ at Chaumont and briefed Colonel Eltinge on their findings. Eltinge, still temporarily in charge of the AEF tank project, instructed them to prepare a formal written report. The pair set about drafting it and, on 5 December, Patton wrote his wife that he was excited because "no one knows any thing about the subject except me. I am certainly in on the ground floor. If they [the tanks] are a success I may have the chance I have always been looking for."[10]

Patton submitted a double-spaced, fifty-eight-page report on 12 December 1917. Later, while organizing his files, he penciled on the envelope containing the paper: "Original Tank Report. The Basis of the U.S. Tank Corps. Very Important. GSP."[11] He was right. The document served as the foundation for subsequent tank developments in the AEF. At least one of his recommendations (a proposal that tanks be organized in platoons of five tanks, with three platoons to a company and three tank companies to a battalion) survived as part of American tank organization until the early 1980s.

The report, divided into four sections, includes a detailed mechanical description of the Renault light tank, recommendations for the organization and equipping of light tank units, a discussion of tank tactics and doctrine, and proposed methods for the conduct of drill and instruction.

The most engaging part of the report is Patton's discussion of tactics. He envisioned several missions for the tanks in their infantry support role: (1) clearing wire obstacles, (2) suppressing enemy crew-served weapons and preventing the enemy's infantry from manning the parapets after the preparatory artillery barrage lifted, (3) helping the doughboys mop up on the objective, (4) guarding against counterattack by patrolling ahead of the most advanced infantry positions, and (5) exploiting the attack supported by reserve infantry, seeking "every opportunity to become pursuit cavalry."

He concludes his discussion of tank tactics with the observation that heavy tanks were more independent and should thus precede light tanks in the attack—especially when no artillery preparation was em-

Lt. Col. George S. Patton Jr. poses in front of a Renault Char FT light tank at the AEF's light tank school at Bourg, France, in July 1918. National Archives.

ployed—capitalizing on the heavy tanks' superior ability to cut wire. Nevertheless, he thought light tanks held an advantage in mobility because they could be easily transported by truck or trailer, whereas the heavy tanks could only be moved by rail.[12]

Subsequent operations proved Patton's ideas at least partially correct. Employed almost exactly as he had proposed, the light tanks had a difficult time keeping pace with the infantry at Saint-Mihiel because of poor ground conditions and the rapidity of the German retreat. However, they were a valuable asset in support of the Meuse-Argonne offensive. His innovative ideas on operational mobility were never tested because the AEF lacked sufficient trucks to transport the light tank force. Instead, they had to be moved by rail to both the Saint-Mihiel and Meuse-Argonne sectors, then under their own power during operations. This meant conducting long road marches to link up with units at the front, contributing to a high mechanical failure rate as the tanks were forced to conduct extended operations without benefit of overhaul.

The report was so thorough and Patton's proposals so well reasoned that the majority of them were enacted. As commander of the light tank training center and school at Bourg, he was able to implement the training program he devised. It was a simple plan calling for recruitment of sufficient men to fill two companies, training them to a battle-ready standard, then using them as cadre to train additional companies. His rationale for this program was that not only would the men learn the skills needed for combat but they would have time to develop unit cohesion and esprit de corps—attributes he considered essential for battlefield success. The relative ease with which he was able to recruit and train two battalions of light tank troops (the 326th and 327th[13]) for combat in less than five months bears witness to the soundness of his ideas.[14]

Development of heavy tank training and tactics followed a different track. The focus within the AEF was on light tanks, so it was left to the Tank Service in the United States, a separate entity that came into being in January 1918, to provide the first heavy tank battalion—the 301st. To ensure that the 301st's officers and men would be ready for combat by the time the AEF needed them, Col. Samuel D. Rockenbach (who had been appointed Chief of the AEF Tank Corps in late-December 1917) was ordered to form a heavy tank training center at Bovington Camp in England, adjacent to the British tank school at Wool. British tankers conducted most of the Americans' initial training. Then, when the unit was ready to deploy to France, a number of officers and men from the 301st were ordered to remain at Bovington to form a cadre to

Brig. Gen. Samuel D. Rockenbach, commander of the AEF Tank Corps. National Archives.

train additional heavy tank units as fast as they could be recruited and shipped to England.[15]

Aside from volunteers recruited by Patton in France, nearly all Tank Corps personnel volunteered in the United States and were shipped to one of several tank centers before deploying overseas. The task of training the initial influx of volunteers fell to Capt. Dwight D. Eisenhower, who had been ordered to Gettysburg, Pennsylvania, in March 1918 to establish a training center called Camp Colt and prepare for their arrival. Additional Tank Corps training centers were established at Camps Summerall and Tobyhanna, also in Pennsylvania, and Greene and Polk in North Carolina that summer. Colt remained the largest, and by war's end Eisenhower was a lieutenant colonel. He commanded a force of more than ten thousand officers and men when Colt's personnel strength reached its peak in September 1918.

Col. Ira C. Welborn, Commander of the Tank Service in the United States, was so impressed by Eisenhower's administrative ability—particularly his creation of a meaningful training program despite the lack of resources—that he recommended "Ike" for the Distinguished Service Medal and offered him promotion to colonel if he would agree to remain stateside. Eisenhower refused. More than anything, he wanted to get into the fight in France. Unfortunately for Ike, the Armistice was signed before he could deploy.[16]

Getting vehicles into the hands of the men who would ride them into battle was much more difficult than originally envisioned. In the fall of 1917, Majs. James A. Drain and Herbert W. Alden were detailed by the Chief of Ordnance in Washington to visit France and England and determine how best to produce vehicles for the AEF. They quickly reached the conclusion that licensing U.S. manufacturers to build copies of the Renault was the only viable solution. They further concluded that none of the existing British heavy tank designs were satisfactory and recommended that a joint British-American effort be made to design a suitable vehicle. They further recommended that the approved design should be assembled at a factory in France, preferably close to a major port and rail center, with engines and automotive parts from the United States, and armor plate, weapons, and ammunition provided by the British.[17]

Their proposals were subsequently approved. Drain was ordered to remain in France, where he represented the United States on the Inter-Allied Tank Commission. Working closely with the British (the French showed no interest in the project), Drain developed specifications for a thirty-five-ton heavy tank measuring thirty-four feet long, twelve and a half feet wide, and nearly ten feet high. The tank's armor was to be of sufficient thickness to protect the crew and internal components from all small-arms bullets—including armor piercing. The American V-12 Liberty aircraft engine, which was chosen to power the vehicle, was expected to move the behemoth at speeds of up to six miles per hour. It also lent its name to the vehicle, which was designated the Mark VIII Liberty tank. Armament was to consist of seven machine guns and two six-pounder (57mm) cannon mounted in retractable sponsons. The components would be built in the United States and Britain and shipped to Neuvy-Pailloux, where an assembly plant was to be erected.

Unfortunately, the Liberty required numerous minor modifications before full-scale production could begin, and the fledgling Air Service siphoned off finished engines as fast as they came off the assembly lines—which was not fast enough for any of the parties concerned. In the end, not one Liberty engine—or any of the other components needed for Mark VIII production—made it to France from the United States. Frustrated, the British and French (who belatedly demanded six hundred of the vehicles), washed their hands of the project and Pershing ordered work on the assembly plant halted on the eve of its completion in late November 1918. The project's materiel assets were eventually shipped to Rock Island Arsenal, Illinois, and a hundred Mark VIIIs were assembled there by June 1920.[18]

Efforts to produce an American-made copy of the Renault light tank were slightly more successful but infinitely more frustrating for the officers involved in the project. As noted earlier, the French were eager to help the Americans begin production. Little work, however, was accomplished until Lieutenant Braine was detached from Patton's staff and sent back to the United States to serve as the AEF Tank Corps's liaison with the light tank production effort there. Although the French originally promised to provide two production copies of the Renault and complete sets of the plans (in metric, not English measure), all that Braine was able to obtain to take with him when he departed in February 1918 was a turret, a 37mm gun, and gun mounts.

When he arrived in New York on 13 March, Braine encountered the first in a series of bureaucratic obstacles that would plague the light tank production program throughout the remainder of 1918. No one expected him, and it took most of the day to find a berth for his shipment. At one point, Braine later wrote, it looked like his mission might end in failure when his precious cargo "almost landed in the bottom of the bay."[19]

Given this inauspicious start, it is not surprising that Braine became increasingly bitter as he shuttled back and forth between Washington and the Ordnance Department's Motor Equipment Section in Dayton, Ohio. It seemed to him that the left hand did not know what the right was doing. Design engineers in Washington and Dayton were working on a turret design independent of each other. Furthermore, Renault's plans had to be recast with English measurements. Confusion arose over whether vehicle speed should be measured in miles or kilometers per hour. The solution was to build speedometers that measured speed in miles per hour and odometers that measured distance in kilometers. Conflict arose over whether the vehicle should be equipped with the French Poteaux 37mm gun and Hotchkiss machine gun or U.S.-built weapons. Braine favored the former but his Ordnance Department bosses ruled in favor of the latter. Ironically, the machine gun decided upon—the Marlin-Rockwell aircraft weapon—proved unsuitable for use in tanks after more than three months of attempts to modify it for that purpose. In the end, more than twenty independent contractors were involved in producing the M1917, as the vehicle became known. It was, wrote Braine, enough to "necessitate [having] a Philadelphia lawyer to keep track" of it all.

While all of this was going on, the Ordnance Department was carefully controlling the flow of information back to AEF headquarters. Braine was instructed not to communicate with his superiors in France. Furthermore, all of his official and personal mail was opened before he received it. When he complained, Braine's Ordnance Department superiors "emphatically informed me that [the] people in France were fully advised as to the progress and situation at all times."[20]

But they were not. Patton and Rockenbach elatedly operated on the assumption that the M1917 light tank would be in full production by late spring—an assumption based on information provided to them by the Ordnance Department. By the early summer of 1918, when it was clear there would be no tanks produced in the United States in time to support First Army's initial operations, Rockenbach was forced to ask the Allies to equip the AEF's three battalions. Although faced with their own vehicle shortages, the British agreed to provide forty-seven Mark V heavy tanks—but only if the 301st was attached to the British Fourth Army—and the French promised 144 Renaults for Patton's two light tank battalions.[21] Fortunately, both allies delivered the required vehicles in time—something the Ordnance Department, for whatever reasons, was unable to do.

Braine, thoroughly disgusted with his treatment by Ordnance Department officers, discreetly sought to make contact with Benedict Crowell, the assistant secretary of war. Thanks to the help of a friend in the Army's New York recruiting office, Braine was able to indirectly get word of the light tank fiasco to Crowell, who showed personal interest in tank production and spent five days investigating Braine's charges. At about the same time, Lieutenant Colonel Drain returned from Paris and was appalled by the situation he found. According to Braine, Drain played a key role in securing the appointment of Louis J. Horowitz as the civilian head of tank production—a move Braine had earlier recommended. He was convinced that Horowitz deserved credit for breaking the bureaucratic logjam.[22]

A possible explanation for the Ordnance Department's efforts to keep Braine from communicating with Patton and Rockenbach may be the Ford Motor Company's effort to produce a competing vehicle. A number of ordnance officers were impressed by the Ford design—a two-man, three-ton light tank—and especially by Ford's promise to quickly produce a large number of the vehicles. Hoping to gain Braine's ap-

First Army Plan of Attack, 12 September 1918

Patton's men were ready. Some had been in training at Bourg since late January, and all had been imbued with their fiery commander's fighting spirit. They sported a "pyramid of power" on the left shoulder of their tunics: a triangular patch divided equally into portions bearing the colors of the infantry, cavalry, and artillery branches—blue, yellow, and red. The patch, designed by 2d Lt. Will G. Robinson, was but one of many manifestations of the light tankers' esprit. According to Robinson, Patton sought "to make the Tank Corps tougher than the Marines and more spectacular than the Matterhorn. That triangle was the first step."[26]

The brigade's officers knew better than anyone that Patton would accept nothing less than success on the battlefield. His fiery exhortations left 2d Lt. Julian K. Morrison, a platoon leader in Company A, 344th Tank Battalion,[27] with no doubt as to what was expected of him: "[I] was made to understand by the Colonel that a Tank Officer was meant to die." Morrison added that Patton's favorite message to his proval, they invited him to Detroit to see a prototype demonstration. He was baffled. The Ford technicians were unable to get the engine, a complicated affair consisting of two standard Ford motors hooked up in tandem, to start. In addition, the tank lacked a tailpiece—something Braine had learned from his experience in France was needed for trench-crossing operations. He later learned that the Ordnance Department had advised the AEF Tank Corps that he approved of the design, "which was not so."[23]

Acting independently of the AEF's tank experts, the War Department contracted with Ford to produce 15,015 of the vehicles at $4,000 each. By war's end, despite the company's highly touted assembly-line prowess, Ford managed to produce only fifteen. The remainder of the order was canceled. Although he does not directly accuse anyone in the Ordnance Department of wrongdoing, it is clear from Braine's postwar account that he was incensed by his superiors' actions, especially the decision to actively pursue the inadequate Ford design.

By early autumn the M1917 production line began to move and the first light tanks began rolling off in October 1918. Braine, anxious to get back to France, encountered resistance to his departure from the Ordnance Department, so he turned to Colonel Welborn for help. The aging Spanish-American War Medal of Honor recipient issued orders for Braine's return that same month. He barely beat the tanks he had spent so long trying to build. On 20 November 1918, almost nine months to the day after Braine left France on the USS *Apples*, and nine days after the Armistice, two M1917s arrived at Bourg. Eight more of the light tanks followed in December, bringing to ten the total delivered to the AEF.[24]

On the night of 11 September 1918, 144 light tank crews in Patton's 1st Tank Brigade were poised in position or making their way through a steady rain to the jump-off point for their first attack. They were to support the advance of the U.S. IV Corps, which was slated to assault the southeast face of the German salient at Saint-Mihiel the following morning. They were joined by 275 Schneiders, St.-Chamonds, and Renaults from Lt. Col. Emile Wahl's French 1st Assault Artillery Brigade. That brought to 419 the number of tanks on hand to support the U.S. First Army's combat debut. Twenty-four of the Schneiders were attached to Patton's 1st Tank Brigade in the IV Corps sector. The remainder (a mix of 35 Schneiders and St.-Chamonds and 216 Renaults) were assigned to support the U.S. I Corps, attacking on IV Corps's right. Lt. Col. Daniel D. Pullen's 3d Tank Brigade headquarters was assigned to serve as the liaison between I Corps and the 1st Assault Artillery Brigade.[25]

officers was "Go forward, go forward. If your tank breaks down go forward with the Infantry. There will be no excuse for your failure in this, and if I find any tank officer behind the front line of infantry I will [probably 'shoot him']."[28] Morrison must have taken the message to heart, for he was one of four Tank Corps officers—all from Maj. Sereno E. Brett's 344th Tank Battalion—to receive two Distinguished Service Crosses for heroism.

Patton took the personalities of his battalion commanders and their men into account when formulating his battle plan. He assigned the aggressive Brett the mission of supporting the battle-wise 1st Infantry Division's advance, giving him the simple mission of getting out in front of the doughboys and leading them to their objectives. Patton was much more controlling with Capt. Ranulf Compton, commander of the 345th Battalion. Compton had just taken over the battalion from Maj. Joseph W. Viner, who was ordered to remain at Bourg and run the tank center. That made Compton a comparatively unknown quantity, a fact that was reflected in Patton's instructions to him. Uncertain of Compton's ability, Patton transferred sixteen of the 345th's tanks to the brigade reserve. He tasked Compton to follow the initial advance of the 42d Infantry Division's 167th Regiment with his remaining vehicles, and then pass through and lead the way into Essey and Pannes. Several rain-swollen streams and other obstacles crossed that part of the battlefield, and Patton, expecting the Schneiders to have problems, instructed Maj. C.M.M. Chanoine, their Commander, to divide his force and have half follow the 42d's 165th Regiment and the other half follow the 166th Regiment.[29]

The attack kicked off with both of Patton's battalions employing the tactics they had worked out at Bourg. They moved out from the line of departure in a formation that allowed them a depth of at least three lines of tanks. This was accomplished by deploying two companies in the lead echelon, each company leading with two platoons abreast and holding the third platoon in support. The third company in each battalion was retained in battalion reserve. Brett had the added support of the brigade reserve in his sector.[30]

While it all may have looked good on chalkboards and sand tables, the situation deteriorated rapidly on the battlefield. The Germans already had begun infiltrating units to the rear, and they put up only token resistance once the assault started. The rapidity of their withdrawal, coupled with the influence of "General Mud," caused the neat formations to fall apart.

Patton was able to observe Brett's initial advance in the 1st Division sector but quickly lost sight of Compton's tanks. Concerned about what might be happening in the 42d Division zone, he abandoned his command post and set off in search of Compton. Although he failed to find his beleaguered subordinate, Patton encountered numerous tanks mired, broken down, or struggling singly or in small groups to keep up with the advancing doughboys. Despite the difficulties, Compton's tankers managed to advance up to fifteen kilometers while materially assisting the infantry in capturing "great quantities of machine-guns, field pieces of large and small caliber, and immense quantities of stores and supplies in towns."[31]

Brett fared little better that first day. When Patton caught up with him outside of Nonsard, the slightly wounded battalion Commander was sobbing with frustration. Only twenty-five of Brett's tanks had made it to the objective, and they were all out of fuel, so he was unable to continue the attack. Patton consoled Brett as best he could before setting off in search of more gasoline.[32]

The fuel situation remained critical throughout the rest of the first day and into the next. The thick mud had caused fuel consumption rates to soar, and clogged roads kept fuel trucks from getting forward in a timely manner. By the time Compton was able to refuel the majority of his tanks on 13 September, the 42d Division no longer needed their services. Compton and his crews spent the balance of the operation performing minor repairs, recovering mired vehicles, and tuning up in preparation for movement to the railhead late on the fourteenth.

Brett was similarly hampered. However, he had lost contact with all 1st Division units, so Patton ordered him to continue forward on the fourteenth until contact was reestablished. When the brigade commander finally caught up with Brett, his column of fifty-one tanks was halted on the outskirts of Woël. The infantry still was nowhere in sight. While Patton was talking the situation over with Brett, he recognized Brig. Gen. Dennis E. Nolan, Pershing's intelligence officer, passing by in an automobile. He flagged Nolan down and told him they were looking both "for a fight" and for the 1st Division. Nolan replied that the Germans had already evacuated Woël and said it was being held by a platoon of French infantry.

Concerned by the presence of German aircraft, which sporadically appeared and fired on the column, Patton ordered Brett to conceal his tanks in the foliage along the road. The brigade commander then sent a message to Rockenbach and the IV Corps and 1st Division commanders requesting further instructions. While awaiting a reply, he sent four officers south, using captured German horses as mounts, in search of American infantry. They found no one. He also dispatched a patrol of

Maj. Sereno E. Brett, commander of the 344th Tank Battalion, in a Renault Char FT light tank. National Archives.

three tanks and five dismounted Soldiers through Woël and on toward Saint-Benoit.[33]

At about 1:30 P.M., 2d Lt. Edwin A. McClure of Company A, the patrol leader, reported that the town was clear and that he was returning. About a half-hour later the patrol was attacked by a force estimated to be a battalion of infantry supported by a battery of 77mm guns. McClure sent a messenger back to battalion with word that he was attacking. Patton responded by having Brett order 2d Lt. Gordon M. Grant's platoon, also from Company A, to go to McClure's aid. Together, their eight tanks—still without infantry support—drove the Germans six kilometers to the outskirts of Jonville. During the running battle the tankers killed or silenced more than a dozen machine-gun crews. They also captured four 77mm guns. The Germans shelled them while they were trying to hook the artillery pieces up for towing, and McClure, Grant, and four of their men were wounded. Two tanks were disabled and the guns had to be abandoned so the damaged tanks could be towed to the rear. They managed to get clear of the area just before the Germans hammered it again with a barrage of 150mm artillery fire.[34]

McClure's and Grant's fight marked the end of Tank Corps combat in the Saint-Mihiel campaign. Orders soon followed to withdraw all tanks to the vicinity of the Bois de la Hazelle to prepare for railloading and transport to the Meuse-Argonne sector. Moving under cover of darkness, the French and American tank units closed on their original assembly areas by the evening of the sixteenth.

Casualties in both the French and American tank units were light. Only four U.S. officers were wounded and five enlisted men killed and another fifteen wounded. French losses were five officers wounded and six enlisted men killed and twenty-three wounded. Vehicle losses were similarly light: only three tanks in Patton's brigade were destroyed, all by artillery fire. However, recovery operations were conducted around the clock for mired vehicles and those disabled by mechanical failures.[35]

Although Patton bemoaned the fact that the Germans' rapid withdrawal meant "the full value of tanks in this operation was not possible to demonstrate," he could not complain about the aggressive spirit shown by his subordinate leaders and men.[36] But Rockenbach did. The AEF Tank Corps Commander was incensed by Patton's failure to maintain contact with higher headquarters. Brigade Commanders belonged in command posts and battalion Commanders belonged with their reserves, he chastised. Only company Commanders and platoon leaders should position themselves with forward elements. Finally, Rockenbach sarcastically advised Patton that although he appreciated the men's esprit de corps and personal valor, he wanted them to know "that they are fighting [in] tanks, they are not Infantry, and any man who abandons his Tank will in the future be tried [by court-martial]."[37]

Patton dutifully passed Rockenbach's message along to his subordinate officers and men but failed to take the instructions to heart.

While American and French tankers worked feverishly to prepare their vehicles for the rail move to the new sector, First Army planners busily put the finishing touches to the most ambitious military effort mounted by U.S. forces to that point in their history. More than a million Americans, most of them without prior combat experience, participated in the month-and-a-half-long U.S. offensive launched in the sector between the Meuse River and the French Fourth Army boundary west of the Argonne Forest on 26 September 1918. That operation was part of a series of coordinated Allied offensives that would hopefully bring the war to an end.

Before the offensive could begin, however, more than 220,000 French troops had to clear out of the area, to be replaced by more than half a million Americans making the sixty-mile trek from the Saint-Mihiel region under cover of darkness. That operation, skillfully coordinated by Col. George C. Marshall of the First Army's operations staff, employed several ruses aimed at making the Germans believe the Americans were planning continued action in the Saint-Mihiel sector. Two of those were demonstrations by fourteen 345th Battalion Renaults led by 1st Lt. Ernest A. Higgins, on the nights of 22-23 and 23-24 September. They rejoined their parent battalion near Boureuilles on the twenty-seventh, the second day of the Meuse-Argonne campaign.[38]

The First Army operation plan called for the attack to kick off at 5:30 A.M. on the twenty-sixth after a three-hour artillery preparation. The sector was divided into three roughly equal corps zones. The III Corps sector was on the right, bounded by the Meuse River in the east and in the west by a line drawn just east of the high ground at Montfaucon. The V Corps was assigned the center sector, sharing its eastern boundary with III Corps on the right and its western boundary—a north-south line east of Exermont, Cheppy, and Neuilly—with I Corps on the left. The I Corps sector included the Aire River and Argonne Forest in the western third of the First Army zone.[39]

General Rockenbach, after reviewing a reconnaissance report prepared by one of Lt. Col. Wahl's officers, allocated Pullen's larger 3d Brigade to V Corps, whose sector had two avenues of approach for tanks, and Patton's 1st Brigade to I Corps, which had only one clearly definable avenue for tanks. Pullen, in turn, instructed Wahl and his 1st Assault Artillery Brigade to support the advance of the 37th and 79th Divisions as soon as they were north of the Bois de Montfaucon. The French subsequently added the 17th Light Tank Battalion with forty-eight Renaults to the 505th Regiment and Pullen ordered it to operate in the 91st Division sector on the V Corps left.[40]

Patton's plan called for employment of his brigade in depth across a mile-and-a-half-wide corridor extending north from Clermont and bounded by the Argonne Forest in the west and the Bois de Cheppy near Varennes and Boureuilles in the east. The I Corps assault divisions were the 28th and 35th, and their sectors were divided by the Aire River. Patton's natural inclination was to assign a single battalion to support each division. However, because of the river's position in the corps zone, that option was not viable. The 35th Division sector could support just two companies, and the 28th's sector had maneuver space for only one. Patton called on Brett and the 344th to take the lead, with A and C Companies trailing lead elements of the 35th Division and B Company supporting the 28th Division's advance. Compton was instructed to array his companies in similar fashion about a mile behind Brett's battalion, and Chanoine's Schneiders were given the mission of trailing a little more than a mile behind the 345th Battalion until the ground opened

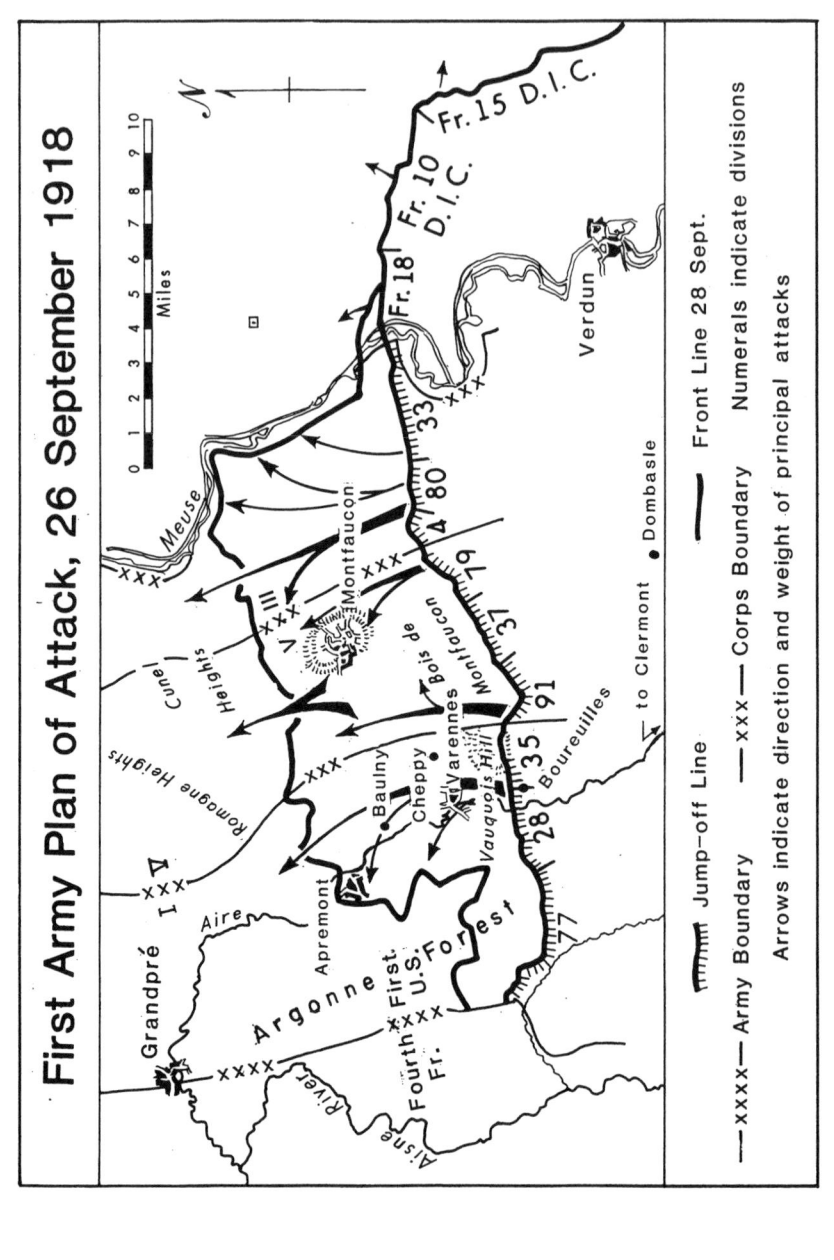

First Army Plan of Attack, 26 September 1918

Miles
0 1 2 3 4 5 6 7 8 9 10

N

Fr. 15 D.I.C.

Fr. 10 D.I.C.

Fr. 18

Verdun

Meuse

Montfaucon

Bois de Montfaucon

33

80

4

79

37

91

35

28

77

Cunel Heights

Romagne Heights

Baulny

Cheppy

Varennes

Vauquois Hill

Boureuilles

• Dombasle

to Clermont

Grandpré

Aire

Apremont

Argonne Forest

First U.S.

Fourth Fr.

Aisne River

First

V

I

V

III

XXX

XXXX

XXX — Corps Boundary   Numerals indicate divisions

Arrows indicate direction and weight of principal attacks

Jump-off Line

Army Boundary

Front Line 28 Sept.

up. Patton's intent was for Brett to lead the way as far as the first day's objective, then have Compton leapfrog forward and support the advance to the next objective.[41]

Still smarting from the vision of a teary-eyed Brett bemoaning the shortage of fuel that halted his advance at Nonsard on the first day of the Saint-Mihiel operation, Patton ordered that two twenty-liter gas cans be tied to the tail of every tank. Although this "created some danger from fire, . . . the risk was thought preferable to the lack of gas."[42]

The brigade's command post was established at the 35th Division's advance headquarters in the Côtes de Forimont. But Patton, Rockenbach's earlier admonishment notwithstanding, had no intention of remaining there. He planned to maintain contact with Tank Corps headquarters by taking nearly a dozen runners with him as he advanced up Route Nationale 46 with the leading tanks.

Major repair operations were to be handled by the 321st Repair and Salvage Company at Camp Fourgous, with forward maintenance support provided by a tank from each company designated to carry extra fan belts and other spare parts plus a mechanic from battalion. The latter served as the genesis for the tank company maintenance teams that were employed with such success in subsequent wars and which have remained a part of the U.S. Army's tank battalion organization ever since. The idea came from one of the brigade's privates, who observed during the Saint-Mihiel campaign that tanks suffering from minor malfunctions such as broken fan belts had to be evacuated to the rear for repair. That the idea came from a Soldier in the ranks should not be seen as unusual. Capt. Harry H. Semmes, commander of Company A, 344th Tank Battalion, recalls that Patton regularly sought suggestions from junior officers and enlisted men "because he knew that those intimately working with and fighting tanks would furnish worthwhile ideas."[43]

On the eve of battle, Patton could count on Chanoine's 28 Schneiders, 69 Renault's in Brett's 344th Battalion, and 58 in the 345th (Compton was still waiting on the return of Higgins's 14 Renaults from the demonstrations in the Saint-Mihiel sector). Pullen had a total of 250 French-manned vehicles available to support V Corps. This brought the total number of tanks supporting First Army back to 419 despite the losses suffered at Saint-Mihiel.[44]

Unlike in the Saint-Mihiel sector, the Germans were prepared to make a stand in the Meuse-Argonne. Their positions incorporated all of the tactical innovations they had developed the year before into a massive, five-layer deep defensive belt. When the sound of whistles split the air at 5:30 A.M. on 26 September and the doughboys in the assault divisions began streaming out of their trenches, it became quickly apparent that this offensive would be every bit as difficult as the Saint-Mihiel affair had been easy. Although initial German resistance was light—as expected—the advancing Americans were plagued by dense fog, shell-pocked terrain, tangled undergrowth, barbed wire, and their own inexperience. All of these factors combined to make it difficult for the attackers to maintain contact with the protective artillery barrages rolling inexorably toward the German rear.

Brett's company Commanders and platoon leaders, unable to see, were forced to dismount and precede their units on foot, picking their way through an eerie moonscape shrouded by fog and smoke that was occasionally disturbed by machine-gun fire and shell bursts. Although mud did not hinder maneuver as badly as it had at Saint-Mihiel, increasing German resistance—especially direct fire from 77mm guns and heavy machine-gun fire—began to exact a toll on the advancing tankers. Within a matter of a few hours Captain Semmes was wounded and evacuated and Capt. Newell P. Weed, the B Company Commander, was captured as he worked his way forward through the German trenches. Fortunately for Weed, the Germans who held him panicked and ran when a tank caught up with them. He was soon reunited with his company.

Capt. Dean M. Gilfillan, Commander of Company A, 345th Battalion, which trailed Weed's company in the 28th Division sector, was wounded by machine-gun fire but remained with his tank despite its being hit twice by artillery fire. As flames spread through the vehicle during an advance on Varennes, Gilfillan was finally forced to get out. He was wounded a second time by fragments from a nearby shell burst as he struggled to get clear before the tank exploded.

A platoon leader in C Company, 345th Battalion—2d Lt. Guy Chamberlain—became the first Tank Corps officer to die in action when he was shot while walking in front of his tanks, trying to lead them through difficult terrain in the vicinity of Vauquois Heights.[45]

Patton, too, became a casualty on the morning of the twenty-sixth. Ignoring Rockenbach's orders to remain at his command post, the brigade Commander set out on foot with his detachment of runners and a handful of staff officers. He caught up with the lead elements just south of Cheppy, where the 35th Division's attack had faltered. Confusion reigned and Patton was seriously wounded by a machine-gun bullet in the thigh while trying to restore control and knock out the machine-gun emplacements in and around Cheppy.[46]

Major Brett, who took over the brigade after Patton's evacuation, soon discovered that his new command had been decimated. Forty-three vehicles were lost on the first day and the number of operational vehicles in the 1st Tank Brigade dwindled to 83 by the morning of the third day of the offensive. That figure included the 14 Renaults that had been left in the Saint-Mihiel sector to perform demonstrations and which rejoined the rest of the brigade on the night of the twenty-seventh. Although far more vehicles had fallen prey to German guns than had at Saint-Mihiel, the tanks' worst enemies continued to be mechanical failure, mud, and the deep German trenches.

Chanoine's Schneiders, which had reached their limit mechanically, were withdrawn from the I Corps sector on 29 September, leaving Brett and Compton with a force of just 55 Renaults. The rest of First Army was in similar condition and Pershing ordered a halt so his forces could regroup and prepare for a renewed assault on the Hindenburg Line on 4 October.[47]

The tankers' success had not come cheaply. In just four days of fighting the 1st Tank Brigade suffered 53 percent of its officers and 25 percent of its enlisted men killed or wounded. The brief respite permitted mechanics to restore the brigade to 50 percent of its force operationally ready. However, Rockenbach informed Pershing that, assuming a five-kilometer-per-day advance when operations resumed on the fourth, Brett's brigade would be combat-ineffective by the tenth.[48]

The situation was nearly as grim in the French tank units supporting V Corps. Although they had been less a factor on the first day of the offensive because of the thick woods in the forward part of the V Corps zone, the fighting around Montfaucon took an equally heavy toll of French men and vehicles. Rockenbach issued a similarly negative forecast for their performance once operations resumed.[49]

Rockenbach's two brigades retained their original assignments when the offensive began again on 4 October: Brett's 1st Brigade supporting I Corps on the left and Pullen's 3d Brigade continuing to act as liaison to the French tank units supporting III and V Corps operations. Brett, with eighty-nine tanks operational, allocated two companies to the 1st Division and one company to the 28th Division west of the Aire River. The remaining 1st Brigade tanks were held in reserve.[50] Pullen's 3d Tank Brigade was able to muster just nine St.-Chamonds to support the III Corps attack. Two French Renault-equipped battalions were committed in the V Corps zone—the 15th in support of the 3d Division and the 17th with the 32d Division.[51]

The first day of the attack proved disappointing at best. III Corps units were stymied by effective fire from the commanding high ground east of the Meuse River, and fierce German resistance slowed movement in the other two corps sectors. Only the 1st Division made any

appreciable gains, advancing about a mile and a half to the outskirts of
Fléville, where enfilading fire from high ground west of the Aire River
forced the division to halt. Withering fire from massed German machine guns, antitank rifles, and 77mm guns—often firing at close range
over open sights—took a heavy toll of men and machines. Mines also
accounted for at least some vehicle losses. By the morning of 5 October,
Brett could count only thirty tanks ready for action. Although Brett
divided his remaining vehicles equally between the two divisions, only
a single platoon saw action that day. Working in support of a 28th
Division assault on German positions in Le Chêne Tondu—a rocky out-
cropping on the west side of the Argonne Forest offering commanding
fields of fire over the I Corps zone—1st Lt. Harry E. Gibbs and his men
were met by a fierce blast of artillery, mortar, and machine-gun fire.
Three of his five tanks were knocked out almost immediately and the
assault ground to a halt after gaining just fifty yards. Gibbs was wounded
but remained in his tank and helped repulse a German counterattack
before being evacuated.

No 1st Brigade tanks saw action on the sixth, a handful supported
the 28th Division the next day, and none were employed from 8-10
October. By the eleventh, only forty-eight Renaults were still operable
in the 1st Brigade. A force of twenty-three under the command of Capt.
Courtney H. Barnard of the 345th Tank Battalion was sent to support an
attack by the 82d Division's 164th Brigade. The division had asked for
only five tanks, but Captain Compton sent as many as possible to ensure the required platoon made it to Fléville from Varennes. As it was,
only four tanks managed to link up with the 164th, and two of those
broke down shortly after the attack began, forcing the remainder to
withdraw as Rockenbach had issued an order that tanks were not to go
into action in less than platoon strength.

That brought an end to the 345th Battalion's participation in the
campaign. Compton was relieved of his duties as Commander of the
brigade's forward elements on 12 October and ordered to assemble his
men at Varennes in preparation for movement back to the light tank
center at Bourg. Similar instructions were issued to the 344th Battalion.
However, Brett and a number of men from both battalions remained
behind to support I Corps until reinforcements from Bourg arrived on
the scene. Maintenance crews from the 321st Repair and Salvage Company worked around the clock to restore as many tanks as possible to
operating condition.[52]

Similar failures dogged French tank efforts in support of III and V
Corps attacks on the heights at Cunel and Romagne. Operations there
were further marred by the assault divisions' lack of experience with

tanks and the language barrier, which made liaison difficult. On 10 October, after a week of disappointing results, the 1st Assault Artillery Brigade was withdrawn from First Army control. French losses during the campaign were 1 officer and 19 enlisted men dead, 1 officer and 21 men missing, and 16 officers and 109 men wounded. They lost 25 tanks to artillery fire, 2 to mines, and abandoned 22 more behind German lines.[53]

Brett's provisional tank company, led by Captain Barnard, saw action on two more occasions: 14 October, the last First Army operation conducted under Pershing's command, and 1 November. The latter operation was so successful that the mechanically spent Renaults were unable to keep up with the rapidly advancing doughboys and the war ended without any more tanks seeing action.

The AEF's tank brigades were redesignated on 6 November to bring them in line with the War Department's numbering system. The 1st Brigade became the 304th Tank Brigade, the 3d Brigade became the 306th Tank Brigade, and the newly organized 2d and 4th Brigades became the 305th and 307th Tank Brigades.[54]

The 304th Brigade's casualty toll in the Meuse-Argonne campaign was 3 officers and 16 enlisted men killed, and 23 officers and 131 enlisted men wounded. These numbers represent nearly 43 percent of the officers and 21 percent of the enlisted men participating. Vehicle losses to all causes amounted to 123 percent. Despite organizational maintenance problems encountered at battalion level, the fact that mechanics were able to recover, repair, and return tanks to action at the rate they did reflects well on their efforts. A total of eighteen American-manned Renaults were destroyed by German fire. Many more were damaged and subsequently restored to action.[55]

The AEF tankers' aggressive spirit is further confirmed by the comparatively high number of valor awards made to officers and men in the 304th and 306th Tank Brigades during the Meuse-Argonne offensive: two Medals of Honor and twenty-one Distinguished Service Crosses.[56]

The 301st Tank Battalion deployed to France from Bovington, England, on 23 August 1918. There the battalion drew a mix of forty-six Mark V and Mark V Star heavy tanks from the British on the thirtieth and began performing maintenance on the vehicles while waiting for orders to move to the front.[57]

The Mark V Star tank was more than thirty-two feet long and weighed about thirty-seven tons. It came in two versions: a "male" mounting two six-pounder cannon and five 8mm machine guns, and a "female" sporting seven 8mm machine guns. The Mark V measured six feet shorter and weighed in at about thirty-two tons. It came in three versions: a male and female model each mounting one less machine gun than its Star tank counterpart, and a "hermaphrodite," which carried a single six-pounder cannon and five 8mm machine guns.[58]

The battalion finally got its marching orders on 19 September. It was to be attached to the British 4th Tank Brigade, which would support the Fourth Army's Australian Corps in its assault on the Hindenburg Line in the vicinity of the Saint-Quentin Canal on 29 September. The U.S. II Corps, which had been placed under the operational control of the Australian Corps, would provide the assault divisions—the 27th and 30th—giving the Aussie infantry a much-needed break.

The 301st would be going into battle with a new Commander: Maj. Ralph I. Sasse, an aggressive cavalryman recruited by Patton in France. Sasse joined the battalion at Bovington Camp as tactical (operations) officer. However, shortly after they deployed to France, Rockenbach tapped Lt. Col. Henry E. Mitchell, the battalion Commander, to take command of the soon-to-be-activated 2d Tank Brigade and Sasse replaced him at the helm of the 301st. It was a move that sat well with both leaders. The tankers had developed great respect for Sasse, whereas Mitchell was held in "low esteem."[59]

Sasse devised a simple scheme of maneuver for his battalion: Each company was assigned to lead one of the 27th Division's regiments, and the companies allocated a platoon to each of the regiments' battalions. The tanks would precede the doughboys by about a hundred yards, clearing wire obstacles and destroying German machine-gun and artillery emplacements.[60] This plan was in perfect harmony with approved doctrine. Nevertheless, Sasse lacked confidence in the infantrymen who would be following them. The 27th Division had no previous experience working with tanks, and Sasse, who had no opportunity to rehearse with his infantry counterparts, later observed that they simply "did not seem to grasp the idea of Tanks co-operating with Infantry."[61]

Compounding their problems was the fact that they had to move their tanks more than thirty miles to reach the front lines. On the twenty-seventh the British gave Sasse a heavy tank equipped with radio gear and instructions to employ it in the center of the sector behind the lead battalions in the two assault regiments. That night, the 301st left its initial assembly area and moved twenty-five miles to a forward assembly area eight miles from the jump-off line. After refueling on the twenty-eighth, a painstaking operation that involved emptying two-gallon gas cans into the vehicles' huge tanks, the heavy tankers moved out under cover of darkness for the final leg of their trip to the front.

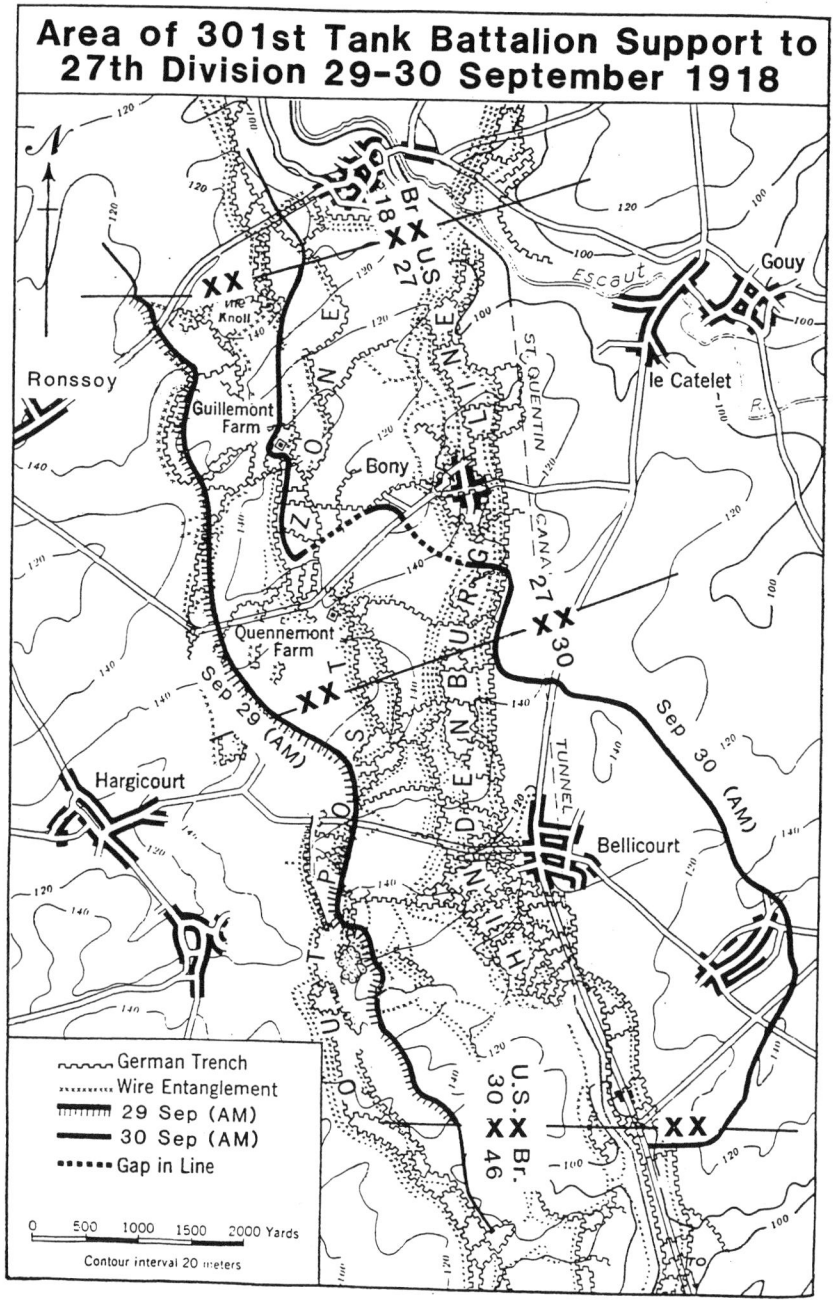

Area of 301st Tank Battalion Support to 27th Division 29-30 September 1918

Ronssoy

Gouy

le Catelet

Guillemont Farm

Bony

Quennemont Farm

Hargicourt

Bellicourt

Sep 29 (AM)

Sep 30 (AM)

TUNNEL

German Trench
Wire Entanglement
29 Sep (AM)
30 Sep (AM)
Gap in Line

0   500   1000   1500   2000 Yards

Contour interval 20 meters

Captain Ralph deP. Clarke, the C Company Commander, said the approach march was particularly hard on the officers as they preceded their units on foot, stumbling through the pitch dark over sunken roads and muddy fields, through barbed-wire entanglements, and across shell holes and abandoned trenches. Platoon leaders ran up and down the columns in an effort to keep the lines of vehicles moving steadily forward, and individual tank Commanders dismounted and guided their drivers forward, communicating instructions by frantically waving a glowing cigarette.[62]

The tanks reached the jump-off line at 5 A.M., the new H hour. Unfortunately, only A Company received infantry support as it ground forward through the fog and crushing German machine-gun and artillery fires. C Company, which was operating in advance of the 107th Infantry Regiment, got a one-hour head start because the regimental Commander refused to advance until the originally scheduled H hour. The doughboys in that sector never closed the gap and the operation was a disaster. The inexperienced doughboys fought their way forward aggressively enough. However, they failed to mop up pockets of German resistance they bypassed. Many more Germans escaped into the Saint-Quentin Canal tunnel, emerging onto the battlefield from hidden passageways after the Americans passed.

Both the 301st and the 27th Division suffered frightful casualties—without taking a single one of the day's objectives. In some places, attacking units barely covered a thousand yards. Lt. Gen. Sir John Monash, the Australian Corps Commander, realizing the severity of the situation, pulled the badly mauled U.S. divisions out of the line and ordered his Australians to continue the attack the next day.

It was impossible to get an accurate count of American infantry casualties because of the number of men who became separated from their units, either isolated behind German lines or intermingled with the Australians moving forward to assume the lead. Estimates of 107th Regiment casualties range as high as 50 percent in the assault battalions—about a thousand men. The 301st's casualty toll was less difficult to tally: 3 officers and 17 enlisted men killed, 15 officers and 70 enlisted men wounded, and 7 enlisted men missing. Of the 40 tanks that began the operation (7 were held in brigade reserve), 2 were destroyed by mines the British left in an unmarked minefield, 16 were lost to artillery fire, and the remainder—except for the radio tank—were either ditched in trenches or had fallen prey to mechanical problems.[63]

The heavy tankers spent the next week working around the clock in an effort to get as many of their vehicles ready for action as possible. By 5 October, when Sasse received word to prepare for another attack—

this time in support of the 30th Division—twenty-two of his tanks were operational.

The Australians had finished clearing the Germans out of their Hindenburg Line positions in the Saint-Quentin Canal region and captured the Beaurevoir Line before handing the sector over to Maj. Gen. George W. Read's U.S. II Corps on the sixth. Read's instructions were simple: keep up the pressure on the Germans. He did this by ordering the 30th Division to seize Brancourt on 8 October.

Unlike the Saint-Quentin Canal operation, the Brancourt attack was launched without an artillery preparation. However, the plan did call for a creeping barrage to commence at H hour, so Sasse ordered his tanks forward in the predawn darkness to ensure they reached the start line when the artillery fires began. The Germans cooperated by not shelling the area and by illuminating a searchlight in the objective area, giving the tankers something to guide on.

It was apparent from the beginning that the battalion's luck had changed. The terrain was more favorable and trafficability was ideal. The rolling barrage pounded the enemy's forward positions and a favorable wind carried an artillery smoke screen into the German trenches, blinding machine gunners and forward artillery observers. Finally, the doughboys—who had lost confidence in the British tankers supporting them at Saint-Quentin—fought their way forward with their aggressive Yank counterparts in the 301st. After capturing Brancourt, several tank Commanders, caught up in the excitement of their success, disregarded orders to halt and continued on to the second objective with the exploitation force.

The operation was a morale-boosting success for the men of the 301st. Although nine of the twenty vehicles that began the attack fell prey to various mechanical problems or German artillery fire, casualties were relatively light: two enlisted men killed and three officers and thirteen men wounded. Sasse gave high marks to the doughboys for their support, and they in turn praised the tankers. The presence on the battlefield of large numbers of German bodies riddled with case shot from the tanks' six-pounder guns bore mute testimony to their effectiveness.[64]

The 301st participated in two more attacks before its vehicles reached their mechanical limits. The first of these, in the vicinity of the Selle River on 18 October, was part of a general offensive conducted by the Fourth Army with three corps abreast. The heavy tankers again supported II Corps, with A Company operating ten tanks in the 27th Division sector and B and C companies consolidating with the battalion's remaining fifteen tanks in the 30th Division zone. As they had on 29

September, fog and mechanical problems combined to hinder the tanks' effectiveness. Only one vehicle, a tank commanded by 2d Lt. William C. Rock of A Company, was lost to enemy fire. Rock's tank was hit by a trench mortar, which set it aflame and blew off a track. The intrepid officer helped drag wounded crewmen to safety. Then, armed only with a pistol, he charged a machine-gun position but was killed before he could reach the Germans. Casualties in the battalion were light. In addition to Rock, one enlisted man died and eight more were wounded—mostly by gas. Only three of the twenty-five tanks that began the assault reached the objective. By the end of the day, the battalion had just seventeen vehicles operational—all in need of a major overhaul. Although the rest of II Corps was withdrawn for much-needed rest and reorganization, Sasse and his men remained with the British 4th Tank Brigade to help Fourth Army maintain pressure on the reeling Germans.[65]

The battalion was given a few days to rest and refit in Busigny before again being alerted for what proved to be its final operation: a night attack in support of the British 1st and 6th Divisions on 22-23 October in the vicinity of the Sambre Canal. Like Brett in the Meuse-Argonne sector, Sasse was forced to consolidate his few remaining men and machines into a provisional company. Captain Clarke was given command of this force, which mustered twelve tanks divided equally into three platoons. Sasse, who placed considerable stock in reconnaissance and coordination, had to make do with little of either. Reconnaissance efforts were hampered by the lack of time, constant changes in the frontline trace, and the lack of a detailed operations plan. The battalion's shortage of motor transport and the distance to the supported infantry brigades' headquarters limited coordination to a single briefing for platoon leaders by their respective brigade Commanders on the evening of the attack. Despite these handicaps, the operation went smoothly and the battalion escaped without casualties. Nevertheless, both men and machines had reached the limits of their endurance; the 301st was finally withdrawn from the line for good.[66]

The Tank Corps's personnel strength peaked with the signing of the Armistice. The AEF Tank Corps counted 752 officers and 11,277 enlisted men in its ranks, and an additional 483 officers and 7,700 men were in stateside units—divided about equally between Camps Colt and Polk.[67] Congress wasted no time reducing the Tank Corps's authorized strength to 300 officers and 5,000 men. Rapid demobilization quickly followed.

Lieutenant Colonel Eisenhower took his remaining troops to Camp Dix, New Jersey, for demobilization in early December and was instructed to discharge all but 250 enlisted men. Col. William H. Clopton

Jr., Commander of Tank Corps units at Camp Polk, was instructed to retain two hundred men and ship them to either Camp Dix or Camp Benning, Georgia. Only thirty-one Tank Corps officers in the United States were to be retained on active duty.[68]

Brigadier General Rockenbach presided over a similarly rapid dismemberment of his organization. Although a few tank battalions were assigned to perform occupation duty, the majority gathered at the 302d Tank Center at Bourg to prepare for their return to the United States and demobilization or reassignment to Camp Meade, which was to become the new "home" of the Tank Corps. In March 1919, Tank Corps units from overseas began arriving at Meade, where they were joined by Eisenhower and Clopton and their small contingents. While Rockenbach remained in France watching over the redeployment of the remnants of the AEF Tank Corps, Colonel Welborn retained his position in Washington as the stateside Tank Corps director. Colonel Clopton, the ranking Tank Corps officer at Camp Meade, became post Commander and served as Welborn's deputy.

Tank production efforts, which had failed miserably during the war, soon began to bear fruit. As noted earlier, components for the Mark VIII Liberty tank were shipped to Rock Island Arsenal for assembly. Meanwhile, the Maxwell Motor Company and affiliated contractors, who had completed 64 copies of the Renault light tank by the time the Armistice was signed, built an additional 714 by 31 March 1919. Total

A Mark VIII Liberty heavy tank at Camp Meade, Maryland. National Archives.

production of the M1917 reached 952 before contracts were canceled later that year. In addition, Tank Corps units redeploying from France brought back 218 French-built Renaults and 28 British Mark Vs.[69]

While their men kept busy with maintenance, training, and demonstrations, Tank Corps officers entered into a protracted campaign to ensure the survival of their organization as a separate combat arm. The battle began in the summer of 1919 and continued through the first half of 1920. Congress, responding to the national backlash against war and all things military, fired the opening salvo by passing a law in July 1919 that included further reduction in the Tank Corps's strength to 154 officers and 2,508 enlisted men. Rumors quickly began circulating that the next step would be a major, across-the-board reduction in force that would include releasing the majority of officers commissioned from the enlisted ranks or from officer training schools during the war and restoring regular officers to their permanent ranks.

A month later, Rockenbach was ordered to Camp Meade to assume control of the remnants of the tank force. As Commandant, he was in a particularly strong position to fight for the Tank Corps's survival. Unfortunately, he proved to be the wrong man for the job. Although Rockenbach spoke often and wrote several articles and reports on tank tactics and operations, he seemed more interested in maintaining the status quo than in promoting research, development, and rigorous training—three essentials for the creation of a vigorous, improving force. Maj. Gen. Charles P. Summerall, whose 1st Infantry Division and V Corps had been beneficiaries of tank support in the Meuse-Argonne campaign, was especially critical of the Tank Corps's chief. Commenting on Rockenbach's 1919 lecture to the General Staff, Summerall advised Rockenbach that although he had a sympathetic audience, his presentation had been far too conservative and did little to bolster his case.[70]

It thus fell on the younger tank officers to take up the cudgel on behalf of their branch. This they did—with a relish. In 1920, when the debate over whether tanks should be retained as a separate combat arm or given to the infantry, both Patton and Eisenhower wrote and published articles expounding the idea that tanks were far more than just an infantry-support weapon. The thrust of their argument was that large fleets of light, fast tanks employed in mass rather than scattered about the battlefield and tied to the pace of the infantry's advance could break into the enemy's rear and unhinge his defensive positions.

Patton's article, which was highly provocative in both language and conception, appeared in the *Infantry Journal* in May. He concluded it with a vain plea to save the Tank Corps, which he believed would, if "grafted on infantry, cavalry, artillery or engineers . . . be like the third leg to a duck—worthless for control, for combat impotent."[71]

While the Army's senior leadership debated the Tank Corps's future, an increasingly penurious Congress viewed it as a matter of simple economics: The nation could ill afford a separate organization for tanks and the concurrent overhead it would engender. With that in mind, Congress passed the National Defense Act on 2 June 1920. Section 17 of that legislation dissolved the Tank Corps and assigned all tank units to the infantry. It was a bitter pill for the tankers, many of whom had come from the cavalry and viewed affiliation with the infantry as anathema. It was especially bitter because the flamboyant Brig. Gen. William "Billy" Mitchell, the leading proponent of airpower, had overshadowed the lackluster Rockenbach and succeeded in convincing Congress to establish a separate Air Service. The National Defense Act also codified earlier rumors about reductions in the Officer Corps: Regulars would revert to their permanent grades at the end of June, and boards were established to identify all but a handful of temporary officers for separation.

Tank Corps morale plummeted with the collapsing rank structure. Rockenbach lost his star, and many officers—among them Patton and Eisenhower—traded in silver eagles and oak leaves for captains' bars.

From left, Ford three-ton light tank, M1917 light tank, Mark VIII Liberty heavy tank. National Archives.

Capt. Dwight D. Eisenhower poses beside an M1917 light tank at Camp Meade, Maryland, in 1920. National Archives.

An exodus soon followed. Among the first to leave was Ralph Sasse, who returned to the cavalry in June. Patton watched the departures of Henry Mitchell and Daniel Pullen, his fellow tank brigade Commanders, before finally deciding in mid-August that he had only a limited future with tanks. He petitioned Rockenbach for reassignment to the cavalry and was allowed to make the move at the end of September.

Shortly after Patton's departure, Eisenhower's article on tanks appeared in the *Infantry Journal.* Although his arguments followed lines similar to Patton's, they were couched in characteristically more tactful terms.[72] Nevertheless, they angered the chief of infantry, who called Ike on the carpet and ordered him to keep his views to himself. Not long after that, Eisenhower petitioned Rockenbach for a transfer. The top tanker dutifully passed the request along to the War Department with an endorsement asking that it be denied. Brig. Gen. Fox Conner finally rescued Eisenhower from his career dilemma by getting Pershing to personally approve Ike's transfer to Conner's brigade staff in Panama in January 1922.

Only Sereno Brett, of all the WWI Tank Corps's leaders, remained committed to tanks during the interwar years. Brett commanded the Expeditionary Tank Force in Panama in 1923-24 and served as executive officer of the Experimental Mechanized Force created at Meade in 1930-31. He remained a vocal proponent of mechanization throughout those long, lean years.[73]

NOTES

1. Brig. Gen. Samuel D. Rockenbach, "Report of Chief of the Tank Corps," 13 October 1919, copy in U.S. Army Military History Institute, Carlisle Barracks, Pa. (hereafter USAMHI), p. 4375.

2. Two of Patton's biographers, Martin Blumenson and Carlo D'Este, examined Patton's WWI Tank Corps experiences in great detail. Dwight D. Eisenhower's biographers devote little space to his experiences training tank crewmen at Camp Meade, Md., and Camp Colt, Pa. The only detailed narrative account of Tank Corps activities in the United States and training and operations in the American Expeditionary Forces in France is Dale E. Wilson's *Treat 'Em Rough!: The Birth of American Armor, 1917-20* (Novato, Calif.: Presidio, 1989).

3. Martin Blumenson, *The Patton Papers, 1885-1940* (Boston: Houghton-Mifflin, 1972), p. 437, and Timothy K. Nenninger, "The World War I Experience," *ARMOR* 78, no. 1 (Jan.-Feb. 1969): p. 47.

4. Samuel D. Rockenbach, "Operations of the Tank Corps, A.E.F., with the 1st American Army," Dec. 1918, copy in USAMHI, p. 2-3.

5. Wilson, *Treat 'Em Rough!*, pp. 10-11.

6. Patton, report, "Subject: Light Tanks," 12 Dec. 1917, George S. Patton Jr. Collection, Patton Chronological File, 11-12 Dec. 1917, Library of Congress, Washington, D.C. (hereafter cited as Patton Collection); Patton, "History of the 304th (1st) Brigade Tank Corps," undated, Patton Writings, Box 49, Military Writings 1918-20 File, Patton Collection, 1; and Braine, "Personal Experience Report," 22 Dec. 1918, Patton Military Papers, Box 47, Personal Experience Reports of Tank Operations-1918 File, Patton Collection.

7. Blumenson, *Patton Papers*, pp. 444-46.

8. Patton, "Light Tanks," p. 1; Rockenbach, "Tank Corps Operations," p. 4.

9. Patton, ibid.

10. As quoted in Blumenson, *Patton Papers*, p. 447.

11. Patton, "Light Tanks," p. 1.

12. Ibid., "Attachment C, Tactical."

13. Keeping track of Tank Corps unit designations is a difficult task. Once a table of organization for tanks in the AEF was approved, Rockenbach issued an order delineating how units would be designated. Light tank battalions were to be designated 1 to 40 and heavy battalions 41 to 50 in order of activation. Companies would be designated A, B, and C within each battalion. Tank centers were to be numbered 1st, 2d, and so on as they

were formed. Repair and salvage companies would take on the numerical designation of the centers to which they were allocated. Meanwhile, in the United States, the first heavy tank unit was originally designated the 1st Separate Tank Battalion, Heavy Tank Service, 65th Engineers. It was subsequently redesignated the 41st Tank Battalion upon arrival at the 2d Tank Center in Bovington, England. The AEF's numbering system was scrapped in June 1918 to bring it in line with the system devised by the War Department earlier in the year. The 1st and 2d Tank Battalions thus became the 326th and 327th, and the 41st Tank Battalion became the 301st. Patton's 1st Tank Center became the 311th and the 2d Tank Center at Bovington became the 312th. See General Orders no. 5 (dated 18 Apr. 1918) and no. 10 (dated 8 June 1918), General Tank Headquarters, AEF, Modern Military Records Division, RG 165, Entry 310, Box 446, File 66-32.13, National Archives, Washington, D.C. (hereafter cited as NA); and *Order of Battle of the United States Land Forces in the World War (1917-19), Zone of the Interior,* vol. 3, part 2 (Washington: GPO, 1949), pp. 1543-44.

14. Patton, "Light Tanks, Attachment D, Instruction."

15. Wilson, *Treat 'Em Rough!,* pp. 52, 54-55.

16. Ibid., Chapter 4.

17. Drain and Alden, "Report of Investigations by Majors Drain and Alden," 10 Nov. 1917, included as an enclosure to Maj. Raymond E. Carlson, "Memorandum on the Development of Tanks," 15 Mar. 1921, USAMHI.

18. The saga of the Mark VIII can be traced through a series of cables included as enclosures to Lt. Col. Robert L. Collins's undated "Report on the Development of the Tank Corps," Modern Military Records Division, RG 120, Entry 22, Folder 387, NA.

19. Braine, "Personal Experience Report."

20. Ibid.

21. Rockenbach, "Tank Corps Operations," p. 7.

22. Braine, "Personal Experience Report."

23. Ibid.

24. See the table included as an enclosure to Collins, "Development of the Tank Corps."

25. Rockenbach, "Tank Corps Operations," 10, and Lt. Col. Emile Wahl, "Operations of the French Tank Corps with the First American Army," included as Appendix 7 to Rockenbach, "Tank Corps Operations," p. 1.

26. Will G. Robinson to Lt. Col. Arthur J. Jacobsen, 6 July 1961, Patton Military Papers, Box 45, Miscellaneous Military Papers File, Patton Collection.

27. Anyone examining the records of the Saint-Mihiel campaign is apt to wonder just how many light tank battalions were involved in the battle. The problem stems from the fact that the AEF learned on the eve of the attack that two of the light tank battalions en route from the United States bore the same unit designation as Patton's battalions. On 12 September,

Rockenbach's headquarters issued General Order no. 16 calling for the redesignation of the 326th Tank Battalion as the 344th and the 327th as the 345th. The same order called for the 311th Tank Center at Bourg to become the 302d. Unfortunately, the change in designations was slow to be implemented and was not consistently applied until the beginning of the Meuse-Argonne campaign. See Modern Military Records Division, RG 120, Entry 1296, NA.

28. As quoted in Blumenson, *Patton Papers*, p. 599.

29. Rockenbach, "Tank Corps Operations," pp. 3, 12.

30. Patton, "Operations of the 304th Tank Brigade, September 12th to 15th, 1918, Saint-Mihiel Salient," included as Appendix 4 to Rockenbach, "Tank Corps Operations," p. 4.

31. Capt. Ranulf Compton, "War Diary of the 345th Tank Battalion," World War I Survey Collection, Archives, USAMHI.

32. Blumenson, *Patton Papers*, pp. 589-90.

33. Wilson, *Treat 'Em Rough!* pp. 115-16.

34. Blumenson, *Patton Papers*, 594; 1st Lt. Edwin A. McClure, "Personal Experience Report," undated, Patton Military Papers, Box 47, Personal Experience Reports of Tank Operations—1918, Patton Collection.

35. Wilson, *Treat 'Em Rough!*, p. 117.

36. Patton, "Appendix 4, Tank Corps Operations," p. 6.

37. Brig. Gen. Samuel D. Rockenbach to commanding officers of American Tank Brigades, Subject: Notes on American Tanks, 14 Sept. 1918, Patton Chronological Files, Box 10, 13-16 Sept. 1918, Patton Collection.

38. Wilson, *Treat 'Em Rough!*, pp. 124-29.

39. Edward M. Coffman, *The War to End All Wars: The American Military Experience in World War I* (New York: Oxford University Press, 1968; reprint, Lexington: University Press of Kentucky, 1998), pp. 304-305.

40. Rockenbach, "Tank Corps Operations," p. 18.

41. Patton, "304th Brigade History," pp. 29-30 and "Operations of the 304th Brigade, Tank Corps from September 26th to October 15th 1918," 18 Nov. 1918, Patton Chronological Files, Box 11, 7-20 Nov. 1918, Patton Collection, p. 3.

42. Patton, "304th Brigade History," p. 31.

43. Harry H. Semmes, *Portrait of Patton* (New York: Appleton-Century-Crofts, 1955), p. 52.

44. Wilson, *Treat 'Em Rough!*, pp. 136, 160 n31.

45. Ibid., pp. 136-39.

46. Ibid., pp. 141-44.

47. Ibid., pp. 147, 151-53, 156.

48. Rockenbach, "Tank Corps Operations," p. 23.

49. Ibid., pp. 23-24, and Wilson, *Treat 'Em Rough!*, p. 147.

50. Brett, "Operations Report of the 1st (304th) Tank Brigade from Noon,

September 26th, 1918, to November 10, 1918," included as Appendix 6 to Rockenbach, "Tank Corps Operations," p. 3.

51. Rockenbach, ibid., pp. 31-32.

52. Wilson, *Treat 'Em Rough!*, pp. 165-71.

53. Wahl, Appendix 7 to Rockenbach, "Tank Corps Operations," pp. 11-12.

54. General Orders no. 21, General Headquarters, Tank Corps, AEF, 6 Nov. 1918, Modern Military Records Division, RG 120, Entry 1296, NA.

55. Brett, Appendix 6 to Rockenbach, "Tank Corps Operations," p. 11.

56. Wilson, *Treat 'Em Rough!*, Appendix B.

57. Ibid., pp. 58-59.

58. Ibid., pp. 236-37.

59. Ibid., 187, and Questionnaire, Sgt. David A. Pyle, 301st Tank Battalion, World War I Survey Collection, Archives, USAMHI.

60. Sasse, "Report on Operations: Sept. 27th-Oct. 1st, 1918," included as Appendix 8 in Rockenbach, "Tank Corps Operations," Appendix B.

61. Ibid., p. 6.

62. Capt. R. deP. Clarke, "Personal Narrative," undated, Patton Military Papers, Box 47, Personal Experience Reports of Tank Operations—1918, Patton Collection.

63. Wilson, *Treat 'Em Rough!*, pp. 196-204.

64. Sasse, "Report on Operations, October 8, 1918," reprinted in A Company, 301st Tank Battalion History (Philadelphia: Wright, 1919), pp. 96-102.

65. Sasse, "Report on Operations, October 17, 1918," reprinted in ibid., pp. 112-14.

66. Sasse, "Report of Action on October 23, 1918," reprinted in ibid., pp. 132-34.

67. Rockenbach, "Report of Chief of the Tank Corps," 13 Oct. 1919, USAMHI, p. 4375.

68. The adjutant general to director, Tank Corps, "Subject: Discharge, Enlisted Men, Tank Corps," 26 Nov. 1918, RG 165, Entry 310, Box 216, File 7-61.13/6, NA.

69. Wilson, *Treat 'Em Rough!*, pp. 221-22.

70. Nenninger, "The Tank Corps Reorganized," *ARMOR* 78, no. 2 (Mar.-Apr. 1969): pp. 34-35.

71. Patton, "Tanks in Future Wars," *Infantry Journal* (May 1920): p. 962.

72. Eisenhower, "A Tank Discussion," *Infantry Journal* (Nov. 1920): pp. 453-58.

73. Wilson, *Treat 'Em Rough!*, pp. 227-31.

2

Organizational Milestones in the Development of American Armor, 1920-40

Timothy K. Nenninger

After 11 November 1918 the Tank Corps, like the rest of the U.S. Army, rapidly demobilized. The future of the tank, to say nothing of the continued existence of the Tank Corps, was in question. Despite stalwart service and heroic deeds during World War I (two Medals of Honor and 39 Distinguished Service Crosses), the Tank Corps decreased from more than twenty thousand men in the United States and France at the time of the Armistice to less than 10 percent of that one year later.

Beginning early in 1919—as demobilization accelerated—the War Department closed stateside tank training centers, such as Camp Colt and Tobyhanna in Pennsylvania, and concentrated the remaining tank personnel, units, and equipment, at Camp Meade, Maryland, just north of Washington, D.C. By March of 1919 tank troops from overseas also began to arrive with their equipment. Eventually, nearly seven hundred light tanks, both French- and American-made Renaults, and more than a hundred British-built heavies were shipped to Camp Meade. When the AEF Tank Corps completed its demobilization in August 1919, the War Department ordered its former Commander, Brig. Gen. Samuel

D. Rockenbach, to return from overseas and take command of tank activities at Camp Meade.[1]

The task facing the U.S. Army after the Armistice involved not simply demobilizing from one war, but determining what might be learned from the experience to prepare for the next armed conflict. In order to consider the basic organizational and tactical lessons of the war, Gen. John J. Pershing appointed the AEF Board on Organization and Tactics, which consisted of AEF senior commanders and staff officers. The board met from April to June 1919, analyzing each of the combatant arms and support services, how they should separately be organized and function, and especially how they should operate as combined arms in a field army. The board devoted but two pages in its 184-page final report to tanks, concluding that they "should be recognized as an infantry supporting weapon incapable of independent decisive action." Tanks in the field should be part of the General Headquarters reserve, with individual tank battalions and companies attached to corps or divisions as required by the tactical situation.[2] Pershing, in his endorsement to the report, stressed that while tanks should be concentrated at a decisive point, they also should remain closely associated with infantry. Acknowledging their potential, he predicted the future employment of tanks would be increased "many fold."[3] With few exceptions, senior leaders in the U.S. Army valued what tanks had accomplished during the war, anticipated future improvements, yet recognized their limitations.

The immediate postwar question was not about the value of tanks, but whether the Tank Corps should remain a separate service as it had been during the war. Not surprisingly, Rockenbach advocated autonomy. So did army Chief of Staff Gen. Peyton C. March, who observed that "the Tank Corps is technical enough and important enough to keep it a separate corps."[4] The Superior Board, however, had recommended consolidation with the infantry, as did Pershing and other senior AEF Commanders, like Maj. Gen. Charles P. Summerall. He believed "the only way which has ever been discovered of making this service a maximum is to link the arm permanently and closely to the infantry."[5] However, the organizational debate ultimately did not revolve around tactical doctrine or past performance but on more fundamental issues best articulated by one who had responsibility for helping to resolve it. During the 1919 hearings on army reorganization, Rep. Henry E. Hull declared: "I can see how perhaps in case of war there might be some need of a separate organization for tanks, but I am unable to see any reason absolutely during peacetime for the creation of the overhead that would have to be established to give you a separate organization."[6]

The 1920 amendments to the National Defense Act abolished the Tank Corps as a separate arm and assigned responsibility for tanks to the infantry (Section 17). Although principally concerned with determining national military policy, defining the "Army of the United States," and creating mechanisms for manpower mobilization during times of emergency, this legislation also largely established the organizational hierarchy that would govern the peacetime army for the next twenty years. The 1920 act restored much of the autonomy to War Department technical and administrative bureaus (e.g., quartermaster, adjutant general, and ordnance) that they had lost to the General Staff during the World War. In addition, it extended the bureau system to the maneuver combat arms by creating positions for chiefs of the infantry and cavalry branches. Moreover, restrictions were imposed on the General Staff and its ability to assert effective control over the bureaus. The legislation limited the General Staff to a planning and coordinating body, rather than as an operating agency that could impose control over the bureaus as it had during the war. As a consequence, responsibility and authority for such matters as development of doctrine, procurement of equipment, and determination of proper tactical organization, was diffuse and shared among a number of agencies. This would have a marked impact on how and even whether the U.S. Army proceeded in the field of mechanization during the interwar years.[7]

For most of the 1920s tanks remained an infantry weapon organizationally and doctrinally; their principal role was to support infantrymen in the attack. Nevertheless, studies within the offices of the Chiefs of Infantry and Cavalry, at the Command and General Staff School, and by the War Department General Staff, took growing cognizance of developments overseas and led some to conceive of more independent and mobile roles for mechanized units. Such concepts gained greater credence as the mechanical capabilities of tanks, particularly speed, range, and reliability, improved. Even some cavalrymen saw their potential because tanks neutralized the defensive advantage of the machine gun and forced combat into the open. In 1927, for instance, the Chief of Cavalry recommended attaching tank companies to cavalry divisions and assigning antitank weapons to cavalry regiments.[8] General Rockenbach had proposed that the cavalry and other branches, in addition to the infantry, contribute to tank development. He declared, "I submit that the recent developments by the British will have an effect in modifying our ideas in regard to tanks and that the role of tanks is no longer a special weapon for infantry, but that it is just as important to cavalry division, corps, and the Army."[9]

Early in 1927 events got a boost when Secretary of War Dwight

Davis witnessed the maneuvers of the British Experimental Mechanized Force on Salisbury Plain. The performance of this combined arms force impressed him.[10] Later that year Davis ordered the organization of a similar American unit to serve as a military laboratory. Ultimately, the objective was for the Experimental Mechanized Force (EMF), which included infantry, cavalry, field artillery, tanks, armored cars, and supporting services, to be self-sufficient. The service tests conducted by the EMF helped to determine future equipment needs and the degree of self-sufficiency and capability of mechanized units. On 30 December 1927, the Chief of Staff, General Summerall, approved a War Department General Staff plan for the organization of the experimental force.[11]

By 3 July 1928 the components assigned to the EMF had assembled at Camp Meade under the command of an infantry officer and former Commandant of the Tank School, Col. Oliver Eskridge. His force consisted of two tank battalions, an artillery battalion, an infantry battalion, an armored car troop, and assorted engineer, ordnance, and quartermaster units for support. Initial training conducted from 9 to 14 July consisted of an equipment inspection and instruction on procedures for road marches. Following this preliminary training, the EMF made a five-day march to Aberdeen Proving Ground, Maryland; Carlisle Barracks, Pennsylvania; and back to Camp Meade. The exercise provided valuable experience in moving a large motorized and mechanized formation. Proper grouping in march columns—such as economical rates of march; means of command, supply, and reconnaissance while on the march; and methods of conducting night marches—were techniques that had to be perfected. During much of July and August the unit developed and practiced offensive tactics for mobile operations. From late August to mid-September, the EMF applied these in solving a number of tactical problems in the field.[12]

Throughout its existence the EMF faced some significant difficulties. Obsolete wartime equipment, which often broke down, proved the greatest handicap. Insufficient equipment and improper balance, notably lack of cross-country mechanized equipment other than tanks, limited the force as an effective experimental test unit. Despite its shortcomings, the work of the EMF did provide useful technical and tactical information to those on the General Staff who prepared subsequent mechanization studies. By the end of September 1928, the force had accomplished its mission. Consequently, on 19 September the Assistant Chief of Staff for Operations and Training, Brig. Gen. Frank Parker, recommended to the Chief of Staff that the EMF be disbanded as originally planned. Summerall approved the recommendation the next day and after 1 October 1928 all component units of the EMF returned to their home stations.[13]

During the spring of 1928, while organizing the EMF, the War Department simultaneously began planning for a long-range mechanization program. General Parker submitted a report in March emphasizing that success in modern warfare required a combination of firepower and mobility, attributes he believed tanks possessed. Parker looked beyond the use of tanks in direct support of infantry to utilizing their shock effect and mobility as an arm of strategic decision. Tanks providing the principal striking power for combined arms forces was what Parker envisaged for the immediate future.[14]

Tactical missions contributing to operational or strategic goals that mechanized forces were uniquely capable of performing included serving as the spearhead of an important attack, as a counterattack force, and as the advance or flank guard for strategic formations. Tank companies would comprise the principal power of the mechanized force. Light tanks would lead an assault, seeking weak points in the defense or vulnerable enemy flanks. Self-propelled field artillery and medium tanks would support the advance by overcoming strong points and widening gaps in the enemy line. Infantry, carried forward in cross-country capable vehicles, would consolidate the ground gained by the leading tank units. Supply, maintenance, and other support elements would also require mechanized transport in order to keep up with the advance.[15]

Parker made several recommendations for the long-range development of mechanization in the U.S. Army. Beginning with fiscal year (FY) 1930, he proposed an ambitious procurement program that included light and medium tanks, a reconnaissance car, cross-country vehicles for infantry and support units, and self-propelled artillery. As this procurement progressed, he envisaged the gradual phasing out of the large stocks of obsolete equipment mostly left over from the World War. Finally, he recommended that Congress be asked to establish a permanent mechanized unit in FY 1931. Secretary of War Davis approved Parker's report as the basis for future development and appointed a board of General Staff officers to prepare the details for implementation.[16]

The appointment of the board was crucial, for among its members was Maj. Adna R. Chaffee Jr., a cavalryman and member of the G3 Organization and Training Section of the General Staff. From the time of his assignment to the G3 in June 1927 until his death on 22 August 1941, Chaffee was the most effective American advocate of mechanization. Prior to 1927 he knew nothing of the subject. Aware that the G3 was initiating studies on mechanization, Chaffee sought information on the subject from diverse sources. He personally witnessed the demonstrations and test of new tank models, including a Christie that could go forty-two miles an hour. These demonstrations, as well as his cavalry background, convinced Chaffee that tanks should not be tied to infantry advancing at a walking pace. He received information on recent maneuvers of British mechanized units from a friend, Maj. Charles G. Mettler, then an assistant military attaché in Great Britain.[17]

The eleven-man mechanization board on which Chaffee served first met on 15 May 1928. Members of the board, representing all branches of the army, attended proving ground demonstrations of tanks, studied the exercises of the EMF, and in October 1928 issued their own report with recommendations about mechanization similar to those in General Parker's report six months earlier. The mechanization board concluded that the army should establish a permanent mechanized force with tanks as the backbone. Like Parker, the board outlined a development program that included establishing a unit similar to the recently disbanded EMF to serve as a technical and tactical laboratory. Furthermore, the board had the details: a force of 131 officers and 1,896 enlisted men, organized into a headquarters, one light tank battalion, one field artillery battalion, two mechanized infantry battalions, an engineer company, and an attached medical company. In order for tactical doctrine and organization to keep pace with technology, the board recommended supplying the force with the most modern equipment. Responsibility for future development affecting the force would be with its Commander. Rather than being a recommendation to create a new, separate branch, it emphasized the need to move beyond traditional branch rivalries and competition for resources to make progress in mechanization. On 31 October 1928 the Secretary of War approved these recommendations, but postponed organization of a mechanized force from FY 1930 to FY 1931 for budgetary reasons.[18]

The Chief of Infantry, Maj. Gen. Stephen O. Fuqua, was the sole dissident among the branch chiefs. Fuqua's criticism of the mechanization board's recommendations stemmed in part from two reasons. He believed that a separate mechanized force contradicted the provisions of the 1920 National Defense Act, which had allocated tanks only to the infantry. In addition, Fuqua's criticism stemmed from branch rivalry and competition for funds. A separate mechanized force threatened his branch's complete control over tanks. Fuqua protested: "The tendency in this study to set up another branch of the service with the tanks as its nucleus is heartily opposed. It is as unsound as was the attempt by the Air Corps to separate itself from the rest of the army. The tank is a weapon and as such it is an auxiliary to the Infantryman, as is every other arm or weapon that exists."[19] Despite Fuqua's protests, the War Department proceeded with plans for a mechanization program.

Col. James K. Parsons initially had the responsibility for "field development" of the Mechanized Force, as well as continuing to serve as Commandant of the Tank School. These responsibilities led Parsons to draft a plan for organizing a tank division, eventually to be expanded to a force of six divisions. Parsons's was the most ambitious, far-reaching mechanization plan proposed for the U.S. Army between the wars. His tank divisions were to be self-contained combined arms units suited for extended operations. Parsons determined that tank divisions were not assault units. He concluded their mobility was ideal for covering the advance or retirement of an army, attacking an enemy flank or rear, exploiting a breakthrough, seizing a strategic position, or filling a gap in the line. Although cavalry traditionally had performed these roles, Parsons was politically careful to state that these divisions would not replace or restrict development of any existing arm or unit.[20]

The divisions Parsons proposed required procurement of considerable amounts of equipment. Each would have 486 combat cars, 172 reconnaissance tanks, and eighty-seven command tanks. The total cost of organizing and equipping six divisions, Parsons estimated, would be $270 million. After the War Department General Staff analyzed Parsons's proposal, General Summerall commended his "vision, initiative, and energy" in preparing the study. However, he concluded that given the realities of the Army bureaucracy, the Army budget, and other modernization needs, a plan for immediate development of mechanization needed to be less costly and more modest.[21]

In his 1930 annual report Chief of Staff Summerall reaffirmed the Army's commitment to form a mechanized force, "From being an immediate auxiliary of the infantry the tank will become a weapon exercising offensive power in its own right."[22] Acknowledging the importance of a suitable tank force, Summerall on 19 August 1930 ordered the creation of a permanent Mechanized Force. The force was to be based, at least in part, on the experience of the EMF, Parsons's report, and the General Staff studies on mechanization.[23]

However, the development program soon ran into difficulties. Production of a faster, more reliable tank to replace the obsolete surplus wartime models then largely in use was especially critical. Unfortunately, the retrenchment and stabilization of military budgets made a modernization and reequipment program very difficult. For example, priorities for the FY 1932 budget included $2.4 million for limited service tests and procurement of a new semiautomatic rifle, a 3-inch antiaircraft gun, and new tanks. When the final War Department budget directive reduced the amount to $1 million, the General Staff decided to use most of the money for the highest priority item—the rifle. It provided funds for only a few tanks to serve mainly as pilot models for testing tactics and keeping pace with the latest technology.[24] Consequently, the Mechanized Force—when fully organized at Fort Eustis, Virginia, in November 1930—continued to use mostly obsolete equipment. One example is Company A, 1st Tank Regiment, which was equipped with six World War vintage M1917 light tanks, five modernized M1917s, and four newer Cunningham tanks, including three light T1 variants and a T2 medium. These tanks formed the nucleus of the force. An armored car troop of ten vehicles was the reconnaissance element. One battery from the 6th Field Artillery, its guns towed by trucks, provided fire support. Equipment problems also plagued the force's engineer company, whose transportation initially consisted of horse-drawn wagons. The original force included fifteen light tanks, ten armored cars, seven tractors, sixty-six trucks, twenty-two automobiles, and less than six hundred men.[25]

General Summerall selected a cavalryman, Col. Daniel Van Voorhis, who had no previous service with tanks, to command the force. Maj. Sereno Brett, an infantry officer and former wartime commander of the 304th Tank Brigade, was the executive officer. In September 1930 Van Voorhis, Brett, and Chaffee, the latter now head of the General Staff section responsible for troop training, visited Aberdeen Proving Ground, Holabird Quartermaster Depot near Baltimore, and Fort Eustis, where they conferred about equipment and organization for the Mechanized Force.[26] The immediate tactical and training missions of the force stemmed from their deliberations.

In combat the Mechanized Force was to utilize mobility and striking power to perform both tactical and strategic missions. Initial training was to determine the proper tactics involved in the operation of fast tanks with other mechanized and motorized elements. From 1 November 1930 until 13 June 1931, the force conducted a brief period of organization and individual training, followed by drills and exercises to perfect tactical teamwork, which led to more elaborate maneuvers and field training. Among these exercises were night and strategic road marches, offensive combat against entrenched infantry, operations against an opposing mechanized force, attacks involving wide turning movements, and operations as a covering force for a larger unit.[27]

The training and exercises of the Mechanized Force emphasized mobility. Although cavalry traditionally was the branch of mobility, Summerall's successor as Chief of Staff, Gen. Douglas MacArthur, who assumed his duties on 21 November 1930, changed that. He disbanded

Civil War veterans visit the Mechanized Force and pose with two Cunningham light tanks. Patton Museum.

Four Cunninghams and three 6-ton M1917A1s, including a signal tank from A Company, 1st Tank Regiment, prepare to mount truck carriers during march operations. This tank company was the principal combat organization for the Mechanized Force. Patton Museum.

the Mechanized Force and directed all ground combat branches, particularly the cavalry, to mechanize so far as possible.

Issued 1 May 1931 as "General Principles to Govern Mechanization and Motorization throughout the Army," MacArthur's new policy largely determined the Army's mechanization program for the next decade. MacArthur believed that missions rather than equipment should dictate organization, which had not necessarily been the existing practice. In short, the principal combat arms all would use mechanized vehicles insofar as it was practical to carry out their missions. Tanks would continue to support the infantry; however they would also be used as mechanized cavalry in performing the traditional reconnaissance, security, exploitation, and pursuit roles of that arm. The field artillery would also mechanize by developing some cross-country capability.

The new policy brought the most significant change to the cavalry. In order to assist the cavalry in developing its organization and equipment for modern warfare, MacArthur ordered the Mechanized Force disbanded and some of its equipment and personnel used to create a mechanized cavalry regiment. Under the new policy tanks operating with the mechanized cavalry became "combat cars" to distinguish

Wives of Mechanized Force officers pose beside a Cunningham T1E3 light tank in 1931. Withers Collection, Patton Museum.

between cavalry and infantry tanks. This was done to bypass the National Defense Act provision that assigned "tanks" exclusively to the infantry.[28]

MacArthur's program encountered opposition from several quarters. Van Voorhis, the Mechanized Force Commander, feared that branch rivalry would disrupt the proposed program and advocated strengthening his force, thus continuing mechanization development independent of any branch control.[29] The Chief of Infantry, General Fuqua, who had previously opposed the formation of the Mechanized Force, also disagreed with the new policy, which he argued violated the spirit of the National Defense Act of 1920. Furthermore, in an argument that was repeated regularly over the next ten years, Fuqua stated that neither the cavalry nor the infantry should strip personnel from established formations to create new mechanized units.[30]

The cavalry's reaction to the new policy was mixed. Maj. Gen. Guy V. Henry Jr., the Chief of Cavalry, welcomed the addition of mechanized units to his arm. Mechanized cavalry, in his view, would in no way be analogous to the infantry tank units. He believed that combat cars could replace horses without changing the essential role of cavalry; in fact, they would enhance the effectiveness and capability of units employing them. Henry believed that more mechanically reliable vehicles with greater mobility would be needed if the cavalry was to perform its traditional missions with mechanized and motorized vehicles.[31] However, not all cavalry officers shared Henry's enthusiasm for mechanization. Many could not accept the fact that the horse had a dwindling role on the modern battlefield. Yet others did recognize the need to fight from something more substantial that the back of a horse, and mechanized cavalry offered a logical alternative to the horse.[32]

Some Congressmen also had reservations about MacArthur's change in mechanization policy. There were the usual "pork barrel" concerns. When MacArthur terminated all tank activities at Fort Meade, Maryland, he moved the Tank School to Fort Benning, Georgia, where it became part of the Infantry School. Other tank units moved from Fort Meade to Camp Knox, Kentucky, the site of the soon-to-be-established mechanized cavalry regiment. Maryland politicians protested these moves.[33] However, there also were congressional supporters of an aggressive mechanization program, such as Rep. Ross Collins, chairman of a military affairs subcommittee of the House Appropriations Committee. He considered MacArthur's policy reactionary. Collins, an advocate of a small but modernized army, criticized the Chief of Staff for abolishing the Mechanized Force.[34]

Instead of spending money for a number of new tanks, complete mechanized units, and immediately creating a thoroughly modernized Army, MacArthur advocated a more gradual approach. His future Army was to be structured by maintaining a strong personnel base, particularly among the professional officer corps. He believed that progress in mechanization should consist of producing the best pilot vehicles, procuring only a sufficient number of tanks for tactical tests to develop doctrine, thoroughly testing and improving them mechanically, planning for expedited production in time of emergency, and indoctrinating the entire Army as to the capabilities of mechanized units. MacArthur thought that the immediate production of large numbers of new tanks was an unnecessary expense. After testing, he reasoned, many new models might quickly become obsolete. He hoped, however, that tank technology would eventually become more stable and the operational life of tanks would increase, at which point additional investments in mechanization would be warranted.[35]

In his *Reminiscences,* MacArthur claimed he "stormed, begged, ranted, and roared; I almost licked the boots of certain gentlemen to get funds for motorization and mechanization and air power." As Chief of Staff he did strongly support mechanization in theory, and his revision of mechanization policy was not necessarily a backward step as has sometimes been claimed. Faced with a choice between using funds for mechanized equipment and money to support personnel costs—as for the reserve components and officer education—MacArthur chose the latter.[36]

Representative Collins had a different vision. He believed the best defense for the least expense could be provided by a comparatively small Army of well-trained experts using the newest concepts of warfare and modern equipment. Collins wanted to develop fully weapons such as tanks and airplanes. According to him the Army, especially the War Department General Staff, was "utterly unable to lift themselves out of the rut and apply new principles to military science in the United States." Because MacArthur put preservation before progress, Collins opposed increasing War Department appropriations.[37]

Budgetary limitations doomed any prospects for a large mechanization project during the early 1930s. Until the limited wartime mobilization in 1940, when the Roosevelt Administration began spending earnestly for defense, American mechanized units consisted only of two regiments of mechanized cavalry and the infantry tank units remaining from the 1920s. Because of insufficient equipment, the War Department actually skeletonized several infantry tank companies during the early 1930s. Conforming to the policy dictated by MacArthur, the Army concentrated on developing tactical doctrine for mechanized combat and on producing a few pilot models to improve the mechanical capabilities of tanks.

Although the War Department mechanization policy of 1931 directed all branches to mechanize, it primarily concerned the cavalry and infantry. During the 1930s infantry tanks functioned much as they had in the past, with their primary mission being to assist the advance of the rifleman. The infantry used tanks in two different roles. "Leading tanks" would precede the main assault force, break into the enemy line, and penetrate deeply to disrupt the defense. "Accompanying tanks," normally organized in light tank platoons and attached to infantry battalions, were to reduce points of resistance encountered in front of or to the flanks of the infantry units they directly supported.[38]

Infantry tank units during the 1930s—often skeletonized and understrength—were stationed at posts that stretched from Miller Field, New York, to Schofield Barracks, Territory of Hawaii. There were two tank regiments, the 66th and 67th Infantry (Light Tanks), but the only active company from either regiment was Company F of the 67th at Fort Benning. There were also seven organic divisional tank companies, which served at the home stations of their parent units. About two thousand enlisted men and 120 officers served in the two regiments and the divisional companies.[39]

In 1939 the Army changed basic infantry doctrine to utilize tanks in mass for infantry support. Organizationally this resulted in elimination of the light tank company as a unit organic to infantry divisions. The Chief of Infantry, Maj. Gen. George A. Lynch, ordered the organization of existing light tank companies into battalions. Only rarely would tanks be used in units smaller than a battalion.[40] Redesignated the 68th Infantry, all the former divisional tank companies concentrated at Fort Benning in January 1940. Combined with the other infantry tank units, the 68th participated in the 1940 Carolina and Louisiana maneuvers as part of a provisional tank brigade. During the maneuvers, the infantry tanks and the mechanized cavalry began to become closely associated.

The most immediate organizational effect of the 1931 policy-change initiated by MacArthur was the mechanization of one cavalry regiment. The regiment selected, the 1st Cavalry from Fort D.A. Russell at Marfa, Texas, had a long and distinguished lineage and some regrets over losing its familiar mounts for ones less devoted. General Henry, the Chief of Cavalry, assured the regimental Commander: "It is with a feeling of sadness that we see this change in our oldest mounted organization with memories of a century's service as such."[41] To provide personnel skilled in mechanical activities, the War Department replaced some horse cavalrymen with troops from the recently disbanded Mechanized Force.

Vehicles, equipment, and maintenance and service detachments from the force also joined the cavalry regiment.[42]

As early as 1930 General Staff officers had declared Fort Eustis, the home of the Mechanized Force, an unacceptable site for mechanized training because it had only one small area suitable for field maneuvers. In December 1930 Chaffee and Van Voorhis traveled to Camp Knox (the War Department upgraded it to the status of a "fort" on 24 December 1931), near Louisville, Kentucky, and determined that the thirty-three thousand acres of rolling, rugged terrain was well suited for mechanized unit training.[43] Fort Knox became not only the initial station of the 1st Cavalry (Mechanized), but eventually the permanent home of American armor.

The first elements of the new organization, consisting of personnel and equipment detached from the disbanded Mechanized Force from Fort Eustis, arrived at Camp Knox in November 1931. The Mechanized Force commander, Van Voorhis, took command of the 1st Cavalry, with Chaffee as his executive officer. The post commander was Brig. Gen. Julian R. Lindsey, another cavalryman. All three officers concentrated their immediate efforts on preparing the post for the arrival of the cavalry from Texas. The major problem was procuring housing for troops. Because the bulk of the regiment remained at Marfa, the few troops at Knox had a limited training schedule during 1931 and 1932. There were drills, road marches, classroom lectures, and instruction on the use of equipment, with particular emphasis placed on developing mechanical skills. With few troops, little equipment, and no fully organized units at Fort Knox, there was no tactical training in the field. Training on this limited scale continued through November 1932, when the 1st Cavalry began its move from Texas to Fort Knox.[44]

Although the War Department announced its intention to mechanize the 1st Cavalry in May 1931, little happened for a year and a half. Politics played a role in creating this impasse. The Texas congressional delegation opposed the transfer, pressuring the Army to keep the 1st Cavalry on the border. Eventually Chaffee began working on the problem; first going to Washington in September 1932 to determine what the Army planned to do, then trying to end the impasse. In a letter dated 5 October 1932 he wrote, "I was given to understand . . . that the War Department would take up, following a political promise that had been made, a proposition to furnish the personnel of this regiment (mechanized) by abandoning D.A. Russell after the election; and they were hoping this would be organized by the 1st of January."[45] Had the plan to abandon Fort D.A. Russell failed, the mechanized cavalry probably would have moved to Texas. There even was the possibility the War

Department might have scrapped the mechanized cavalry idea entirely.

Eventually the army proceeded with the original plans. In November 1932 Van Voorhis received orders to proceed from Fort Knox to Texas with sufficient motor transport to move the 1st Cavalry to Kentucky. The convoy left for Marfa on December 17 and returned with the new mechanized cavalry troopers shortly after the New Year.[46]

The cavalrymen's mechanization training began in early 1933. Training was progressive and very similar to the previous years' work with the initial detachment. With development of individual skills and equipment familiarization underway, command post exercises and field problems began in earnest. From the beginning officers stressed the *cavalry* aspect of mechanized cavalry. Van Voorhis, particularly, emphasized cavalry "traditions, esprit, and smartness." There were tactical lessons to be learned from the horse cavalry. For example, years later officers who served in the early mechanized cavalry units recalled they were taught to "think mounted," that is to make quick tactical estimates

Members of the 1st Cavalry (Mechanized) at Fort Knox, Kentucky, in February, 1934. Col. Daniel Van Voorhis is in front, flanked by (from left) Lt. Col. Adna R. Chaffee Jr., Maj. Robert W. Grow, and other members of the regimental staff. Note the four CCT3 combat cars immediately behind the troop formation. Patton Museum.

and decisions, that there are no "foxholes for horses" (or tanks), and that mounted troops always have an advantage over dismounted opponents.[47]

Two cavalrymen, Van Voorhis and Chaffee, came to dominate the activities of the regiment. According to officers who served under him, Van Voorhis was principally an administrator, but the "ideal regimental commander." By contrast, Chaffee, outgoing, gregarious, and charismatic, "was an outstanding, forward-thinking officer who had a major influence on the development of armored doctrine, tactics, and equipment." Chaffee was a master tactician who advocated that "the mission of cavalry [both horse and mechanized] is to fight."[48] His foresight and breadth of vision inspired subordinates. This, along with his accomplishments on the General Staff in Washington and with the mechanized cavalry at Fort Knox, arguably made him the leading pioneer of American armor.

There were several important changes among the mechanized cavalry personnel at Fort Knox during the mid-1930s. When Van Voorhis departed in 1934 for a command position in the Hawaiian Department, Chaffee became regimental Commander. By late 1934, however, he too had transferred back to the War Department General Staff. Colonel Bruce Palmer, former assistant commandant at the Cavalry School, took command of the 1st Cavalry, with Col. Henry Baird as his executive officer. Other new faces arrived; among them was Lt. Col. Willis D. Crittenberger, who as regimental operations officer (S3) played an important role in the development of mechanized cavalry tactics. He later moved to the Office of the Chief of Cavalry as its mechanization expert. General Lindsey retired in 1934 and was replaced as Fort Knox Commander by General Henry, the former Chief of Cavalry and a strong advocate of mechanization. Van Voorhis returned from Hawaii in 1936 to succeed Henry upon the latter's retirement.

Because the *mechanized* aspect of mechanized cavalry was so new, following initial formation of the regiment the War Department anticipated organizational modifications in light of training and exercises. Within four months of the 1st Cavalry's arrival at Fort Knox, the Chief of Cavalry and officers of the regiment recognized the need for changes. General Henry suggested a major revision in the regimental tables of organization and equipment (TO&E), principally increasing the combat car squadrons from one to three, while other officers recommended removing supply vehicles from the combat car squadron and creating a separate regimental service troop.[49] More changes came when maneuvers at Fort Riley in the spring of 1934 revealed deficiencies in flexibility and offensive power. The ability of platoon and troop Commanders to control fast-moving small-unit actions was one noticeable problem. On 26 April 1935 the War Department approved a new regimental TO&E that reduced the number of combat cars per platoon from five to three. Moreover, it increased the regimental combat-car total from forty-two to fifty-six and raised the number of personnel from forty-two to forty-six officers and 610 to 749 enlisted men. Fire support for the mechanized cavalry also increased with the addition of 4.2-inch mortar and infantry units to each squadron, and attachment to the regiment of a battalion from the 68th Field Artillery.[50]

Chaffee's move back to the War Department in July 1934 was especially propitious both in terms of time and place. He became chief of the Budget and Legislative Liaison Branch of the General Staff, a position from which he materially could promote mechanization. When in late 1934, for example, the War Department received an additional $45 million, Chaffee saw to it that $5 million went for procurement of tanks and other mechanized vehicles. Expanding and augmenting the mechanized cavalry, originally anticipated in MacArthur's 1931 policy, also cost money. One obstacle to sending auxiliary units to Fort Knox was the lack of sufficient housing for the troops. Yet Chaffee pushed housing projects for Fort Knox as a priority in War Department budget requests. By 1936, signal, ordnance, quartermaster, and more field artillery units had been added to support the mechanized cavalry in Kentucky.[51]

Of even greater importance was the effort to mechanize a second cavalry regiment, an effort in which Chaffee exerted considerable influence. Insufficient funds, lack of housing, interbranch jealousy, and politics had hindered progress. Expansion of mechanized forces at the expense of existing units, essentially "dehorsing" the cavalry, became a particularly difficult problem. Chaffee first recommended using a four-hundred-man increase to the cavalry, which was allotted in the 1935 army budget. The plan was to mechanize the 4th Cavalry at Fort Meade, South Dakota. Because the loss of horse troops would be offset by the four-hundred-man increase, it was palatable within the Army. However, political pressures prevented the regiment's transfer to Fort Knox. Despite this setback Chaffee continued his efforts to get another mechanized regiment. Eventually MacArthur's successor as Chief of Staff, Gen. Malin Craig, himself a former Chief of Cavalry, suggested mechanizing the 13th Cavalry at Fort Riley, offsetting its loss to Kansas by using the increased cavalry personnel allocation to bring an undermanned regiment at that post, the 2d Cavalry, up to war strength. Consequently, the Secretary of War on 27 May 1936 approved the mechanization of the 13th Cavalry, which in September of that year moved from Kansas to

Kentucky. Col. Charles L. Scott, a former director of instruction at the Cavalry School, became the regimental Commander.[52]

In June 1937 General Craig ordered a reexamination of the Army's mechanization policy. He had come to believe that mechanized cavalry was developing into something tactically different from cavalry. Recent experience in training at Fort Knox and Fort Riley and in the 1936 Second Army maneuvers demonstrated to Craig the offensive potential of mechanized cavalry. Yet he was concerned that the Chiefs of Cavalry and Infantry, with their parochial branch interests, would inhibit its development. To assure its full potential, the Chief of Staff indicated he would again "inaugurate a mechanized force without regard to arm of service."[53] Dividing responsibility between the infantry and cavalry, according to General Craig, led to many technical and tactical ideas but no clear-cut program of expansion and development. He directed the G3 Section of the General Staff again to study mechanization policy.[54]

The G3 report completed on 25 October 1937 under the direction of Brig. Gen. George P. Tyner, began: "Experience has shown that the older arms will fight in their traditional way and that, except for the mechanized reconnaissance detachments of the Cavalry, mechanization can be applied only through what it is in effect, if not in name, a new arm." The report concluded by recommending rescinding the 1931 policy of general mechanization throughout the Army and supported reestablishing an independent mechanized force.[55] There was, however, opposition to these conclusions, principally by the Chief of Infantry, General Lynch. He argued there was no justification for restricting mechanization to certain designated units and noted that "Tanks should be used whenever and wherever they will contribute to the effectiveness of the unit considered."[56]

In part because of this opposition General Craig in March 1938 modified only the War Department policy of decentralized mechanization. He restated the essence of the 1931 policy that "mechanization will be applied to certain cavalry and infantry units to the extent necessary to enable these arms to better carry out their prescribed tactical functions." Most importantly, both the infantry and cavalry retained control of their mechanized units. The new policy also eliminated the concept of "leading tanks" operating beyond the range of the infantry. More importantly, however, it broke with the 1931 policy of mechanized cavalry merely performing traditional cavalry roles. The 1938 directive recognized that independent, self-contained mechanized cavalry forces were capable of distant strategic employment and in essence sanctioned development of a mechanized cavalry division capable of such independent missions.[57]

Chaffee and Van Voorhis had for some time advocated independent missions for mechanized cavalry. In a 1937 Army War College lecture, Van Voorhis proposed immediately organizing a heavy mechanized striking force, a division consisting of tanks supported by other arms.[58] Also in a War College address, Chaffee recommended expanded development of mobile mechanized units—one mechanized cavalry division for each of the four field armies. Experience in recent maneuvers had convinced Chaffee that independent mechanized cavalry units, if sufficiently large, could move beyond traditional cavalry roles to engage in heavy combat and deliver decisive offensive blows.[59] As Commander of the 7th Cavalry Brigade (Mechanized) from 1938 until the creation of the Armored Force in 1940, Chaffee shaped the mechanized cavalry to become such an independent striking force. Although mechanized cavalry leaders could influence doctrinal developments, there were events beyond their control, such as budgets and opposition within the War Department. These conditions continued to limit the expansion of mechanized units.

Several proposals for mechanized cavalry expansion—beyond the existing two regiments at Fort Knox—surfaced during the late 1930s. The October 1937 G3 report recommended creating a self-contained tank division of three mechanized cavalry regiments.[60] In a more conservative vein—also in 1937—Van Voorhis advocated expanding the mechanized brigade at Fort Knox to division size by adding reconnaissance and support elements.[61] Although the chief of cavalry, Maj. Gen. Leon Kromer, supported Van Voorhis's idea, the War Department approved a plan by Kromer for organizing a three-thousand-man mechanized cavalry division "for planning purposes only."[62] Nevertheless, Van Voorhis tried again in May 1939. By then a Major General and Commander of the V Corps Area, he repeated the recommendation to expand mechanized cavalry from brigade to division size by augmenting it with support and reconnaissance elements while retaining the tactical integrity of the mechanized cavalry regiments.[63] Despite support from the G3, Maj. Gen. Robert M. Beck (Tyner's successor), nothing happened as a result of Van Voorhis's persistent efforts.

Maj. Gen. John K. Herr, chief of cavalry from 26 March 1938, to 9 March 1942, was much less friendly to mechanization than his immediate predecessors. The demise of the horse was anathema to Herr. As late as 1953 he contended, "One basic and immutable truth stands out through all our wars. Sometimes our commanders have to learn it the hard way: There is no substitute for cavalry!" Herr meant his beloved *horse* cavalry.[64] He told the chief of staff in 1938, "It [mechanized cavalry] has not yet reached the place in which it can be relied upon to displace horse cavalry. For a considerable period of time it is bound to play an important but minor role while the horse cavalry plays the major role so far as our country is concerned."[65] Herr would not sacrifice a single horse or man from horse cavalry regiments to organize any mechanized units. An officer who served in the Office of the Chief of Cavalry recalled of Herr, "I can hear him in the corridors of the old

General Herr's beloved "water-cooled transportation" was gradually replaced by mechanization in the 1930s and faded into history in the early days of World War II. Patton Museum.

Munitions Bldg.: 'Not one horse will I give up.'"[66] In 1939 Herr ranked the expansion of the 7th Cavalry Brigade (Mechanized) fourth out of five on a list of priorities for overall cavalry enlargement.[67]

Horses had friends in the halls of Congress as well. During Senate hearings on the 1938 Army appropriations bill, Sen. William G. McAdoo asked General Craig whether the mechanized cavalry had superseded horse units and relegated them to insignificance. When the Chief of Staff replied that horse units performed a necessary role in a modern army, the senator responded, "I am glad to hear that."[68] It is hardly surprising that, with such opposition, funds for mechanized cavalry expansion still did not materialize during 1938 or 1939.

The German invasion of Poland on 1 September 1939 forced a reorientation of national priorities, including a limited expansion of the Army. The rapid and decisive German success in Poland served to demonstrate the capability of mechanized units. Chaffee, using the occasion to further the idea of expansion of American mechanized forces, pointed out that the main components of panzer divisions—mechanized cavalry, mobile infantry, and medium tanks—already existed in the U.S. Army. Expansion of mechanized units, in his view, could best be affected by building on the 7th Brigade. Chaffee recommended the immediate organization of one mechanized cavalry division and the procurement of equipment and personnel to establish cadres for two more.[69] Even General Herr, influenced by events in Europe, advocated expanding mechanized cavalry by turning the brigade into a ten-thousand-man division. Moreover, Herr thought that each of the four field armies should possess a mechanized division. He still demurred in using personnel from horse units, which he also wanted to strengthen.[70]

Personnel for mechanized cavalry remained scarce. The president's declaration of a "limited national emergency" on 8 September 1939 resulted in an army troop increase of "only 17,000 men."[71] Even in the wake of the German panzer successes and seeming acceptance of augmentation of American mechanized units, interbranch and intrabranch rivalries continued to delay expansion. The Chief of Infantry—like the Chief of Cavalry—opposed expanding mechanized units at the expense of existing organizations. These combat-arms chiefs traditionally competed with one another to secure for their branch funding and any expansion that did occur.[72] Once again competition within the War Department thwarted mechanization.

However, a solution began to unfold during the spring of 1940 when Maj. Gen. Stanley D. Embick commanded the Third Army maneuvers in Texas and Louisiana. Formerly Chaffee's superior at the War Department Budget and Legislative Liaison Branch, Embick was par-

The 7th Cavalry Brigade (Mechanized) forms on the grounds of the United States Military Academy at West Point, New York, in August 1939. Patton Museum.

ticularly enthusiastic to test the possibilities of mechanization by having the 7th Brigade participate in the maneuvers. At meetings in Washington, Embick, Chaffee, and the General Staff agreed to organize a provisional mechanized division for the maneuvers by attaching an infantry regiment to the mechanized cavalry brigade. As Chaffee noted on 9 April: "A month ago . . . as an upshot [to] some missionary work in the War Department I will have the Sixth Infantry in trucks attached to the Brigade for the period of the Third Army Maneuvers. . . . So little by little we are getting the troops if not the name and dignity of a mechanized division."[73]

The Provisional Tank Brigade, organized in January 1940 at Fort Benning from infantry tank units, also participated in the maneuvers. Commanded by Brig. Gen. Bruce Magruder, it consisted of one regiment and two separate battalions of light tanks and one company of mediums. Virtually all the tank and mechanized units in the army were in Louisiana. During the maneuvers Embick used the infantry tanks and mechanized cavalry in various combinations. In the second phase they were combined in a single unit. It was reported that "the makeshift force worked smoothly and inspired the leading officers to take thought of the future organization of such a unit."[74]

Once again the Germans provided a timely example of the potential decisiveness of mechanized warfare. Spearheaded by panzer forces, the Wehrmacht ended the so-called sitzkrieg in the West on 10 May 1940, attacking through Holland, Belgium, and France and within ten days reached the English Channel. Magruder's executive officer and a future Armored Force Commander, Maj. Gen. Alvan C. Gillem, recalled that the German success and the experience of combining the American mechanized cavalry and infantry tank units resulted in discussions among officers at the maneuvers "as to the need for a Force composed of Infantry, Cavalry, and Artillery. The thought was not along the lines of the old Infantry and Cavalry tanks but of a Force separate from all current Arms and one that combined not only the combat elements but the essential support units to make a tactical entity."[75] On 25 May 1940 several officers, including Chaffee, Magruder, Gillem, Sereno Brett, Col. George Patton (an observer at the maneuvers), and Maj. Gen. Frank M. Andrews, the Assistant Chief of Staff for Operations and Training (G3), met in the basement of the high school in Alexandria, Louisiana, to discuss the future of mechanization. They agreed the army could no longer delay development of mechanization under a unified command free from cavalry or infantry control. Andrews conveyed this sentiment to the Chief of Staff, Gen. George C. Marshall.[76]

Despite continuing bickering from the cavalry and infantry, Marshall on 10 June convened at the War Department a meeting with the branch chiefs, representatives from the General Staff, and leaders of the mechanized cavalry and infantry tank units. His purpose was to outline plans for organizing a separate armored force. The word *armored* symbolized the break with the past. The conferees eventually agreed that German successes demonstrated the value of armored forces. Moreover, those at the meeting acknowledged that American tactical and technical developments had previously proceeded along sound lines but too modestly. To become more combat effective, they agreed that all mechanized resources now had to be consolidated into a single command to obtain maximum value from the limited equipment and personnel. The meeting concluded with emphasis on the idea that further expansion of armored units was imperative.[77]

On 10 July 1940 the War Department issued a directive that declared: "For purposes of a service test, an Armored Force is created." Chaffee became Chief of the Armored Force, responsible for tactical and training doctrine; research, development, and procurement of equipment; and command of "all armored corps and divisions, and all GHQ [General Headquarters] Reserve tank units."[78] Although the implementing instructions explicitly denied it, Chaffee was, in effect, chief of a branch just like the cavalry, infantry, and field artillery. The mechanized cavalry brigade formed the nucleus of the 1st Armored Division with Magruder as its Commander. General Scott became Commanding General of the 2d Armored Division, which was organized from the infantry tank units at Fort Benning. American armored formations during World War II evolved from this 1940 Armored Force structure.

The military situation in 1940 required creation of a separate Armored Force. Because the army's leadership had been unable to overcome branch rivalries that inhibited development during the previous decade, the bureaucratic imperatives now required a radical organizational solution more than ever. Perhaps most importantly, the entire context had changed. Within a matter of weeks during the spring of 1940, administration and congressional sentiment markedly shifted on the question of rearmament and significant mobilization of military manpower, and by mid-June the General Staff was beginning to plan for a four-million-man Army.[79] With the worsening international situation, the administration and Congress became more willing to provide for a general expansion of the Army, which freed resources for the expanded mechanized program.

NOTES

The development of American armor during the interwar years is the subject of several monographs, each with a particular strength of research, coverage, or analysis. Mildred Hanson Gillie, *Forging the Thunderbolt: A History of the Development of the Armored Force* (Harrisburg, Pa.: Military Service Publishing Co., 1947), is more a biography of Adna R. Chaffee Jr. than an organizational history of armor; Gillie had access to the apparently no longer extant Chaffee papers. George M. Shuffer Jr., "Development of the U.S. Armored Force: Its Doctrine and Tactics, 1916-40" (master's thesis, University of Maryland, 1959), is a brief yet reasonably well-rounded account based on official records and professional journals that puts U.S. developments into an international context. Timothy K. Nenninger, "The Development of American Armor, 1917-40" (master's thesis, University of Wisconsin, 1968), makes extensive use of Army records at the National Archives, and correspondence with many armor officers who participated in interwar developments highlights its research. David E. Johnson, "Fast Tanks and Heavy Bombers: The United States Army and the Development of Armor and Aviation Doctrines and Technologies, 1917-1945" (Ph.D. diss., Duke University, 1990), is a useful study with insightful analysis comparing the technology and doctrine of two emerging weapons systems. John L.S. Daly, "From Theory to Practice: Tanks, Doctrine, and the U.S. Army, 1916-1940" (Ph.D. diss., Kent State University, 1993) is a huge (over a thousand pages), often wordy dissertation that offers some excellent description and analysis of tank doctrine at the small-unit level. Robert S. Cameron, "Americanizing the Tank: U.S. Army Administration and Mechanized Development within the Army, 1917-1943" (Ph.D. diss., Temple University, 1994), is another lengthy (over nine hundred pages) study. Nevertheless, it is perhaps the best organizational history of the subject, and makes excellent use of the personal papers collections at the USAMHI. It also includes a clear description of what the Army learned of foreign mechanization developments and how that information was used. Vincent J. Tedesco III, "'Greasy Automatons' and 'The Horsey Set': The U.S. Cavalry and Mechanization, 1928-1940" (master's thesis, Pennsylvania State University, 1995), is a sympathetic study of how the horse cavalry eventually learned to live with the tank. The author emphasizes that there were degrees of acceptance and rejection, not simply "progressives" and "reactionaries," within the cavalry.

1. U.S. House of Representatives, 66th Cong., 2d sess., *House Documents,* vol. IX (Washington, DC: GPO, 1920), p. 253; U.S. Senate, 66th Cong., 1st sess., Subcommittee of the Committee on Military Affairs, *Reorganization of the Army,* vol. I (Washington: GPO, 1919), p. 773.

2. Report of the Superior Board on Organization and Tactics, 1 July 1919, pp. 29-31, RG 120, National Archives (hereafter NA).

3. Pershing, 16 June 1920, Endorsement to Report of the AEF Superior Board on Organization and Tactics, RG 120, NA.

4. U.S. Senate, 66th Cong., 1st sess., Subcommittee of the Committee on Military Affairs, *Reorganization of the Army,* vol. I (Washington: GPO, 1919), statement of Rockenbach, pp. 761-85; U.S. House of Representatives, 66th Cong., 1st sess., Committee on Military Affairs, *Army Reorganization,* vol. I (Washington: GPO, 1919), statement of Peyton C. March, p. 59.

5. Summerall to Rockenbach, 13 Jan. 1919, War College Division File 8479-86, RG 165, NA.

6. U.S. House of Representatives, 66th Cong., 1st sess., Committee on Military Affairs, *Army Reorganization,* vol. I (Washington: GPO, 1919), p. 285.

7. James E. Hewes Jr., *From Root to McNamara: Army Organization and Administration, 1900-1963* (Washington: Department of the Army, 1975), pp. 50-6; Mark Watson, *Chief of Staff: Prewar Plans and Preparations: United States Army in World War II* (Washington: Department of the Army, 1950), pp. 19-28; and Cameron, "Americanizing the Tank," pp. 16-27.

8. Chief of Cavalry (hereafter ChCav) to the Adjutant General (hereafter TAG), 12 Apr. 1927, AG 320.2 (4-12-27), RG 407, NA.

9. Rockenbach to TAG, 7 Dec. 1926, Chief of Infantry (hereafter ChInf) Correspondence, 470.8/7251-B, RG 177, NA.

10. Gillie, *Forging the Thunderbolt*, p. 20. Although I have uncovered no corroborating evidence from contemporary documentation that Davis actually attended these maneuvers, neither is there any reason to doubt Gillie. The Military Intelligence Division records, RG 165, NA, however, include many reports on the British mechanized force, yet contain nothing on Davis.

11. Chief of Staff (hereafter C/S) to Assistant Chief of Staff for Organization and Training (hereafter G3), 7 Nov. 1927, AG 354.2 (11-7-27) (1) Sec. 1; and G3 to C/S, 29 Dec. 1927, AG 354.2 (11-7-27) (1) Sec. 1, RG 407, NA.

12. Training Memo #1, Experimental Mechanized Force, 21 June 1928; TAG Memo, 11 July 1928; G3 to C/S, 31 Aug. 1928, AG 354.2 (12-21-27) (1) Sec. 2, RG 407, NA.

13. G3 to C/S, 19 Sept. 1928, AG 354.2 (12-21-27) (1), Sec. 2, RG 407, NA.

14. G3 to C/S, "A Mechanized Force," 20 Mar. 1926, AG 537.3 (4-14-28) (1) Sec. 3, RG 407, NA. Although Parker signed the memo, Maj. Adna R. Chaffee Jr. actually drafted it, as he did nearly all other memos relating to mechanization during the time he served in G3 from July 1927 to November 1930.

15. Ibid.

16. Ibid.

17. Gillie, *Forging the Thunderbolt*, pp. 20-44; Charles G. Mettler, "Adna Romanza Chaffee," *Assembly: Association of Graduates, U.S. Military Academy* (obituary, April 1942), pp. 13-15, and Timothy K. Nenninger, "Chaffee, Adna Romanza, Jr.," in Roger Spiller, ed., *Dictionary of American Military Biography* (Westport, Conn.: Greenwood Press, 1984), pp. 164-67.

18. "Proceedings of the Board of Officers to Make Recommendations for the Development of a Mechanized Force," 1 Oct. 1928, AG 537.3 (4-14-28) (1) Sec. 3, RG 407, NA.

19. ChInf to G3, 26 Mar. 1928, ibid.

20. Parsons to TAG, 17 Apr. 1930, AG 537.3 (2-21-30) (1) Sec. 1, RG 407, NA. Parsons's responsibility for "field development" put him in a somewhat ambiguous situation. He was responsible for preliminary development of the force, including preparing a systematic, comprehensive study. Although he was not named Commander of the force, seemingly implied in his status was the possibility that eventually he would. TAG to Parsons, 30 Apr. 1930, ibid.

21. Parsons to Maj. T.R. Sims, 28 Aug. 1943, Background File to Study 27, Army Ground Forces Historical Studies, RG 337, NA; Summerall to Parsons, 20 May 1930 included therein.

22. "Report of the Chief of Staff," *War Department Annual Reports, 1930* (hereafter *WDAR*) (Washington: GPO, 1930), p. 125.

23. C/S to G3, 19 Aug. 1930, AG 537.3 (2-21-30) (1) Sec. 1, RG 407, NA.

24. John W. Killigrew, "The Impact of the Great Depression on the Army, 1929-1936," (Ph.D. diss., Indiana University, 1960), p. III-4.

25. Acting G3 to C/S, 24 Sept. 1930, AG 537.3 (2-21-30) (1) Sec. 2, RG 407, NA.

26. TAG to Van Voorhis, Brett, and Chaffee, 16 Sept. 1930, ibid.

27. Acting G3 to C/S, 20 Oct. 1930, ibid.; Van Voorhis, "Consolidated Report of Operations," 1 July 1931, AG 537.3, (2-21-30) (1) Sec. 4, RG 407, NA.

28. "General Principles to Govern Mechanization and Motorization Throughout the Army," 1 May 1931, copy filed AG 537.3 (6-20-37), RG 407, NA. The Deputy Chief of Staff, Maj. Gen. George Van Horn Moseley, actually drafted the document. MacArthur's rationale for the new policy also is explained in "Report of the Chief of Staff," *WDAR, 1931* (Washington: GPO, 1931), pp. 42-43, and "Report of the Chief of Staff," *WDAR, 1932* (Washington: GPO, 1932), p 82.

29. Historical Section, Army Ground Forces, *History of the Armored Force, Command, and Center,* Study 27 (Army Ground Forces, 1946), p. 3.

30. ChInf to Deputy C/S, 27 Apr. 1931 and 13 May 1931, AG 537.3 (2-21-30) (1) Sec. 3, RG 407, NA.

31. ChCav to Deputy C/S, 3 July 1931, ibid.

32. Maj. Gen. Guy V. Henry to author, 11 June 1967. The best, most sophisticated discussion of the battle for the hearts and minds of the cavalry over the question of mechanization is Tedesco, "'Greasy Automatons and the 'Horsey Set'."

33. Sen. P. L. Goldsborough to TAG, 6 Nov. 1931, AG 352 Tank School (2-1-26), RG 407, NA.

34. U.S. House of Representatives, 72d Cong., 1st sess., Subcommittee of the House Committee on Appropriations, *War Department Appropriations Bill, 1933* (Washington: GPO, 1933), MacArthur testimony on 21 Dec. 1931, pp. 17-19.

35. U.S. Senate, 73d Cong., 1st sess., *Hearings Before the Committee on Military Affairs* (Washington: GPO, 1933), MacArthur testimony on 26 Apr. 1933, pp. 28-32.

36. Douglas MacArthur, *Reminiscences* (New York: McGraw Hill Books Co., 1964), p. 99. Also see D. Clayton James, *The Years of MacArthur,* vol. I, *1880-1941* (Boston: Houghton Mifflin Co., 1970), pp. 354-63.

37. 72d Cong., 1st sess., *Congressional Record,* vol. 75, pt. 9, pp. 9932-35, as quoted in Killigrew, "Impact of the Great Depression on the Army," pp. V-22-5.

38. U.S. Army, *Infantry Field Manual,* vol. II, *Tank Units* (Washington: GPO, 1931), p. 181.

39. U.S. House of Representatives, 71st Cong., 3d sess., Hearings Before a Subcommittee of the House Appropriations Committee, *War Department Bill, 1932* (Washington: GPO, 1930), Chief of Infantry testimony on 2 Dec. 1930, pp. 600-606.

40. ChInf to TAG, 31 Aug. 1939, ChInf Correspondence, 320.3/4370, RG 177, NA.

41. ChCav to Commanding Officer, 1st Cavalry, 17 November 1931, ChCav Correspondence, 332.02, RG 177, NA.

42. TAG to Corps Area Commanders, etc., 13 May 1931, AG 537.3 (2-21-30) (1) Sec. 3, RG 407, NA.

43. G3 to C/S, 5 Dec. 1930, AG 537.3 (12-5-30), RG 407, NA. On 24 Dec. 1931 the War Department redesignated the installation Fort Knox; C/S to G3, 24 Dec. 1931, AG 680.9 Ft. Knox (12-24-31), RG 407, NA.

44. "Training Program, Mechanized Cavalry Detachment, 1 February— 1 June 1932"; TAG to CO Mechanized Cavalry Regiment, 25 Jan. 1932; AG 353 (12-28-31); CO Mechanized Cavalry Regiment to TAG, AG 353 (1-25-32); and Maj. Robert W. Grow to TAG, 2 Sept. 1932, AG 353 (9-10-31) (1) Sec. 1, RG 407, NA.

45. Chaffee, as quoted in Gillie, *Forging the Thunderbolt*, p. 56. As former chief of the Budget and Legislative Liaison Branch on the General Staff, Chaffee was well suited for this role.

46. Ibid., 58.

47. Lt. Gen. Geoffrey Keyes, taped reply to questionnaire submitted by author (received by author 23 Sept. 1967 six days after Keyes died); Maj. Gen. Robert W. Grow, 10 June 1967, correspondence with author; and Lt. Gen. W.H.S. Wright, correspondence with author, 23 June 1967.

48. Lt. Gen. Charles G. Dodge, 3 July 1967, correspondence with author, and Grow correspondence.

49. ChCav to CG Fort Knox, 7 Apr 1933, AG 320.2 Cavalry (4-7-33), and CG Fort Knox to CG V Corps Area 1 May 1933, (3d Endorsement to previous memo), RG 407, NA.

50. "Continuation of Development of Mechanized Cavalry," ChCav to TAG, 28 Mar. 1935, ChCav Correspondence, 322.02, RG 177, NA.

51. Gillie, *Forging the Thunderbolt*, pp. 91-92, and "Additions to the 7th Cavalry Brigade (Mechanized)," Maj. LeRoy Martin to ChCav, 16 Dec. 1936, ChCav Correspondence, 322.02, RG 177, NA.

52. Gillie, *Forging the Thunderbolt*, p. 92, and ChCav to G3, 30 Mar. 1936, AG 320.2 (3-30-36) and ChCav to G3, 13 May 1936, AG 320.2 (5-13-36), RG 407, NA.

53. "Mechanized Forces," C/S to DC/S, 28 June 1937, AG 537.3 (6-28-37), RG 407, NA.

54. C/S to AC/S G3, 14 Oct. 1937, AG 537.3 (10-14-37), RG 407, NA.

55. "Tanks and Mechanized Units," AC/S G3 to C/S, 25 Oct. 1937, ChInf Correspondence, 470.8/550-B, RG 177, NA.

56. ChInf to AC/S G3, 5 Nov. 1937, ChInf Correspondence, 451/9264-B, RG 177, NA.

57. "Policies Governing Mechanization," TAG to Branch Chiefs, etc., 6 Apr. 1938, AG 537.3 (6-28-37), RG 407, NA.

58. "Mechanization," Van Voorhis address to the Army War College

(hereafter AWC), 13 Oct. 1937, Mechanized Cavalry Board: Maneuvers, 1929-39, RG 177, NA.

59. "Mechanized Cavalry," Chaffee address to AWC, 29 Sept. 1939, Mechanized Cavalry Board: Maneuvers, 1929-39, RG 177, NA.

60. AC/S to C/S, 25 Oct. 1937, ChInf Correspondence 470.8/550-B, RG 177, NA.

61. Van Voorhis AWC address, 13 Oct. 1937, RG 177, NA.

62. ChCav to TAG, 8 July 1937, AG 320.2 (7-8-37), RG 407, NA.

63. CG V Corps Area (Van Voorhis) to C/S, 17 May 1939, AG 537.3 (5-17-39), RG 407, NA.

64. John K. Herr and Edward S. Wallace, *The Story of the U.S. Cavalry* (Boston: Little, Brown, and Co., 1953), p. 261.

65. ChCav to C/S, 17 Oct. 1937, AG 320.2 (10-17-38), RG 407, NA.

66. Grow correspondence. For a full account of Herr's attitude see, Maj. Gen. Robert W. Grow (Ret.) (edited by Capt. Peter R. Mansoor), "The Ten Lean Years, Part 4" *ARMOR* (Jan.-Feb. 1987): pp. 34-42.

67. ChCav to C/S, 5 Sept. 1939, AG 320.2 (9-5-39), RG 407, NA.

68. U.S. Senate, 75th Cong., 1st sess., Subcommittee of the Committee on Appropriations, *War Department Appropriations Bill, 1938* (Washington: GPO, 1937), p. 55.

69. "Observations and Recommendations on Expansion and Development of Mechanized Cavalry," Chaffee to TAG, 15 Sept. 1939, AG 320.2 (9-15-39), RG 407, NA.

70. ChCav to C/S, 3 Oct. 1939, ChCav Correspondence, 322.2, RG 177, NA.

71. Watson, *Chief of Staff*, p. 156.

72. *History of the Armored Force, Command, and Center*, p. 5.

73. Chaffee to Scott, 9 Apr. 1940, quoted in ibid., p. 6. Also see, Gillie, *Forging the Thunderbolt*, pp. 146-48.

74. *History of the Armored Force, Command, and Center*, pp. 6-7. The Louisiana maneuvers were not principally about mechanization. The corps on corps exercises were first a test of the new tactical organization (the triangular formations) and then practice in operational movements, moving corps and divisions to the battlefield. See Jean R. Moenk, "A History of Large-Scale Army Maneuvers in the United States, 1935-1964," Historical Branch, Continental Army Command, 1969, pp. 26-33.

75. Lt. Gen. Alvan C. Gillem, 26 July 1967, 22 and 27 Sept. 1967, correspondence with author.

76. AC/S G3 to C/S, May 29, 1940, G3 41665, as quoted in *History of the Armored Force, Command, and Center*, p. 8. On the "schoolhouse meeting" also see the notes of an interview of Brig. Gen. Thomas J. Camp by Lt. Kenneth Hechler, 3 Aug. 1943, and excerpts from memo by Camp, 5 Aug. 1943, in Background File to Study 27, Army Ground Forces Historical Studies, RG 337, NA; Camp attended the meeting.

77. Gillie, *Forging the Thunderbolt*, pp. 165-67; *History of the Armored Force, Command, and Center*, p. 8; and statement by Chaffee in U.S. House of Representatives, 76th Cong., 1st sess., Subcommittee of the Committee on Appropriations, *Military Establishment Appropriations Bill for 1942* (Washington: GPO, 1941), p. 558.

78. "Organization of Armored Forces," TAG to Commanding Generals of all Armies, etc., 10 July 1940, AG 320.2 (7-5-40), RG 407, NA.

79. Watson, *Chief of Staff*, pp. 166-77. The link between expansion of mechanization (creation of the Armored Force) and general army expansion is clearly documented in the 23 June 1940, G3 study for the chief of staff outlining his options on "Mechanization." Completion of the proposed four-division Armored Force is explicitly linked to the recently approved 375,000-man Army expansion with equipment needs being met by the supplemental 1941 budget estimates; G3 to C/S, 23 June 1940, AG 320.2 (6-5-40) (3) Sec. 1, RG 407, NA.

3

The Marine Corps's First Experience with an Amphibious Tank

George F. Hofmann

Naval strategist RAdm. Alfred Thayer Mahan wrote in 1889 that changes in tactics historically have not kept pace with advances in weapons technology. He attributed this to the inertia of a conservative military class, thus causing an unduly long developmental period. The advantage, he wrote, lies with those who recognize each change and study the qualities each new weapon presents. He maintained that this understanding would lead to a change in tactics, thus giving an advantage to those going to battle. Nevertheless, Mahan was not too optimistic, claiming "history shows that it is vain to hope that military men generally will be at pains to do this."[1] Admiral Mahan credited improved weapons technology to the energy of one or two men. J. Walter Christie was one of these men. A number of Marines understood the potential impact that Christie's new technology would have on tactics. During the early post–World War I period there were a few Marines who took advantage of the qualities and limitations of new weapons and attempted to apply them to the Corps's perceived mission. These Marines attempted to disprove Admiral Mahan's predictions.

This chapter examines the history of the first U.S.-designed assault amphibian tank, its controversial inventor, and its influence on the operational and tactical aspects of post–World War I Marine Corps amphibious doctrine.[2] It also considers the visionary ideas of postwar Marine Corps leaders, which were overwhelmed by the lean years of the 1920s, when severe budget restraints, disarmament, retrenchment and personnel turbulence—and a national indifference toward military matters—threatened the Corps's very existence.

A brilliant and farsighted Marine at the U.S. Naval War College, Earl H. Ellis, wrote a series of papers between 1911 and 1913 dealing with advanced base operations as a result of his perception of potential problems with Japanese expansion in the Pacific. In 1921, as the experience of World War I unfolded and the strategic balance shifted to Japan, Ellis produced a paper titled "Advanced Base Operations in Micronesia," which outlined the Marine Corps's operational plan for seizing hostile bases in the Pacific.[3] The document was approved in July 1921 by Maj. Gen. John Archer Lejeune, the Major General Commandant of the Marine Corps. This revolutionary plan became the linchpin of Marine Corps amphibious doctrine in the event of a Pacific war with Japan. Influenced by the prewar War Plan Orange, Ellis formulated elements for an amphibious assault doctrine that depended on extensive postwar training, the maturation of meticulous logistical and tactical planning, and the utilization of special landing vehicles. His plan had all the elements of technology, ideas, and reality. Although Ellis did discuss the problems of landing over barrier and fringing reefs, he made no reference to the role advanced technology would play in such an operation.

Then, in December 1922, General Lejeune received a report from a Marine Corps observer, a *New York Times* article, and a picture of a Christie amphibian that could operate either on its tracks or wheels. This convertible principle was an engineering achievement that promised to revolutionize cross-country transportation by eliminating carriers and bridging equipment. The report noted that the vehicle had been unveiled with a great deal of public notoriety. Unlike recent times, publicity surrounding new tank designs was rarely controlled. Most information on new military designs and displays were for public consumption and virtually never classified. The *Times* article reported the Christie was "expected to revolutionize modern warfare."[4] The Marine Corps observer advised the Commandant that "the use of this Mount seems to be indicated in landing operations of the Advance Base Force."[5] Thus, by the end of 1922, certain leaders in the Marine Corps not only demonstrated future operational direction but they had begun to consider the application of advanced technology to improve the operational capabilities and tactical mobility of the Expeditionary Force. This action was going against an attitude that began to prevail in military thinking during the period of postwar retrenchment. Maj. Gen. Clarence C. Williams, the Army's Chief of Ordnance, summed up this attitude when he paraphrased Admiral Mahan's earlier comment, remarking that the

Maj. Gen. John A. Lejeune, USMC Commandant, 1920-29. Marine Corps
University Archives.

military mind is a rigid one and does not lend itself to keeping pace with advance technology and its impact on tactics.[6]

J. Walter Christie, the creator of this amphibious wonder, was a self-made automotive engineer who approached mechanical problems by experience and application. He was from the school of men like Thomas Edison, who believed in empirical relevance to surmount technical difficulties. Christie had a propensity for developing ideas spontaneously and then exhibiting a passionate desire to put them to use with a great deal of public fanfare. He was an inventor rather than an engineer. Unlike Henry Ford, he preferred to work with prototypes rather than engage in the essentials of production. Nevertheless, Christie was a true proponent of mobility. He saw the need for developing a fighting vehicle that could negotiate water obstacles, especially when a mobile force had to deal with river a crossing.

When the United States entered the war in 1917, Christie began working on the convertible principle for armored fighting vehicles, which at the time was considered a revolution in automotive design. This track-wheel capability had the potential for enhancing the operational and tactical role of military vehicles. The AEF's Tank Corps and Army ordnance officers were very interested in applying this principle to fighting vehicle designs. As a result, the Ordnance Department issued work orders for a number of self-propelled artillery mounts and one tank. However, all Christie projects were terminated by 1924 due to numerous mechanical faults, major redesigns, poor Army service tests, an unclear tank development policy, and budgetary limitations.[7]

It was during this period when the Army was uncertain about its future tank policy that Christie conceived the idea of building a convertible amphibian gun mount. The project was an outgrowth of his experience gained building self-propelled gun mounts for the Army toward the end of the war. It is possible Christie was also familiar with foreign developments on an amphibian vehicle, such as the British Medium D tank. Major Raymond E. Carlson, an Army ordnance officer close to the Christie tank and self-propelled gun projects and a member of the Inter-Allied Tank Commission during the war, was aware of British attempts to develop an amphibian tank. The idea for this tank was perpetuated in Col. J.F.C. Fuller's revolutionary "Plan 1919," which was designed to break the static nature of trench warfare by executing a deep tank attack into the enemy's control and communications center. Fuller, the Royal Tank Corps's operations officer, was among the first to visualize a framework that associated speed, mobility, and extended range with a mechanized force. Plan 1919 became the guiding principle for proponents of mechanized warfare during the 1920s and 1930s. In its theoretical form, according to Fuller, the plan was designed to replace body warfare with brain warfare, with mechanical penetrations deep into the enemy's territory causing the dislocation of the enemy's forces by shutting down his ability to command. Fuller believed that if the enemy's will was paralyzed, the body then became inarticulate. He envisioned a force of fast Medium D tanks with an amphibious capability to execute his revolutionary plan. The Armistice eliminated the implementation of Plan 1919; nevertheless, Medium D development continued until 1922. The tank, however, was no better than a track-laying vehicle with an enhanced fording ability.[8] When he returned to the United States, Major Carlson suggested that future tanks should have the ability to negotiate land and water obstacles with impunity.[9]

In the December 1921 edition of the *Tank Corps Journal* (published at Bovington Camp, England) two interesting articles appeared. Fuller's "Tanks in Future Warfare" described a future scenario in which a "floating mechanical Army" could be launched from a seaborne mechanical base. This future scenario originated with his controversial lecture at the Royal United Service Institution on 11 February 1920 entitled: "The Development of Sea Warfare on Land and Its Influence on Future Naval Operations," which was published that year in the Institution's journal. Fuller's scenario envisioned an amphibian tank being launched from a "submarine tank carrier." Within a few years the U.S. Navy and Marine Corps attempted to make Fuller's projection a reality off the coast of Culebra Island near Puerto Rico. The other article, "An Amphibious Vehicle. Its Construction and Possible Uses," related to an earlier French experiment with a tracked amphibian. Its designer, M. Foenguinos, predicted the vehicle could be used "as land and water tank for attacking purpose."

In June 1921 the Christie vehicle was reported to have successfully negotiated the Hudson River. Christie used balsa wood floats encased in sheet metal to enhance flotation. The amphibian could operate on tracks or wheels. The vehicle was propelled through water by engaging two propellers and steered by varying the pitch of the propellers and speed of the tracks. This model had an open top that gave it the appearance of a self-propelled gun carriage. Brig. Gen. Samuel D. Rockenbach, Commander of the AEF Tank Corps in World War I, witnessed the exhibition. He felt the vehicle was not a tank but a gun mount. He believed additional details had to be worked out before it could be converted into a tank. Rockenbach theorized the flotation gear was too costly in weight for a tank. Nevertheless, he strongly endorsed Christie's accomplishments.[10] Christie, however, was not satisfied, and the following year he completed his second model using parts from the first amphibian. It was this vehicle that drew the Marine Corps's attention. *Automotive Industries* described the Christie as a combination of three types of an automotive vehicle in one: a wheeled motor truck, a crawler, and a motor boat.[11] *Scientific American* called the vehicle an amphibious military tank, designed to extend the mobility of field artillery by its ability to cross water obstacles.[12] *Army Ordnance* and the *Field Artillery Journal* also reported on the demonstration, placing emphasis on the dimensions and technical characteristics of the amphibian.[13] Considering the technology of the period, Christie's convertible design was quite an accomplishment. General Lejeune and Brig. Gen. Smedley D. Butler, the Commanding General of the Marine Barracks at Quantico, Virginia, took notice. They seriously began to view its potential role as an assault tank for sea-to-shore and overland operations for their Expeditionary Force.

During the fleet maneuvers that ran from December 1923 through February 1924, the Marine Corps Expeditionary Force participated for the first time in large-scale landing operations. One was scheduled for Culebra. Plans were made to evaluate two landing craft: the Christie

The second prototype Christie amphibious vehicle. National Archives.

Brig. Gen. Smedley Butler inside the fourth prototype Christie amphibious vehicle. Icks Collection, Patton Museum.

amphibian and a fifty-foot twin-engine landing barge with partial armor protection.[14] Ironically, Marine Corps interest in Christie's amphibian heightened at a time when the Army, due to budget constraints and technical problems, was losing interest in developing Christie-designed tanks and self-propelled artillery.

Soliciting financial aid from the Sun Shipbuilding Company of Chester, Pennsylvania, Christie introduced another modified version of his amphibian in November 1923.[15] It was this vehicle that General Lejeune instructed General Butler on 7 December to acquire for a test and included as part of the equipment for the 5th Regiment, the assaulting force.[16] General Butler, who was the designated commander of the Expeditionary Force, immediately contacted Christie at Sun Shipbuilding. Christie had earlier indicated he would lend the Marine Corps his vehicle at no cost, hoping that its performance would inspire the government to purchase it for the Marines. A written loan request was made contingent on Christie's ability to deliver the Marine tank to New York for a brief trial by General Butler. If it proved satisfactory to all concerned, Butler contemplated placing the vehicle aboard a battleship by the end of the year for transit to the Caribbean. General Butler advised Christie to keep the project secret because he wished it to be a surprise element in the landing operations. He concluded with a request that any written response be marked "Personal."[17]

On 11 December General Butler contacted the commander of the Scouting Fleet, VAdm. N.A. McCully, and outlined the Christie project. In the 1920s the Marines used motor-powered open landing barges for ship-to-shore transportation.[18] In his letter to Admiral McCully, General Butler proposed to reduce the Marines' exposure to hostile fire by employing the Christie, "a tank which is equally effective both on land and sea." General Butler again requested the "tank business" be kept secret not only because of the anticipated surprise but to demonstrate that landings could be made with minimal losses. Meanwhile, General Lejeune had contacted the Navy Department and received approval to include the Christie vehicle.[19] At 4:31 P.M. on 27 December, the boxlike Christie steel amphibian was hoisted aboard the port side of the USS *Wyoming* and logged in as the "U.S. Marine Corps Tank, GC2." Christie furnished one of his mechanic-drivers to operate the vehicle during the proposed surprise landing. General Lejeune arranged for the Marine Corps quartermaster to pay the mechanic $150 per month.

The maneuvers were broken up into a number of fleet problems, one of which involved Marine landings on Culebra. It was during this

The fourth Christie amphibian aboard the USS *Wyoming* in the Brooklyn Naval Yard, December 1923. National Archives.

The fourth Christie amphibian is loaded onto the deck of the submarine *S-20* off Culebra Island on 21 February 1924. National Archives.

problem that Christie's convertible amphibian was to execute General Butler's surprise assault. However, his enthusiasm for this revolutionary landing was short-lived. Shortly before the USS *Wyoming* sailed, Butler was replaced by Brig. Gen. Eli E. Cole, who was chosen to command the Expeditionary Force that would execute the Culebra landing. Finally, on 21 February, the Marine tank—mounting a 75-mm gun—was transferred to a submarine for launching. The idea of launching an amphibious track-laying vehicle from a submarine was not unusual at the time. Colonel Fuller had speculated that in future sea-to-shore warfare tanks would-be transported close to the enemy's beaches on submarines. The main advantage of such an operation, according to Fuller, was the ability of amphibian tanks to negotiate beaches and travel inland, thus causing command and control paralysis. It was the U.S. Marine Corps and Navy that first attempted to execute Fuller's idea in February 1924.[20]

The Christie reportedly swam in the sea successfully for a few miles. Though capable of swimming the Potomac and Hudson Rivers, it was unable to reach the beach because of the surf, thus proving the Marine tank to be unseaworthy. The driver, afraid that the vehicle might sink, returned to the launching area and it eventually ended up in Chiriqui Lagoon on Culebra, where Marines from the 8th Company, 5th Regiment, unofficially tested it. The overall assessment of the maneuvers was that they were a near disaster. General Cole called the operation chaotic. Nevertheless, Lt. Col. Kenneth J. Clifford noted years later in his important study the importance of the knowledge gained by experiencing the negatives of a large-scale landing operation and the "many recommendations to improve future landings."[21]

In May General Lejeune presented his report on the Christie tests at Culebra to the War Department. Showing exceptional political adroitness, the Commandant observed that it was not Marine Corps policy to undertake tests of or adopt weapons that differed from those in use by the Army or Navy. Therefore, he noted, there was no specific board of officers on hand to observe and report on the Marine tank's demonstration off Culebra. Nevertheless, a number of officers who participated in the maneuvers submitted unofficial reports on the Christie, which were included in the Commandant's correspondence to the War Department.

General Cole said he considered the tank "capable of being developed into an extremely valuable weapon of war, not only in connection with landing operations, but in a war of movement." He visualized its importance in crossing rivers and neutralizing the enemy's gun positions. He also viewed it as a potential supply vehicle. These visionary comments were expressed years before the threat of war in the Pacific made the search for such a vehicle a matter of prime importance.

Other observations reflected the officers' rank and experience. General Cole's Chief of Staff, Maj. James J. Meade, thought the tank had a future but said he believed it required remodeling and additional demonstrations. The Expeditionary Force's Air Service liaison officer, Capt. Robert J. Archibald, also agreed on the vehicle's potential, suggesting continued development, especially regarding ship transport for amphibious tanks and "their possible use for day or night operations." Capt. David L.S. Brewster of the 5th Regiment was part of a group of Marines who tested the Christie tank at Chiriqui Lagoon after it failed to negotiate the heavy surf. He was more critical, noting that in its present state it could not affect a landing in rough water. Furthermore, he was concerned that the two propellers might be sheared off when the vehicle was crossing rivers or streams with high banks. However, he accepted the idea of an amphibian tank as fundamentally sound and said an improved design would be valuable "not only in landing operations but also as accompanying guns for an infantry regiment, especially where a river might have to be crossed." Lt. Carlos H. McCullough, also from the 5th Regiment, believed that "The amphibious tank has numerous possibilities if an effort is made to carry on experiments."[22]

Overall, comments on the Christie tested at Chiriqui Lagoon were positive and visionary. The Marines were convinced that an amphibian assault tank could revolutionize the expeditionary role of the Marine

Marines testing the Christie amphibious tank at Chiriqui Lagoon on Culebra Island, February 1924. Icks Collection, Patton Museum.

Corps provided that continued tests and improvements were made to enhance its amphibious assault capabilities. This was a problem during the 1920s and 1930s because appropriations severely limited amphibious training and experiments with new landing equipment.

Years later, there were conflicting reports, articles, and studies on the Christie Marine tank tested at Culebra. Many of these studies added to the mystique of Christie and his armored fighting vehicles. For example, the charge that the vehicle was "unseaworthy" was made by Jeter A. Isely and Philip A. Crowl in *U.S. Marines and Amphibious War* (1951), and a December 1957 report by the Ingersoll Kalamazoo Division of Borg-Warner Corporation entitled "Research, Investigation and Experimentation in the Field of Amphibian Vehicles for U.S. Marine Corps."[23] An umpire at the 1924 maneuvers—Maj. Holland M. Smith—wrote later in the *Marine Corps Gazette* (1946) that the Christie "demonstrated a singular lack of seaworthiness." In his autobiography, *Coral and Brass* (1949), General Smith noted the Christie "worked fairly well in the water." He claimed the captain of the USS *Wyoming* became unnerved over the experiment. Consequently he aborted the test by placing out the boom and hoisting the Marine tank back on deck.[24] On the other hand, the Marine School's Series No. 18 on *Amphibious Operations: Employment of Tanks* reported the Christie "swam ashore from the USS *Wyoming*."[25] This observation was also noted earlier in Ralph E. Jones, George H. Rarey, and Robert J. Icks, *The Fighting Tanks Since 1916* (1933), and later in Richard M. Ogorkiewicz, *Armoured Forces* (1970), where it was stated that Christie "gave the first practical demonstration when it swam ashore from the USS *Wisconsin*."[26] J. Edward Christie—J. Walter's adopted son—in his controversial memoir, *Steel Steeds Christie* (1985), claimed the amphibian eventually made a safe ship-to-shore landing.[27]

In *Soldiers of the Sea* (1962), author Robert Heinl claimed the Christie suspension employed on the Marine tank was purchased by the Soviet Union after the U.S. government failed to accept it.[28] In fact, the Red Army acquired two Christie convertible nonamphibian tanks in 1930, and in 1931, the U.S. Army contracted for seven. The Christie system, however, was never standardized by the U.S. military. Nevertheless, throughout the 1930s and early 1940s the U.S. Army—in spite of a lack of funds during the depression era—continued to develop, modify, and experiment with the Christie system. The large road wheels with rubber tires mounted on coil springs (three of the four pairs of wheels were supported on springs) used in the 1924 Marine tank were not the type of suspension arrangement used on Red Army Christies as maintained by Robert Heinl. In 1928 Christie introduced a remarkable combat vehicle whose suspension system employed long, individually sprung helical springs for each road wheel. This arrangement allowed for greater wheel amplification and depression than did the suspension systems in existing armored tracked fighting vehicles. It was this design that the Red Army and the U.S. Army were interested in developing, because it provided greater speed and a more stable firing platform.[29] However, Christie's new design was not suited for amphibious operations.

General Lejeune's 1924 report on the Christie was circulated for comments to the War Department and then routed to the Commandant of the Army's Tank School, and the Army's Ordnance Chief. The Tank School Commandant advised that with suitable armor "the vehicle would not float," and that "the problem should be to make a fighting tank float and not try to use something because it will float." The Ordnance Department's position was that the amphibious tank was not in a sense a tank. In addition, the running gear, wheels, and track were too light. The Army ultimately concluded that the vehicle could not endure the demands of combat, and staff comments were eventually forwarded to General Lejeune after the decision not to buy the Christie had already been made.[30]

Meanwhile, Christie contacted General Lejeune, offering to sell his amphibian for the manufacturing price of $30,000 cash. At first he attempted to sell the Marine Corps the vehicle for $120,000, a figure that included the total expense of manufacturing, additional experimental costs, drawings, patterns, testing, and demonstrations. The commandant responded by advising the inventor that Congress funded all weapons purchases. In the end, because the amphibious tank project was considered a private venture and had not been authorized congressional funding or formally authorized by the Department of the Navy, the Marine Corps decided not to acquire the vehicle.[31] A major determining factor in this decision was a study by the Marine Corps's Division of Operations and Training that indicated the Christie Marine tank was still experimental and that acquiring repair parts would pose a major problem. Furthermore, the operations staff saw little benefit from the vehicle "except as a means of experiment." They concluded their report with a recommendation that the Commandant advise Christie that the Marine Corps simply had no money to buy his amphibious tank.[32]

On 27 June General Lejeune wrote Christie thanking him for the loan. He observed that, while the amphibian tank was not yet perfected, he believed it was capable of being developed into a valuable assault weapon.[33] In his annual report to the Secretary of the Navy, the Commandant played down the Marine Corps's experience with the Christie, mentioning that "two special types of boats for landing operations were used experimentally with interesting, although not decisive, results."[34]

Christie's failure to interest the Marine Corps and U.S. Army in his armored fighting vehicles, coupled with a need for financial aid, led him to open negotiations with the Japanese government in April 1925. The Japanese Imperial Army became familiar with Christie's amphibian tank through observations and reports from Maj. Ko Daikaku, an artillery officer engaged in ordnance designs and intelligence activities in the United States. Major Daikaku participated in demonstrations of the amphibian at Chester, Pennsylvania, after an introduction arranged by the Okura Shoji Trading Company of New York. Upon his return to Japan he reported the demonstrations to Lt. Gen. Katsuichi Ogata, an artillery officer also engaged in ordnance designs. Early in 1925 General Ogata was sent to the United States as chief of purchasing. While in that post he negotiated with Christie for the sale of certain patent rights with the possibility of building combat tanks for the Imperial Army. The financially strapped Christie had neither prototypes nor manufacturing facilities, so he sold skeleton drawings of his amphibian tank to the Okura Shoji Trading Company, acting on behalf of the Japanese government.[35] Christie, through his attorney, subsequently advised the Army's Ordnance Department of his negotiations with the Japanese. He sarcastically informed the Chief of Ordnance that he felt "obliged" to accept funds from the Japanese government "and upon your shoulders [Ordnance Department] will rest the responsibility."[36]

Maj. Tomio Hara, who was personally acquainted with General Ogata and Major Daikaku, was engaged in ordnance designs and intelligence work during the interwar period. In January 1931 he met Christie at Aberdeen Proving Ground and viewed the M1930 tank, which at the time was undergoing U.S. Army acceptance tests. A Red Army representative was also in attendance. The Japanese attempted to resume negotiations, but after January Christie cut off contact with them and focused his efforts on selling tanks to the Soviet Union, Poland, and the United States. However, the Japanese, continued to show interest in amphibious developments in the Soviet Union and England. This was due to the introduction of the British Vickers Carden-Lloyd amphibian tank, which was bought, copied, and then employed by the Red Army. By 1942 Japan had developed its own amphibious tank program, which was initially divided between the Army and Navy.[37]

The Japanese and Soviets were not the only ones interested in Christie's amphibian tank. The Italian naval military attaché pressed the director of Naval Intelligence for information "whether these tanks have been tested by the Navy, whether their nautical and performing qualities have been found satisfactory, and whether the U.S. Navy intends adopting them." The attaché was only provided with information on the 1922 Hudson River test. No information was provided on the Culebra experience even though unofficial comments had been circulated among U.S. military agencies. The attaché was advised to contact Christie. Whether he did or not is unclear. What is clear is that Christie's designs had no influence on the Italian amphibious vehicle development program.[38]

In Germany the Reichswehr also took notice of Christie's amphibious designs. The *Technische Mitteilugen* (1926) credited Christie with solving the amphibian tank problem. The first major postwar German armor tactician, Ernst Volckheim, who was influenced by Fuller and familiar with Christie's armor vehicle designs, considered amphibian tanks necessary to the Reichswehr's developing doctrine of mechanized warfare. Later, in the early 1930s, when Christie's fortunes were on the decline and developmental costs were rising, the Reichswehr's Weapons Office lost interest in the project. By the mid-1930s Germany had developed its own tank program.[39]

The Marine Corps's interest in Christie's tank did not end with the commandant's reluctant rejection. While the Marines continued their commitment to the limited conflicts in Central America called the "Banana Wars," the Corps also was called upon to suppress unrest in China. General Butler, who had borrowed Christie's Marine tank for the 1924 experiment, commanded the Marine Corps Expeditionary Force in China. In April 1927, during the Nanking affair, he sent an urgent telegram to the Navy Department requesting immediate shipment of six Christie amphibians or armored cars to China. General Butler was advised that there were no amphibian tanks or armored cars available. Instead, a platoon of obsolete six-ton M1917s was sent as infantry support weapons for the 4th Marines. In discussing General Butler's request for the Christies or armored cars, the Division of Operations and Training staff advised the Commandant that Christie since 1924 had not attempted to improve on or manufacture additional amphibian tanks. They further explained that neither the Army nor the Marine Corps had adopted or standardized any of his armored vehicles for military use.[40]

The idea of using a tracked amphibian tank for military operations did not subside with the Marine Corps's and Christie's withdrawal from its development. In December 1929 Capt. George H. Rarey—at the time a respected tank historian—was teaching a course on tank design at the Army's Tank School. He requested from the Commandant information on the Marine Corps's experience with the Christie Marine tank. He was advised of the unofficial test held at Culebra and the comments made by General Cole and the officers who tested the vehicle in Chiriqui Lagoon. In addition, Rarey was advised to discuss Christie with Army ordnance personnel because of their extensive experience with his designs since World War I. In his response, Captain Rarey opined that the value of amphibian tanks for ship-to-shore and overland operations should have high priority in any mechanized development program. Captain Rarey, who had access to plans for a joint Army and Navy operation, also suggested that amphibian tanks be incorporated in future joint operations. These suggestions fell on deaf ears since the written material on joint Army and Navy operations in 1931 made no reference to amphibian tanks, although a passing comment was made about the problem of getting nonamphibian tanks to a hostile shore.[41]

After their brief experience with the Christie, the Marine Corps lost interest in the development of a convertible Marine tank. This was primarily a result of budgetary and mechanical problems with the Christie vehicle, especially regarding funds for research and development. During the interwar period the U.S. military was subjected to draconian budget cuts and suffered from a national lack of interest in military affairs, a characteristic that normally accompanied periods of peace after American wars. In addition, there was a preference for developing tank carriers and bridging equipment rather than providing a developmental program for amphibian tanks.[42] The only country to extensively pursue amphibian tank development was the Soviet Union. During the 1930s light amphibian tanks based on the British Vickers Carden-Lloyd design were employed throughout the Red Army as reconnaissance tanks for mechanized units and in mixed tank battalions in infantry divisions. This model, the T37, advanced the Red Army's level of engagement with a mechanized force capable of deep land operations.[43] The T37, however, was not designed for ship-to-shore operations.

Further Japanese expansion and aggression in China caused a renewed interest in an amphibian assault vehicle. This situation caused the Marine Corps to devote considerable effort at their Quantico schools to analyzing amphibious warfare doctrine. Meanwhile, in September 1937, 1st Lt. Victor H. Krulak, an intelligence officer with the 4th Marines in Shanghai, was provided a rare opportunity to observe a Japanese amphibious assault on Chinese positions on the Yangtze River. The future lieutenant general was impressed with a Japanese ramp-bow craft capable of landing infantry heavy support vehicles directly on the beaches. At the time Krulak believed the Japanese were ahead of the navy and Marines in landing craft development.[44] In the United States at the same time a similar invention was found that resurrected the Christie idea. It resulted in solving the need for a specialized troop landing craft for ship-to-shore assault operations.

On 4 October 1937 *Life* magazine featured a story on a nonmilitary amphibian tractor used for rescue operations in the Florida Everglades. Designed and built by Donald Roebling in 1935, the "Swamp Gator" caught the attention of naval and Marine Corps officers.[45] Unlike the twin removable propellers used on the Christie, the Roebling tractor was propelled by a paddle-tread track. Officers who had feared the Christie's propellers were vulnerable to damage saw an answer to the Pacific coral reef problem in the Roebling design. This greatly enhanced its military potential.

It was ironic that during one demonstration in 1940 the Roebling prototype was alongside the old battleship USS *Wyoming*, which had launched the Christie in February 1924.[46]

The Roebling design was the basis for the Landing Vehicle, Tracked (LVT) so successfully used by the Navy, Marine Corps, and some Army units during World War II. More than eighteen thousand LVTs had been built in various configurations by the end of the conflict.[47]

When World War I ended, rapid demobilization, retrenchment, and a contracting economy affected the Marine Corps, creating personnel turbulence and uncertainty about the future. Disarmament was in vogue, and the country—consumed with a false sense of national security—became indifferent to military matters. After all, was not World War I the "war to end all wars"? General Lejeune, the Commandant during the 1920s, was aware of the Corps's precarious situation and moved to reconnect it with the navy. He recorded in his *Reminiscences* (1930) that a major, and many times strenuous, effort was directed to defending the existence and integrity of the Marine Corps.[48] The Republican administrations of the 1920s were committed to controlling government expenditures, especially military expenditures. This austerity program rested very heavily on the military, and for a time the Commandant was concerned that the Corps might be eliminated.

It was evident during the experiment with the first Marine Corps amphibian tank in 1924 that the conflict between technology, ideas, and reality of the times generated tensions between tradition and modernity at a time when the Corps was faced with force reduction. The Christie convertible Marine tank was a new weapon of war that emerged out of wartime technology into the idea of an assault amphibian tank.

This idea found its clearest expression in the minds of such visionaries as Generals Lejeune, Cole, Butler, and other officers at Culebra who unofficially tested the vehicle. The Marine Corps and its Commandant attempted to graft this idea of a Marine tank to the Corps's perceived assault mission. However, lack of funds and the halcyon years of the 1920s prevented further technology investigations, resulting in the inability of the Marine Corps to gauge the influence of an assault vehicle on tactics. As for J. Walter Christie, he can be credited with making the Marine Corps aware in the interwar years that an amphibian assault vehicle existed. However, his convertible principle was finally discarded in the late 1930s as an unworkable engineering compromise.

Nevertheless, the Christie experience demonstrated that the Marine Corps was serious in linking new technology with its "Advanced Base Operations" plan, but it was stifled by political and civilian indifference to military matters. Furthermore, the Corps was influenced by a naval disarmament mood and an Army tank policy that lacked direction. Through most of the interwar period U.S. military ground forces used obsolete World War I equipment in spite of the growth in weapons technology. In addition, the Marines were governed by a policy that prevented adopting weapons that differed from those in use by the Army and Navy. The Corps thus was never provided the opportunity to form an experimental unit to further evaluate the Christie Marine tank as a ship-to-shore assault vehicle. The doctrine and early technology were evident; however, there was no opportunity to develop operational concepts and training in tactics for assault vehicles. Realistically, there was no immediate need for an amphibian tank during the 1920s, when the Marine Corps was mainly engaged in traditional political-military actions that were never challenged by hostile forces during landing operations. Only when the conversion factor was added—that is, Japan's power projection into Asia and the Pacific in the 1930s—did the Navy and Marine Corps realize that tracked assault vehicles were needed to engage a formidable expansionist foe.

Notes

This article is a refinement of my Christie amphibian tank articles that appeared in the *Marine Corps Gazette* in September 1977, and in Donald F. Bittner, ed., *Perspectives on Warfighting Number Three,* Marine Corps University (May 1994). In crafting these articles, I was fortunate to have the generous cooperation of the following: U.S. Marine Corps (USMC) University Research Center Archives Branch, USMC Development and Education Center, and the *Marine Corps Gazette* and the Marine Corps Association,

Quantico, Va.; USMC History and Museums Division, Washington, D.C.; Naval War College, Newport, R.I.; Patton Museum of Cavalry and Armor, U.S. Army Armor Center, Fort Knox, Ky.; and the National Archives in Washington, D.C., and Suitland, Md.

1. Alfred Thayer Mahan, *The Influence of Sea Power Upon History 1660-1783* (Boston: Little, Brown and Company, 1890), pp. 9-10; and J.K. Christmas, "Tanks: The Ideal Combination of Fire Power, Mobility and Protection," *Army Ordnance* (Jan.-Feb. 1941): p. 22.

2. Recent scholarship has paid little attention to Christie's influence on and relationship with the Marine Corps. Merrill L. Bartlett, Lejeune's biographer, does not mention the Commandant's support of the Christie amphibian in the assault mission. See *Lejeune: A Marine's Life, 1867-1942* (Columbia: University of South Carolina Press, 1991), pp. 146-71, 197. An examination of *John Archer Lejeune 1869-1942 Register of His Personal Papers* (Washington: USMC History and Museums Division, 1988) compiled by Bartlett, failed to turn up correspondence between the Commandant and Christie. General Butler's biographer, Hans Schmidt, does note briefly Butler's role in arranging to have the Christie tank secretly introduced at Culebra early in 1924. See *Maverick Marine: General Smedley D. Butler and the Contradictions of American Military History* (Lexington: University of Kentucky Press, 1987), pp. 128, 142. Anne Cipriano Venzon, ed., *General Smedley Darlington Butler: The Letters of a Leatherneck, 1898-1931* (New York: Praeger Publishers, 1992), does not list correspondence between Butler and Christie. Victor H. Krulak, *First to Fight: An Inside View of the U.S. Marine Corps* (Annapolis: Naval Institute Press, 1984), depicts a picture of the experimental Christie at the 1924 Culebra maneuvers but makes no mention on its contribution.

More recently, William P. McLaughlin, "The Assault Amphibian Vehicle (AAV): Its Past, Present and Future," *ARMOR* (Apr.-May 1993), wrote that the Christie was rejected because of "poor water speed and buoyancy." This article, however, failed to identify the major historical issues that influenced the development of the first amphibious assault vehicle. See pp. 8-17. In Allan R. Millett's comprehensive history, *Semper Fidelis: The History of the United States Marine Corps, A Revised and Expanded Edition* (New York: The Free Press, 1991), makes a passing reference to the Christie experiment. See pp. 327, 341. Earlier, Jeter A. Isely and Philip A. Crowl wrote briefly on the innovative Christie tank, commenting that it was the forerunner of the remarkable World War II LVT. See *The U.S. Marines and Amphibious War: Its Theory and Its Practice in the Pacific* (Princeton, N.J.: Princeton University Press, 1951), pp. 31-32, 68-69. By far the most interesting accounts of the Christie Marine tank and the 1924 maneuvers are Kenneth J. Clifford, *Progress and Purpose: A Developmental History of the U.S. Marine Corps, 1900-1970* (Washington: History and Museums Division, 1973), pp. 34-35; and George F. Hofmann, "A self-made automotive engineer finally convinced the military that an LVT existed in the 1920s," *Marine Corps Gazette* (Sept. 1977), pp. 42-50. Two Occasional Papers for limited distribution prepared by the USMC History and Museums Division made brief comments on the Christie experiment. See Alfred Dunlop Bailey, *Alligators, Buffaloes, and Bushmaster: The History of the Development of the LVT Through World War II* (Washington: Headquarters, Marine Corps [hereafter HQMC], 1986), pp. 17-19; and Holland M. Smith, *The Development of Amphibious Tactics in the U.S. Navy* (Washington: HQMC, 1992), pp. 21-22.

3. For a recent study on Ellis, see Dirk Anthony Ballendorf and Merrill Lewis Bartlett, *Pete Ellis: An Amphibious Warfare Prophet, 1880-1923* (Annapolis, Md.: Naval Institute Press, 1997). Also see, John J. Reber, "Pete Ellis: Amphibious Warfare Prophet," *U.S. Naval Institute Proceedings* (Nov. 1977), pp. 53-64, and Ballendorf, "Earl Hancock Ellis: The Man and His Mission," *U.S. Naval Institute Proceedings* (Nov. 1983), pp. 53-60; and Frank O. Hough, Verle E. Ludwig, and Henry I. Shaw Jr., "Evolution of Modern Amphibious Warfare, 1920-1941," in *Pearl Harbor to Guadalcanal: History of U.S. Marine Corps Operations in World War II* (Washington: GPO, 1958), pp. 8-11. Also see, Fleet Marine Force Reference Publication (FMFRP) 12-46, *Advanced Base Operations in Micronesia* (Washington: HQMC, 1992).

4. "Tank Swims Hudson, Mounts Palisades," *New York Times,* 6 Dec. 1922; and "To See Web-Footed Tank: Armored Truck which Swam Hudson to be Shown in Washington," *New York Times,* 7 Dec. 1922.

5. Maj. Robert W. Voeth, USMC, to the Maj. Gen. Comd't., 6 Dec. 1922, Records of the USMC, Adjutant and Inspector's Office, General Correspondence, 1913-1932, Box 23, RG 127, NA. (Hereafter cited as USMC Gen. Corr.)

6. Williams, "Automotive Ordnance Equipment," *Journal of the Society of Automotive Engineers* (Oct. 1919), p. 304.

7. For additional elaboration on the problem of mental rigidity, integrating ideas and technology, and a lack of policy direction, see George F. Hofmann, "The Demise of the U.S. Tank Corps and Medium Tank Development Program," *Military Affairs* (Feb. 1973): pp. 20-25.

8. Maj. Gen. J.F.C. Fuller, *Memoirs of an Unconventional Soldier* (London: Nicholson and Watson, 1936), pp. 318-41; Brian Holden Reid, *J.F.C. Fuller: Military Thinker* (New York: St. Martin's Press, 1987), pp. 48-55, 56-57, 74, 154, 231; and Anthony John Trythall, *'Boney' Fuller: The Intellectual General, 1878-1966* (London: Cassell & Co., Ltd., 1977), pp. 60-64. Fuller's plan was entitled "The Tactics of the Attack as affected by the Speed and Circuit of the Medium D Tank."

9. Maj. Raymond E. Carlson, "Papers on the Development of Tanks," 16 Mar. 1921, Ordnance Technical Intelligence Files, (OKD), 451.25/56.1, RG 156, NA, pp. 8-9, (Hereafter cited as Ordnance Gen. Corr.); Also see "The Tanks of the Future," *Infantry Journal* (Jan. 1921): p. 83; "Tanks of the Future," *Infantry Journal* (Feb. 1921): p. 182; Chief of Field Artillery to Dir. of Military Intelligence, Subject: Tanks, 14 Jan. 1920; Dir. of Military Intel-

ligence to American Attaché London, England, Subject: British Tanks, 19 Jan. 1920, Military Intelligence Division Correspondence, 1917-1941, Report No. 10318-249, Records of the War Department General and Special Staffs, RG 165, NA. (Hereafter cited as G2 Intell. Files.)

10. Rockenbach's comments in Ralph E. Jones, George H. Rarey, and Robert J. Icks, *The Fighting Tanks Since 1916* (Old Greenwich, Conn.: WE Inc., 1933), pp. 303-304.

11. "Amphibious Tank, Novel Engineering Development, Swims Hudson River," *Automotive Industries* (14 Dec. 1923): pp. 1164-65.

12. "An Amphibious Military Tank," *Scientific American* (Feb. 1923): p. 90.

13. "Demonstration of Land and Water Type of Motor Gun Carriage by the Front Drive Motor Co.," *Field Artillery Journal* (Jan.-Feb. 1923): pp. 85-87—also published in *Army Ordnance* (Jan.-Feb. 1923).

14. Robert D. Heinl Jr., *Soldiers of the Sea: The United States Marine Corps, 1775-1962* (Annapolis, Md.: Naval Institute Press, 1962), p. 259.

15. John J. Jordan, Dir. of Public Relations, Sun Shipbuilding and Dry Dock Co., to author, 8 June 1971. This letter contained comments from William Smith who worked in the shipyard when Christie was there in 1923 redesigning his second model. According to Smith, Sun Shipbuilding's arrangement with Christie was financial along with providing its facilities.

16. Maj. Gen. Comd't., to CG, Quantico, Subject: "Christie Tank-Test of and use of in Culebra during Winter Maneuvers," 7 Dec. 1923, USMC Gen. Corr. The Marines associated with the project referred to the vehicle as the Christie Convertible Marine tank.

17. Gen. Butler to Christie, 7 Dec. 1923, USMC Gen. Corr.

18. Rufus H. Lane, "The Mission and Doctrine of the Marine Corps," *Marine Corps Gazette* (Mar. 1923): p. 11.

19. Gen. Butler to VAdm. McCully, 11 Dec. 1923; and Gen. Butler to Christie, 11 Dec. 1923, USMC Gen. Corr.

20. Maj. Gen. J.F.C. Fuller, *The Reformation of War* (London: Hutchinson and Co., 1923), pp. 182-83, and "The Development of Sea Warfare on Land and Its Influence on Future Naval Operations," *Journal of the Royal United Service Institution* (May 1920): p. 293. In Duncan Crow and Robert J. Icks, *Encyclopedia of Tanks* (Secaucus, N.J.: Chartwell Books, Inc., 1975), p. 19, a Japanese amphibian tank is pictured on the deck of a submarine. This was the Christie on a U.S. submarine, the *S-20*, which was moored alongside the USS *California* until the morning of 21 Feb. when it apparently sailed next to the USS *Wyoming* to take aboard the Marine tank. Later that afternoon the *S-20* departed for the Virgin Islands. Neither the *California*, the *Wyoming*, nor the *S-20* logged experiments with the Christie vehicle.

21. Clifford, *Progress and Purpose*, p. 32, and Heinl, *Soldier of the Sea*, p. 259.

22. Maj. Gen. Comd't., to Adjutant Gen., U.S. Army, War Department,

Subject: Christie amphibious tank, 27 May 1924, USMC Gen. Corr.

23. Isely and Crowl, *The U.S. Marines and Amphibious War*, p. 31; and Final Report, Dec. 1957, USMC Contract MOm 66245, James Carson Breckinridge Library, USMC Development and Education Command, Quantico, Va., pp. 40-42. Isely's and Crowl's source was Frederick L. Wieseman's "Proper Design and Employment of a Marine Corps Tank for Landing Operations," a lecture he gave at the Marine Corps School in Mar. 1949. See p. 593 n28.

24. Holland M. Smith, "The Development of Amphibious Tactics in the U.S. Navy," *Marine Corps Gazette* (Aug. 1946), p. 21; and Holland M. Smith with Percy Finch, *Coral and Brass* (New York: Charles Scribner's Sons, 1949), p. 97.

25. Historical Amphibious File, USMC Educational Center, Breckinridge Library.

26. Jones, et al., *Fighting Tanks*, p. 304; and Richard M. Ogorkiewicz, *Armoured Forces: A History of Armoured Forces and Their Vehicles* (London: Arms and Armour Press, 1970), pp. 409-10.

The most reliable primary sources providing historical data dealing with Christie and the tests at Culebra were located at the time in the Records of the Marine Corps, Adjutant and Inspector's Office, General Correspondence, 1913-1932, RG 127, NA. It is doubtful the authors listed above had seen these files because this rather large collection was frequently overlooked at the time their comments on the Christie test at Culebra were recorded. Only when Colonel Clifford began his research on the developmental history were the sources dealing with the Culebra maneuvers of 1924 and the Christie experiment adequately noted. See Clifford, *Progress and Purpose*, pp. 32-35, n117.

27. For a discussion of the controversy, see J. Edward Christie, *Steel Steeds Christie* (Manhattan, Kans.: Sunflower University Press, 1985), pp. 24-26; *Steel Steeds Christie* reviewed by Col. Leo D. Johns, *ARMOR* (Jan.-Feb. 1986): p. 52; "Crossfire: Review of Christie Biography Draws Comments from Author, Readers," *ARMOR* (July-Aug. 1986): pp. 2-3; "The Christie Biography: Conflict Revisited," *ARMOR* (Nov.-Dec. 1986): pp. 3-4; and "More on Christie's Battles with Ordnance," *ARMOR* (Sept.-Oct. 1990): p. 3. Also see "Book Review" by D.P. Dyer, *AFV News* (Jan.-Apr. 1986): p. 20; Konrad F. Schreier Jr., "Christie Revived," *AFV News* (Jan.-Apr. 1986): p. 18; D.P. Dyer, "Steel Steeds Christie (Part umpety umpth)," *AFV News* (Sept.-Dec. 1986): p. 22; "Christie Drawings," *AFV News* (Jan.-Apr. 1987): p. 17; D.P. Dyer, "Comments Regarding Article by Konrad F. Schreier, Jr.," and "J.E. Christie Responds," *AFV News* (May-Aug. 1987): pp. 15, 19.

28. Heinl, *Soldiers of the Sea*, p. 259. For additional misinterpretations of the 1924 Christie experiment, see Richard M. Ogorkiewicz, "Evolution of the Amphibious Tank," *Marine Corps Gazette* (Aug. 1957): p. 22; and James D. Mason, "Tracked Landing Vehicles," *Ordnance* (Jan.-Feb. 1972): p. 314.

29. On the Christie-Red Army affair, see George F. Hofmann, "Doctrine, Tank Technology, and Execution: I.A. Khalepskii and the Red Army's Fulfillment of Deep Offensive Operations," *Journal of Slavic Military Studies* (June 1996). On Christie's relationship with the U.S. Army, see George F. Hofmann, "The Troubled History of the Christie Tank," *ARMY* (May 1986): pp. 54-65. The last U.S. military evaluation of the Christie suspension occurred early in World War II when Army tank destroyer doctrine was being formulated. See Hofmann, "Christie's Last Hurrah," *ARMOR* (Nov.-Dec. 1991): pp. 14-19.

30. Adjutant General's Office to Maj. Gen. Comd't., Subject: Amphibious Tank, 7 July 1924, USMC Gen. Corr.

31. Christie to Gen. Lejeune, 10 June 1924; Law Offices of George K. Perkins to Gen. Lejeune, 21 June 1924; P.H. Gill and Sons Forge and Machine Works to Gen. Lejeune, 24 June 1924; and Gen. Lejeune to P.H. Gill Jr., Pres., 27 June 1924, USMC Gen. Corr.

32. Memorandum for General Feland, HQMC, O and T Div., 25 June 1924, USMC Gen. Corr.

33. Gen. Lejeune to Walter Christie, 27 June 1924, USMC Gen. Corr.

34. Maj. Gen. Comd't. Report in the Report of the Secretary of the Navy for 1924, Historical Division, HQMC, Quantico, Va.

35. Gen. Tomio Hara to author, 21 Sept. and 25 Nov. 1970, and 29 Jan. 1971, author's files.

36. J.R. Tiffany to Gen. Clarence C. Williams, 27 Apr. 1925, OO 072/4999, Ordnance Gen. Corr.

37. Hara and Akira Takeuchi, *Japanese Tanks and Armored Vehicles* (Tokyo: Shuppan Kyodo Sha, 1961), pp. 93-108, and Hofmann, "Doctrine, Tank Technology, and Execution," p. 298. Also see "Confidential Report," Lt. Col. Fred Cardway, USAR, to Maj. Walter O. Boswell, 30 Apr. 1925, G2 Intelligence Files, MID 2281-H-22, RG 165, NA.

38. Office of Naval Intelligence (ONI) to Director of Military Intelligence (DMI), 5 May 1926, OO 402.2/133 Ordnance Gen. Corr.; ONI to DMI, Subject: Information Relative to Amphibious Tanks of the Wheel Track Layer Corporation Desired by the Italian Naval Attaché, 21 May 1926; and War Dept., Ordnance Office to Office, Assistant Chief of Staff, G2, 4 June 1926, G2 Intelligence Files No. MID 301-V-33, RG 165, NA. Also see DMI to Director, ONI, Subject: Christie Wheel Track Light Marine tank, 27 Aug. 1925, ONI, File No. MID 315-V-1, RG 165, NA; and Memorandum for Chief of Bureau, Subject: Test of Marine tank, 4 June 1924, File No. 40438, Navy Ordnance Bureau Records, RG 74, NA. The latter source indicated the navy was somewhat interested in the Christie. The memorandum reported on a public demonstration of the Christie amphibian after the tests at Culebra. It noted that the Marine tank maneuvered in and around the muddy surface extension of Potomac Park adjoining the river. The demonstration was reported as satisfactory; how-

ever, no naval interest developed at the time except responding to foreign attachés' questions.

39. Walter J. Spielberger to author, 12 Jan. 1970, author's files. Spielberger, a noted German tank authority and historian, was involved with the engineering and development side of German armor in World War II. Also see Walter Nehring, *Die Geschichte der deutschen Panzerwaffe 1916 bis 1945* (Berlin: Propyaen Verlag, 1974), pp. 39-51; and James S. Corum, *The Roots of Blitzkrieg: Hans von Seecht and German Military Reform* (Lawrence: University Press of Kansas, 1992), pp. 113-14, 130-31. Corum's source is Ernst Volckheim's *Der Kampfwagen in der heutigen Kriegfuhrung. Ein handbuch fur alle Waffen* (Berlin: Mittler & Sohn, 1924), p. 83.

40. CING ASIATIC to Adjutant and Insp. Department, 2 Apr. 1927; Dir., Div. of Operations and Training to Commandant, Subject: Amphibian Tanks (Christie) or Armored Cars for Service in China, 2 Apr. 1927, USMC Gen. Corr.; and Maj. Gen. Comd't. *Annual Report 1927*, Historical Division, HQMC, p. 1193. Victor J. Croizat in "The Marines' Amphibians," *Marine Corps Gazette* (June 1953) stated that in 1927 six modified Christies were used in China. See p. 42. His source was the Secretary of the Navy Continuing Board for Development of LVT, *History of Landing Vehicle, Tracked* (1 Dec. 1945). Other authors subsequently repeated the use of Christie amphibians in China. See Ogorkiewicz, *Armoured Forces*, p. 409; and Mason, "Tracked Landing Vehicles," p. 314. One author went so far as to write that Christie's amphibians were developed by the Japanese and used in their 1941 seaborne offensive. See John Wheldon, *Machine Age Armies* (London: Abelard-Schuman, 1968), p. 221. Seven years later Chris Ellis and Peter Chamberlain in *The Great Tanks* (London: Hamlyn Publishing Group Ltd., 1975) wrote that Christie won an order for six amphibians from the Marine Corps and that an additional one was bought by Japan. See p. 40.

41. Rarey to Maj. Gen. Comd't., 15 Dec. 1929; Dir., Div. of Operations and Training to Maj. Gen. Comd't., Subject: Test of Christie Amphibian Tank, 20 Dec. 1929; Maj. Gen. Comd't. to Rarey, 23 Dec. 1929; Rarey to Maj. Gen. Comd't., 18 Mar. 1930, USMC Gen. Corr., File No. 2455-75-35, RG 127, NA; and "Joint Army and Navy Operations 1931: The Conduct of a Forced Landing," Naval War College, p. 74.

42. Ingersoll Kalamazoo Division of Borg-Warner Corporation Report, "Research, Investigation and Experimentation in the Field of Amphibian Vehicles for U.S. Marine Corps," (Dec. 1957), Breckinridge Library, p. 40-42.

43. In Feb. 1932 the Soviet Union ordered eight light amphibian tanks from Vickers. Like the Christie tested in 1924, the T37s were propeller driven in water. In 1933 the Red Army ordered its tank industry to furnish 1,200 T37s. In 1938 the T37 light amphibious tank fulfillment program was concluded for an advanced model. See Order No. TD 908: "8 Vickers Carden Lloyd Light Amphibious Tanks Complete with Turret & Block Mountings," 5 Feb. 1932, The Tank Museum, Bovington, Dorset, England, pp. 1-2; To:

People's Commissar of Army and Navy and the Chairman of RMC [Revolutionary Military Council], USSR Mr. Voroshilov, From: Griazov, MMC [Directorate of Mechanization and Motorization of Red Army], 15 Mar. 1933, *f.* (fond) 9, *op.* (inventory) 14, *d.* (item) 717, *l.* (folio or page number) 51; and To: People's Commissar of Army and Navy and the Chairman of RMC USSR Mr. Voroshilov, From: Pavlov, Chairman of ABTU [Directorate of Motor and Armored Troops of Red Army], 21 Feb. 1938, *f.* 9, *op.* 19, *d.* 55, *l.* 5, Russian State Military Archives [*Rossiskii Gosudarstvennii Voennyi Arkhiv-*RGVA], Moscow.

44. Krulak, *First to Fight,* pp. 90-91.

45. "Science & Industry: Roebling's 'Alligator' for Florida Rescues," *Life,* 4 Oct. 1937: pp. 94-95; and Holland M. Smith, "The Development of Amphibious Tactics in the U.S. Navy," *Marine Corps Gazette* (Oct. 1946): pp. 32-33.

46. Krulak, *First to Fight,* pp. 103-104.

47. Clifford, *Progress and Purpose,* pp. 54-57; and Robert J. Icks, "Landing Vehicles Tracked," *AFV Profiles 16* (Windsor, England: Profile Publications, Ltd., 1972), p. 13. Also see Croizat, "The Marines' Amphibians," pp. 40-49.

48. Maj. Gen. John A. Lejeune, *The Reminiscences of a Marine* (Philadelphia: Dorrance and Co., 1930), pp. 461-62, 473. Also see Bartlett, *Lejeune,* pp. 146-68.

4

Army Doctrine and the Christie Tank

Failing to Exploit the Operational Level of War

George F. Hofmann

During the interwar period the U.S. Army engaged in a heated doctrinal dispute that prevented the tank, especially the Christie tank, from becoming the foundation for the service's approach to the operational level of war—the theory of larger-unit operations in which combined arms elements fight a series of battles known as campaigns. With the emergence of the tank as the main maneuver element for a mechanized force, the potential existed for the Army to embrace a level of war between strategy and tactics rather than emphasizing specific techniques of firepower and attrition warfare. However, this issue was obscured by the combatant branch chief organization dominated by the infantry. Although the specific-mission echelons of corps, field Army, and Army group were studied at the Command and General Staff School at Fort Leavenworth,[1] the format was controlled by a traditional infantry orientation. This bias was sustained by the 1923 *Field Service Regulation: Operations,* which stated: "The coordinating principle which underlines the employment of the combined arms is that the mission of the infantry is the general mission of the entire force. The special missions of other arms are derived from their powers to contribute to the execution of the infantry mission."[2] As a result, this orientation became a heated issue during the interwar period because it prevented doctrine from moving from infantry tactical considerations to an operational focus centered on a combined arms mechanized force independent of the traditional combatant branches.

To understand the branch bias that led to an imprecise tank policy and the resulting failure to integrate it into an operational level of war theory, a series of historical issues will be analyzed that centered on doctrine[3] and the dispute over Christie tanks. This analysis will examine the influence tank designer J. Walter Christie had on U.S. tank policy and the potential his vehicles presented for creating a link between maneuver and the operational level of war. Hopefully the reader will gain some insight into the Army's inability to adopt an operational level of war theory during the interwar period—in spite of the advances in tank and aviation technology. It was not until the early 1980s that the idea of the operational level of war was formally embraced by the U.S. Army.

In July 1917 Gen. John J. Pershing, commander of the AEF in France, asked the Army Chief of Staff to detail a board of selected officers "to make a careful and confidential study of this new tank." Two of the members detailed to the AEF Tank Board were Maj. Frank Parker (Cavalry) and Lt. Col. Clarence Williams (Ordnance); both would play a key role in postwar tank development and doctrine. The board's "Report on Tanks" outlined the classes and tactical employment of heavy British and light French tanks and suggested their employment with the AEF, and that immediate steps be taken to produce both models.[4]

The most innovative portion of the report was an attachment to the appendix titled "The Tactical Employment of Tanks in 1918." It was prepared for the AEF Tank Board by a British General Staff officer assigned to the Royal Tank Corps, Lt. Col. J.F.C. Fuller.[5] What had become evident, wrote Fuller, is that "mechanical warfare is going to supersede muscular warfare." Therefore, the Allies should prepare to raise mechanical (the term used at the time) driven armies to expand the base of attack and restore mobility to the battlefield. The projected main maneuver element, the tank, was considered a time saving machine that altered infantry tactics. Fuller considered it an armored horse in the sense that its mobility led to mass, surprise, and security. The main problem, he explained, was that the tanks currently in use were unable to keep pace with infantry because of the tactical conditions caused by a greater defense in depth and ground obliteration due to all types of artillery bombardment. He criticized these "blast tactics" as removing one obstacle and replacing it with another.[6]

The solution, Fuller suggested, must consider two grand tactical acts of battle: penetration and envelopment. He claimed the tank was ideal to carry out these tactics. Regarding the tactics of penetration, he recommended eliminating the use of heavy artillery fire on the ground to be traversed by the infantry except for counterbattery fire. The defense in depth, Fuller maintained, can be overcome by a continued advance, not only in "the very van of the battle itself," but "behind the battle front." This could only be accomplished, he wrote, with the creation of a mechanical Army. He noted that "The horse's legs have been replaced by mechanical forces," and therefore, "so should men's legs. . . . We stand on the brink of one of the greatest epochs in the art of war, . . . [and a] mechanical force is going to supplement muscular force as regards movement." Fuller proposed creating a Mechanical Striking Force consisting of a breaking and exploiting component. The breaking component, composed of heavy tanks with infantry and mechanically transported field artillery, would penetrate the first two lines of the enemy's defenses. This action would be followed by light tanks, infantry, and field artillery combining to disorganize the enemy, forcing him farther to the rear. This exploiting component would consist of light tanks followed by infantry, cavalry, and artillery supplied by mechanical carriers. The most innovative aspect of Fuller's Mechanical Striking Force was that it was an organization built on a combined arms grouping with tanks forming the basic maneuver element. This was a major problem, he realized, because of a lack of tactical experience. Nevertheless, "the idea of mechanizing the infantry and horse-drawn guns on the same footing of tanks should not be abandoned," he argued. Fuller maintained that once the enemy's lines were ruptured and the mechanized force had moved to battle depth the tactics of envelopment would take advantage of the morale and bewilderment by encircling the forming reserves.[7]

Fuller's "Tactical Employment of Tanks" also emphasized a combined arms tank raid (a forerunner of today's mechanized or armored cavalry) designed to demoralize, disorganize, confuse, and fix the enemy through "exceedingly mobile" actions.

Thus by August 1917 Fuller had suggested to the AEF Tank Board a new method for employing a combined arms mechanized force capable of penetrating deep into the Germans' defense in depth. His proposed force was designed both to restore mobility to the battlefield and create psychological panic through deep operations at an operational level of war. "Success in war greatly depends on mobility, mobility on time," he concluded.[8]

Colonel Fuller, who emerged from the war as one of the most

controversial and innovative proponents of armored warfare, later recalled meeting Major Parker from the AEF in August 1917. Fuller claimed Parker held "ultramodern views." Apparently, after witnessing the position warfare in France and engaging in discussions with Fuller, Parker argued that the only way to break the stalemate was with a combined mechanized force supported by aviation that was capable of widening the breach and then penetrating deep into the enemy's defense in depth. Fuller later quoted Parker, whom he called a "veritable he-man," as believing his views would win World War I because the Germans were not capable nor had they the resources "to adopt such a plan." Fuller gave credit to Parker's views (no doubt because they agreed with his) and indicated they were not put into practice until 1939—and then by the Germans in Poland—when they became known as "blitzkrieg," a term reportedly coined by *Time* magazine.[9]

Since the tank was a new weapon, its tactical use on many occasions met with poor results. With greater use and experience, the Allies were able by November 1918 to formulate certain principles for their employment. The heavy tanks were used to break the way for the infantry, and light tanks accompanied the attacking infantry. However, the war ended before Fuller's radical conception of deep mechanized operations, called Plan 1919, could be executed with newly designed medium tanks as the main maneuver component.

By October 1918, Brig. Gen. Samuel D. Rockenbach, the AEF Tank Corps Commander, had begun to reassess the role of tanks. He accepted the prevailing doctrine that heavy and light tanks should be used to assist the infantry, and shunned the idea of using them in an independent tank attack. Yet Rockenbach saw the need for a medium tank, which he called a "raiding tank" similar to that recommended in Fuller's "Tactical Employment." Such a tank, based upon British experience with the "Whippet," must be fast and capable of destroying and confusing forces in the enemy's rear, similar to the use of cavalry in the exploitation of a successful penetration, he argued.[10]

Towards the end of the war the Tank Corps underlined the need for a breakthrough, an accompanying, and a raider tank. However, the tactical conception and execution finally adopted by General Headquarters, AEF, after the Armistice was based on existing British heavy and French light tank models, positioning them as infantry supporting weapons. Only on rare occasions was it suggested they be used independently, meaning attacking a disorganized retreating enemy or for the purpose of increasing confusion in the enemy's rear.[11] No further consideration was given to developing a raider or cavalry tank. No doubt this was due to the British failure to allocate medium Whippets to the horse cavalry during the battle of Amiens in August 1918, resulting in a poorly executed operation. It was subsequently determined the Whippets, operating independently, managed to penetrate the defense-covering machine guns, whereas the horse troops were neutralized by the enemy's remaining machine guns and reserves.[12]

After 11 November the Allies still lacked consensus on the mission of light, medium, and heavy tanks, the type desired, and a definition of their characteristics. The U.S. Army also exhibited confusion over a tank development policy. Nevertheless, during the postwar period the Army did accept the role of tanks in altering infantry tactics. Many prominent senior officers, however, had questioned their value, while others held various ideas regarding their mission and tactical employment.

No sooner had the war ended than the Ordnance Department conceived the idea of developing an all-purpose tank—a medium similar to the British Medium D, which was designed to be the main tank for executing Plan 1919. Lt. Col. Herbert W. Alden, assistant to the Chief of Ordnance in charge of tank engineering and a member of the wartime Inter-Allied Tank Commission created to develop the Mark VIII heavy tank, believed future tank development presented great possibilities. Though accepting the need for light and heavy tanks, he preferred an all-purpose fast medium capable of extending the radius of action.[13] Work on an American twenty-three-ton medium tank began in June 1919 without consulting the Tank Corps. One month later, the Ordnance Department placed an order for a revolutionary combination wheel and caterpillar fifteen-ton medium tank designed by eccentric automobile inventor J. Walter Christie.[14] Shortly before the war ended, Christie had built a prototype self-propelled gun mount for an 8-inch howitzer that incorporated the convertible design. This type of vehicle, Christie said, would eliminate railroad facilities for gun handling. "The speed that can be made in moving from point to point is so great," he wrote, "that it gives value from an economic and strategic military standpoint."[15] There was considerable validity to Christie's lofty idea of operational mobility, because the majority of tank casualties during the war were due to mechanical problems. The convertible idea had the potential for revolutionizing the transportation of armored fighting vehicles as it greatly improved the operational and tactical mobility of tanks by allowing for rapid deployment and employment. With wartime demands and encouragement from the Ordnance Department, Christie also built self-propelled gun mounts and antiaircraft gun carriages.

The driving force behind the postwar medium tank program was Maj. Raymond E. Carlson of the Ordnance Department's Tank, Tractor,

The Christie 8-inch howitzer. Icks Collection, Patton Museum.

and Trailer Division in the Automotive Section of the Artillery Division Manufacturing Service. Like Alden, he had been a member of the Inter-Allied Tank Commission. He believed that future mechanical wars required fast medium tanks with amphibious capabilities, greater mobility, and extended range operating as armored cavalry in close cooperation with aviation. Even though tanks and airplanes were new weapons of warfare, they were never fully developed as a tactical entity, he argued. Carlson was explicit about the need for greater cooperation between tanks and aviation. He also argued that a convertible medium tank of the type developed by Christie offered both strategic mobility (by deploying to a theater of operations in the wheel mode) and tactical mobility (battlefield employment in the tracked mode).[16] The idea of a convertible tank had earlier been under consideration. Commenting on General Rockenbach's lecture on tank lessons presented at the AEF Tank School on 20 October 1918, Maj. Gen. Charles Summerall, whose division had the most experience with tanks, voiced his concern over the useless wear and tear caused by driving tanks long distances to the point of engagement.[17] The Ordnance Department found the convertible principle "an extremely attractive idea" for resolving the problem of strategic mobility. There was caution, however, because dual-purpose equipment was considered a general violation of good engineering practices.[18] Accordingly, at the end of the war the Ordnance Department offered a strong but cautious endorsement for developing a fast convertible tank capable of deep independent operations.

General Rockenbach at first questioned the need for developing a medium tank. However, in a paper titled "Tanks—Functions in Relation to Design," Rockenbach did an about-face and came out in support of the development of mediums along the lines outlined by Carlson.[19] By early November 1919, two concepts were under consideration: the M1921, an Ordnance design similar to the British Medium D, and the Christie convertible Model 1919.

However, this joint development by the Tank Corps and the Ordnance Department met with disapproval from the chief of staff's office. The primary objection was that the medium tank program was being developed independent of the combatant arms and the General Staff. Because tanks were viewed solely as a means of preparing the way for an infantry attack, Rockenbach was advised that responsibility for carrying out the tank program was to be determined by the General Staff and the Army War Plans Division, with the latter having responsibility for establishing general principles to be followed by the Tanks Corps and the Ordnance Department.[20]

Meanwhile, the AEF Superior Board was established to consider lessons from the war and how they affected the tactics and organization of the combatant arms. The board declared that tanks were infantry accompanying weapons and inadequate for independent operations. Commenting on the tactical lessons, the board predicted that future wars would follow tactics similar to those of 1918—that is, a combination of position and open warfare with the latter being more in line with U.S. doctrine.[21] The Superior Board thus established the doctrinal tone for the peacetime Army. General Pershing, despite endorsements for an independent Tanks Corps by General Rockenbach, Chief of Staff Peyton C. March, and Secretary of War Newton D. Baker, supported the board's recommendation during the 1919 congressional hearings.[22] Congress subsequently passed the 1920 National Defense Act, which restored the autonomy of the bureaus, abolished the World War I Tank Corps, and assigned tanks to the infantry. This action in effect limited postwar medium tank development by making the tank an arm of the infantry, which saw no need for such a vehicle.

The Superior Board also ratified the traditional role of the cavalry, noting that the war furnished few reasons for changing cavalry doctrine, which called for defeating the enemy's cavalry, breaking up his communications, providing flank and rear security, and conducting pursuit, harassment, and reconnaissance operations. For the time being the horse was saved, but it was apparent to some officers that mechanization spelled the end of horse-mounted cavalry. In 1921, Maj. Bradford Chynoweth argued in the *Cavalry Journal* for cavalry tanks, claiming they offered a greater balance of mobility and firepower. However, it would be more than ten years before the cavalry received its own tanks, which would be called "combat cars."

Before the Superior Board rendered its conclusions, the chief of ordnance, Maj. Gen. Clarence Williams, suggested that General Rockenbach appoint a Tank Board to recommend a permanent tank development policy. This recommendation was based on the Westervelt Board or Caliber Board, which was in the process of formulating a future policy and program for artillery designs that emphasized the military importance of mechanical transport.[23] The Tank Corps's conservative chief did not embrace General Williams's suggestion for a similar policy. Rockenbach later sent Williams an article that won the *Journal of the Royal United Services Institution* (RUSI) 1919 Gold Medal Prize.[24] The article was Fuller's revolutionary call for analyzing the increasingly technological character of land warfare. In future wars, he argued, a mechanical Army scientifically trained and armed with tanks would replace the traditional combatant arms. Commenting on Fuller's article, Rockenbach admitted that the British were ahead of the United States in tank development.[25] Yet Rockenbach was never able to crystallize a concept that included the role of tanks in independent deep operations as envisioned by Fuller. Although he supported a separate tank branch, it was one tied to the tactically narrow experiences of the World War. Meanwhile, the Ordnance Department was more focused on avoiding the design dilemmas of the past by pushing research and development in preparation for future armed conflicts. This caused General Williams to request clarification from the War Department in May 1921 regarding a development policy and tactical requirements for all tanks.[26]

The following July the Chief of Infantry, Maj. Gen. Charles S. Farnsworth, recommended to the War Department that tanks continue to be assigned to the infantry and that the service's inventory of Mark VIII heavy tanks and American-built M1917 Renault-type light tanks be maintained and modified. It was also expected a future tank development program would be affected by the cost of construction and limited funds, which already had become complicated by investigating various tank accessories for Ordnance-designed medium tanks as suggested by the infantry. In addition, Ordnance claimed there was practically no private armor plate industry. This, it was noted, was a "'choke point' in war plans for the procurement of tanks." The War Department subsequently established development programs calling for a light tank to weigh no more than five tons and a medium to weigh no more than fifteen tons. These weight limitations were adopted because they were within the tolerances of the country's and the Army's pontoon bridges and motor transport for light tanks. The Ordnance Department later argued that the weight limit caused a great deal of policy uncertainty and paralysis during a crucial period in tank development. Ordnance officers also found demands by the infantry regarding armor, weight, and speed "mutually incompatible and unattainable."[27]

To make matters even more unsettling, General Farnsworth's successor, Maj. Gen. Robert H. Allen, disagreed with the postwar Ordnance tank program. He blamed the confusion over a medium tank policy on General Rockenbach and his insistence on controlling development. The idea of combining the characteristics of light and heavy tanks into a medium tank was unsound, he argued. He was also very critical of the Ordnance Department's Christie tank project, which he called "chasing a rainbow." Furthermore, he questioned the mechanical difficulties of the wheel-track principle, claiming that it would provide "beautiful flash demonstrations, but [it is] not dependable." General Allen instead focused his energies on endorsing the development of light tanks because they were more portable and made ideal accompanying weapons for the infantry. The Commandant of the Army War College, whose function was to provide midcareer officers an environment for studying the art of war, including operational analysis and plans, supported Allen.[28] As a result, the Ordnance Committee (composed of members of the using services and Ordnance personnel), working with the Ordnance Advisory Committee of the Society of Automotive Engineers, designed a new infantry light tank. In April 1927 the firm of James Cunningham and Sons was awarded a contract to manufacture the light tank T1, thus beginning another controversial period in design and development that again pitted the Ordnance Department against maverick tank designer J. Walter Christie.

Another factor that had a lasting effect on medium tank development was Christie's influence. The contract for the M1919 included specifications provided by the Ordnance Department's technical staff regarding the design and manufacture of military caterpillars and a combined wheel-track tank. During an inspection, Ordnance personnel discovered that Christie had failed to comply with several essential specifications as defined in the contract. Christie's position—indicative of his arrogance and condescending attitude that would plague his relations with the military for years—was that he knew every phase of tank design and construction. He therefore saw no need to follow con-

Christie Model 1919. Icks Collection, Patton Museum.

tract specifications. That did not sit well with Ordnance officers, who insisted that the specifications listed in the contract must be followed.[29]

It took Christie until January 1921 to finally offer the M1919 for a preliminary demonstration. Meanwhile, during hearings before the House Armed Services Committee, Rockenbach, who questioned the utility of the British Medium D, strongly supported Christie's tank, telling the members it would "revolutionize cross country transportation and fighting."[30]

The M1919's preliminary demonstration in January 1920 proved unsatisfactory. This prompted Christie to make changes. However, he did not provide the Ordnance Department with details of his proposed changes. He advised Ordnance he should "be left largely to his own resources in the matter [of changes] and not bound by detailed specifications."[31] On 5 February Christie delivered the M1919 to Aberdeen Proving Ground for a test, where it was discovered he had not corrected the defects apparent in January.[32] Again, test results indicated the tank did not meet specifications due to numerous mechanical problems. Frustrated, Christie contemplated testing his tank with an Army agency

Christie Model 1921. National Archives.

other than the Ordnance Department.[33] He also attempted applying political pressure at a time when wartime funding and political support had evaporated. His request for appropriations with which to continue his design effort, submitted to Sen. Joseph S. Frelinghuysen, was denied.[34]

During the spring of 1922 Christie's M1921 (a turretless design rebuilt from the M1919) was tested at Aberdeen Proving Ground and the Infantry Tank School at Camp Meade. Serious mechanical problems were again experienced. The Tank School concluded the vehicle was not developed to the point where it could function satisfactorily. Besides design and engineering problems, the Ordnance Department found Christie difficult to work with because of his informal approach and unwillingness to conform to the written specifications in his contracts. Consequently, the Christie medium tank project was concluded. Nevertheless, there were still officers who believed his convertible principle had some commendable features for strategic deployment and tactical employment.[35]

While the infantry was struggling to develop a tank policy, the field artillery—as a result of the Westervelt Board report—began examining the impact motorization and mechanization had on its branch.

The idea of self-propelled artillery first surfaced during the war and reached its highest development in the form of Christie mounts intended for division and corps artillery that were produced after the Westervelt Board report was issued. The board had set three categories of field artillery: light division artillery to accompany infantry, medium corps artillery to engage in counterbattery fire, and heavy Army artillery for interdiction, neutralization, and destruction beyond the corps artillery's capabilities. Christie designed a convertible gun carriage with interchangeable capabilities for mounting a 75mm gun or a 105mm howitzer for division light accompanying artillery, and his four convertible self-propelled 155mm G.P.F. (*Grande Puissance Filloux*) guns were detailed for corps artillery use. The *Field Artillery Journal* wrote that Christie's experimental designs were promising and had the potential to become the self-propelled mount of the future. The chief of field artillery, Maj. Gen. William J. Snow, said the Christie mounts were of "exceptional value."[36]

However, by 1924 the branch's enthusiasm for mechanization had waned. The Field Artillery Board that tested the convertible Christie 75mm-105mm howitzer mount and a track-type Holt 75mm gun mount concluded they were "devoid of tactical usefulness for light guns and howitzers." The board noted that the Christie mount was unsuitable, mechanically unreliable, too heavy and large, and reflected a crude design and imperfect construction.[37] Christie's 155mm self-propelled guns presented the same mechanical problems and were determined to be equally unsatisfactory.[38] The field artillery's apathy for developing self-propelled gun carriages during the 1920s was also driven by limited funds, a lack of interest, and a belief that they were too heavy, unreliable, and offered easy targets for counterbattery fire due to their large silhouettes.[39]

By the end of 1924 the Army had invested almost a million dollars, an amount it could hardly afford at the time, in Christie's various military vehicles without establishing a clear need for them.

In the mid-1920s tank development was firmly under the direction of the Chief of Infantry, and the less than imaginative doctrine coming from Fort Benning inhibited the full realization of technical opportunities, many of which had already been demonstrated. For example, in 1923 the Army's first postwar *Field Service Regulation: Operations* (*FSR*) was adopted. This regulation, describing how the Army was to fight the next war, echoed French doctrine: *Instruction provisoire sur l'emploi tactique des grandes unites* (1921), which emphasized firepower and the "methodical" battle that made infantry the governing arm of combat. Reflecting French doctrine, tanks were prescribed in the U.S. Army's 1923 *FSR* as accompanying weapons for infantry. They were tactically ordained as assault weapons, providing firepower and shock.[40]

Christie 75mm/105mm howitzer motor carriage. National Archives.

Christie 155mm self-propelled gun mount. National Archives.

The 1923 *FSR* called for an inflexible, mechanistic tactical approach that preserved the primacy of infantry linear tactics and discounted the potential independent role of tanks. In 1925 the General Service Schools at Fort Leavenworth, Kansas, produced a text entitled *The Employment of Tanks in Combat.* This book upheld the decree of the 1923 *FSR* that the tactical role of tanks was the province of the Infantry Branch. It also reiterated the World War I definition of light accompanying and heavy breakthrough tanks, while discounting the potential role for medium tanks.[41]

In May 1927 a controversial article on mechanization written by J.F.C. Fuller appeared in the *Infantry Journal.* The reaction to his article demonstrated the deep concern infantry officers had with reconciling doctrine and tank development and defending their institution. Fuller argued that progress in weapons power threw tactical ideas out of focus. Tactics in the last war had ceased to be an art because of attrition, limited penetration, and powerful defensive salients. To restore the art of fighting, he argued, "We must get the present form of war out of our heads." By form he meant battles waged separately by the traditional combatant arms—cavalry, infantry, and artillery. "We must forget these arms, and no longer be chloroformed by their names, or organization," he wrote. "We must cease to think in names and must learn to think in the terms of tactical functions, such as finding, holding, hitting, protecting, and smashing." This, he maintained, was the first rational step toward mechanization. He concluded with the observation that if the military persisted in thinking out tactical problems in terms of the traditional combatant arms "then we shall render our minds rigid to all new ideas."[42] These comments were heresy to the U.S. combat arms branch system, most especially the infantry.

In discussions following the publication of Fuller's article, General Rockenbach called him a prophet, a romantic, and an extremist. The former chief of the Tank Corps reiterated the established doctrine that "all arms and services are auxiliaries to the infantry and must be trained to facilitate the advances of the infantry." The assistant commandant at the Infantry School stated, "We must . . . look for a wonderful development in infantry auxiliaries, especially tanks." The chief of the War Plans Section in General Allen's office claimed Fuller had abandoned the cult of the bayonet and predicted tank technology and capabilities would be met by corresponding advances in antitank weapons. An instructor at the Army War College commented, "We cannot subscribe to [Fuller's] view of a tank or mechanical Army." The only member of the article's discussion group to disagree was an instructor from the Infantry School, Maj. Merrill E. Spaulding. "We should not lightly dismiss [Fuller's] ideas," wrote Spaulding. "We ignore the opinions of those most competent to speak." Spaulding concluded with a challenge: Should the U.S. Army study and investigate the potential of mechanical warfare, as Fuller suggested, or continue "to loiter sleepily in the past?"[43]

The Army answered Spaulding's challenge in 1928 when, at the insistence of the War Department General Staff, it embarked on the turbulent road to mechanization and the Armored Force. Impressed by the British Mechanized Force established in 1927, Secretary of War Dwight D. Davis made two important decisions. First, he ordered the creation of an experimental mechanized force during the summer of 1928. Second, he directed the Chief of Staff, General Summerall, to study the employment of a mechanized force on the future battlefield and determine how the country could effectively prepare for such an eventuality.

Early in America's involvement in World War I, Summerall, who became an innovator in artillery tactics, was a member of the Baker Commission, which drew up plans to organize a force with modern weapons, including tanks. Considering his wartime experiences, the future Chief of Staff became a firm believer in the infantry mission of tanks. During the 1920s he was viewed as an enthusiastic supporter of tanks. Summerall attempted to place their development first on his list of priorities. He reminded students at the Army War College that the United States always entered wars unprepared. The next war would be different, he warned, and cautioned his audience not to rely on their World War I experience when viewing military problems. The Army's last war experience, Summerall noted, was "a special case that cannot be repeated."[44]

Summerall ordered Brig. Gen. Frank Parker, the Assistant Chief of Staff for Training and Operations (G3), to conduct the study. Parker, who had impressed Fuller during their discussions in 1917, had developed his own ideas on mechanization. He had questioned the necessity of the frontal attacks so common on the western front, and had suggested cultivating some other offensive means than human. Parker was now directed to study all mechanical and accessory fires as a preliminary to the advancing infantry. He anticipated that the tank, as the maneuver element, along with airpower and artillery, would provide the means to restore mobility to the battlefield. Considering the obsolete equipment then available, he believed that mechanization could not be carried forward to any great extent, mainly because of its unwieldiness and size.[45]

Meanwhile, Parker directed members of his staff to execute the

Secretary of War's order. The chief architect of the study, "A Mechanized Force," was Maj. Adna R. Chaffee Jr., who had joined the General Staff in June 1927.

The most innovative conclusion from Parker's G3 study called for an evaluation of the role of tanks in deep offensive operations. Recall that in August 1917 Parker and Fuller had shared their views on deep operations with a mechanical force. The 1928 G3 study called for a self-contained, highly mobile mechanized force capable of spearheading an attack and holding "distant key positions." Regarding tactics and techniques, the study viewed the mechanized force as reflecting more the cavalry's spirit of mobility than the infantry's spirit of close combat. The most controversial part of "A Mechanized Force" was a plan calling for a balanced grouping of light and medium tanks, self-propelled field artillery, mechanized infantry, engineers, air support, and a service detachment. This organization differed from the tank-heavy mechanized force assembled on the Salisbury Plain in England in 1927. "A Mechanized Force" was the first rational attempt to encroach on the autonomy of the combatant arms chiefs by advocating a true combined arms organization. However, the study was vague about air support, noting only that "Appropriate squadrons of air corps may assist in overcoming enemy resistance, both ground and aerial."

General Parker's study met with approval from the Secretary of War, the G1, the G2, the G4, and the chief of the War Plans Division. In addition, all of the branch chiefs except the Chief of Infantry, General Allen, concurred.[46]

General Allen vehemently objected to setting up another branch with the tank as its focus. Instead, he recommended that control of tanks remain with the infantry and that control of armored cars and mechanized artillery reside with their respective arms.[47] The rationale for his objection was found in the Superior Board report and the 1920 National Defense Act. Parker responded to Allen's disapproval by calling attention to the slow speed of World War I tanks, which made them auxiliaries to the infantry, and to the increased speed of postwar tanks, which allowed them a greater radius of action and mobility. This situation, Parker reasoned, "forces the consideration of [tanks] as a principle arm under certain circumstances, as well as auxiliaries of the infantry." By continuing to acknowledge that the Chief of Infantry was responsible for tank development, Parker concluded that tanks would remain tied to that branch and to the speed of the foot soldier.[48] Accepting that tanks were adjuncts of the infantry also obstructed the creation of a new doctrine that could have been instrumental in formulating a combined arms mechanized force, rather than a fragmented combatant arms policy as prescribed in the 1923 *FSR*. Parker believed that a combined arms team could form the cutting edge in future operations and that "their efficiency depends upon their ability to move, coordinate, and direct their various weapons."[49]

Shortly after the G3 study was completed the War Department directed that a board of officers from the various branches be appointed "to make recommendations for the development of a mechanized force within the Army and to study questions of defense against such forces." One of the eleven officers detailed to the board was Major Chaffee. The board summarized its results by endorsing a combined force with tanks forming the main maneuver element for the attack. The board also proposed that mechanized infantry be included in the force and that self-propelled field artillery be incorporated for mobile fire support. In addition, it was suggested the mechanized force be set up as a tactical laboratory for analyzing the role of fast tanks in a combined arms operation. However, in an apparent compromise with the Chief of Infantry, the board recommended a separate branch not be established.[50]

Maj. Levin H. Campbell Jr.—one of the officers appointed in October 1928 to make recommendations regarding the development of a mechanized force, and who became the Army's Chief of Ordnance in World War II—at first supported Christie and his tanks, arguing that mechanization was of immense importance in securing mobility for the modern U.S. Army. He predicted that self-propelled artillery had a future. Unfortunately, he wrote, its development was "lying dormant, largely because of funds." Nevertheless, Campbell called for complete Army reorganization based upon mechanization and its potential for "great mobility." This was feasible, he contended, because the country possessed the largest automobile industry in the world. He was disturbed by the fact that much of the Army's current equipment was of World War I design and he concluded that this situation precluded a correct picture of the capabilities of a modern Army.[51]

About this time J. Walter Christie introduced his "National Defense Machine"—the M1928. It was referred to as the "M1940" because it was considered a ten-year advance in tank technology. Major C.C. Benson, a cavalry officer, wrote that the Christie system, with its innovative long independent helical spring suspension, provided greater compression and extension amplitude for its large road wheels. This noticeably enhanced stability of the firing platform and speed of the vehicle.[52] The design represented a major breakthrough because its chassis had the potential of greatly increasing the tactical and operational mobility of armored fighting vehicles. General Summerall, who supported the development of light tanks, was so impressed with its

J. Walter Christie, displaying, as usual, his proclivity for publicity, points to the front of his turretless M1928 for the benefit of a congressman. Icks Collection, Patton Museum.

potential that he circumvented the Ordnance Department and ordered the Infantry Tank Board to test the M1928.[53]

The Infantry Tank Board (established by the Chief of Infantry in October 1924 to make studies, tests, and recommendations on tanks and related equipment) started its evaluation of the chassis in November 1928. After a number of modifications by Christie during the test period, the Tank Board concluded the program in January 1929 and then loaned the chassis to the Cavalry Board for further evaluation. In August 1929 the board recommended that the M1928 be included in the current production program.[54] Maj. Gen. Stephen O. Fuqua, the new Chief of Infantry, responded in December 1929 with a request for five M1928 tanks for tactical experimentation and one for mechanical and performance tests.[55] The cavalry also expressed interest in acquiring one for testing.

Between the time the Tank and Cavalry Boards completed their tests and General Fuqua submitted his request for six Christie tanks, a dispute arose between the Tank School, the Chief of Cavalry, and the Ordnance Department over the technical and tactical value of the M1928. A report by the Ordnance Committee's technical staff in March 1929 outlined past difficulties with Christie, citing poor design, engineering, and workmanship. The report called attention to 1924, when the Tank School and the Chief of Infantry had favored terminating all Christie projects. The report also noted that "continuing work on the Christie mounts was not done without some political pressure on the Army. It ceased when funds began to be curtailed."[56] General Fuqua took issue with the report, and directed that Col. H. L. Cooper, the assistant commandant of the Tank School and a member of the board that tested the M1928, respond. After noting that there was no comparison between the M1928 and earlier Christie designs, Cooper sent a strong endorsement to the office of the Chief of Cavalry for comment. The cavalry's reply was prepared by Maj. George S. Patton Jr.[57] Patton saw the potential of the vehicle for cavalry operations. However, he found a few limitations with the M1928 and observed that "Mr. Christie's histrionic inclinations lead him to over stress speed and ability to cross fallen logs." Fuqua, in turn, instructed Cooper to respond to Patton's concerns. This was addressed again in favor of the M1928.[58] Subsequently, Fuqua decided to conclude the Ordnance Department–sponsored Cunningham light tank project in favor of the Christie vehicle.[59]

Christie, however, again proved difficult, insisting the Army pay for development plus the cost of the M1928. General Williams, the chief of ordnance, considered this financially impractical because of the limited funds authorized by Congress. Nevertheless, he was willing to work with Christie. The Army Appropriations Act for Fiscal Year 1930 included $250,000 for the manufacture of Cunningham light tanks. Under pressure from Generals Williams and Allen—the latter disliked the unwieldy Cunninghams—light tank appropriations were redirected to buy six Christies.[60] Meanwhile, Christie had discreetly begun negotiations to sell his tanks to the Soviet Union and Poland. His main client, however, was the Red Army, which at first was interested in the Ordnance-designed Cunninghams but then found them tactically cumbersome and mechanically impractical.

One of Christie's critics, and a strong supporter of the Cunningham tank series, was Capt. John K. Christmas of the Automotive Section in the Office of the Chief of Ordnance. During the 1930s, Christmas argued that tactically tanks were much cheaper than men. He was one of the Army's leading experts on tank technology, and, like Summerall, Williams, Parker, Chaffee, Campbell, and Alden and Carlson earlier, had foreseen the importance of mechanization in future wars. In 1929 Christmas wrote that peacetime armies "endeavor to develop in conformity with the lesson of the last war." The problem, however, was that "such developments are often deflected by traditions, immediate expediencies, peace requirements, and the political and financial factors involved."[61] He believed the tank was the central element of mechanization. Regarding the mission and tactics of a mechanized force, Christmas understood the necessity for establishing a permanent, self-contained, and highly mobile offensive force capable of fighting independently. He argued that mechanization not only would save lives but that it should come easy for the United States because "we are the world's leading industrial nation." Regarding preparing for the future, he suggested that it was advisable for the Army to continue to pursue mechanization to work out tactics. However, he viewed it practical at the time not to mechanize the whole Army, but to develop a nucleus for expansion in time of war. Christmas believed the sooner the country increased its degree of mechanization "the better it will be . . . for other leading nations are already adopting this new form of warfare." In conclusion, he stressed that American industry was mechanized, the Navy was mechanized, the Army Air Corps was mechanized, and that mechanization was rapidly gaining acceptance within the American home. "Only the Army remains to increase greatly its degree of mechanization."[62]

At the time Christmas's article appeared, Chaffee, who was quite aware of the potential of the new Christie, delivered an impressive lecture at the Army War College entitled "The Status of the Mechanized Combat Organization and the Desired Trend in the Future." The lecture was an elaboration on his "A Mechanized Force" report. He held that future offensive operations in modern war required a self-contained, highly mobile, mechanized corps with great distance striking power. Chaffee understood and had analyzed Fuller's ideas on a mechanized force; however, as did Parker, he questioned its heavy dependence on tanks mixed with armored cars, motorized machine guns, artillery, and engineers at the expense of mechanized infantry or a more balanced force.

Chaffee's thinking challenged the infantry's linear doctrine of firepower and attrition warfare.[63] Parker and Chaffee agreed that a rapid and deep attack by fast-moving tanks supported by a balanced combined arms team including mechanized infantry and self-propelled field artillery was the proper and logical doctrine for bringing the Army into the future. This suggestion, which was in opposition to infantry doctrine, provided the Army with an opportunity to move to an operational level of war at the corps level, the corps being a combined arms, operational-tactical organization. This was the concept of large-unit operations with a mechanized force capable of deep independent operations with Christie tanks as the main maneuver element.

In response to these new ideas, Col. James K. Parsons, the infantry officer in charge of field development for the Experimental Mechanized Force and commandant of the Tank School, completed a rudimentary but revolutionary staff study in April 1930 calling for the creation of six tank divisions to be included with the mobilization of six field armies. He foresaw the use of mechanized forces in the next war and proposed using a Christie tank chassis mounting a 47mm gun as the key vehicle for extended maneuvers. Since the last war, Parsons argued, the speed, radius of operations, and mechanical dependability of tanks had increased, thus making it possible to consider tanks for operations other than close cooperation with the foot troops. He emphasized extended maneuver, such as deep operations, because the mobility and radius of action of tanks made them an operational level force. Parsons also proposed using self-propelled accompanying artillery and tactical aviation. His study also offered the Army an opportunity to move to an operational level of war. However, he did not include infantry foot troops in his organization because they needlessly limited the radius of operations and mobility of tank divisions. Like Campbell and Christmas, Parsons argued that the United States had the necessary production capacity. He concluded with the observation that the cost would be small compared to the postwar development of the Army Air Corps.[64]

As expected, Parsons's proposal for six tank divisions did not excite the branch chiefs. Most critical was Major Patton, who prepared his objections for chief of cavalry, Maj. Gen. Guy V. Henry. Patton maintained that mobile, deep independent operations with tanks would end in disaster. "A stalled tank is junk," he wrote. Furthermore, he noted, Parsons's paper suggested that tank forces could replace the other combat arms. He criticized Parsons for excluding foot infantry in his organization. Parsons's "opinion is as dangerous as it is impossible," Patton argued. One of his main objections was a logistics issue. For example, deployment of overseas transport requirements for tanks could materially reduce troop-carrying transport capacity either to Europe or Asia. Patton also cited terrain features that could hinder tank operations, such as mountains, jungles, and deserts. He then proposed producing more machine guns capable of firing armor-piercing ammunition. Improvements in this type of ammunition, he noted, made tanks more vulnerable than in the past. He believed tanks would thus become less effective in a future war. Nevertheless, he did support creating a small combined arms mechanized force.[65]

Meanwhile, the Christie conflict continued. To formulate a tactical doctrine for fast breakthrough tanks, Fuqua pressed for resolution of the Christie matter. In January 1930 he forwarded to the Chief of Ordnance tank specifications prepared by the Tank Board. After a series of specification changes and problems with Christie's failure to provide suitable drawings, it was evident by May that the purchase of six tanks and one combat car for the cavalry was doubtful.[66] The incoming Chief of Ordnance, Maj. Gen. Samuel Hof, challenged the proposed contract. He was familiar with efforts to buy the M1928 for a pilot test and Christie's refusal to sell it at a reasonable price. General Hof was upset because the Ordnance Department was not given the opportunity to test the M1928 as were the Infantry and Cavalry Boards. This was a valid point. It was War Department policy that a new tank project could not be started or money spent until it had written approval from the Ordnance Committee, whose members were representatives from the using services and technical sections of the Office of the Chief of Ordnance. In their enthusiasm to acquire the M1928, the Chief of Staff and the Chiefs of the Infantry and Cavalry branches ignored standard procurement policy. Furthermore, General Hof questioned the practicability of the convertible principle, claiming it had never been conclusively demonstrated. There was also considerable doubt over the weight of the tank once a turret and armament were added. For light tanks, this exceeded the prescribed weight. Finally, Hof added that "too much stress was being placed on speed at the expense of ruggedness and fighting ability."[67]

Negotiations with Christie now reached a critical point. Because of the Army's past dissatisfaction with the tank designer and the fact that he had concluded a contract to sell two tank chassis to the Red Army, General Hof again expressed his opinion that it would be prudent to purchase only one Christie tank for test and evaluation before releasing additional funds.[68] General Summerall, irritated over the conflict, called for a conference. After a heated debate, it was decided to reject Christie's bid for six tanks and instead acquire just one. As usual the decision inflamed Christie. At a subsequent conference to resolve the difficulties, he made a statement that further alienated Ordnance officers, especially Captain Christmas. Christie referred to spies or agents in all branches of the Army who kept him posted on tank development. Furthermore, if the Ordnance Department decided to build a convertible-type vehicle, he would make every effort to stop it. Then he made a threatening statement: He, Christie, would apply political pressure if negotiations failed. Summerall thereupon ordered all parties involved to get together, set specifications, and acquire one Christie fitted with a turret.[69] Finally, on 28 June, an agreement was reached calling for delivery of one Model 1930 on 1 September 1930 at a cost of $55,000.

Before leaving office in late 1930, Summerall ordered the creation of a permanent mechanized force to be established at Fort Eustis, Virginia, to be commanded by a cavalry officer, Col. Daniel Van Voorhis. The force was organized on the theory that modern fast tanks were capable of extended maneuver beyond the immediate support of divisional infantry. The principal role was that of a mobile force to execute operational and tactical missions requiring deep operations with fast, hard-hitting striking power. It would be a force skilled in enveloping movements. If possible, the mechanized force was to avoid highly organized areas of resistance, leaving those to assaulting infantry tanks. Because of its speed, the G1, Brig. Gen. Campbell King, like Colonel Parsons, visualized the new Christie tank chassis as the basic maneuver weapon for the mechanized force.[70] There was considerable interest in the General Staff for developing the Christie tank for deep offensive operations. However, the dispute with Christie, ongoing since late 1928, prevented the Mechanized Force at Fort Eustis from investigating the tank's potential. Unfortunately, the Mechanized Force had to experiment with the Cunninghams and obsolete equipment.

Meanwhile, the debate continued over the need for creating an independent mechanized force. Although costs were a constraint in creating such a force, the main obstruction came from the Chief of Infantry, General Fuqua. He rejected the notion that cavalry, because of its mobility, was more suitable for forming a mechanized force. "There is no such animal as 'armored Cavalry' in these modern days. Remove the 'horse' and there is no Cavalry," was his comment. General Fuqua, in

A brief visit with the Mechanized Force: A Christie M1930 with A Troop, 2nd Armored Car Squadron, in column march with a movie vehicle, Fort Eustis, March 1931. Patton Museum.

a highly charged memorandum to the Deputy Chief of Staff, stated: "I am trying to lead infantry thought into the same doctrine of open warfare" that was adopted in France by General Pershing. "The dehorsing of these units [due to mechanization]," he added, "will mean an irretrievable loss to the Cavalry." General Fuqua claimed that fire-power and maneuver were the infantry's mode of attack, and that fast tanks should be used to execute wide turning movements, exploit break-throughs, engage in flanking movements, and provide close combat support for the attacking foot soldier.[71] Following Fuqua's reasoning, the infantry, armed with fast tanks, could take over some of the cavalry's traditional missions. Thus, by 1931, the argument about how to execute the doctrine of open warfare conducted by fire and movement had resulted in a disagreement between the Army Staff in the War Depart-ment and the branch chiefs. The infantry believed open warfare was best facilitated by placing tanks with its linear attacking force, whereas the Army Staff concluded that the optimum solution was to create a combined arms mechanized force designed to engage in deep opera-tions.

Finally, in January 1931—after a more than four-month delay caused by Christie's commitment to the Soviet Amtorg Corporation—a board of officers from the using branches and the Ordnance Depart-ment began acceptance tests with the Christie Model 1930. The test vehicle was fitted with an Ordnance-designed turret and a government-furnished V-12 Liberty engine. The board included Van Voorhis, Chaffee, and Patton. At the end of February, due to repeated delays as a result of repairs and modifications caused by breakage, the M1930 finally passed all the acceptance tests except for a cross-country run on tracks. General Hof consequently offered Christie $54,000 for the tank, explain-

From left: Cunningham T1E1 and T1E2 light tanks, Cunningham T2 medium tank, and two Christie light tanks. National Archives.

ing that the $1,000 reduction from the agreed price reflected the facts that the tank did not pass all the acceptance tests and that it failed to meet the specifications set forth in the contract. True to form, the headstrong Christie became disjointed, accusing General Hof of breaking faith. Hof responded that Christie should have restricted his activities to working with the proper contract authorities instead of lobbying and engaging in general publicity.[72]

Hof was angry because Christie had circumvented the Ordnance Department by writing to the chairman of the House Subcommittee for the War Department Committee on Appropriations, Henry E. Barbour. Barbour in turn took up the matter with the new Chief of Staff, Gen. Douglas MacArthur.[73] Not satisfied with letting matters rest there, Christie had contacted Secretary of War Patrick J. Hurley.[74] The Ordnance Department, responding to strong political pressure and continued interest in the Christie tank by the infantry and cavalry, conducted a final brief acceptance test and finally signed a contract with Christie in June calling for the production of seven Model 1931 tanks. Due to Van Voorhis's influence four, designated Combat Car (CC) T1, went to the mechanized cavalry at Fort Knox. The chief of infantry responded by demanding that three, designated T3, be shipped to Fort Benning for testing in an infantry support role. The Army never bought the M1930. In November 1936, Christie sold that tank chassis to the British government.[75]

Years later, Chaffee credited General Summerall and Parker's G3 Division for getting the Army to start thinking about mechanization.[76] The doctrine that emerged from the General Staff in 1928 and which Chaffee embraced broke from the 1923 *FSR*, which gave primacy to the infantry over other branches. Instead, Summerall's staff perceived a future war driven by mechanization and organized on a combined arms team configured for deep offensive operations with the tank as the primary maneuver element. This new doctrine rejected the traditional principle of open warfare shaped by firepower and attrition and a linear orientation. The Army Staff supported this principle rather than the more extreme mechanization ideas advanced by Fuller. The commander of the permanent mechanized force assembled at Fort Eustis in October 1930 and later commander of the mechanized cavalry at Fort Knox, Colonel Van Voorhis, added that the mechanized cavalry's characteristic of mobility, integrated with fire and movement, was its strength. Its operational concept, however, embraced depth and speed rather than infantry linear tactics.

By the time General MacArthur became Chief of Staff in late 1930, Congress and the president had determined that the only way to attain financial and economic stability was by balancing the federal budget. That meant the funds needed to move the Army into the future simply were not available. As a result, the Mechanized Force created at Fort Eustis was short-lived.[77] General MacArthur ordered its deactivation and directed all branches to adapt mechanization and motorization to their traditional roles.[78]

MacArthur's directive made the issue of a combined arms mechanized force for deep offensive operations moot by requiring each arm to mechanize according to its prescribed mission as defined in the 1923 *FSR*. MacArthur's policy reinforced the traditional autonomy of the combatant arms branch chiefs and made it even more difficult to formulate a new doctrine centered on an operational level of war, which would better prepare it for a threatening international situation, such as the rise of fascist regimes. As a result of MacArthur's directive, branch mechanization prevented the planning, conducting, and sustaining of larger combined arms units to obtain strategic goals within a potential theater of operations. Van Voorhis objected to MacArthur's directive, believing the only way to effectively achieve mechanization was through a coordinated effort.[79]

During MacArthur's first year in office the *Infantry Journal* published one of the first articles to suggest the creation of tank divisions. Major Benson, who earlier wrote the buoyant article introducing the new Christie high-speed tank chassis, presented a clear thought for the future. Like Parsons, he offered a conceptual opportunity to move the Army to think in terms of the operational level of war.[80] Unfortunately, it proved to be a lost opportunity. The Chief of Staff's decision to decentralize mechanization caused the branch chiefs, especially the infantry, to focus on their traditional missions and combat tactics as outlined in the 1923 *FSR*. That regulation, in effect until World War II, stated that the combined employment of all arms was essential to success. However, the mission of the infantry was the general mission of the entire Army.[81] Benson's and Parsons's tank divisions did not materialize until the creation of an Armored Force in 1940. In the meantime, Van Voorhis and Chaffee had to settle for a decentralized effort to be determined by their branch chief. Consequently, their only road to furthering mechanized doctrine was through the cavalry.

Meanwhile, one of the supporters of the Christie chassis, Major Patton, was assigned to the War College in September 1931. That month he chaired a G3 committee that analyzed mechanized units. Writing for the committee, Patton reemphasized his earlier objections that were a replay of those in his paper written in opposition to Parsons's paper. Conforming to school custom of supporting the Chief of Staff, the committee upheld MacArthur's directive decentralizing mechanization. The group made an extensive examination of the limitations of track-laying vehicles, including terrain, antitank fire, and mechanical and supply difficulties. These limitations, it was noted, "forbid the organization of large units." Patton's committee concluded that mechanized units should not operate beyond infantry support. It also recommended supplementing horse cavalry with mechanized cavalry.[82] For whatever reason, until the eve of World War II, Patton never published ideas that strayed beyond the established doctrine of traditional cavalry. It can be argued that later he executed maneuvers at the operational level of war while commanding the Third Army. Somewhere along the way he completely reversed his opinion on the importance of tanks in deep operations.[83]

During MacArthur's tour as chief of staff two distinct opinions emerged from the Infantry School regarding the tactical employment of tanks: Were they to be used as accompanying or leading tanks? The conservative consensus was that they should accompany the infantry in the attack. On the other hand, the progressives who worked to overcome the prejudice of combatant arms partisanship argued for leading tanks because their mobility and improved speed—as with the Christie type tank—made them capable of independent operations. The conservatives were firmly opposed to the independent use of tanks. They rejected the tactical employment of leading tanks because it would reduce the support from other arms.[84] The Christies were to operate as supporting tanks in conjunction with the infantry's linear tactical doctrine of firepower and maneuver by moving fast through the breakthrough and then engaging in flanking movements. It was still expected that heavy tanks, such as the obsolete Mark VIII, would make the initial assault.

In February 1932 a tactical combat car platoon that would employ the soon-to-arrive Christie CCT1s was organized in Fort Knox, Kentucky. The unit was commanded by Capt. H.H.D. Heiberg, who served with the mechanized force at Fort Knox until 1936 and again in 1939 as Chaffee's aide. Heiberg—as were most cavalrymen at Fort Knox—was impressed with the eleven-ton CCT1, especially its stable gun platform produced by an enhanced road wheel articulation and its convertible capabilities that improved operational mobility. Soon, however, mechanical problems began to plague the Christies. The worst, according to Heiberg, was the power train. Although the V-12 Liberty engine gave little trouble, the separately mounted transmission and double steel clutches caused misalignment in shifting, resulting in gear failures.[85]

Maj. Robert W. Grow, the executive officer and S3 of the mechanized cavalry, believed the CCT1 was not built as a fighting vehicle but as a "mobile cradle" for its V-12 Liberty engine. As early as 1934 he preferred an all track rather than a convertible vehicle, a decision the Cavalry Board finally made in October 1939.[86]

While the mechanized cavalry at Fort Knox was developing a doctrine for fluid deep operations based upon their mechanical mounts, U.S. Diplomats at the 1932 Geneva Disarmament Conference proposed "the total abolition of tanks and all heavy mobile land artillery over 155mm in caliber."[87] General MacArthur concurred. He was ready to give up tanks because they were considered offensive weapons of war.[88] Not only were the Chief of Staff's directive and opinions troublesome for establishing a mechanization policy, but an "absolutely essential" order by the Secretary of War in April 1933 furthered impeded conditions for establishing a balanced doctrine. The order limited the development of tank technology by cutting funds and limiting the weight of tanks and combat cars to 7.5 tons.[89] It was evident the Army was not only subject to constant budget restraints but to a protracted and indecisive tank development program orchestrated by its leaders. The Army's situation was also heavily influenced by the idea that Japanese aggression in Asia and the rise of fascist regimes in Europe were not threats to the country's security.

The Soviet Union was not the only power to take an interest in American doctrine and tank development. Colonel Van Voorhis recalled

German interest in developments at Fort Knox in 1933: "They were not particularly interested in our equipment. . . . They were keenly interested in our views on the proper tactical and strategic employment of mechanized forces."[90] Major Grow also recalled those evenings in 1933 with German staff officers at the Doe Run Inn near Fort Knox. The visiting officers stated the U.S. Mechanized Cavalry was ahead of them in tactical employment of self-contained fighting units but that they, the Germans, were more advanced in the development of vehicular equipment.[91]

In spite of the mechanized cavalry's advanced thinking regarding deep operations, agility, initiative, and attempt at combined arms, the Army's leadership still inhibited innovations. For example, one of the lessons learned from the 1934 Fort Riley maneuvers was that they demonstrated the conflict between tradition and modernity caused by General MacArthur's directive. The 1934 maneuvers were designed to determine how far the cavalry had progressed with mechanization, motorization, and new weapons development. The 1st Cavalry (Mechanized), commanded by Chaffee, demonstrated its ability during the maneuvers to extend its "sphere of action" within the cavalry's prescribed mission. Chaffee's force carried out such traditional cavalry missions as "reconnaissance and counter-reconnaissance, seizing and holding positions, flank cooperation, and delaying action." More important was the unit's overland movement to the maneuvers. It demonstrated the operational mobility of the mechanized force.

The mechanized force's performance during the maneuvers made it obvious to many cavalrymen at Fort Riley that the horse was in decline and mechanization was in ascendance.[92]

Nevertheless, after extensively evaluating the maneuvers, the Cavalry School's Academic Division recommended further participation with horsed and mechanized units.[93] The Infantry Board observer claimed the purpose of the exercise was to determine "first and foremost, whether or not mechanized cavalry could entirely replace horsed cavalry." He concluded that the mechanized cavalry's principal role was to supplement the mission of horse cavalry, noting that rarely would an independent mission be assigned to a mechanized force.[94]

Before the 1934 maneuvers, a new Ordnance-developed combat car, the CCT4, based on Christie's convertible wheel-and-track and helical suspension system, was briefly tested at Fort Knox. The test committee recommended that the vehicle, with modifications, be declared standard. It was a decision Chaffee strongly supported based on his earlier experience observing the Christie tank acceptance tests and comparing those with the CCT4's operational mobility and speed.

CCT4 combat car at Fort Riley, Kansas, in May 1934. Patton Museum.

CCT5 combat car at Fort Riley, Kansas, in May 1934. Patton Museum.

During service tests following the maneuvers, the CCT4 outperformed an Ordnance-designed seven-ton CCT5, which displayed a double "Mae West" turret and a new, more rigid suspension system. The CCT5 was a radical departure from the Christie design, being full tracked and nonconvertible and employing a volute spring or bogey suspension system and a divided power train. During the tests the Christie-type suspension system provided a more stable gun platform with better ditch crossing capabilities, while the Ordnance-designed vehicle was more maneuverable but so choppy when moving cross-country that accurate marching fire was impossible. Understandably, observers at Fort Riley did not favorably view the Mae West profile.[95]

At year's end, Chaffee, a proponent of the 9.5-ton convertible CCT4, was overruled. The decision was made to acquire a modified CCT5 (minus the Mae West dual-turret configuration) for the cavalry. Generally, combat car proponents at the user level favored the CCT4. At the staff level, the War Department favored the seven-ton weight and lower cost of the CCT5, thus taking advantage of the opportunity to produce a less expensive vehicle manufactured at Rock Island Arsenal. Also, the CCT5 avoided the engineering dilemma imposed by the wheel-track convertible design. Heiberg recalled that the decision to adopt the CCT5 was made in the War Department "by officers [who had] probably never ridden in a tank, much less fired from one."[96] The CCT5 was standardized for production as the Combat Car M1. The vehicle reflected certain features, such as the Ordnance-designed volute suspension system, which remained characteristic of all U.S. tanks until late in World War II.[97]

In November 1936, a German tank expert wrote in the military weekly *Militar Wochenblatt* that, in spite of its high speed, the CCT5 was a "perfect example of bad construction." The Americans, he noted, had "an ambition to repeat all the mistakes the European tank and armored car constructors already have behind them. . . . Their armor is held to be too light to resist modern weapons." In response, the War Department praised its mechanized equipment, claiming it compared favorably with that of any nation.[98]

As a result of the 1934 maneuvers, the Army's branch leadership determined that combat cars—the cavalry's tanks—should be tied to horse units just as tanks were tied to infantry. This unfortunate situation continued to stifle the Army's effort to develop a new doctrine for deep, fluid offensive operations by a combined arms team. Events at Fort Riley that spring nevertheless convinced the Fort Knox contingent that a self-contained unit with new equipment, organized as a mechanized division, could carry out the cavalry role and fight independently.

When the mechanized cavalry returned to Fort Knox, two towed field artillery firing batteries were added to the force. According to Grow, the mechanized force preferred its accompanying artillery to be self-propelled. Unfortunately for mechanization proponents, World War I field pieces still dominated the inventory. In addition, the Artillery Branch continued to express a preference for towed over self-propelled artillery.[99] In 1934, the Chief of Field Artillery stated that the reason for "retaining horses for division light artillery was that they were available in sufficient quantities and suitable tractors were not." He did admit that a wide difference of opinion existed regarding the soundness of replacing horses with tractors.[100]

Between 1936 and 1937 the Command and General Staff School at Fort Leavenworth published an instructional text describing the organization and tactical employment of a mechanized division. In this text the mechanized force was described as "all arms," self-contained and capable of deep independent operations, leading to pursuit and exploitation of success. Adding to force mobility, the text saw the use of aviation for command control, reconnaissance, and tactical support.[101] However, the marriage of tactical aviation with a mechanized force was a problem. During the interwar period, ground support attack aviation did not develop as expected late in World War I because of neglect, technical problems, and the controversy over mission and air tactics. The 1923 *FSR* directed that one of the missions of aviation units was to attack hostile ground forces and their supporting units—including supply columns. No direction was given regarding a tactical effort against enemy tanks or in support of an infantry assault with breakthrough and accompanying tanks. This was due in part to the influence of Brig. Gen. William "Billy" Mitchell, the controversial airman who had questioned the future application of ground attack aircraft because he believed that airpower should focus on deep strategic operations against the enemy's supply concentrations and manufacturing areas. By the mid-1930s, ground attack aviation emphasis give way to high-speed, long-range heavy bombers.[102]

Brig. Gen. Walter Krueger, chief of the War Department General Staff's War Plans Division, also opposed creating a mechanized division, claiming it was too large and costly.[103] General Krueger's opinion was debatable. Major Christmas of the Ordnance Department, for one, persuasively demonstrated that the cost of maintaining and employing a mechanized cavalry unit was considerably less than that of a horse cavalry unit.[104]

Shortly before his retirement in late 1935, General MacArthur proposed a model five-year plan that would take the Army to 1940. His solution for the future was a machine-gun Army. MacArthur's plan was to increase the Army's motorization capabilities and use tanks to enhance the mobility of infantry. He proposed using aviation for deep striking power, a move he linked with "the more slowly delivered but powerful and sustained blows of the bulk of the Army."[105] One of the more ill-conceived experiments during this period, a reflection of MacArthur's recommendation for a machine-gun Army, involved replacing tank and combat car turrets with a fixed barbette superstructure loaded with machine guns. This made the vehicles look more like movable pillboxes.

During the Spanish civil war (1936-39), Army attachés stationed in Europe reported that lightly armored tanks armed only with machine guns were unable to overcome determined enemy fire, especially tanks armed with 45mm cannon supplied to the Republican forces by the Soviet Union. These lessons were misread in the United States. For example, in 1939, the short-lived, underpowered, and underarmored infantry M2 medium tank was introduced. It was armed with one high velocity 37mm tank gun and eight .30-caliber machine guns. The M2 was essentially an enlarged version of the CCM1 and M2A1 light tank.

Meanwhile, in the Soviet Union, the Red Army had developed the Christie system into the BT (*Bystrokhodnii Tahk*) fast tank series, which was intended to be the key fighting vehicle in deep offensive operations doctrine. BTs would be the main maneuver element in the Soviet combined arms force. At the Kiev military district maneuvers in 1935, the Red Army demonstrated an operational capability that included mechanization and a combined arms force.[106] This was largely made possible by product improvements and extensive production of the BT series. Later, numerous BTs armed with 45mm tank guns were sent to Republican Spain, where observers considered them the best tank used by any belligerent. Apparently arms development was less restricted in Stalin's centrally controlled government.

Attaché reports from Spain reinforced the parochial attitude of the Army's leadership, especially that of MacArthur's successor, General Malin Craig, and the ground combatant arm branches. General Craig noted that a balanced Army operating in any theater of operations could never "dispense with a proper proportion of mounted cavalry and horse-drawn artillery."[107] The chief of field artillery added that in spite of the tremendous improvements in mechanization and transportation, "horse-drawn is a little better than motor-drawn" artillery.[108] The field artillery also continued to view the tank as an infantry accompanying weapon, an idea that had not appreciably changed since 1918.[109]

Concerning tank development, General Craig recommended "a

type suitable for close support [with the] infantry."[110] He summarized his feelings in testimony before a military affairs congressional subcommittee, where he suggested future military operations "be carried out by the traditional arms; that well-trained infantry and artillery form the bulk of armies. Air and mechanized troops are valuable auxiliaries." Regarding military operations in Spain, the Chief of Staff observed that tanks were not successful due to antitank weapons, insufficient armor protection, mechanical defects, tactical errors in their employment en masse, and inadequate support from artillery and tactical aviation.[111]

One of the officers influencing General Craig, the General Staff, the Command and General Staff School, and the Army War College was the former Chief of Infantry, General Fuqua, who was the military attaché in Spain during that country's civil war. His intelligence reports were circulated and analyzed by the Army's General Staff and school system. It was his opinion—as well as the opinion of most of his peers in England and France—that tanks did not prove themselves in separate offensive operations in Spain because they were effectively challenged by antitank guns. They concluded that tanks were only useful in support of attacking infantry.[112]

In April 1938 the War Department issued an important but reactionary policy governing mechanization and the tactical employment of mechanized forces. The policy avowed that recent operations in Spain demonstrated that "combatant arms will fight in their traditional roles." It further emphasized that the mechanized cavalry was to adhere to its traditional mission of exploiting the infantry's success.[113] The policy also resolved the issue of independent versus supporting tanks by endorsing the latter.

The Chief of Infantry, Maj. Gen. George Lynch, one of the authors of the 1923 *FSR*, responded by ordering a board of officers to rewrite the Army's tank manual. He instructed the authors to take into account the fact that the accepted use of tanks, as observed in Spain, had been largely discredited.[114]

An article in *Army Ordnance* noted that "independent tank forces are a delusion," and suggested that tanks be heavily armored and function as mobile supporting artillery or as accompanying artillery for the attacking infantry.[115]

In the meantime, Chaffee was also paying close attention to events in Spain. A General Staff intelligence memorandum he digested reported that tanks used in Spain by the belligerents were unsuccessful in almost all operations. The problems identified were numerous, such as inadequate crew training and poor discipline, mechanical deficiencies, insufficient terrain reconnaissance, lack of infantry and artillery support, the questionable use of tanks against strong obstacles and villages, inadequate numbers, and the reported superiority of antitank guns. As far as the mechanized cavalry was concerned, the Spanish civil war provided ample evidence of what not to do.[116]

Unfortunately, the U.S. Army's leadership began to emphasize the development of antitank weapons at the same time mechanized cavalry was even more firmly embracing the concept of a mobile combined arms force capable of executing deep operations. Mechanized cavalry pioneers at Fort Knox believed that the new weapons of war—combat cars, self-propelled artillery, and mechanized infantry vehicles—required new mission-oriented tactics rather than the tank tactics inherited from World War I and demonstrated in Spain. The consensus at Fort Knox was that the tank tactics used in the civil war were unsound and that tanks were improperly used.

Partly as a result of antitank reports from Spain, a board of officers from the 7th Cavalry Brigade (Mechanized) recommended self-propelled guns for the field artillery unit with the mechanized force. The board believed self-propelled artillery was necessary to neutralize antitank weapons, while providing general supporting fire for combat cars. However, the Chief of Field Artillery disagreed. He supported towed artillery because, he believed, it could deliver far more supporting fire. He also regarded the mechanized cavalry's appeal for self-propelled artillery as no more than a request for a vehicle with the essential characteristics and limitations of a tank. The solution, he argued, was a combat car armed with a cannon and sufficiently armored to withstand shelling from anti-mechanized weapons.[117]

Nevertheless, with support from the Chief of Ordnance, C.M. Wesson, a 75mm pack howitzer was mounted on a CCM1 and classified as the T3 75mm Howitzer Motor Carriage.[118] The field artillery, however, considered the T3 unsuitable because of limited crew space. As a result, no additional T3s were built for the mechanized force.[119]

In spite of antitank reports from Spain and problems acquiring self-propelled artillery, Chaffee's 7th Cavalry Brigade (Mechanized) continued to test and expand its operational and tactical mobility. During the Plattsburg, New York, maneuvers in August 1939, the brigade executed a wide enveloping movement and completed a successful deep operation, leading to Chaffee's recommendation for an armored division. This occurred before the Germans had launched their blitzkrieg against Poland. The following May, at the Louisiana maneuvers, the reinforced Mechanized Brigade participated for the first time in large-unit operations that included a corps and three divisions. It was again evident to Chaffee and a few others who evaluated the maneuvers that, considering the German blitzkrieg, U.S. armored divisions should be created without delay.[120]

This action was contrary to the thinking of the Chiefs of Infantry, Cavalry, and Field Artillery, who unanimously objected to organizing around new concepts and weapons. They continued to espouse adapting weapons to the traditional missions of their respective branches. For example, before the German invasion of Poland in September 1939, Chief of Cavalry John K. Herr was adamant about his preference for the horse. His attitude reflected his inherent romantic love for horses and traditional cavalry methods.[121] While General Herr had initially supported the creation of a mechanized cavalry division, he later changed his mind and refused to mechanize his horse units. Grow, who served in Herr's office, claimed he "lost it all."[122] Shortly after the German blitzkrieg consumed Poland, Herr, whose only commitment to mechanization was its use with the horse cavalry, told attendees at the War College it was obvious "that the machine cannot eliminate the horse."[123] With this public pronouncement, mechanization passed Herr; Chaffee and the forces at Fort Knox had finally prevailed. The Mechanized Cavalry would have a new name—the Armored Force. According to Grow, the Armored Force was created not because a new combatant arm was necessary, but because Herr and the cavalry could not grasp the concept of mechanization in a future war.[124]

A few days after Herr's lecture at the War College, the Chief of Field Artillery, Maj. Gen. Robert M. Danford, told attendees that the horse was still the prime mover for division light artillery. Although he cautiously supported motorization and preferred towed artillery, he refused to introduce the self-propelled artillery that was critically needed by the Mechanized Cavalry Brigade.[125] On the eve of the war this apathy made it almost impossible to develop a field artillery doctrine for a combined arms mechanized force. Chaffee thus was never able to develop mobile self-propelled artillery to accompany his fast-moving combat cars. As a result, little attention was given to self-propelled mounts until Maj. Gen. Jacob L. Devers succeeded Chaffee as Chief of the Armored Force in August 1941.[126]

General Lynch, the chief of infantry, also made no secret that his first love was for the foot soldier. Before the creation of the Armored Force his branch bias caused him to veto a proposal to convert foot troops to tank units.[127]

When the G3, Maj. Gen. Frank Andrews, recommended in November 1940 to Chief of Staff George C. Marshall that the Armored Force created in June be legally established as a separate combatant arm, the proposal was strongly opposed by Generals Lynch and Herr. Lynch responded by requesting that his tank units be returned to infantry control.[128] Herr called the G3's recommendation a "petty effort," and argued that "the Armored Force had been violating the terms of the National Defense Act of 1920 in creating non-infantry and non-cavalry armored units." He reasoned that whatever the Armored Force was intended for "could have been accomplished equally well" by the established branches.[129]

It was plainly evident that the self-serving autonomy of the combatant arms branch chiefs prevented consensus not only on how to deal with operational complexities but how best to exploit the operational and tactical opportunities offered by mechanization. With operational concepts, doctrine, tactics, organization, and training methods deeply rooted in the bias of the Army's branch system, it became impossible to work through joint commands and joint combined commands.[130]

This attitude prevailed during the 1930s even though the U.S. Army was aware of developments in mechanization at an operational level in other countries, especially in Germany and the Soviet Union. An Army officer who had completed a tour at the Kriegsakademie was greeted with apathy when he reported to Chief of Staff Malin Craig on Germany's development of a combined air-ground mechanized force capable of deep operations.[131]

In 1937 Van Voorhis reported on a Fort Knox visit by a German staff officer who observed that, unlike the U.S. Army, the Germans placed mechanization and motorization under a single proponent organization: the Panzer Corps. This, the German officer noted, eliminated duplication of development efforts and reduced costs.[132]

For ten years the Army's leaders failed to create an operational vision due to their fixation on the traditional branch organization and the branch chiefs' desire to defend their microinstitutions. The chiefs thus became inflexible toward new ideas that could have moved the Army to change. A student at the Army War College summed up this attitude: "I hold here a pamphlet, 'Tactical Employment of the Mechanized Division,' used as a text at Leavenworth during the past few years. The April directive [the 1938 War Department policy on mechanization] consigns the booklet to the school archives. There will be no Panzer Division in our Army."[133]

In an attempt to provide a capability to defeat the tanks that spearheaded the German blitzkrieg, the Army Ground Forces commander, Lt. Gen. Lesley J. McNair, who had earlier questioned the cost of funding an armored force, prescribed a mobile tank destroyer force. The search for an ideal tank destroyer again spurred reconsideration of Christie's suspension system. A special War Department G3 planning

T49 gun motor carriage. Icks Collection, Patton Museum.

T67 gun motor carriage. Icks Collection, Patton Museum.

group eventually urged development of a lightweight tracked vehicle possessing speed and maneuverability, capable of carrying an adequate gun, and lighter than a tank at the expense of some armor protection. Lt. Col. L.D. Tharp, a G3 staff officer who in 1930 served as a test officer from the Tank Board and a strong supporter of the Christie tank, influenced the planning group to consider Christie's independent helical spring system. The Ordnance Department reluctantly agreed. However, ordnance officers refused to work with the strong-willed, eccentric Christie and instead contracted with Buick Motors to design and build two Gun Motor Carriages using a modified helical spring suspension system.

An important technological change incorporated into the T70 tank destroyer design led to the rejection of the Christie suspension by January 1943. The T70, which became the M18 Hellcat, employed a torsion-bar suspension system requiring less hull space than the Christie design.[134] Ironically, the torsion-bar suspension system outlived the family of vehicles for which it was designed because the tank destroyer concept had been abandoned by the war's end.[135]

Absent a well-thought-out doctrine for an operational level of war with a large mechanized force, the Army performed poorly early in World War II. Under the branch organization, senior Army leaders lacked experience in commanding and controlling large units of any type, let alone a new force bringing new technology to the field as part of a combined arms mechanized force.

Although it eventually defeated the depleted German forces in western Europe, it is arguable whether the Army's armor force could have prevailed in the predominately tank-on-tank combat in the east. There, major engagements were initially characterized by the Wehrmacht's blitzkrieg, then by the Red Army's reintroduction of deep offensive operations with tanks providing the main maneuver element for combined arms mechanized forces.

One of history's great ironies is that the vehicle so critical to the execution of the Red Army's version of the operational level of war was the T-34 medium tank, which, through continued product improvement, was the final development of the Christie BT.[136] The T-34's resounding success must have provided some small satisfaction to J. Walter Christie, who, embittered and impoverished, died on 11 November 1944.

Supported by the productivity of American industry, the Army was able to meet the challenges of World War II, but only at great cost. The U.S. Army fought with both infantry and armored divisions. Infantry divisions were assigned separate tank battalions to assist with infantry attacks. Armored divisions expanded on the cavalry's traditional missions of pursuit and exploitation, fighting as combined arms teams with the tank as the main maneuver element. The Army's main battle tank, the M4 Sherman, was an evolution of the M2 medium, and it continued to compromise on two basic design features: firepower and protection. Although the M4 was fairly reliable, it was underarmored compared to the German panzers. The Sherman's greatest strength was that it was manufactured in staggering quantities.

The lean years of the 1930s should not have hampered the Army from thinking about concepts of warfare at the operational level with a tank-based mechanized force. Even though budgetary restrictions limited the expansion of mechanization, they did not preclude the Army's leadership from modernizing the mind. Other than the Mechanized Cavalry, there was little interest in developing such a progressive doctrine because the Army's leaders limited their efforts to improving past performance rather than learning from and building on present achievements, applying new technology, and developing new doctrine. In addition, initiative was discouraged by the country's climate of isolationism, rendering a lack of interest in military matters by civilians and politicians alike. Taking the initiative in such an environment was politically dangerous. Thus, as the Army moved through the tumultuous 1930s, its leadership exhibited too much caution; there were too many obstacles in the path of progress toward mechanization. As a result, the combat arms failed to cooperate in establishing the organization, technology, and doctrine to support a tank-based mechanized division.[137]

Even during the early months of World War II, the Army's leadership was still unable to establish a consensus regarding the concepts for operations at the tactical and operational levels of warfare. General MacArthur's policy of decentralizing mechanization and the subsequent 1938 War Department policy on mechanization had intensified the autonomy of the combat arms branches and reinforced their traditional tactical orientation. The Army's branch arrangement also encouraged fragmentation. There were many divergent ideas regarding a future tank policy as to type and technology. This situation required compromises with the Ordnance Department regarding such design features as speed, armor protection, armament, and reliability. Chaffee blamed this lack of direction on costs, pacifist tendencies, differences of opinion, and especially, a lack of branch chief awareness. He admitted that the United States "failed to evaluate properly the importance of combined arms in armored units."[138] As a result, when U.S. ground forces finally entered World War II they engaged the Wehrmacht with inferior tanks, equipment, and doctrine.

NOTES

1. See *Tactical and Strategic Studies . . . 1923-1927,* vol. 1, *The Detached Corps;* vol. 2, *Corps and Army;* and vol. 3, *A Group of Armies* (Fort Leavenworth: The General Service Schools Press, 1923-27). This series was used as an instructional aid for the faculty and Army officers to understand specific levels and missions at various echelons. At the Army War College students were prepared for high commands, such as the conduct of operations by Army and higher echelons. By the late 1920s the G3 section of the General Staff was completing *Field Service Regulations: Employment of Large Units and Administration of Large Units.* For an excellent examination of the school system, see Timothy K. Nenninger, "Leavenworth and Its Critics: The U.S. Army Command and General Staff School, 1920-1940," *Journal of Military History* (Apr. 1994): pp. 199-231. For a historical bibliography, see Elizabeth R. Snoke, *The Operational Level of War,* Combat Studies Institute, U.S. Army Command and General Staff College, Fort Leavenworth, Kans. (Dec. 1985). Also see Michael R. Matheny, "The Development of the Theory and Doctrine of Operational Art in the American Army, 1920-1940," Monograph (Fort Leavenworth, Kans.: School of Advanced Military Studies [SAMS], 1988).

2. (Washington: GPO, 1924), p. 11.

3. This argument is not as simplistic as one may think. Carl von Clausewitz recognized the operational level of war in his time-honored classic, *Vom Kriege* (1832). It can also be traced to the grand tactics of the Napoleonic era, as described in Baron Antoine Henri de Jomini's *Precis de l'Art de Guerre* (1838) and recently reproduced as the *Art of War* (1992). In the introduction, British military historian, Charles Messenger, claims operational art was not new to the U.S. military. For example, B. H. Liddell Hart's *Sherman: Soldier, Realist, American* (1930) is an analysis of the general as a planner and executor of operational art. Recently James J. Schneider provided a number of examples of operational art associated with protracted campaigns. One example, he notes, is the force the Industrial Revolution had on the events of the American Civil War. As a result, the essence of operational art, which he describes as distributed free maneuver, was demonstrated in Lt. Gen. U.S. Grant's 1864 campaign. See Schneider, "Theoretical Implication of Operational Art," in Clayton R. Newell and Michael D. Krause, eds., *On Operational Art* (Washington: Center of Military History [hereafter CMH], 1994). The distinction in these examples is that operational art—the execution of learned methods at higher echelons—is a prerequisite for a proactive approach to an operational level of war.

4. "Report on Tanks," "Classes," AEF GHQ, 17 July 1917, G3 Reports, Tank Corps Folder, Records of the American Expeditionary Forces (WWI), 1917-23, RG 120, NA, (hereafter AEF, RG 120), pp. 1-7.

5. Fuller, "The Tactical Employment of Tanks in 1918," Appendix 6, in "Report on Tanks," 8 Aug. 1917, AEF, RG 120, pp. 3-4, 8-10.

6. Ibid., p. 1-22.

7. Ibid., pp. 17-20.

8. Ibid., p. 21.

9. Fuller, *Decisive Battles of the U.S.A.* (New York: Thomas Yoseloff, Inc., 1942), p. 398 n6. At the time Fuller could not recall if the name was Parker or Palmer. However, it could only have been Parker since he was responsible for most of the reports, including the "Tactical Employment of Tanks." Fuller's observation occurred in Aug. 1917. Also see Fuller, *Memoirs of an Unconventional Soldier* (London: Ivor Nicholson and Watson, Ltd., 1936), p. 158; *Tanks in the Great War* (London: John Murray, 1920), pp. 277-78; and I.B. Holley Jr., *General John M. Palmer, Citizen Soldiers, and the Army of a Democracy* (Westport, Conn.: Greenwood Press, 1982), p. 407. Holley claims Palmer, who was Pershing's G3 at the time, did not understand the potentialities of tanks in warfare.

The *Time* article called the head of the Wehrmacht, Gen. Walter von Brauchitsch, the "Blitzkrieger" who had fostered, planned, and led the lightning war against Poland. As opposed to the war of position that dominated World War I, the article defined blitzkrieg as merely a war of movement executed by a new kind of cavalry: fast tanks, airpower, and infantry transported in armored trucks. See "Polish Theater: Blitzkrieger," *Time*, 25 Sept. 1939: pp. 25-27.

10. Brig. Gen. Samuel D. Rockenbach, "The Role of Tanks in Modern Warfare," lecture, AEF, France, 20 Oct. 1918, File No. 42-3, USAMHI, pp. 1-6.

11. GHQ AEF, *Tanks, Organization and Tactics*, Dec. 1918, AEF No. 1432, RG 120, NA, p. 6.

12. "Notes on the Relation Between Tanks and Cavalry," Jan. 1921, in Maj. Raymond E. Carlson, "Papers on Development of Tanks," 16 Mar. 1921, Ordnance Technical Intelligence Files (OKD) OO 451.25/56.1, Records of the Chief of Ordnance, RG 156, NA, pp. 1-2.

13. Lt. Col. Herbert W. Alden, "Tanks," *Journal of the Society of Automotive Engineers* (*SAE*) (May 1919): pp. 406-407.

14. Carlson, "Development Record Medium Tank," 12 Jan. 1921, in "Papers on Development of Tanks," p. 1.

15. J. Walter Christie, "Christie Wheel Caterpillar Gun Mount," 18 Nov. 1918, Ordnance Office (OO), OO 472/178, RG 156, NA, pp. 1-2.

16. Carlson, "Memorandum on Development of Tanks," 1 Aug. 1919, in "Papers on Development of Tanks," pp. 1-3, and "Chronology of Medium Tank Development," in "History of Medium Tank Development," 16 Aug. 1929, Ordnance Committee (Technical Staff) Minutes (OCM), Item 7814, RG 156, NA, pp. 4-5.

17. Maj. Gen. Charles P. Summerall to Brig. Gen. Samuel D. Rockenbach, Chief of Tank Corps, 13 Jan. 1919, "Tanks," File No. 42-1, USAMHI, p. 2.

18. "Christie Wheel-Track Tank," in "History of Medium Tank Development," p. 9.

19. Brig. Gen. Samuel D. Rockenbach, "Tanks—Functions in Relation to Design," 18 Aug. 1919, in Part 6, "Tanks with Infantry," 1 April 1932, Command and General Staff College (CGSC) Library, Fort Leavenworth, Kans., pp. 1-17.

20. Memorandum for the Chief of Staff, Subject: Tank Development, 26 Nov. 1919, OO 451.25/1017, RG 156, NA, p. 1.

21. "Report of Superior Board on Organization and Tactics," General Orders no. 68, AEF-GHQ, Chaumont, France, 19 Apr. 1919, CGSC Library, pp. 64-77.

22. *Historical Documents Relating to the Reorganization Plans of the War Department and to the Present National Defense Act*, pt. 1 (Washington: GPO, 1927), pp. 365-66, 404-405.

23. "Report of Board of Officers," Special Orders No. 289-0, 11 Dec. 1918, War Department, Washington, OKD 334.3/1.17, RG 156, NA, p. 24. Brig. Gen. William I. Westervelt, who headed the board, stated the report produced a noticeable look of amazement on the faces of ranking officers, especially regarding the board's ideal of complete motorization. See Westervelt, "A Challenge to American Engineers," *Army Ordnance*, 1 (1920): p. 60.

24. Carlson, "Development Record Medium Tank," p. 4.

25. Maj. Gen. J.F.C. Fuller, "The Application of Recent Developments in Mechanics and Other Scientific Knowledge to Preparation and Training for Future War on Land," (May 1920), pp. 239-74. Also see Anthony John Trythall, `Boney' Fuller: The Intellectual General* (London: Cassell & Co., Ltd., 1977), pp. 86-8.

26. Daniel Chase, "The Development Record in Combat Vehicles," in vol. II "Research and Development: History of the Ordnance Department in World War II," (unpublished manuscript, Aberdeen Proving Ground, Md., 1947), Robert J. Icks Collection, Patton Museum of Cavalry and Armor, Fort Knox, Ky., pp. 13-15.

27. Ibid., and "History of Medium Tank Development," Aug. 1929, OCM, Item 7814, RG 156, NA, pp. 14, 17, 21, 31-35, 45. Accessories mentioned were a pneumatic control system, odograph, stroboscope, compasses, and intratank phones.

28. "Approval of Pilot Tank Chassis by the Infantry," in "History of the Development of the Light Tank," 16 July 1929, OCM, Item 7786, RG 156, pp. 14-16; and Maj. Gen. Robert H. Allen, "A Resume of Tank Development in the United States Army," lecture, Army War College (AWC), 27 Oct. 1927, USAMHI, pp. 1-4, 9.

29. Ordnance Office, Tank, Tractor and Trailer Division to Ordnance

Technical Staff, Subject: Departure from Specifications in Manufacture of a combined wheel and caterpillar Tank, 20 May 1920, OO 451.25/916, RG 156, NA, pp. 1-2.

30. Carlson, "Development Record Medium Tank," p. 4.

31. Memorandum to Technical Staff, 20 Jan. 1921, OO 451.25/1023, RG 156, NA, p. 1; and Christie to Chief, Tank, Tractor and Trailer Division, 29 Apr. 1921, OO 451.25/1331, ibid., p. 2.

32. Memorandum to Technical Staff, 5 May 1921, OO 451.25/1338, RG 156, NA, p. 3.

33. Memorandum to Chief of Manufacturing, Subject: Test of Vehicles Manufactured by Christie, 19 Jan. 1922, OO 451.25/1567, RG 156, NA, p. 1.

34. Christie letters, 13 Mar., 23 Mar., and 12 Apr. 1922, OO 160/21162, 21236, 21394, RG 156, NA.

35. War Department, Office of Chief of Infantry to CO, Tank School, 22 May 1924, OO 451.25/2078, RG 156, NA, p. 1.

36. "Current Field Artillery Notes" (Sept. 1919): pp. 603-604; and Memorandum, Maj. Gen. William J. Snow to C.C. Williams, Chief of Ordnance, 20 Mar. 1920, OO 072/44, RG 156, NA.

37. "Horses, Tractors and Self-Propelled Mounts," *Field Artillery Journal* (Nov.-Dec. 1923): pp. 491-92; and Sub-Committee on Mobile Artillery to: Ordnance Committee, Technical Staff, 16 Nov. 1923, Subject: 75mm Gun-105mm Howitzer Motor Carriage (Christie), OCM, Item 3391, RG 156, NA, pp. 1-5.

38. Sub-Committee on Mobile Artillery to Ordnance Committee, Technical Staff, 6 Jan. 1924, Subject: 155mm Gun Motor Carriage (Christie), OCM, Item 3486, RG 156, NA, pp. 1-2.

39. Boyd L. Dastrup, *King of Battle: A Branch History of the U.S. Army's Field Artillery* (Fort Monroe, Va.: U.S. Army Training and Doctrine Command, 1992), pp. 189-90.

40. War Department, *Field Service Regulations United States Army 1923: Operations* (Washington: GPO, 1924), pp. 17-18. For an excellent study on the disparity between the 1923 *FSR* and technological advances, see William O. Odom, "The Rise and Fall of United States Army Doctrine, 1918-1939," (Ph.D. diss., Ohio State University, 1995). On French doctrine, see Robert A. Doughty, *The Seeds of Disaster: the Development of French Army Doctrine 1919-1939* (Hamden, Conn.: Archon Books, 1985), p. 91.

41. *The Employment of Tanks in Combat* (Fort Leavenworth, Kans.: The General Service Schools Press, 1925), pp. 7-8.

42. Fuller, "Tactics and Mechanization," (May 1927): pp. 457-65.

43. Ibid., "Discussions," pp. 465-76.

44. Summerall to Rockenbach, 13 Jan. 1919, p. 1; "Summerall Paints War of the Future: It will conserve soldiers and use plans and tanks in

attack," *New York Times*, 2 Sept. 1927: p. 18; and Allen, "A Resume of Tank Development in the United States Army," pp. 1-2.

45. Brig. Gen. Frank Parker, "The G3 Division of the War Department General Staff and Its Problems," lecture, AWC, 11 Sept. 1928, USAMHI, pp. 6-7.

46. Brig. Gen. Frank Parker, "A Mechanized Force," 20 Mar. 1928, AG 537-3 (3-20-28), Records of the Adjutant General's Office, RG 407, NA, pp. 1-4, 8. (Hereafter RG 407.)

47. Willey Howell for the Chief of Infantry, memorandum for the Assistant Chief of Staff, G3, 26 Mar. 1928, RG 407, p. 3.

48. Brig. Gen. Frank Parker, "A Mechanized Unit," 2 Apr. 1928, RG 407, NA, pp. 3-4.

49. Brig. Gen. Frank Parker to Brig. Gen. James C. Dozier, AG, State of South Carolina, 9 Jan. 1929, Box 3, Folder 58, Parker Papers, Southern Historical Collection, University of North Carolina at Chapel Hill.

50. Proceedings of a Board of Officers, Subject: A Mechanized Force, 1 Oct. 1928, CI 537.3/7884-B, Records of the Chief of Arms, RG 177, NA, pp. 4, 7-10, 15-16, 18, 35. Unfortunately the Proceedings and the G3 study paid no attention to tactical air other than observation and reconnaissance missions in support of a mechanized force.

51. "Special Automotive Equipment of the Army," *SAE* (Sept. 1928): p. 299. Campbell later became disenchanted with Christie because of the designer's tempestuous personality. Campbell, interview with author, Annapolis, Md., 13 Aug. 1973.

52. Maj. C.C. Benson, "The New Christie, Model 1940: An Estimate of the Tank Armored Car Combined," *Infantry Journal* (Sept.-Oct. 1929): pp. 255-61; and *Army Ordnance* (Sept.-Oct. 1929): pp. 34-41.

53. Resume of Tank Board's Report on Christie Tank Chassis, 20 May 1930, OO 451.25/3387, RG 156, NA, p. 1.

54. First Partial Report of Test of Christie Tank Chassis, The Tank Board, Fort Meade, Md., Test No. 29, 22 Aug. 1929, TB 470.8/233, RG 177, NA, pp. 1-2.

55. Maj. Gen. Stephen O. Fuqua to The Adjutant General, 30 Dec. 1929, Subject: Purchase of Christie Tanks, CI 470.8/550-B-P II, RG 177, NA, pp. 1-3.

56. Procurement of Christie Armored Car, 5 Mar. 1929, OCM, Item 7522, RG 156, NA, pp. 3-14.

57. Maj. George S. Patton Jr. to Commandant, The Tank School, 28 Oct. 1929, Subject: Christie Tank, TS 470.8 (Christie), RG 177, NA, pp. 1-6. According to Patton's grandson, the future general often invited Christie to his Washington home. Patton claimed the Christie tank was "the future"; however, he soon lost interest in its development. See Robert H. Patton, *The Pattons: A Personal History of an American Family* (New York: Crown Publishing, Inc., 1994), p. 238. Martin Blumenson suggested Patton might per-

sonally have funded Christie. See Blumenson, *The Patton Papers, 1885-1940* (Boston: Houghton Mifflin Co., 1972), p. 868.

58. Col. H.L. Cooper to Chief of Cavalry, 22 Nov. 1929, Subject: Comments on Tank School Report to the Chief of Infantry, Office of the Chief of Cavalry, RG 177, NA, pp. 1-3.

59. Maj. Gen. Stephen O. Fuqua to Commandant, The Tank School, 26 Nov. 1929, Subject: Christie Tank, TS 470.8 (Christie), RG 177, NA, pp. 1-2.

60. Procurement of Christie Armored Car, pp. 3-4; and Chronology of Christie Tank Procurement, 1 June 1931, OO 451.25/3693, RG 156, NA, p. 1.

61. Capt. John K. Christmas, "The Mechanization of Armies," *Military Engineer* (July-Aug. 1929): p. 340. Also see Christmas, "Mechanization in Our Army Today," *Army Ordnance* (July-Aug. 1932): pp. 11-17; "Tanks and Tactics," Christmas, *Army Ordnance* (Jan.-Feb. 1937): pp. 208-14; and Christmas, "The Manufacture of High-Speed Tanks," *Mechanical Engineering* (Jan. 1939): p. 139.

62. Capt. John K. Christmas, "The Mechanization of Armies," *Military Engineer* (Sept.-Oct. 1929): pp. 452-57.

63. Maj. Adna R. Chaffee Jr., "The Status of the Mechanized Combat Organization and the Desired Trend in the Future," 19 Sept. 1929, USAMHI, pp. 2, 10.

64. Col. James K. Parsons to The Adjutant General, Washington (Through: The Commanding General Third Corps Area), 17 Apr. 1930, Subject: Mechanized Forces, AG 573.3 (5-15-30), RG 407, pp. 1-10.

65. Maj. George S. Patton Jr., memorandum for the Chief of Cavalry, Subject: Study of Mechanized Forces by Colonel James Kelly Parsons, 19 May 1930, Ibid., pp. 1-7. Also see John Daley, "Patton Versus the 'Motor Maniacs': An Inter-War Defense of Horse Cavalry," *ARMOR* (Mar.-Apr. 1997): pp. 12-15.

66. "Chronology of Christie Tank Procurement," 1 June 1931, OO 451.25/3693, RG 156, pp. 2-7.

67. "Reasons Why General Hof Opposed the Initial Purchase of Six Christie Tanks," in "Chronology of Christie Tank Procurement," 1 June 1931, OO 160/5435, RG 156, NA, p. 1.

68. Maj. Gen. Samuel Hof, memorandum for Brig. Gen. W.H. Tschappat, 5 Mar. 1931, OO 451.25/3660, RG 156, NA, p.1; "Chronology of Christie Tank Procurement," 1 June 1931, OO 451.25/3693, RG 156, NA, pp. 1-3; and "Resume of the Christie Situation," 14 Jan. 1932, OO 451.25/4404, RG 156, NA, p. 1. Also see Hearings before Subcommittee of House Committee on Appropriations, *War Department Appropriations Bill for 1933: Military Activities*, 71st Cong., 3d sess. (Washington: GPO, 1931), pp. 537-629.

69. Capt. John K. Christmas, "Report on Conference," 18 June 1930, OO 451.25/3387, RG 156, NA, p. 1.

70. Brig. Gen. Campbell King, memorandum for General Moseley,

Subject: Organization—Mechanized Force, 27 Apr. 1931, CI 537.3/7884-B, RG 177, NA, pp. 2-3; and Report of Mechanized Force 1931, HQ, Fort Eustis, Va., 1 July 1931, Patton Museum, p. 1. The author of the report was Maj. Robert W. Grow.

71. Maj. Gen. Stephen O. Fuqua, "An analysis on the chronological history of progress toward the formation of a Mechanized Force in the United States Army and the purpose of the organization," 24 Mar. 1931, CI 537.3/7884-B, RG 177, NA, pp. 4-5.

72. J. Edward Christie wrote a biography of his father that was very critical of the Ordnance Department. See *Steel Steeds Christie* (Manhattan, Kans.: Sunflower University Press, 1985). The book was panned in reviews because the author altered historical events to fit the would-be imperatives of his family's past. See Hofmann, "The Christie Biography: Conflict Revisited," *ARMOR* (Nov.-Dec. 1986): pp. 3-4.

73. Henry E. Barbour to Gen. Douglas MacArthur, 7 Mar. 1931, OO 451.25/3666, RG 156, NA, p. 1, and MacArthur to Barbour, 9 Mar. 1931, OO 451.25/3667, RG 156, NA, p. 1.

74. J. Walter Christie to Patrick J. Hurley, 11 Mar. 1931, OO 451.24/290, RG 156, NA, p. 1.

75. For an excellent British view of Christie and his tank chassis, see David Fletcher, *Mechanised Force: British Tanks between the Wars* (London: HMSO, 1991), pp. 2, 50, 120-23.

76. Brig. Gen. Adna R. Chaffee Jr., "Mechanized Cavalry," lecture, AWC, 19 Sept. 1939, USAMHI, p. 1.

77. John W. Killigrew, "The Impact of the Great Depression on the Army, 1929-1936," (Ph.D. diss., Indiana University, 1960), pp. IV:13-19.

78. MacArthur biographer D. Clayton James maintained that a lack of funds during the 1930s prevented the progression of a tank development program. See James, *The Years of MacArthur 1880-1941*, vol. 1 (Boston: Houghton Mifflin Co. 1970), p. 358.

79. "Prelude to Armor," in *Armored Force Command and Center,* Study No. 27, Historical Section, 1946, Army Ground Forces, RG 407, NA, p.3.

80. Maj. C.C. Benson, "Tank Divisions," in the *Infantry Journal Reader* (Garden City, N.Y.: Doubleday, Doran & Co., 1944), pp. 36-39.

81. *FSR 1923*, pp. 11, 13, 88-9.

82. "Mechanized Units," G3 Course, AWC, 1931-1932, Report of Committee No. 6, 28 Sept. 1931, USAMHI, pp. 1-4.

83. Steve D. Dietrich, "The Professional Readings of General George S. Patton, Jr.," (paper, American Military Institute, Lexington, Va., 15 Apr. 1989), pp. 36-37, and Kris P. Thompson, "Trends in Mounted Warfare: Blitzkrieg and the Operational Level of War," *ARMOR* (July-Aug. 1998): pp. 57-58.

84. William C. Lee, "Fast Tanks as Leading Tanks and Exploiting Tanks," lecture, Tank School, 12 Jan. 1932, *Tank Notes* (Fort Meade, Md.: Mar. 1932), USAMHI, p.14; and Bradford Chynoweth to author, 17 Jan. and 25 Feb.

1972; and S.R. Hinds to Chynoweth, 7 Jan. 1972, USAMHI. Generals Chynoweth, who in 1921 had argued for cavalry tanks, and Hinds, who would serve with distinction in the 2d Armored Division during World War II, were both considered progressives at the time and referred to themselves as dissenters because they challenged established infantry doctrine that tied the tank to the pace of the foot soldier.

85. Capt. H.H.D. Heiberg, "Organize a Mechanize Force," (manuscript, Patton Museum, Fort Knox, Ky., n.d.), pp. 5-8. The manuscript is part of the Heiberg Collection in the Patton Museum. An edited version of his experiences with the mechanized force during the 1930s, "Organize a Mechanize Force," appeared in *ARMOR* (Sept.-Oct. 1976): pp. 8-11, 48-51. Heiberg recalled many evenings gathering with his unit in the Central Mess singing the newly composed "Lament of the Cavalry Tanker," sung to the tune of the artillery's caisson song (now the Army's official song):

> First on wheels then on tracks.
> As we break our bloody backs,
> Keep those Christies a-rolling along
> In and out, mostly out,
> While you hear the Colonel shout
> Keep those Christies a-rolling along.
> For its Hi-Hi-Hee
> In the Horse-Tank Cavalree
> Heave on your clutches hard and strong
> Where e're you go
> You will always know
> That those Christies are rolling along
> Lord keep them rolling
> Keep those Christies a-rolling along

86. Maj. Gen. Robert W. Grow, "Ten Lean Years. From the Mechanized Force (1930) to the Armored Force (1940)," (manuscript, Falls Church, Virginia, 1969), pp. 35, 57. Copy on file at the Patton Museum. An edited version of "Ten Lean Years" appeared in a four-part series in *ARMOR* in 1987.

87. U.S. Department of State, *Foreign Relations of the United States, 1932,* vol. 1 (Washington: GPO, 1948), pp. 65-67, 70, 180-82.

88. Ibid., p. 65.

89. Directive for the Future Development of Combat Cars and Tanks, 29 Apr. 1933, OO 451.24/622, RG 156, NA, p. 1.

90. Van Voorhis, as quoted in "Prelude to Armor," p. 5.

91. Grow, "Ten Lean Years," pp. 55-56.

92. Maj. Adna R. Chaffee Jr., "Report of Maneuvers," The Cavalry School, Fort Riley, Kans., 1 Oct. 1934, Grow Files, p. 419, and Heiberg, "Organize a Mechanize Force," p. 13.

93. Academic Division, "Report on Maneuvers," The Cavalry School, Fort Riley, Kans., 1 Oct. 1934, Grow Files, p. 419. (The documents in this file were given to the author by General Grow and hereafter are referred to as Grow Files.)

94. Jesse A. Ladd, "Report of Observations of Fort Riley Maneuvers (May 14 to May 26 1934)," Grow Files, pp. 1, 10.

95. Report of Technical Committee, 24 Mar., and Proceedings of Board of Officers, 25 Mar. 1934, HQ, 1st Cavalry (Mechanized), OO 451.24/1789, RG 156, NA, pp. 1-5 and 1-4; and The Daily Log of Combat Cars T4 and T5, During Test at Fort Riley, 8-21 May 1934, Grow Files. Grow was a member of the technical committee at Fort Knox that recommended the CCT4 be declared standard and procured. Also see Heiberg, "Organize a Mechanized Force," pp. 13-15 and "Mechanization in the Army," lecture, Society of Mechanical Engineers, Pittsburgh, Pa., 23 Apr. 1940, Heiberg Collection, pp. 11-13; Daniel Chase, "The Developmental Record in Combat Vehicles," pp. 12-21; and Richard P. Hunnicutt, *Stuart: A History of the Light Tank*, vol. 1 (Novato, Calif.: Presidio Press, 1992), pp. 55-112.

The Mae West arrangement was due to the divided power train where the engine was mounted in the rear and the transmission in the front of the CCT5. A long driveshaft connected the two units. This was a problem because the tunnel that enclosed the driveshaft bisected the crew, causing an obstruction, thus the two side-by-side mounted turrets.

96. Heiberg, "Organize a Mechanize Force," pp. 13-15.

97. On the evolution of the volute suspension system, see Chase, "Combat Car," in "The Development Record in Combat Vehicles," in vol. II "Research and Development," pp. 18-21; Capt. Thomas H. Nixon, Ordnance Department, memorandum for the Chief of Staff, Subject: Volute Suspension for the Light Tank, T2, 25 Apr. 1934, OO 451.25/4716, RG 156, NA; Sub-Committee on Automotive Equipment to Ordnance Committee Technical Staff, 1 May 1934, Subject: Light Tank T2-Application of Volute Spring Type Suspension; and Maj. Gen. Samuel Hof to Adjutant General, 4 May 1934, Subject: Light Tank T2, OO 451.25/4728, RG 156, NA.

98. "German Expert Finds U.S. Tanks Would not Stand Test of War," *New York Times*, 21 Nov. 1936, pp. 1-2, and "Army Denies Tanks are Second Rate," *New York Times*, 22 Nov. 1936, p. 7. The German's observation was correct. Later in Tunisia, when American tanks were first employed, they lacked sufficient armor and armaments to engage German tanks. This disparity was never corrected. See Omar N. Bradley, *A Soldier's Story* (New York: Henry Holt and Co., 1951), pp. 40-41; Blumenson, *The Patton Papers, 1940-1945*, pp. 137-38; and Hamilton H. Howze, *A Cavalryman's Story* (Washington: Smithsonian Institution Press, 1996), pp. 52-53. Also see Harry C. Thomson and Lida Mayo, *The Ordnance Department: Procurement and Supply* (Washington: GPO, 1960), pp. 222-23.

99. Grow, "Ten Lean Years," p. 71, and Dastrup, *King of Battle*, pp. 192-93.

100. Upton Birnie Jr., "Developments in Organization, Armament, and Equipment of the Field Artillery," lecture, AWC, 20 Sept. 1934, USAMHI, pp. 10-11.

101. *Tables of Organization Mechanized Division (Tentative)* (Fort Leavenworth, Kans.: The Command and General Staff School Press, 1936), pp. 3-24; and *Tactical Employment of the Mechanized Division* (Fort Leavenworth, Kans.: The Command and General Staff School Press, 1937), pp. 3-4, 6, 23-24, 31.

102. *FSR 1923*, pp. 21-23; William Mitchell, *Winged Defense: The Development and Possibilities of Modern Air Power—Economic and Military* (New York: G.P. Putnam's Sons, 1925), pp. 188-89; and Thomas H. Greer, *The Development of Air Doctrine in the Army Air Arm 1917-1941* (1955; reprint, Washington: GPO, 1985), pp. 12, 66-67. Another excellent study on the institutional problems regarding the functional duality between the ground and air forces is David Eugene Johnson, "Fast Tanks and Heavy Bombers: The United States Army and the Development of Armor and Aviation Doctrines and Technologies, 1917 to 1945," (Ph.D. diss., Duke University, 1990).

103. Brig. Gen. Walter Krueger, memorandum as quoted in Grow, "Ten Lean Years," p. 88.

104. Maj. John K. Christmas, "Tanks and Tactics," Table II, p. 210. Also see Christmas, "Is Mechanization Expensive? Cost Analysis Reveals Great Economy of Machine Warfare," *Army Ordnance* (Jul.-Aug. 1930): pp. 22-24.

105. "Infantry Modernizes," *Army and Navy Register*, 31 Aug. 1935, p. 177; "Chief of Staff Report," *Army and Navy Register*, 28 Sept. 1935, pp. 261-62; "Army Chiefs and House Group, in Secret Talk, Also Urge Stronger Pacific Forts," *New York Times*, 9 Feb. 1935, p. 7; and "Machine Gun Army is Five Year Plan," *New York Times*, 23 Sept. 1935, p. 3.

106. For a lengthy study on Christie's relationship with the Red Army and the development of the M1930 into the BT series for deep operations and battle, see Hofmann, "Doctrine, Tank Technology, and Execution: I.A. Khalepskii and the Red Army's Fulfillment of Deep Offensive Operations," *Journal of Slavic Military Studies* (June 1996): pp. 283-334. On the Kiev maneuvers see pp. 311-12.

107. "Horse-Drawn Artillery," *Army and Navy Register*, 9 Oct. 1937, p. 4.

108. Upton Birnie Jr., "Obsolescence of Horse-Drawn Artillery," *Army and Navy Register*, 16 May 1937, p. 11.

109. Dastrup, *King of Battle*, pp. 200-201.

110. "Report of Chief of Staff," *Army and Navy Register*, 11 Dec. 1937, pp. 1, 21.

111. Gen. Malin Craig, "Mechanization and Tanks," Special Statement in "General Craig's Hearing," *Army and Navy Register*, 26 Mar. 1938, p. 4. The *Register* earlier predicted "the horse may come back." See 5 Mar. 1938, p. 9.

112. "Fuqua, U.S. Mainstay in Spain, is Returning," *New York Times*, 20

Feb. 1938, p. 15; and U.S. Command and General Staff School, "Lessons from the Spanish Civil War," lecture, 1 Dec. 1937, Fort Leavenworth, Kans., pp. 1-14. For a more detailed discussion, see Hofmann, "The Tactical and Strategic Use of Attaché Intelligence: The Spanish Civil War and the U.S. Army's Misguided Quest for a Modern Tank Doctrine," *Journal of Military History* (Jan. 1998): pp. 101-33.

113. Subject: Policies governing mechanization, and tactical employment of mechanized units, 6 Apr. 1938, AG 537.3 (4-6-38), RG 407, NA, pp. 1-4.

114. "Tank Tactics," *Army and Navy Journal,* 4 June 1938: p. 884, and Robert L. Bateman, "Doctrine & Equipment," *ARMY* (Aug. 1997): p. 24.

115. Henry J. Reilly, "Proving Ground in Spain. Armament Trends as Revealed by the Spanish War," *Army Ordnance* (May-June 1939): pp. 333-36.

116. EO, G2, War Department, memorandum for Col. A.R. Chaffee, 27 Jan. 1938, RG 165, NA, pp. 1-2; and Mildred Hanson Gillie, *Forging the Thunderbolt: A History of the Development of the Armored Force* (Harrisburg, Pa.: Military Service Publishing Co., 1947), pp. 101-108.

117. Chief of Field Artillery, Subject: Letters of Transmittal, Re: Proceedings of Board, 27 July 1938; Chief of Cavalry to Chief of Field Artillery, 6 Oct. 1938; Chief of Field Artillery to Chief of Cavalry, 17 Dec. 1938; and Chief of Cavalry to Chief of Field Artillery, 5 Jan. 1939, RG 156, NA, pp. 1-6.

118. Subject: Self-Propelled Mount for a 75mm Howitzer, 8th Endorsement, 17 Apr. 1939, OO 472/3496, RG 156, NA, p. 13.

119. Hunnicutt, *Stuart*, p. 317.

120. Heiberg, "Organize a Mechanize Force," pp. 24-28, and Brig. Gen. Adna R. Chaffee Jr., "The Seventh Cavalry Brigade in the First Army Maneuvers," *Cavalry Journal* (Nov.-Dec. 1939): pp. 450-61.

121. "Prelude to Armor," p. 5.

122. Grow, "Ten Lean Years," pp. 94, 113, 116.

123. Maj. Gen. John K. Herr, "The Cavalry," lecture, AWC, 19 Sept. 1939, USAMHI, p. 13.

124. Grow, "Ten Lean Years," p. 116.

125. Maj. Gen. Robert M. Danford, "Field Artillery," lecture, AWC, 23 Sept. 1939, USAMHI, pp. 1-18. Also see Constance McLaughlin Green, Harry C. Thomson, and Peter C. Roots, *The Ordnance Department: Planning Munitions for War* (Washington: GPO, 1955), pp. 203-204, 314-17.

126. "The Armored Force: Commanders and Principles" in *Armored Command and Center,* p. 18.

127. "Prelude to Armor," p. 5.

128. "Redesignation of the Armored Force" in *Armored Command and Center,* p. 108.

129. Ibid., and Maj. Gen. John K. Herr, "Editorial Comment," *Cavalry Journal* (May-June 1946): p. 38.

130. For a modern conceptualization of an operational level of war, see Glenn K. Otis, "The Ground Commander's View—1," in Clayton R. Newell and Michael D. Krause, eds., *On Operational Art* (Washington: CMH, 1994), pp. 31-46.

131. Albert C. Wedemeyer, *Wedemeyer Reports!* (New York: Henry Holt and Co., 1958), pp. 61-62.

132. Van Voorhis, "Visit of Lieutenant Colonel [Adolf] von Schell, German Army, to Fort Knox, Kentucky," 23 July 1937, MID (Military Intelligence Division) 343-W-97, Records of the War Department General Staff, RG 165, NA, p. 1. At the time, von Schell was Chief of Staff for the Inspector of the Panzer Corps and Wehrmacht Motorization Bureau.

133. Oral Presentation, "Mechanization and Defense Against Aviation," G3 Course, AWC, 1938-1939, Report of Committee No. 5, 7 Oct. 1938, USAMHI, p. 11.

134. George F. Hofmann, "Christie's Last Hurrah. In 1941, the Army Reappraised the Christie Suspension For Use on Tank Destroyers," *ARMOR* (Nov.-Dec. 1991): pp. 14-19.
The Buick design deviated from the CCT1 and T3 Christie suspension by replacing the large road wheels with smaller ones, adding track support rollers, and eliminating the forward bell cranks. However, the modified design did retain the long helical springs so characteristic of Christie's earlier vehicles. In an attempt to modify the wasted hull space (a major problem in all Christie tanks) due to the springs' position between two spaced side plates, the outside plates on each side of the vehicle were eliminated thus exposing the long helical springs.

135. For an excellent study on the flawed antitank doctrine, see Christopher R. Gabel, *Seek, Strike, and Destroy: U.S. Army Tank Destroyer Doctrine in World War II*, Leavenworth Paper No. 12 (Washington: GPO, Sept. 1985). Also see Charles M. Bailey, *Faint Praise: American Tanks and Tank Destroyers during World War II* (Hamden, Conn.: Archon Books, 1983).

136. On the evolution of Soviet tank development, see Hofmann, "Doctrine, Tank Technology, and Execution." For a comparative study, see Arthur J. Alexander, *Armor Development in the Soviet Union and the United States*, R-1860-NA (Santa Monica, Calif.: Rand, Sept. 1976).

137. Nenninger, "Leavenworth and Its Critics," p. 221.

138. Maj. Gen. Adna R. Chaffee Jr., Statement, *Hearings on the Military Establishment Appropriations Bill, 1942*, House of Representatives, Subcommittee of the Committee on Appropriations, 77th Cong., 1st sess., 14 May 1941 (Washington: GPO, 1941), pp. 552-55.

5

World War II Armor Operations in Europe

Christopher R. Gabel

At the time of America's entry into World War II the U.S. Army's Armored Force consisted of one corps headquarters, five divisions in various states of organization, and a handful of nondivisional General Headquarters (GHQ) Reserve tank battalions. The I Armored Corps and the 1st and 2d Armored Divisions had just completed large-scale training maneuvers, and were counted among the Army's most combat-ready forces. The recently activated 3d, 4th, and 5th Armored Divisions had not yet begun large-unit training. Armored Force headquarters at Fort Knox controlled its own schools and replacement system, and even had organizational authority over many of the nontank elements in its divisions. Fort Knox also exercised a considerable degree of latitude in creating doctrine and force structures. Although not yet an official arm of the service, the Armored Force harbored ambitions of becoming a semi-independent branch of the Army like the Army Air Forces.[1]

Plans for Army expansion suggested that the Armored Force would become a dominant player in the nation's military establishment. The 1941 Victory Program forecast a huge wartime armor contingent consisting of 61 armored divisions supported by 51 motorized infantry divisions in a total force of 216 divisions.[2] On 23 May 1942 the Operations Division of the War Department General Staff issued a slightly more modest proposal calling for a 187-division Army that would include 46 armored and 23 motorized divisions. The Armored Force planned to forge these elements into twenty-three armored corps, each with two armored divisions, one motorized division, and a large array of corps troops including military police, engineer, medical, supply, and maintenance units specially tailored for mechanized warfare. The head-quarters for II, III, and IV Armored Corps were activated in 1942.

The Armored Force's plans for autonomy were short-lived. In March 1942 the War Department established the Army Ground Forces to oversee the organization and training of all ground combat elements, including the Armored Force. The head of Army Ground Forces, Lt. Gen. Lesley J. McNair, believed that the infantry-artillery team, and not the tank, was still the centerpiece of ground combat. While not an opponent of armor per se, McNair was unimpressed by claims that the newer elements of warfare, such as armor, were unstoppable. The effec-tiveness of German antitank guns against British tanks in Libya, as well as events in the Army's own 1941 maneuvers, reinforced McNair's con-victions.

Rather than building a blitzkrieg Army centered on large, special-ized armored corps, McNair intended to create a lean, standardized force structure in which special requirements would be met by attaching ad-ditional forces to conventional divisions and corps. In McNair's scheme, the corps would be a combined arms task force that could direct the operations of any and all combat elements, armor included. Perceiving armored divisions to be useful primarily in exploitation missions, McNair recommended that only about ten percent of the divisions activated be armored.[3]

Given McNair's influence, the number of armored divisions pro-jected for the wartime Army declined sharply. War Department esti-mates issued later in 1942 called for twenty-six, then twenty, armored divisions.[4] Meanwhile, the number of nondivisional tank battalions de-signed for infantry support increased. These battalions were to be ad-ministratively self-sufficient forces that could be held in corps pools and attached to infantry or armored divisions as needed. Regimental-level headquarters called tank groups would be used to combine tank battalions in circumstances calling for larger doses of armor support.

One mission that McNair did not intend for tank battalions or armored divisions to perform was antitank combat. For years he had advocated the creation of highly mobile antitank reserves that could be kept in corps and field Army pools, then dispatched to the scene of a hostile armor breakthrough where they would intercept the enemy tanks and defeat them with superior firepower. Experiments with antitank battalions and regiment-sized groups in the 1941 maneuvers seemed to validate this concept. Accordingly, in November 1941 the War Depart-ment ordered the activation of fifty-three tank destroyer battalions to constitute an Armywide antitank pool. Shortly thereafter, the antitank battalions organic to every infantry division were detached from their

parent organizations and added to the tank destroyer force.[5] The Tank
Destroyer Center, where units were to be trained and doctrine formulated, opened at Camp Hood, Texas, in January 1942. By the end of the
year, eighty tank destroyer battalions were active.[6]

A tank destroyer field manual published in 1942 outlined aggressive, even offensive, tactics for tank destroyer battalions and groups.
Their motto was "seek, strike, and destroy." A table of organization
for the tank destroyer battalion, which consisted of thirty-six tank destroyers organized into three companies, received official sanction in
June.[7]

Work also began on the design of a self-propelled tank destroyer
weapons system, which was characterized by high mobility, light armor, and heavy firepower. But in 1942 the tank destroyer battalions had
to make do with expedient designs. The M3 tank destroyer was nothing
more than a standard half-track mounting a Model 1897 75mm gun.
The M6 was an unarmored pickup truck with a 37mm antitank gun
fixed to the bed. Intended for training only, both of these would see
combat. A third expedient design, the M10, was much more battle-
worthy. Based on the chassis of the Sherman tank (which it closely
resembled), the M10 weighed in at thirty-three tons and carried a
maximum of 2.5 inches of armor. The top of its fully rotating turret was
open to allow better visibility for tank hunting. Its main armament was
a three-inch gun with a muzzle velocity of twenty-eight hundred feet
per second and a maximum range of sixteen thousand yards.[8]

The Armored Force was also busy in 1942 with creating doctrine,
designing an efficient force structure, and weapons development. The
divisional structure created in 1940 when the Armored Force was first
established was an item of particular concern. Numbering 11,200 men
and equipped with 287 light tanks and 120 mediums organized into six
light battalions and two medium battalions, the 1940 division possessed
only two battalions of infantry. Control of its three artillery battalions
was divided awkwardly between the armored brigade and division
headquarters.[9] Given the dominance of light tanks in its ranks, it is not
surprising that the division's tactical doctrine was reminiscent of old-
fashioned cavalry, with emphasis on dash and speed rather than combined arms. The limitations of the 1940 armored division became obvious
in the Louisiana and Carolinas maneuvers of 1941 whenever the armored divisions ran up against antitank forces. Because of their inadequate infantry and artillery support, the armored divisions had little
choice but to charge the antitank guns with light tanks, a technique that
worked as poorly in maneuvers as it would have in combat. During
one six-day exercise in the Carolinas maneuvers, the 1st and 2d Armored

Divisions lost a total of 844 tanks—more than their combined TO&E strengths.[10]

The Armored Force addressed these problems in March 1942 with a new division TO&E. (See Table 1.) The proportion of light and medium tanks was essentially reversed, with 232 mediums and 158 lights. The tanks were grouped into two regiments, each having two battalions of medium tanks and one of light tanks. The addition of a third battalion to the armored infantry regiment enhanced the division's capacity for combined arms action, as did the establishment of a division artillery headquarters to control all three battalions of artillery (fifty-four howitzers). In addition, each armored battalion received a platoon of three assault guns and a platoon of three 81mm mortars for use against hostile antitank positions. To improve the division's ability to operate at long distances from field Army supply points, the division trains organization received a large supply battalion to supplement the service company or battery located within each regiment. Division trains personnel accounted for 1,948 of the division's total of 14,618.[11]

The most innovative features of the new armored division were its two combat command headquarters. These brigade-level organizations replaced the single armored brigade headquarters of the 1940 division but possessed no organic troops beyond their headquarters personnel. Combat elements would be assigned to these organizations as required, allowing maximum flexibility in the combination of arms.

Although the 1942 division TO&E suggested that the Armored Force was evolving away from its mechanized cavalry roots and toward a combined arms capability, armor doctrine lagged behind. According to the *Armored Force Field Manual* published in March 1942, the role of armor was "the conduct of highly mobile ground warfare, primarily offensive in character, by self-sustaining units of great power and mobility, composed of specially equipped troops of the required arms and services."[12] The manual emphasized surprise, speed, shock action, and firepower directed against hostile rear areas.[13] The preferred tactics for armored formations were breakthrough, exploitation, encirclement, annihilation, and pursuit.[14] With top priority accorded to mobility, followed by firepower and protective armor,[15] the light tank remained the centerpiece of doctrine. The prize missions of envelopment, pursuit, and the disruption of hostile rear areas belonged to the light tanks. Medium tanks were to assist the light tanks by neutralizing enemy antitank guns and artillery, and by providing fire support.[16] In a typical operation, the armored regiment would attack on a one-thousand- to two-thousand-yard front, with its reconnaissance company in the lead, followed by the light tank battalion and two medium tank battalions in

Table 1
1942 Armored Division

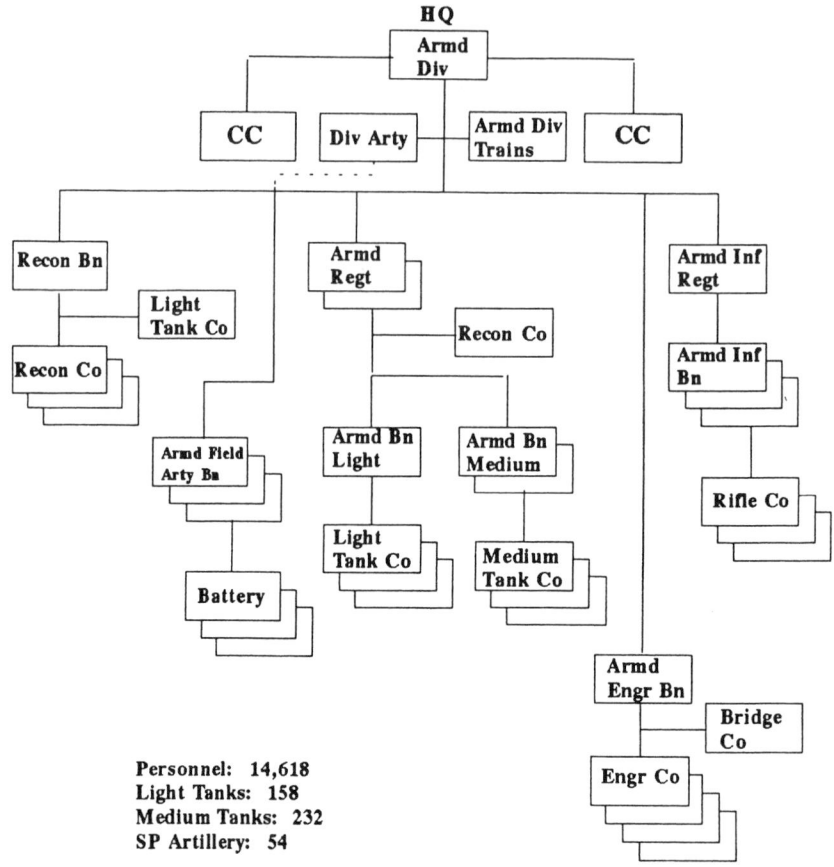

Personnel: 14,618
Light Tanks: 158
Medium Tanks: 232
SP Artillery: 54

Source: Table of Organization 17, March 1, 1942.

column. Artillery and infantry were subordinated to supporting roles—fixing the enemy and occupying captured positions.[17] Battalions were to be employed whole. Combined arms would be achieved by attaching supporting elements to the armored regiment. There are surprisingly few references to the combat commands in this manual, leading one to suspect that the insertion of combat commands into the division TO&E caught the doctrine writers off guard.

Like its doctrine, the Armored Force's equipment in 1942 was not quite up to date. The light tank upon which so much depended was the obsolescent fourteen-ton M3 Stuart. Although its 250-horsepower engine could propel the M3 at speeds of thirty-six miles per hour, its maximum armor thickness was only 1.75 inches. It mounted a 37mm antitank gun that produced a respectable muzzle velocity of twenty-nine hundred feet per second, but the armor-piercing projectile for that weapon weighed only 1.92 pounds. At ranges over a thousand yards, the 37mm gun could not penetrate even two inches of armor.[18]

Medium tank battalions relied on the thirty-ton M3 medium, variously called the Grant or Lee. This was an expedient design based on a 1930's-vintage infantry-support tank sporting a general-purpose 75mm gun in a hull mount. The 75mm fired both armor-piercing and high-explosive rounds. It had a muzzle velocity of two thousand feet per second and a maximum range of fourteen thousand yards. The M3 medium also carried a turret-mounted 37mm gun that was identical to the light tank's main armament. Both the 75mm and 37mm weapons came equipped with gyrostabilizers that were supposed to allow the guns to fire on the move. Unfortunately, the gyrostabilizers never lived up to expectations. The M3 medium carried a respectable 2.25 inches of armor (maximum), but on early models the armor plate was riveted, exposing the crew to the danger of flying rivet heads should heavy projectiles strike the armor.[19] Production began in 1942 on an improved medium tank, the M4 Sherman. The Sherman was built on a chassis similar to the M3 medium's and mounted the same 75mm gun, but in a fully rotating turret.

The T30 howitzer motor carriage, consisting of a 75mm howitzer mounted on a half-track, filled the assault-gun platoon requirement until a better design emerged. The need for self-propelled artillery in the armored field artillery battalions resulted in another expedient design, a standard 105mm howitzer mounted on the chassis of the M3 medium tank. (Later versions employed the Sherman tank's chassis.) Designated the M7, this expedient served quite successfully throughout the war.

The Armored Force needed time to shake out its new TO&E, train units, and upgrade equipment, but the exigencies of war dictated that

armored divisions would be in battle before the first anniversary of Pearl
Harbor. As 1942 progressed, it became increasingly likely that the Armored Force's combat debut would occur in North Africa. To prepare
troops for this hostile environment, the Army opened the Desert Training Center in the spring of 1942. Located at the intersection of California, Nevada, and Arizona, this huge facility was large enough for an
armored corps to conduct maneuvers. A total of three armored corps,
four standard corps, seven armored divisions, three motorized divisions,
and ten infantry divisions, plus a number of nondivisional elements,
trained at the Desert Training Center over the next two years. Ironically,
the two armored divisions that eventually fought in North Africa never
went to the Desert Training Center.[20]

Operation Torch, the Anglo-American invasion of North Africa, was a
hastily-contrived transoceanic amphibious assault directed against
French Morocco and Algeria. Landings began on 8 November 1942. The
2d Armored Division, part of the Western Task Force, landed in Morocco where it met with only sporadic opposition from Vichy French
forces. Elements of the 1st Armored Division, under the control of
Combat Command B, landed in Algeria as part of the Eastern Task Force
and participated in the capture of Oran. As in Morocco, American armor proved more than a match for the outclassed French defenders. On
25 November, however, the Allied forces in Algeria moved east to
Tunisia, where they ran up against the German Fifth Panzer Army and
the German-Italian Panzerarmee Afrika. Combat Command B quickly
discovered that it had much to learn about armored warfare.
    One of the first deficiencies discovered in Tunisia was the inad-
equacy of many expedient weapons systems. The 37mm gun—which
equipped the M3 light tank, the turret of the M3 medium, the M6 pickup-
truck tank destroyer, and the antitank companies in U.S. infantry regiments—was virtually useless against German armor at ranges beyond
five hundred yards. The M3 medium had other problems. Its high profile (ten feet, three inches), coupled with the fact that its main armament
was mounted low in the hull, made it almost impossible for the M3 to
fight from hull-down positions. Fortunately, the remainder of the 1st
Armored Division came with M4 Sherman medium tanks when it joined
Combat Command B in December.
    But bad generalship, not obsolete equipment, was the 1st Armored
Division's greatest problem in Tunisia.[21] Piecemeal employment was
the norm. By early February the 1st Armored Division had been split
into four combat commands, with the division artillery headquarters
and the headquarters of the division's armored infantry regiment dou-

bling as combat command headquarters in addition to the two organic combat commands. Operational control over these four combat commands was divided among the division commander (Maj. Gen. Orlando Ward), II Corps, and the British First Army. The division as a whole occupied sixty miles of front.

On the morning of 14 February 1943 Combat Command A occupied an isolated position at the village of Sidi Bou Zid. Combat Command A consisted of one tank battalion equipped with M4 Shermans, two battalions of infantry from the 34th Infantry Division, an armored field artillery battalion, a battalion of towed 155mm howitzers from II Corps, and a company of M3 tank destroyers (half-tracks with 75mm guns). The II Corps commander, Maj. Gen. Lloyd R. Fredendall, designated the ground that each element was to occupy. These positions were not mutually supporting. At dawn, during a blinding sandstorm, the 10th and 21st Panzer Divisions hit Combat Command A in its front and flank. (The German force included a number of new Tiger tanks carrying fearsome 88mm guns.) The German attack destroyed Combat Command A's tank battalion, overran the 155mm howitzer battalion, and surrounded its two infantry battalions.

Fredendall ordered an immediate counterattack. The only force available was Combat Command C, led by the commander of 1st Armored Division's armored infantry regiment. On 15 February Combat Command C launched its attack to rescue the elements trapped at Sidi Bou Zid. A battalion of M4 Shermans led off arrayed in a column of companies. Two batteries of M7 self-propelled howitzers came next, followed by a battalion of armored infantry riding half-tracks. A company of M3 medium tanks brought up the rear, and a company of M3 tank destroyers covered the column's flanks. There was no other reconnaissance or screening force to the flanks, nor was any deployed in front. Combat Command C, which was smaller than the force destroyed at Sidi Bou Zid the day before, did not even know that it was charging into the teeth of two panzer divisions. However, aside from the lack of intelligence about the enemy and the absence of a reconnaissance element, the attack looked like it came straight out of the Armored Force field manual.

German air attacks struck Combat Command C in its assembly areas and during its thirteen-mile charge over open desert. While still well short of its objective, the column came under fire from German artillery and antitank guns. As the Americans struggled to deploy under fire, German combined-arms battle groups struck both flanks. The leading tank battalion was cut off and destroyed, although most of the infantry and artillery managed to escape.[22]

The twin disasters at Sidi Bou Zid cost the 1st Armored Division ninety-eight medium tanks, fifty-seven half-tracks, and twenty-nine artillery pieces. The Germans followed up with a thrust through Kasserine Pass, and both Ward and Fredendall eventually lost their jobs. A less auspicious battle debut for American armor could scarcely be imagined.

Five weeks later, the tank destroyer force had its own moment of awakening. During an action near El Guettar, the 601st Tank Destroyer Battalion, armed with M3 tank destroyers and reinforced by a company of M10s from the 899th Tank Destroyer Battalion, encountered fifty 10th Panzer Division tanks. Employing the aggressive fire-and-movement tactics prescribed by their doctrine, the tank destroyers turned back a German attack. However, their losses were prohibitive, amounting to twenty of twenty-eight M3s and seven of twelve M10s.[23] This was the only known instance of the entire war in which an American tank destroyer battalion, operating as a unit, attempted to execute its prescribed doctrine.

Subsequent campaigns in 1943 offered few opportunities for either armor or tank destroyers to redeem their doctrines. The 1st Armored Division soldiered on in Tunisia, the 2d helped conquer Sicily, and the 1st landed in Italy. But armor generally operated in subdivisional elements during these campaigns, fighting at the pace of the infantry. Tank destroyer battalions were broken up and distributed among the

An M3 medium tank knocked out by the Germans in North Africa. Patton Museum.

infantry divisions, where they reinforced the inadequate frontline antitank firepower of the infantry regiments, and where they developed secondary missions as assault guns and reinforcing artillery. While there were no stunning armor triumphs in the Mediterranean theater (MTO), at least there were no further catastrophes. American armor learned new ways to wage war.

The doctrine that American armor took to war in 1942 was predicated on the assumption that tanks operated in masses, at their own pace, and that combined arms consisted of attaching supporting, subordinate elements to armored regiments. However, the German panzers in Tunisia had not operated that way, and American armor never really had the opportunity to do so either. Following the early setbacks in Tunisia, new forms of fighting evolved that emphasized intimate cooperation among the arms. Increasingly, tanks and infantry integrated their actions at the small-unit level. Patience and thoroughness in small-unit armor tactics became more important than dash and speed. Infantry learned to guide the tanks over the terrain, leading them around obstacles, pointing out enemy strong points, and actually protecting the tanks from hostile antitank weapons. In return, tanks knocked out automatic weapons strong points located by the infantry and provided close-range fire support when infantry closed with the enemy. Artillery observers advanced with the forward elements, enhancing the effectiveness of artillery fire against antitank positions and hostile batteries.[24]

These techniques worked just as well for infantry divisions as they did for armored divisions. Experience showed that infantry needed tank support routinely, not just on special occasions. But the infantry divisions that fought in North Africa had received little or no training in cooperation with tanks. As a result, Army Ground Forces in 1943 directed that all new infantry divisions train with nondivisional tank battalions before deploying overseas.[25]

The armor establishment reacted to the lessons of Tunisia by drafting a new 1943 armored division TO&E that increased the proportion of infantry to tanks, and which enhanced the ability of division commanders to forge combined arms teams. (See Table 2.) The number of tank battalions dropped from six in the 1942 organization to three in the 1943 division, giving the division equal numbers of tank, infantry, and artillery battalions. Every tank battalion was standardized with three medium and one light tank companies—light tank battalions having been scrapped. The new division had seventy-seven light tanks and 168 mediums. The armored regiment and infantry regiment headquarters

Table 2
1943 Armored Division

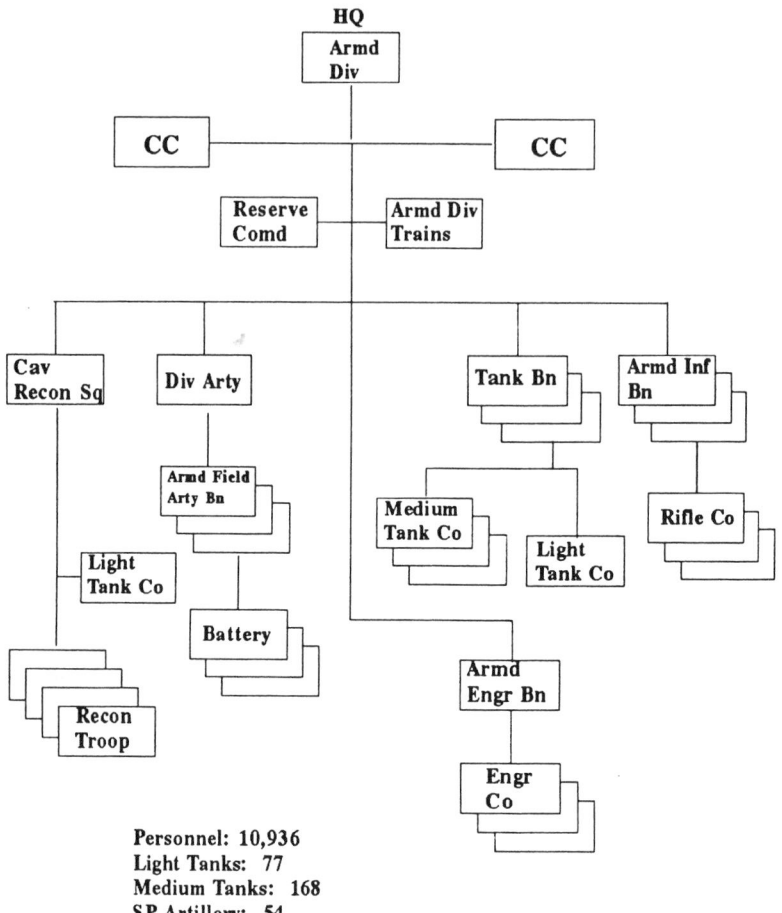

Personnel: 10,936
Light Tanks: 77
Medium Tanks: 168
SP Artillery: 54

Source: Table of Organization 17, September 15, 1943.

were eliminated, leaving the combat command as the only echelon between division and battalion in the tactical chain of command. A Reserve Command was added for the supervision of combat elements in division reserve. This small headquarters consisted of only three officers and five enlisted men.

The 1943 armored division was significantly smaller than the organization it replaced—10,936 personnel as opposed to 14,618 in the 1942 division. Much of the savings came from reductions in the number of headquarters and logistics troops. In addition to eliminating the three regimental headquarters, the new division structure simplified logistics administration. The division trains were reduced, and, with the regimental echelon gone, every battalion in the division was provided with its own service company or battery, making each battalion logistically self-sufficient. Battalions drew their supplies directly from field Army supply points, bypassing the division and corps echelons altogether.[26]

A new field manual published in January 1944 codified the new approach to armored warfare. It stressed the need for timely cooperation among the arms while placing more emphasis on the destruction of enemy forces in contact and less on cavalry-like rampages in hostile rear areas.[27]

The 1943 armored division was not only smaller than its predecessor, but there would be fewer armored divisions than previously anticipated. The War Department Troop Basis adopted in July 1943 finalized the Army's wartime strength at eighty-eight divisions, of which sixteen would be armored. Fourteen of these adopted the 1943 TO&E, while the 2d and 3d Armored Divisions retained a modified version of the 1942 structure and were referred to as heavy armored divisions. The motorized division concept was dropped altogether. When circumstances called for motorized infantry, truck companies would be attached to standard infantry divisions. The armored corps also disappeared from the 1943 troop basis. The four existing armored corps headquarters, none of which saw combat, converted into standard corps or field Army headquarters.[28]

Reducing the size and numbers of armored divisions freed up tank battalions for employment as nondivisional or independent tank battalions that could be held in corps pools and attached to divisions as needed. These battalions were configured almost identically to the armored division's tank battalions. The Army ultimately fielded sixty-five such battalions, a number significantly larger than the fifty-four tank battalions found within the sixteen armored divisions.[29] For those times when it would be necessary to mass two or more independent battalions, the War Department established the armored group head-quarters, the staffing of which was nearly identical to the armored division's combat command. Ten such group headquarters deployed to the European theater (ETO).

All of these changes signaled the ascendancy of General McNair's vision for armored warfare, and a decline in the influence of the armor establishment over its own affairs. In 1943 the Armored Force was downgraded to the status of "Command," and later "Center," with its authority essentially restricted to the confines of Fort Knox. Armor was firmly established as part of the ground forces combined arms team; the prospects of an autonomous armored force had disappeared.

The year 1943 also saw several significant improvements in armor's arsenal of weapons. The M3 light tank was given a larger engine and redesignated the M5. Design also began on an entirely new light tank, the sleek M24, which mounted the same 75mm gun employed by the Sherman. Deliveries of the M24 would begin in late 1944. A new howitzer motor carriage, the M8, replaced the half-track–mounted assault gun used in North Africa. The M8, which utilized the M5 light tank's chassis, mounted a 75mm howitzer in a fully rotating turret. In 1944, yet another weapon would supplant the M8 as the tank battalion's assault gun—a Sherman tank variant armed with a 105mm howitzer.

The M4 Sherman had in fact come of age as the Army's premier armored vehicle. Ultimately, American industry would produce some forty thousand Shermans in a bewildering variety of versions—enough to equip every medium tank company in the Army, not to mention those supplied to the Marine Corps and to lend-lease recipients. Of all the variants produced, the M4A3 can best be considered the Army's standard medium tank of World War II. This model weighed thirty-four tons and featured a welded hull and a cast turret with a maximum armor thickness of two inches. Its power plant was a five-hundred-horsepower Ford V-8 gasoline engine that produced a top speed of twenty-six miles per hour. The M4A3 retained the same 75mm gun featured on the old M3 medium.[30] Ammunition stowage totaled ninety-seven rounds of 75mm ammunition, of which approximately 70 percent would typically be high explosive, 20 percent armor piercing, and 10 percent white phosphorous. The latter served as both an incendiary and a smoke munition.[31]

Just as important as its main armament were the tank's machine guns—one .30-caliber weapon mounted coaxially with the 75mm gun, one .30-caliber in a bow mount (operated by the codriver), and a .50-caliber machine gun adjacent to the commander's hatch on top of the turret. The M4A3 provided stowage for 6,250 rounds of machine-gun ammunition, but crews often crammed in as many as 10,000 or even 13,000 rounds.[32]

When designed in 1941, the Sherman was a world-class tank, and it remained an excellent, reliable weapons system for most of the tasks it was called upon to perform. Nevertheless, by 1943 it was deficient as an antitank weapon. The Sherman could still compete with the older Panzerkampfwagen (Pzkw) IV, but was at a serious disadvantage against the forty-five-ton Pzkw V Panther and fifty-five-ton Tiger I. The Panther featured 4.7 inches of frontal armor and a superb 75mm gun with a muzzle velocity in excess of three thousand feet per second. The Tiger had four inches of frontal armor and mounted the famous 88mm gun, which produced a muzzle velocity of 2,650 feet per second.[33] Due to organizational inefficiency[34] and the constraints of resources and logistics, the U.S. Army failed to keep pace with these developments.

Under American doctrine, the main burden of antitank warfare belonged to the tank destroyers. Unfortunately, the M10 tank destroyer, which replaced the hapless M3 and M6 models in 1943, was only marginally more effective against heavy German armor than the Sherman. A new weapon, the M18, entered production in 1943. The M18 had been designed from the ground up to be the prototypical tank destroyer. Weighing just twenty tons and riding on an innovative torsion-bar suspension, it could travel at fifty miles per hour. Maximum armor thickness was only one inch. The M18 carried a high-velocity 76mm gun in

M18 "Hellcat" tank destroyer. Patton Museum.

a fully rotating turret, and could deliver a projectile with a muzzle velocity of twenty-eight hundred feet per second.[35] But even this gun could not guarantee a kill against Tigers or Panthers. Late in 1944, yet another tank destroyer would appear—the M36—with the chassis of a Sherman tank and a 90mm gun in a fully rotating turret. With a muzzle velocity of twenty-seven hundred feet per second, this weapon was able to deal with even the heaviest German tanks.[36]

Doctrine notwithstanding, both tanks and tank destroyers would have to face enemy armor. To help redress the Sherman's firepower disadvantage, certain variants were equipped with the same 76mm gun as the M18 tank destroyer. In the closing days of the war, an experimental heavy tank mounting a 90mm gun arrived in the ETO. This was the M26 Pershing, ancestor to the M48 and M60 tanks of later years. But the M26 arrived too late and too few in numbers to have much impact. The "crusade in Europe" was the Sherman tank's war.

For most of the U.S. Army, 1943 was a year of preparation and training for the decisive battles ahead. But for two armored divisions it was a year of active campaigning, even as the rest of the armored establishment absorbed the lessons of Tunisia. On 10 July, the 2d Armored Division, under the command of Maj. Gen. Hugh J. Gaffey, participated

M4 and M3 medium tanks on maneuvers at Fort Knox, Kentucky. Patton Museum.

in the Allied invasion of Sicily. Combat Command A was attached to the 3d Infantry Division for the landings, while the remainder of the division assembled in the beachhead as part of the Seventh Army reserve. German counterattacks on succeeding days resulted in the distribution of various armored elements to hard-pressed infantry divisions. Once the beachhead was secure, the division reassembled under a provisional corps headquarters for a quick strike toward Palermo. Facing indifferent opposition, elements of the 2d Armored covered the one hundred miles to Palermo in four days. The division then reverted to reserve status and engaged in occupation duties. The 2d Armored Division saw only twelve days of combat in Sicily, mostly while fragmented among other commands. Four months later the division sailed for England, leaving the 1st Armored as the only American armored division in the MTO.

On 9 September, American forces invaded the Italian mainland at Salerno. The 1st Armored Division, commanded by Maj. Gen. Ernest N. Harmon, did not land until November. By that time, Allied forces were closing in on the German defensive position known as the Winter Line.[37] When the Allied advance stalled, the 1st Armored (less Combat Command B), sailed with the U.S. VI Corps and landed farther up the peninsula at Anzio, well behind the German lines and within striking distance of Rome. Failure to exploit the successful landings at Anzio consigned VI Corps to five months of static warfare within a confined beachhead. Obviously there was no opportunity for the 1st Armored to execute the offensive mission for which it was intended. Instead, the division entered corps reserve, with its armored infantry battalions frequently committed to manning the perimeter. Heavy German counterattacks began in early February and continued for much of the month. Tanks from the 1st Armored were instrumental in blocking German efforts to penetrate the perimeter. During static phases of the battle, the tanks executed indirect fire missions in support of the corps artillery.

Not until May 1944 was Combat Command B brought into the Anzio beachhead and the division given an offensive mission. The division formed up on 23 May with its two combat commands abreast on a three-mile front to lead the breakout from the beachhead. In a week of costly fighting the division succeeded in battering a hole through the German defenses. Then, for a few brief days, the 1st Armored enjoyed the exhilaration of open warfare as it exploited the breakout. On 4 June, mobile columns entered the city of Rome.

After the capture of Rome and another period of exploitation that carried the division well north of the city, the 1st Armored was belatedly reorganized to conform with the 1943 TO&E. Shortly after that the

division was fragmented and its elements parceled out to other formations as the Allies drove north to the Arno River. In September, Allied
forces reached the northern Apennines and a new German defensive
barrier called the Gothic Line, where the Allied advance again stalled.
The 1st Armored spent the winter months manning an essentially static
line. Some tank elements were even dismounted and sent to the front
as infantry. Offensive operations did not resume until mid-April 1945.
In those last weeks of the war the 1st Armored once again enjoyed a
brief interlude of freewheeling offensive action before German forces
in Italy surrendered on 29 April.

The 1st Armored Division's war in Italy was a grueling, frustrating affair that bore little resemblance to prescribed doctrine. The division was used (and abused) in a variety of roles, but rarely operated as
a unit. Only twice did the division's inherent mobility come into play
on an operational scale—during the breakout from Anzio and in the Po
Valley operations at the end of the war. Although the division saw little
in the way of blitzkrieg warfare, the 1st Armored and the separate tank
battalions deployed in Italy proved the value of tanks in a wide variety
of tactical applications.

The U.S. Army's main mission in World War II was to invade northwest
Europe and destroy the German Army. This would involve liberating
the territories that Germany had conquered in its remarkable blitzkrieg
campaign of 1940. Although the ground was the same, and even though
the Americans also planned to conduct a mechanized war, the strategic
context that confronted the Allied invaders in 1944 differed significantly
from that which faced the Germans four years earlier. Germany's desire
to avoid a prolonged war had compelled its armed forces to conduct a
high-risk campaign of annihilation. But in 1944 the Allies felt little need
to run risks. Germany was on the way to defeat thanks to Allied victories in the North Atlantic, in the skies over Europe, and especially on
the eastern front. The Anglo-American forces in northwest Europe could
afford to be conservative. This meant that Allied armor forces would
have few opportunities to replicate the slashing blitzkrieg operations
that had electrified the world from 1939-41.

The campaign got off to a bad start for American armor. Plans for
the invasion of Normandy called for three tank battalions to land in the
initial assault waves—one with each of the U.S. infantry divisions
making the initial landings. These tanks were expected to swim ashore
under their own power and provide fire support from hull-down positions in the water. To accomplish this remarkable feat, the tanks were
fitted with a British invention—a collapsible thirteen-foot-high rubber-

ized canvas screen. When erected, this screen provided enough flotation to support a Sherman tank, even though the tank itself was below water level. (The tank commander used a periscope to see where he was going.) Propulsion came from two steerable twenty-six-inch propellers powered by the tank's engine, hence the designation Duplex Drive (DD) tank. Thus fitted, a Sherman tank was capable of traveling at four knots in the water.[38]

The DD tanks fared poorly on 6 June 1944, when the Normandy landings took place. Waves were higher than the two to three feet of freeboard provided by their canvas screens. Most of the tanks that got ashore did so because their landing craft ran them all the way in to the beach. The 741st Tank Battalion lost twenty-seven tanks in the heavy seas.[39]

Once ashore, the American forces found themselves bogged down in Normandy's hedgerow country. Seven weeks of bloody, grinding, attrition warfare followed. The terrain in the Normandy *bocage* consisted of small fields, roughly two hundred yards on a side, walled in by hedgerows—banks of earth topped by dense vegetation. The hedgerows provided the Germans with a ready-made, compartmentalized defensive battlefield. German troops placed their automatic weapons, mortars, artillery, and antitank guns in concealed positions that afforded interlocking fields of fire across the small open spaces. Standard American fire-and-maneuver tactics fared poorly against such dug-in and camouflaged strong points. It was hard to locate the German positions, hence suppressive fires were generally ineffective. The hedgerows themselves restricted maneuver because any attempt to outflank a German position involved crossing a hedgerow and coming under fire from yet another enemy strong point. Engagements usually degenerated into costly infantry frontal assaults against concealed German machine guns and mortars. Infantry divisions and their attached tank battalions bore the brunt of the fighting on this terrible battlefield.

The tank battalions supporting American infantry were at a serious disadvantage among the hedgerows. German antitank guns, sited to fire along the roads, forced the American tanks into the close quarters of the fields. There they were easy prey for the German version of the bazooka known as the *panzerfaust*. Many of the hedgerows were too tall for tanks to cross, and those that could be traversed exposed the tanks' underbellies to German fire from the hedgerows beyond. Nothing in armor or infantry doctrine explained how to fight under such conditions. Solutions had to be worked out in the field.

First, the hedgerow itself had to be defeated. Engineers could blow gaps in hedgerows with high explosives, but digging a hole in which to plant the explosive was dangerous and time consuming. It was discovered that prongs made of large pipe fitted to the nose of a Sherman tank could punch holes in the base of a hedgerow into which demolitions could be placed. Sometimes the prongs themselves were sufficient to rip through the obstacle. This inspired the development of pointed tusks fabricated from steel beams, which allowed the tank to tear its own hole through a hedgerow. The tusks kept the tank from bellying up and directed its momentum forward. Tanks so equipped were referred to as "Rhinos."[40]

With the tanks now able to maneuver off-road, the next challenge was to develop new tactics for defeating the German strong points. This involved creating small teams of tanks, infantry, and engineers. Infantry served as the eyes and ears for the tanks and protected the tanks from close-range antitank fire. The tanks in turn provided the infantry with physical cover from automatic weapons fire and delivered suppressive fire that enabled the infantry to maneuver and assault strong points. The engineers in turn reduced obstacles. Due to the confined nature of the hedgerow battlefield, these combined arms teams were often very small. In some cases, individual tanks were assigned to infantry squads.[41]

Such tactics called for innovations in battlefield communications. Tank and infantry radios operated on different frequencies. In some cases, tankers and riflemen exchanged man-pack radios so that they could communicate with each other. In others, tank crews installed infantry radios in their tanks. These techniques allowed infantry commanders to ride in the tanks, where they could coordinate face-to-face with the supporting armor commander and still control their foot troops. Another expedient was to attach a field telephone to the rear of the tank and wire it into the tank's intercom system.[42]

The Battle of the Hedgerows confirmed a lesson learned in North Africa and Italy: infantry divisions needed some form of armor support virtually all the time, not just for special operations. Moreover, the techniques of small-unit cooperation developed so painfully among the hedgerows varied from division to division. As a consequence, it became common practice to attach tank battalions to specific infantry divisions and leave them there indefinitely. The divisions in turn proceeded to parcel out individual tank companies to their regiments. For all practical purposes the tank company became an organic part of the infantry regiment.

Unfortunately, there were not enough tank battalions to go around. Forty-two American infantry divisions ultimately participated in the European campaign, but only thirty-seven independent tank battalions.

Tank destroyer battalions helped fill the void. They too were attached to specific divisions on a semipermanent basis, whereupon their constituent parts married up with infantry elements. Tank destroyers performed essentially the same functions as tanks, with the additional mission of reinforcing the division artillery with indirect fires.

Five armored divisions also fought in the Normandy campaign. They too had to correct certain deficiencies in their makeup. Neither the "light" 1943 division nor the older "heavy" division was a properly balanced combined arms force. Both needed additional infantry and artillery on a regular basis. Armored divisions routinely borrowed infantry battalions, or even regiments, from neighboring infantry divisions. Other typical attachments included an armored field artillery battalion and at least one battalion of 155mm howitzers, plus a tank destroyer battalion and an antiaircraft battalion drawn from corps. Engineer bridging elements were also frequently added to the armored division. As was the case with the infantry divisions, such attachments were often more or less permanent.

Most armored divisions usually operated with three full-fledged combat commands, rather than the two provided by their TO&Es. The 2d and 3d (heavy) Armored Divisions employed the headquarters of their armored infantry regiments as a third combat command. The light divisions organized under the 1943 TO&E without regimental headquarters, used armored group headquarters, which were left without a job when the independent tank battalions went off to the infantry divisions.

All of the armored divisions organized task forces within their combat commands. In the two heavy divisions, each of the three combat commands formed two task forces. (See Table 3.) Typically, a task force consisted of one tank battalion, one armored infantry or "leg" infantry battalion attached from an infantry division, one armored artillery battalion, plus some tank destroyers and engineer elements. Task force headquarters personnel came from the headquarters of the two armored regiments and from the tank battalion's headquarters.[43]

Light divisions organized comparable but smaller task forces. (See Table 4.) In a typical arrangement, the three combat commands each received one tank battalion and one armored infantry battalion. The tank battalion traded one tank company for a rifle company from the infantry battalion. Each composite battalion then got a slice of tank destroyer, antiaircraft, and engineer assets, making it a task force. Thus, each combat command controlled a tank-heavy task force and an infantry-heavy task force.[44]

As the war progressed, special circumstances sometimes compelled the armored divisions to organize more than two task forces per com-

Table 3
Typical Combat Command
"Heavy" Armored Division

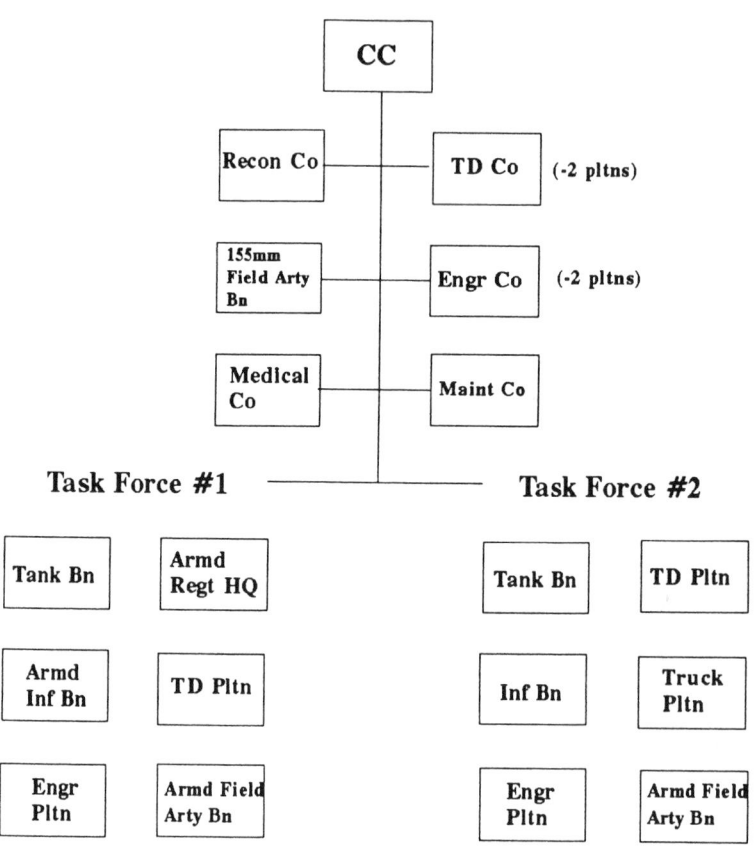

(Combat Command "A," 3d Armored Division, March 25-April 25, 1945).

Source: General Board, Study No. 48, Appendix 1.

Table 4
Typical Combat Command
"Light" Armored Division

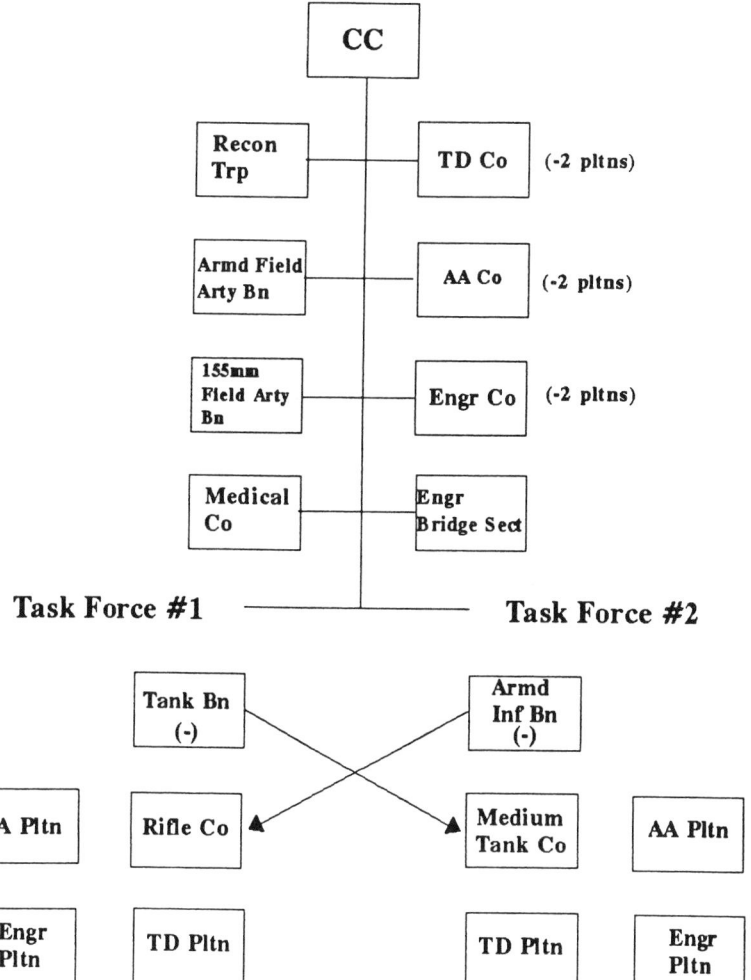

(Combat Command "A," 6th Armored Division, April 1-15, 1945).

Source: General Board, Study No. 48, Appendix 1.

bat command. Late in the war, the 9th Armored Division made use of an attached infantry regiment to organize a total of nine task forces—three in each combat command. Different armored divisions also varied in the permanence of their task force arrangements. The 5th Armored Division, valuing the teamwork that came from long-term associations, seldom if ever altered the composition of its task forces. The 4th and 6th Armored Divisions, which valued flexibility, frequently reconfigured task forces to meet changing battlefield circumstances. The 4th and 6th also differed from other armored divisions in that they did not commonly employ a third combat command as a tactical unit. Moreover, they often assigned artillery directly to their task forces rather than retaining it at the combat command level as was typical in other light armored divisions.[45]

As a school for teaching combined arms, the Normandy *bocage* country provided a bloody but beneficial experience. The operations that followed more closely resembled the type of warfare that American armor leaders wanted to wage.

On 25 July 1944, American forces near Saint-Lô launched Operation Cobra, the breakout from Normandy. Twenty-four hundred aircraft opened the battle with a saturation bombardment of the breakthrough sector. The VII Corps, which massed six divisions on a five-mile front, then ruptured the German lines with two infantry divisions plus one regiment from a third leading the assault. Exploitation began the next day, led by the two heavy armored formations, the 2d and 3d Armored Divisions. The 3d Armored, reinforced by the temporarily motorized 1st Infantry Division, attempted to cut behind and encircle German forces adjacent to the breakthrough zone, but the Germans slipped past the envelopment.

Maj. Gen. Edward H. Brooks's 2d Armored Division met with greater success. Its mission was to drive deep behind enemy lines and protect the 3d Armored's rear by blocking the advance of German reserves. Augmented by a regiment from the 4th Infantry Division, the 2d Armored broke into the clear and conducted a classic armor exploitation. The leading wave, consisting of tanks supported by artillery, found and penetrated the gaps between German elements. A follow-on force of tanks carrying infantry on their decks held open the passage. The third wave of tanks and infantry reduced the strong points with infantry fixing the enemy and leading the tanks in the assault. Rhino tusks enabled the 2d Armored's tanks to maneuver off-road, whereas hedgerows confined German armor to the roads.[46]

The 2d Armored Division got so deep into the enemy's rear that it was in a position to trap German forces retreating past the 3d Armored. Thus the intended encircling force, the 3d Armored Division, became a holding force and the 2d Armored forged an even deeper encirclement with the 2d Armored's reconnaissance battalion racing ahead of the main body and taking up blocking positions in the path of the retreating Germans. Combat Command B, Reserve Command, and a battalion from the 4th Infantry Division followed and sealed off German escape routes. The 2d Armored ultimately killed or captured more than five thousand German troops attempting to break out of the trap.[47]

The Cobra breakthrough ruptured the German front in Normandy beyond hope of repair. The Allies followed up with a pursuit all along their line that lasted more than a month, resulted in the virtual destruction of a German field Army, and liberated most of northern France. The six American armored divisions in theater during the pursuit were in their glory. They operated within corps that consisted, typically, of one armored division and two infantry divisions. The XII Corps, organized just after the Normandy breakout, conformed to this pattern, with the 4th Armored Division leading the pursuit while the 35th and 80th Infantry Divisions kept pace as best they could.

The Commander of the 4th Armored, Maj. Gen. John S. "P" Wood, was perhaps one of the most forceful and daring armored division commanders in the pursuit. Exercising command and control from a liaison aircraft and issuing mission-type orders verbally or over the radio, Wood drove his division more than a thousand road miles in thirty-five days. He reconfigured the division's combat commands and task forces about every three days in response to the constantly changing circumstances of the pursuit. Depending on the situation, a 4th Armored Division combat command might have two, three, or four task forces. Normally, each task force was a mixed column of tanks, infantry, and artillery. Medium tanks, not reconnaissance elements, usually led the way. Medium tanks could cut through most resistance, thus precluding the time-consuming process of deploying from the march column. Wood used his mechanized cavalry reconnaissance forces to screen the division's flanks and maintain contact between task forces. Artillery and engineers with bridging equipment stayed near the head of each column, ready for prompt action. As far as possible, the 4th Armored kept to the side roads as experience showed that these were less likely than the main highways to be defended.

Each combat command positioned its assigned supply and maintenance assets unusually close to the front, where they could service the fighting elements with minimum delay. Under the fluid conditions of

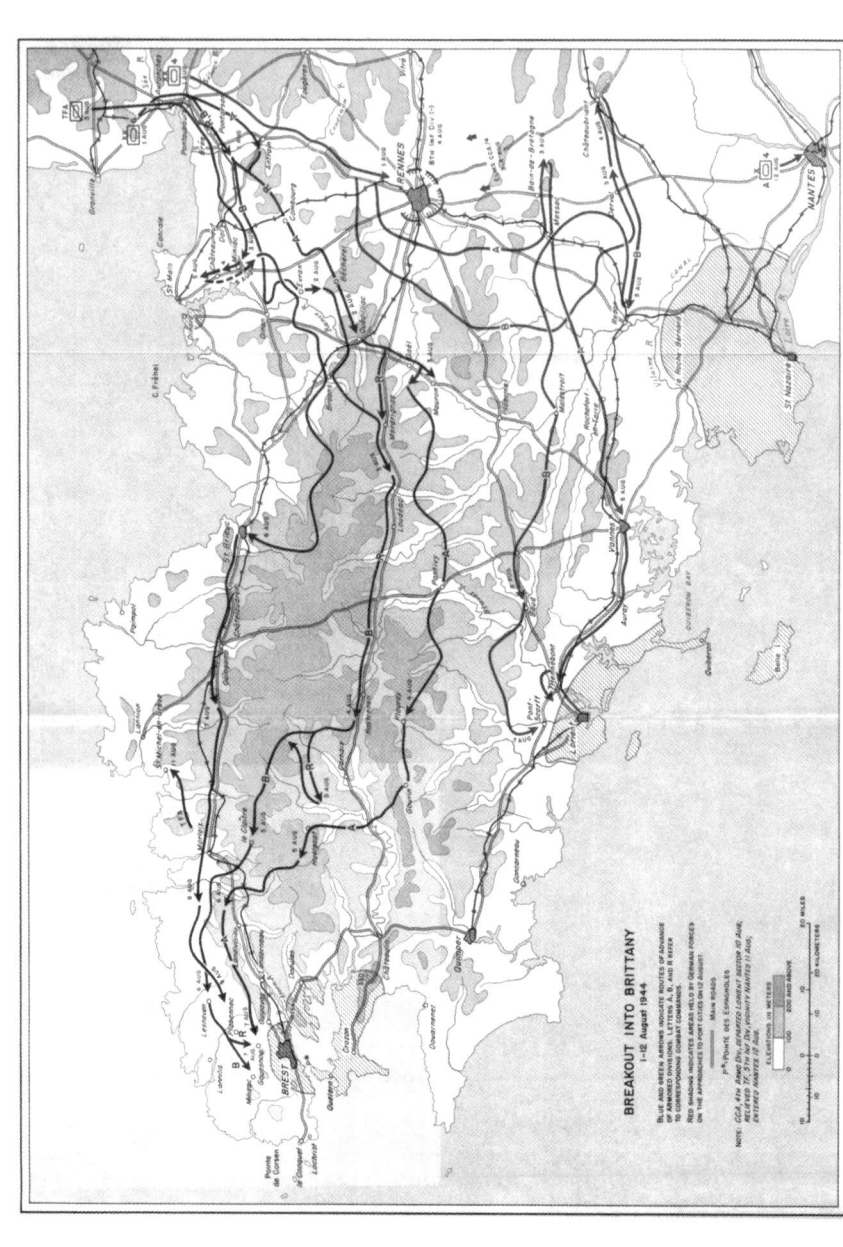

BREAKOUT INTO BRITTANY
1–12 August 1944

BLUE AND GREEN ARROWS INDICATE ROUTES OF ADVANCE
OF ARMORED DIVISIONS, LETTERS A, B, AND R REFER
TO CORRESPONDING COMBAT COMMANDS.

RED SHADING INDICATES AREAS HELD BY GERMAN FORCES
ON THE APPROACHES TO PORT-CITIES ON 12 AUGUST.

━━━━━ MAIN ROADS

━━P━━PONTE DES ESPAGNOLES

NOTE: CCA, 4th ARMD DIV, ADVANCED LORIENT SECTOR 10 AUG,
RELIEVED TF, 5TH INF DIV, VICINITY NANTES 11 AUG,
ENTERED NANTES 12 AUG.

ELEVATIONS IN METERS

the pursuit, Wood believed that the safest place for the supply trains was in the "vacuum" right behind the combat forces. Supply vehicles ran overloaded by fifty percent or more, carrying gasoline for the tanks. Each combat vehicle carried seven days' rations. General Wood did not want logistics to slow down his division.

Aviation played a crucial role in the 4th Armored Division's pursuit across France. The division's liaison aircraft maintained continuous reconnaissance in advance of the task forces. In addition, Ninth Air Force's XIX Tactical Air Command provided a four-ship armed reconnaissance flight over each column. The fighter-bomber pilots, who were in direct radio communication with the commanders of the 4th Armored's combat commands, warned the ground elements of obstacles and enemy strong points, many of which they were able to neutralize before the armored columns arrived.

The secret of Wood's success in this remarkable pursuit was always to stay at least one step ahead of the retreating Germans. The 4th Armored Division preferred to arrive at defensible positions before the enemy was ready to defend them. In this style of fighting, the tank's principal weapon was the machine gun. But on the occasions when the 4th encountered Germans who were prepared to fight, Wood knew the importance of taking time to organize a combined-arms attack in overwhelming strength.[48]

Meanwhile, the commander of the 6th Armored Division, Maj. Gen. Robert Grow, moved his division toward the port city of Brest. In a classic exploitation reminiscent of traditional cavalry, the division disorganized the enemy with a rapid and deep operation.[49] Wood and Grow would emerge, according to General Patton, as the most effective and dependable armored division commanders in his Third Army.

The Allied pursuit across France and Belgium ended in early September. Logistical overextension, geography, weather, and a resurgent German defense all combined to stop the dash toward Germany. For the next several months the Allies were compelled to slog their way through forests, cities, and the fortifications of the Westwall.

Prior to World War II Germany constructed a fortified defensive line called the Westwall along its western frontier. The Allies referred to it as the Siegfried Line. Most of the fortifications within this complex were small pillboxes, each manned by about fifteen defenders, with firing ports for machine guns and light antitank weapons. Their walls consisted of reinforced concrete from three to eight feet thick. Positioned so as to be mutually-supporting, each was flanked by entrenched infantry.[50] Although the Siegfried Line was not much of a defensive position

Maj. Gen. John S. "P" Wood, commander of Third Army's spearhead 4th Armored Division. Patton Museum.

Maj. Gen. Robert W. Grow, Commander of the 6th Armored Division. 6th Armored Division Association.

Lt. Gen. George S. Patton Jr., Maj. Gen. Robert W. Grow, Maj. Gen. Manton S. Eddy, November 1944. 6th Armored Division Association.

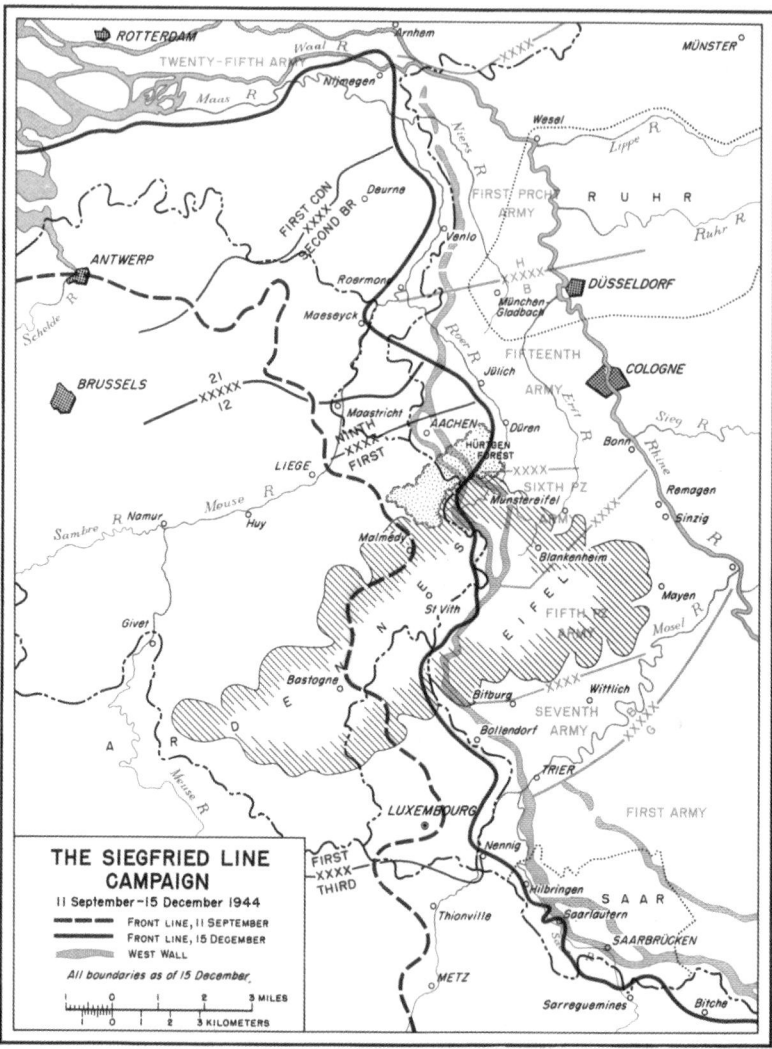

THE SIEGFRIED LINE
CAMPAIGN
11 September–15 December 1944

▬ ▬ ▬ FRONT LINE, 11 SEPTEMBER
━━━ FRONT LINE, 15 DECEMBER
▓▓▓ WEST WALL

All boundaries as of 15 December

by 1944 standards, it proved to be a significant obstacle to the Allies as they attempted to penetrate the German border.

Infantry divisions and their attached tank and tank destroyer battalions again bore the brunt of the fighting along the Siegfried Line, just as they had in the close confines of Normandy's *bocage* country. In fact, the combined arms teams and tactics employed in reducing Siegfried Line pillboxes resembled those used in the hedgerow fighting. For example, the 28th Infantry Division formed assault teams consisting of one rifle platoon, two tanks, and a squad of engineers equipped with demolition charges. Operations were slow and methodical. Artillery pounded the pillbox being attacked and blanketed adjacent positions to suppress flanking fire. Then tanks or tank destroyers advanced to close range and fired at the pillbox's embrasures to drive the defenders from their guns. Infantry then cleared the trenches flanking the pillbox. Finally, infantry-engineer teams worked their way to the rear of the pillbox, where they placed demolition charges against the pillbox door.[51]

When both tanks and tank destroyers were available, tank destroyers, with their lighter armor but higher velocity guns, would provide suppressive fire from overwatch positions while tanks closed on the pillbox with the infantry. Sometimes tank destroyer fire alone could break open a pillbox, particularly when a platoon of four aimed at the same point on the fortification. And if a tank destroyer could manage to maneuver behind the pillbox, its gun could easily penetrate the entry door. One high-explosive round placed through the opening thus created invariably resulted in the surrender of the surviving defenders.[52]

Towns and cities along the German frontier posed their own special problems. During the pursuit across France it had been possible to clear towns rapidly because the Germans rarely had time to organize their defense. The 4th Armored Division would punch a task force into a town on a narrow front at some weakly defended point (usually through yards and lots, rather than along the streets), consolidate at the town center, then attack out in all directions—essentially capturing the town from the inside out.[53]

However, as the Allies neared and entered German territory, urban fighting became more difficult. Experience on the eastern front had provided the Germans with a coherent doctrine for defending cities. It was their practice to site antitank barriers and self-sufficient, interlocking strong points along an irregular line of resistance that was difficult for the attackers to identify. Automatic weapons and antitank guns swept all streets and open areas. Sewers provided covered lines of communication, supply, and infiltration. Frequent counterattacks in unexpected locations kept the attackers off balance and retarded their operations.[54]

American armor units had no training in urban fighting. In fact, doctrine discouraged the use of tanks in cities because of the poor visibility, close quarters, and limited fields of fire.[55] Nonetheless, tanks and tank destroyers proved to be indispensable players in urban combat.

In September 1944 elements of the 1st Infantry Division attacked Aachen, the first German city assaulted by the western Allies. Aachen rested between two belts of the Siegfried Line and the Germans planned a tenacious defense of the city itself. The 1st Infantry Division could spare only two battalions to clear the city. The battalion charged with clearing the city center organized into three teams, each consisting of one rifle company, three tanks or tank destroyers, two 57mm antitank guns, two bazooka teams, two heavy machine guns, and one flame-thrower. These teams employed tactics that were deliberate, carefully orchestrated, and characterized by the profligate expenditure of ammunition.

Tanks and tank destroyers provided a base of fire in an environment that was often too congested for effective artillery support. Sheltering behind buildings already secured, they would nose around the corner of a side street and place fire on enemy strong points in the next block ahead, driving the defenders into the basements. Infantry then cleared the block one building at a time, blowing passages through walls rather than risking exposure in the streets. Once that block was cleared, infantry would place intense fire on every spot in the next block that might harbor hostile antitank weapons, while the armor dashed forward to the cross street just secured by the infantry. Then the process would patiently be repeated. In one instance, the battalion brought up a self-propelled 155mm gun to place direct fire on a suspected enemy bunker. To protect the precious artillery piece, tank destroyers and infantry took up positions from which they commanded every point that might fire on the self-propelled gun. In effect, the tank destroyers protected the artillery piece, while the infantry protected the tank destroyers. Ironically, the "bunker" turned out to be a camouflaged German tank, which succumbed quickly to point-blank 155mm fire.[56] Such teamwork allowed the American forces to secure Aachen in a matter of days with minimal casualties.

Smaller towns posed less of a threat but demanded similar cooperation among the arms. The 2d Armored Division's procedure for capturing a small town began with artillery fire placed on the objective and on all neighboring points that might contain hostile positions. Artillery also isolated the objective by firing high-explosive and smoke rounds beyond it. Tanks and tank destroyers then took up overwatch positions and placed suppressive fire on the village while teams composed of tank and infantry platoons assaulted it.[57]

As a general rule, when tank-infantry forces attacked a strong point (village or otherwise), tanks took the lead when no mines or antitank fire were present, with the infantry following at a distance that precluded risk from hostile artillery fired at the tanks. Infantry led when antitank defenses were significant. When no hostile fire was expected, infantry rode on the backs of tanks.[58]

On 16 December 1944 a German counteroffensive in the Ardennes interrupted the Allied assault on the Siegfried Line. Eight panzer divisions employing fourteen hundred tanks led a twenty-four-division attack that ruptured a thinly held section of the U.S. First Army's line. The American response to the German breakthrough involved holding the shoulders of the penetration and denying enemy access to vital road centers. American armor played a central role in the successful execution of this plan.

The Belgian town of Bastogne was one such road center. Task forces from Combat Command B, 10th Armored Division, staged heroic delaying actions at roadblocks east of the town, allowing time for the 101st Airborne Division to occupy Bastogne. By 20 December German forces surrounded city. Within the encircled position were the airborne division, two battalions of corps artillery, and the 10th Armored Division's Combat Command B and the 9th Armored Division's Reserve Command, with a total of about forty Sherman tanks. The 705th Tank Destroyer Battalion was also present with its M18s.[59] The acting commander of the 101st, Brig. Gen. Anthony C. McAuliffe, organized his forces into regimental task forces, each with a portion of the available tanks, tank destroyers, and artillery. Paratroopers established a defensive perimeter, with the tanks and tank destroyers acting as a mobile reserve to strengthen threatened sections of the line and to snuff out German penetrations.[60]

On 22 December the 4th Armored Division initiated an attack against the southern face of the German "bulge" with the mission of relieving the defenders of Bastogne. Two combat commands attacked abreast on separate routes, but both thrusts bogged down in the face of dogged German resistance. Maj. Gen. Hugh Gaffey, who had succeeded Wood as division commander, committed a third combat command to the fight (an unusual departure from the division's custom of employing just two combat commands). The Reserve Command, composed of one battalion each of tanks, armored infantry, and armored artillery, found a weak spot in the German line and pushed through. On 26 December a task force from the Reserve Command reached Bastogne. Although several days would pass before the corridor was secure, Bastogne remained in American hands.

Another vital road center was the town of Saint-Vith, Belgium. Brig. Gen. Robert W. Hasbrouck's 7th Armored Division conducted a forced march from the Netherlands to the vicinity of Saint-Vith on 17 December, arriving just before the Germans. There Hasbrouck took charge of a regiment from the shattered 106th Infantry Division, Combat Command B of the 9th Armored Division, and elements of three tank destroyer battalions (including some new 90mm M36s).[61] The defense of Saint-Vith differed from that of Bastogne, in that the 7th Armored did

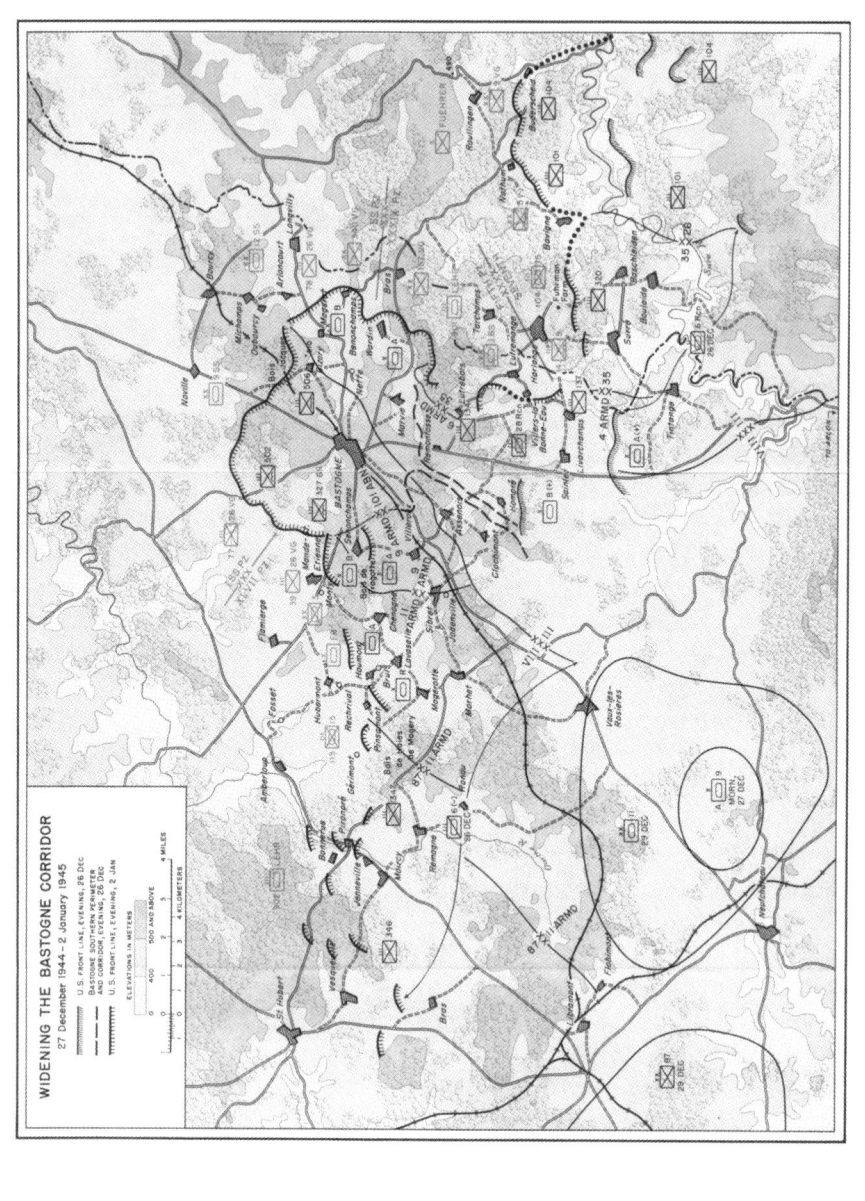

WIDENING THE BASTOGNE CORRIDOR
27 December 1944 – 2 January 1945

U.S. FRONT LINE, EVENING, 26 DEC
BASTOGNE SOUTHERN PERIMETER
AND CORRIDOR, EVENING, 26 DEC
U.S. FRONT LINE, EVENING, 2 JAN

ELEVATIONS IN METERS

400   500 AND ABOVE

0   1   2   3   4 MILES
0   1   2   3   4   5   6 KILOMETERS

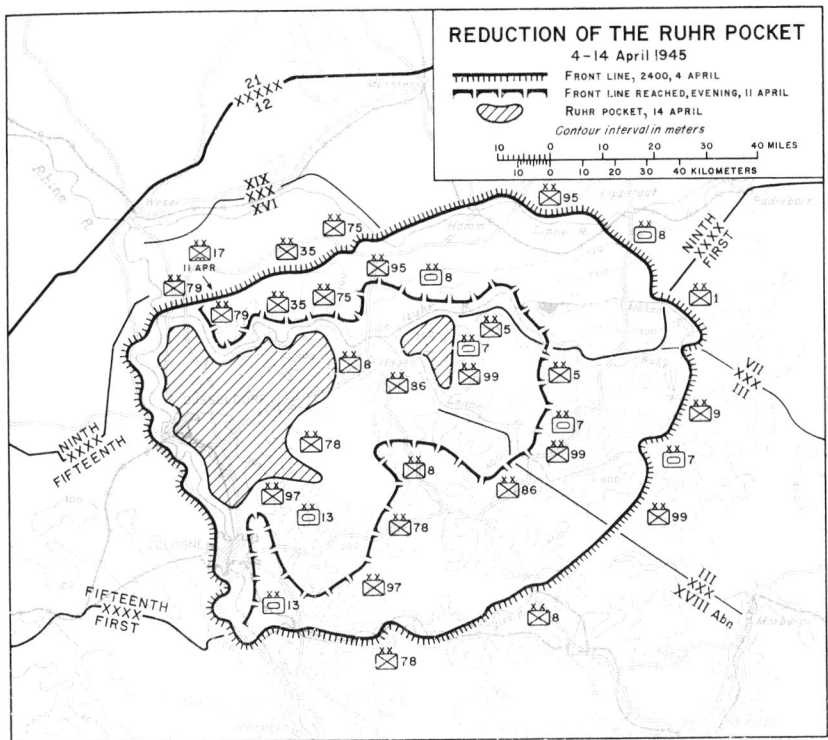

not attempt to hold a rigid perimeter. Instead, it practiced a mobile defense designed to absorb enemy thrusts by giving ground, then counterattacking with a mobile reserve of tanks and tank destroyers.[62] Although ultimately compelled to withdraw from Saint-Vith, the defenders held out for six days, time enough for other divisions to secure the northern shoulder of the bulge and contain the German offensive.

The Ardennes campaign marked Germany's last attempt to regain the initiative in World War II. With its reserves exhausted, the German Army was unable to mount an effective defense of the Rhine River. By late March 1945, the Allies held bridgeheads east of the Rhine and were ready to plunge into the heart of Germany. When the breakout came, American armor was allowed once more to execute the exploitation and pursuit mission at which it excelled.

The main Allied objective within Germany was the Ruhr industrial complex. Two American field armies stood poised to encircle the Ruhr Valley from their bridgeheads along the Rhine—Ninth Army to the north and First Army to the south. First Army erupted from its bridgehead on 25 March. Three corps advanced east, each with an armored division in the lead—VII Corps on the left, III Corps in the center, and V Corps on the right.

Leading the VII Corps advance was Maj. Gen. Maurice Rose's 3d Armored Division. He organized his division into the standard three combat commands, with two task forces per combat command. Each of the six task forces included a tank battalion. Three of the six included an armored infantry battalion apiece, the other three had battalions attached from the 104th Infantry Division. Rose fought four task forces at a time, keeping the other two in reserve. Each of the four task forces engaged with the enemy also included an armored field artillery battalion.[63]

On the day of the breakout, the 3d Armored passed through the lines of the 104th Infantry Division and almost immediately broke free of organized resistance. With swarms of tactical aircraft leading the way, the 3d Armored drove east for four days, scooping up hordes of prisoners. On 28 March alone the division packed fifteen thousand Germans off to prisoner-of-war cages.

First Army wheeled north on 29 March to form the lower jaw of the pincers around the Ruhr Valley. The 3d Armored Division led the way, with the 104th Infantry Division securing its flank and rear. The 3d Armored's advance elements covered forty-five miles that day without incurring a single casualty.

The next day, however, as it neared Paderborn, the 3d Armored ran into an ad hoc defense force composed of students from an SS training establishment. In the hard fighting that ensued, General Rose was killed while personally reconnoitering a route around the German position. On 1 April the 3d Armored Division linked up with the Ninth Army's 2d Armored, completing the Ruhr encirclement.

Organized German resistance in the west virtually ended with the closing of the Ruhr pocket. All along the front armored divisions surged across Germany, to be halted only by encounters with Soviet spearheads and the cessation of hostilities.

The war in Europe ended with sixteen American armored divisions in the field: one in Italy and the other fifteen on the western front. These divisions, together with the forty-four independent tank battalions and sixty-one tank destroyer battalions in Italy and western Europe represented roughly one-fourth of the American ground forces arrayed against Germany. Yet the scope of armor's contribution to victory in Europe is difficult to assess. Certainly tanks did not dominate ground combat in the manner envisaged by the founders of the Armored Force. And if casualty statistics are any measure, armor's role in the war was not particularly prominent. The sixteen armored divisions suffered a combined casualty total of 12,947 killed in action and died of wounds, and another 45,958 wounded.[64] The Army's total losses in the Mediterranean and European theaters (including the Army Air Forces) came to 162,314 dead and 498,040 wounded.[65] The hardest hit armored division was the 3d, with 2,126 dead and 6,963 wounded.[66] By comparison, the 3d Infantry Division, which sustained more casualties than any other Army division in World War II, lost 5,558 dead and 18,766 wounded[67]—approximately 160 percent of its authorized strength.

On the other hand, the historical record shows that armor figured prominently in most of the operations that made up the European campaign—from small-unit combat in the hedgerows and cities to the great encirclements and pursuits. Tanks provided an offensive punch to the American ground forces that was out of proportion to the number of armored units involved or the casualties they sustained.

The European campaign bore out the prewar Army's inability to embrace the concept of operational art—that is, organizing individual engagements and battles into a campaign. On the whole, higher-echelon commanders remained preoccupied with defeating the enemy in contact, and gave only general consideration to the problem of influencing the enemy's ability to wage the *next* battle. As a result, mechanization and motorization were probably not used to their fullest potential. Viewed on a large-scale map, much of this campaign was indistinguishable from World War I. The exceptions, however, were significant. The breakout and pursuit across France, and the encirclement of the Ruhr demonstrated that Allied commanders were beginning to awaken to the potential of the tool within their hands.

The significance of the war in Europe to the evolution of the Army's armor component is beyond question. One might say that the 1944-45 campaign was the defining experience of the armor arm as it exists today. Yet memory tends to be selective. Postwar commentators tended to remember the dashing, blitzkrieg-type operations rather than the slugfests in the Normandy hedgerows or the streets of Aachen. They tended to remember the Sherman tank's weaknesses as an antitank system, while forgetting that 70 percent of the rounds fired by the Sherman were high-explosive, not antitank, ammunition. Day in, day out, armor's chief contributions were in functions that armor doctrine said tanks should avoid: fighting in cities, reducing pillboxes, and generally operating at the pace of the infantry. The blitzkriegs stand out precisely because they were the exception rather than the rule.

But the armor establishment did learn some tangible, enduring lessons in the ETO during World War II. Foremost among these was the value of combined arms. Following the war, the separate tank battalion, which had proved so valuable in a variety of roles during the European campaign, was incorporated as an organic part of the infantry division. Moreover, the campaign showed that tanks could not operate as an independent force on the battlefield. The combat commands that drove across western Europe in 1944 and 1945 live on in the brigades of today's armor forces.

NOTES

1. Kent Roberts Greenfield, Robert R. Palmer, and Bell I. Wiley, *The Organization of Ground Combat Troops,* United States Army in World War II (Washington: Historical Division, U.S. Army, 1947), pp. 56-57, 64.

2. R. Elberton Smith, *The Army and Economic Mobilization,* United States Army in World War II (Washington: Office of the Chief of Military History [hereafter OCMH], 1959), p. 187.

3. Greenfield, et al., *Organization of Ground Combat Troops,* pp. 63, 71, 334.

4. Ibid., p. 161.

5. U.S. Army Tank Destroyer Center, "Tank Destroyer History" (Camp Hood, Tex.: 1945?), pt. 1, chap. 1, p. 16.

6. Greenfield, et al., *Organization of Ground Combat Troops,* p. 161.

7. For an evaluation of the tank destroyer program see Christopher R. Gabel, *Seek, Strike, and Destroy: U.S. Army Tank Destroyer Doctrine in World War II* (Fort Leavenworth, Kans.: U.S. Army Command and General Staff College, 1985).

8. U.S. Army, Office of the Chief of Ordnance Technical Division, *The American Arsenal* [based on Catalog of Standard Ordnance Items, 1944] (Mechanicsburg, Pa.: Stackpole Books, 1996), pp. 51, 180.

9. Army Ground Forces Study No. 27, "The Armored Force, Command, and Center," Historical Section, 1946, Records from the Adjutant General's Office, RG 407, NA, p. 30.

10. Christopher R. Gabel, *The U.S. Army GHQ Maneuvers of 1941* (Washington: CMH, 1991), p. 148.

11. U.S. War Department, "Table of Organization 17," 1 Mar. 1942.

12. U.S. War Department, FM 17-10, *Armored Force Field Manual* (Washington: GPO, 1942), p. 1.

13. Ibid., pp. 85-86.

14. Ibid., pp. 112-15, 117.

15. Ibid., p. 1.

16. Ibid., pp. 90, 247.

17. Ibid., p. 270.

18. Ordnance Technical Division, *American Arsenal,* pp. 13, 177.

19. Ibid., pp. 23, 179.

20. See Army Ground Forces Study No. 15, "The Desert Training Cen-

ter and California-Arizona Maneuver Area," Historical Section, RG 407, NA. For a compilation of division training and combat histories, see Shelby L. Stanton, *Order of Battle, U.S. Army, World War II* (Novato, Calif.: Presidio Press, 1984).

21. George F. Howe, *The Battle History of the 1st Armored Division* (Washington: Combat Forces Press, 1954), pp. 203-204.

22. For the Sidi Bou Zid actions, see George F. Howe, *Northwest Africa: Seizing the Initiative in the West,* United States Army in World War II (Washington: OCMH, 1957), pp. 410-22; and William R. Betson, "Sidi Bou Zid— A Case History of Failure," *ARMOR* (Nov.-Dec. 1982): pp. 38-44.

23. Howe, *Northwest Africa,* 559-60; U.S. Army Tank Destroyer School, "Tank Destroyer Combat," Camp Hood, Tex., n.d., Andrew D. Bruce Papers, USAMHI, Pa., pp. 17-30, and Maj. Allerton Cushman, AGF Observer Report, May 3, 1943 Documents Collection, Combined Arms Research Library, Fort Leavenworth, Kans., p. 1.

24. E.N. Harmon, "Notes on Combat Experience During the Tunisian and African Campaigns," typescript, Library, U.S. Armor School, Fort Knox, Ky.

25. Greenfield, et al., *Organization of Ground Combat Troops,* pp. 1, 416-17.

26. U.S. War Department, "Table of Organization 17" [and accompanying documents], 15 Sept. 1943.

27. U.S. War Department, FM 17-100, *Armored Command Field Manual: The Armored Division* (Washington: GPO, 1944). See especially p. 23.

28. Greenfield, et al., *Organization of Ground Combat Troops,* p. 161.

29. Ibid., p. 333.

30. Ordnance Technical Division, *American Arsenal,* pp. 29, 37.

31. U.S. Forces, European Theater, General Board, Study No. 53, "Tank Gunnery" [1946?], p. 43.

32. Ibid., p. 35 n.

33. Kenneth Macksey and John H. Batchelor, *Tank: A History of the Armoured Fighting Vehicle* (New York: Ballantine, 1970), pp. 89, 118-19, 128-29.

34. See Charles M. Bailey, *Faint Praise: American Tanks and Tank Destroyers during World War II* (Hamden, Conn.: Archon, 1983).

35. Ordnance Technical Division, *American Arsenal,* pp. 58, 180.

36. Ibid., pp. 60, 181.

37. For an excellent history of armor operations during the Italian campaign, see Howe, *Battle History of the 1st Armored Division.* Also see Donald E. Houston, *Hell on Wheels: The 2d Armored Division* (San Rafael, Calif.: Presidio, 1977), pp. 153-95; and Ernest N. Harmon, *Combat Commander: Autobiography of a Soldier* (Englewood Cliffs, N.J: Prentice-Hall, Inc., 1970), pp. 142-203.

38. Peter Chamberlain and Chris Ellis, *British and American Tanks of World War II* (New York: Arco, 1981), pp. 132-33; Richard P. Hunnicutt, *Sherman:*

A History of the American Medium Tank (Novato, Calif.: Presidio, 1978), pp. 422-29.

39. Gordon A. Harrison, *Cross-Channel Attack,* United States Army in World War II (Washington: OCMH, 1951), p. 309.

40. Michael D. Doubler, *Busting the Bocage: American Combined Arms Operations in France, 6 June-31 July 1944* (Fort Leavenworth, Kans.: U.S. Army Command and General Staff College, 1988), pp. 21-34.

41. Ibid., pp. 38-50.

42. U.S. Forces, European Theater, General Board, Study No. 50, "Organization, Equipment, and Tactical Employment of Separate Tank Battalions" [1946?], p. 6.

43. U.S. Forces, European Theater, General Board, Study No. 48, "Organization, Equipment and Tactical Employment of the Armored Division" [1946?], App. 1.

44. Ibid., p. 8.

45. Ibid., and George F. Hofmann, *The Super Sixth: History of 6th Armored Division in World War II and its Post-War Association* (Louisville, Ky.: Sixth Armored Division Association, 1975), pp. 439-42.

46. Doubler, *Busting the Bocage,* pp. 54-58; and Russell F. Weigley, *Eisenhower's Lieutenants: The Campaigns of France and Germany, 1944-1945* (Bloomington: Indiana University Press, 1981), pp. 156, 163. For a full account of Cobra, see Martin Blumenson, *Breakout and Pursuit,* United States Army in World War II (Washington: CMH, 1984 [1961]), pp. 224-80.

47. Blumenson, *Breakout and Pursuit,* pp. 272-80.

48. Headquarters, 4th Armored Division, Training Memorandum 23 Sept. 1944. See also Hal C. Pattison, "The Operation of CCA, 4th Armored Division, Normandy Beachhead to the Meuse River, 28 July-31 August 1944" (Command and General Staff College Report, 1946-47), Documents Collection, Combined Arms Research Library, Fort Leavenworth, Kans.. Also see, the Armored School, Student Research Reports, 1949-50, "Armor in the Exploitation: The 4th Armored Division Across France to the Moselle River," Fort Knox, Ky., Headquarters Army Ground Forces, RG 339, NA.

49. The Armored School, Student Research Reports, 1949-50, "Super Sixth in Exploitation: The 6th Armored Division from Normandy to Brest," Fort Knox, Ky., Army Ground Forces Records, RG 337, NA.

50. Michael D. Doubler, *Closing With the Enemy: How GIs Fought the War in Europe, 1944-1945* (Lawrence: University Press of Kansas, 1994), p. 116.

51. General Board, Study No. 50, p. 7.

52. U.S. Army Ground Forces, Immediate Report No. 58, 16-17 Sept. 1944.

53. Pattison, "Operation of CCA, 4th Armored Division," p. 39.

54. Doubler, *Closing With the Enemy,* pp. 90-91.

55. See, for example, *FM 17-10,* pp. 17, 142, and *FM 17-100,* pp. 91-93.

56. Derrill M. Daniel, "The Capture of Aachen," unpublished lecture,

Documents Collection, Combined Arms Research Library, Fort Leavenworth, Kans.

57. Doubler, *Closing With the Enemy*, pp. 102-104.

58. General Board, Study No. 48, p. 16.

59. Hugh M. Cole, *The Ardennes: Battle of the Bulge,* United States Army in World War II (Washington: OCMH, 1965), pp. 459-60.

60. Ralph M. Mitchell, *The 101st Airborne Division's Defense of Bastogne* (Fort Leavenworth, Kans.: U.S. Army Command and General Staff College, 1986), pp. 13, 58.

61. Cole, *The Ardennes*, p. 395.

62. J.D. Morelock, *Generals of the Ardennes: American Leadership in the Battle of the Bulge* (Washington: National Defense University Press, 1993), pp. 305-6.

63. The Armored School, Student Research Reports, 1949-50, "Armored Encirclement of the Ruhr: The 2d and 3d Armored Divisions," Fort Knox, Ky., RG 337, NA.

64. Compiled from Stanton, *Order of Battle.*

65. *The Army Almanac* (Washington: GPO, 1950), p. 665.

66. Stanton, *Order of Battle*, p. 51.

67. Ibid., p. 79.

SUGGESTED READINGS

There is an abundance of published material on American operations in the European theater of World War II. By far the most important tool for serious students is the series of official histories entitled United States Army in World War II. Individual volumes cover each campaign, plus other studies on War Department activities and on the technical services. Written by professional historians with access to both American and German records, the "greenbooks" set the standard for the genre. Russell F. Weigley drew heavily upon the official histories in writing *Eisenhower's Lieutenants: The Campaigns of France and Germany, 1944-1945* (Bloomington: Indiana University Press, 1981), a thoughtful and insightful analysis of strategy and operations. For a nuts-and-bolts examination of American tactics, techniques, and procedures, see Michael D. Doubler's *Closing With the Enemy: How GIs Fought the War in Europe, 1944-1945* (Lawrence: University Press of Kansas, 1994).

Record Group 407, Records from the Adjutant General's Office, at the National Archives in College Park, Maryland, contains the pertinent Army Ground Forces, Historical Section, Study No. 27, "History of the Armored Force, Command, and Center," 1946. This extensive study was written from the perspective of officers who were directly responsible for the development of American armor before and during World War II. Also at the Archives in College Park are a wide range of documents from the Armored

Center, Armored Board, and Armored School located in Record Group 337, Headquarters Army Ground Forces. In the Armor School, Student Research Reports, dated 1949-50, there are a number of excellent studies regarding armor operations in Europe.

The official lineage and honors for armor units can be found in Mary Lee Stubbs and Stanley R. Connor, *Armor-Cavalry*, pt. 1, *Regular Army and Army Reserve*, Army Lineage Series (Washington: OCMH, 1969). In addition, every armored division (except the 20th) has been the subject of at least one unit history. One of the best is George F. Howe's highly detailed tactical account, *The Battle History of the 1st Armored Division* (Washington: Combat Forces Press, 1954). Donald E. Houston produced a comparable study in *Hell on Wheels: The 2d Armored Division* (San Rafael, Calif.: Presidio, 1977). Combining tactical detail with firsthand color is George F. Hofmann's *The Super Sixth: History of the 6th Armored Division in World War II and its Post-War Association* (Louisville, Ky.: Sixth Armored Division Association, 1975). Unit histories exist for other divisions, and even some of the tank and tank destroyer battalions, but all should be examined with caution— many are essentially yearbooks intended more for unit members than as historical references.

Technical material on tanks and other weapons is also readily available. Armor guidebooks intended for modelers and war gamers abound, and are generally reliable. Peter Chamberlain and Chris Ellis produced one of the handiest reference tools on Allied armored vehicles of all types with their *British and American Tanks of World War II* (New York: Arco, 1969). Richard P. Hunnicutt's exhaustive developmental histories *Stuart: A History of the American Light Tank,* vol. 1 (Novato, Calif.: Presidio Press, 1992) and *Sherman: A History of the American Medium Tank* (Novato, Calif.: Presidio Press, 1978) treat all variants of the premier American tanks of World War II. Christopher R. Gabel's monograph *Seek, Strike and Destroy: U.S. Army Tank Destroyer Doctrine in World War II* (Fort Leavenworth, Kans.: U.S. Army Command and General Staff College, 1985), evaluates doctrine, force structure, and equipment. Charles M. Bailey explores the American failure to keep pace with weapons system evolution in *Faint Praise: American Tanks and Tank Destroyers during World War II* (Hamden, Conn.: Archon, 1983).

6

Marine Corps Armor Operations in World War II

Joseph H. Alexander

The Pacific War had several crucial turning points: Midway, the high tide of Japanese expansion; Guadalcanal, the first Allied offensive; and Saipan, which for the first time brought Tokyo within striking range of American B-29 bombers. Yet it was the bloody battle for Tarawa in November 1943 that proved to be the crossroads of the Pacific War in terms of armor tactics and technology. "Issue in doubt," reported the commanding general of Tarawa's landing force on the afternoon of D day, and indeed the battle hung in the balance for the first thirty hours. Tarawa was the one major battle in the Pacific after Guadalcanal that the Americans could have lost; it was the one full-scale amphibious assault that could well have been thrown back into the sea. Among a handful of decisive factors that swayed the balance was the innovative use of armor assets to overrun Japan's "Gibraltar of the Pacific."[1]

Tarawa represented not only the first combat employment of M4 Sherman medium tanks in the Pacific but also the initial availability of suitable sealift capability and landing craft to deliver them ashore during the critical first hour of the assault. Tarawa also represented the first tactical use of tracked landing vehicles (LVTs or amphibian tractors—"amtracs") to deliver assault troops over barrier reefs to the beachhead.

The lessons of Tarawa were almost prohibitively costly: twelve of fourteen Shermans lost on D day, and ninety of 125 LVTs knocked out in the three-day operation. However, the Marines studied the battle with painstaking honesty, adapted quickly, and generated both the technological and doctrinal fixes necessary to facilitate the larger assaults that soon followed across the breadth of the Pacific.[2]

The Pacific War imposed entirely different requirements upon the armor arm than did the ETO. Imperial Japan's tank units lacked the mass, mobility, and firepower of Germany's heralded panzer divisions. The Japanese 47mm antitank gun, effective enough in cave warfare, could hardly compare with the deadly German 88mm gun. This moderate threat accounts for the reason why American forces could fight the entire Pacific War with light and medium tanks. If the U.S. M4 Sherman proved of increasingly marginal value in shoot-outs with heavier German tanks in Europe, the same combat vehicle ruled Pacific battlefields after Tarawa.

The ETO was given strategic priority by the United States and its Allies. "Germany First" meant that Pacific armored forces typically had to fight with hand-me-down equipment and field improvisations until the America's ordnance productivity peaked in 1944.

However, the principal difference between armored operations in the two theaters was a matter of geography: the vast oceanic expanses and small islands of the Pacific versus the large continental battlefields of Europe and Africa. Japan's concentric rings of fortified islands required strong naval, air, and amphibious forces to overcome.

Because amphibious assaults occur in a prohibitive environment involving the risky buildup of combat power from ground zero and demand extraordinary cooperation between naval, air, and ground commanders, they are frequently described as the most difficult of military operations. As at Tarawa, armored forces would prove decisive in many of the epic amphibious assaults during the final two years of the war, but none of these successes came easily. Indeed, American commanders began the Pacific counteroffensive faced with three daunting requirements in the use of armor in amphibious operations:

How to get assault waves of infantry ashore across barrier reefs under fire

How to get tanks ashore early, in force

How to modify armored vehicles for specialized missions against enemy fortifications (flame units and bulldozer kits, for example)

Solving these problems satisfactorily demanded years of developmental work to produce combat vehicles adapted for the amphibious environment, specialized ships, and landing craft with which to deliver them to the objective, as well as joint doctrine acceptable to each of the armed services. The Marines, increasingly identified with the amphibious assault mission, would take the lead.

The Marine Corps entered World War II without a significant armor or cavalry tradition. A provisional tank platoon of M1917 light tanks served with the Marine Corps Expeditionary Force in the 1920s but folded after five years. Tanks were deployed to Culebra during the 1924 Fleet Winter Maneuvers, but they could only land administratively, dockside, in a distant port. The failure of the Christie amphibious tank to negotiate the surf zone during that same exercise highlighted the extreme difficulties involved in employing combat vehicles in landing operations.[3]

Nevertheless, the seeds of Marine armor operations had been planted. During the austere 1930s—as visionary Marine and Navy officers gathered in Quantico, Virginia, to formulate their seminal doctrine of offensive amphibious assaults—the idea of Marine tanks integral to the landing force remained attractive. The Marines had learned to appreciate the value of combined arms during their recent expeditionary adventures in the Banana Wars of Central America. Marine pilots had devised innovative dive-bombing techniques in support of infantry units engaging guerrillas in Haiti, Santo Domingo, and Nicaragua. Small tanks, planners believed, would add more mobile firepower to lightly armed, seaborne expeditionary forces. Tanks could also augment naval gunfire until field artillery units could be landed.

During these brainstorming sessions, planners were undaunted by the fact that suitable tank lighters did not then exist. The Marines's *Tentative Manual for Landing Operations,* approved by the chief of naval operations (CNO) in 1934, contained these far-sighted roles for landing force tank units: "The primary mission of tanks in the landing operation is to facilitate the passage of infantry through the immediate beach defenses by destroying enemy wire and machine gun defenses at or near the water's edge . . . . In addition tanks in adequate numbers should be provided to support the advance to the final objective. Their speed and maneuverability make them particularly effective for rapid exploitation." Tanks should therefore be landed in the assault waves of infantry, the manual recommended.[4] Drafted eight years before the first Marine tanks rolled ashore at Guadalcanal, this conceptual doctrine proved remarkably valid.

Armed with this new doctrine, and given a new organizational niche with the Navy with the creation of the Fleet Marine Force (FMF) in 1933, the Leathernecks began to gear up for what they perceived to be an inevitable amphibious war against Japan. The Marines established

A Marmon-Harrington Model 1938 two-man turretless light tank under-goes amphibious landing tests. Icks Collection, Patton Museum.

their first tank company in 1937 at Quantico, equipped with the Marmon-Herrington two-man, turretless light tank armed with three fixed machine guns. The tanks were a maintenance nightmare and an operational hazard—the machine guns were aimed by steering the tank at the target—but they were lightweight enough for the primitive amphibious ships of the era, and the Marines were back in the armor business to stay. Significantly, the tank units, as well as aircraft, artillery, and engineers, were integral components of the FMF from the beginning.[5]

The growing insistence of Army and Marine officers for greater ship-to-shore lift capacity to accommodate combat vehicles finally convinced the Navy Department to give more than token attention to amphibious warfare. The Navy Department's preoccupation with building a two-ocean combat fleet had understandably taken priority over what were then classified as auxiliary ships and craft. When experimental boats designed by the Navy swamped or capsized in field tests, the Marines turned to a private boat builder, Andrew J. Higgins of New Orleans, for assistance. Higgins in short order produced the durable Landing Craft, Vehicle, Personnel (LCVP) and Landing Craft Mechanized (LCM)—a heavy-lift, shallow-draft tank lighter—both of which proved to be so useful and reliable that after 1942 they dominated every landing in both theaters of war.[6]

The Marines also sought a lightly armored tracked amphibian capable of operating at sea, traversing the surf zone, and negotiating marginal terrain ashore. In 1937, a brief article in *Life* magazine about inventor Donald Roebling's tracked "Swamp Gator" caught the Marine Corps Commandant's eye. In remarkably short order, the Marines engineered a contract between Roebling and the Food Machinery Corporation of Dunedin, Florida, to produce a militarized version of the "Swamp Gator," which the Marines dubbed the LVT1 Alligator.[7]

In 1940, the Marines procured a number of Army M2A4 light tanks, which mounted a 37mm gun in a centerline turret and proved incomparably more useful than the Marmon-Herringtons. With inadequate numbers of the M2A4s available for Marine use, the initial Marine Corps tank TO&E specified an unwieldy mix of M2s and M3s with the unpopular Marmon-Herringtons.

The Marines thus entered World War II with a far-sighted if unproven doctrine of offensive amphibious assault that integrated the use of combat vehicles in the assault waves and the rudimentary components of the small armored components of the FMF. Two Marine divisions, the first in Corps history, each contained a tank battalion of the combined light vehicles, an amphibian tractor battalion of one hundred LVT1 Alligators, and a weapons company in each infantry regiment that included a pair of 75mm self-propelled guns mounted in half-tracks.

Both divisions (less a provisional brigade in Iceland) were still forming stateside when the war began. The forward-deployed Marine units at Guam, Wake, and in the Philippines—so immediately vulnerable after Pearl Harbor—had no armored units to help repel the Japanese. Not that the Corps's primitive armor units could have stemmed the flood tide of Japanese conquests in those early years. Indeed, the first Marine tanks to deploy to the South Pacific were a handful of toothless old Marmon-Herringtons that went as part of a defense battalion ordered to Samoa in April 1942. A platoon of M2A4s reinforced the Marine defenders at Midway during that pivotal battle the following month.

Shortly after the Battle of Midway, the 1st Marine Division arrived in New Zealand in preparation for their first offensive campaign—a landing on Guadalcanal in the Solomon Islands. The division had just undergone the transition to a new TO&E that eliminated the Marmon-Herringtons, added a tank battalion of seventy-two M2/M3 series light tanks, retained the amphibian tractor battalion, and included a scout company equipped with the White M3A1 armored car. The latter was a 110-horsepower wheeled vehicle acquired during a moment of strategic confusion when it appeared that the Corps might be destined for the North African deserts. The armored cars, totally unsuited for the jungle warfare that ensued, never saw combat. The Marine scouts first traded them to the New Zealand Army for Bren-gun carriers, then adapted jeeps and rubber boats for their missions.[8]

Operation Watchtower was the code name assigned to the Guadalcanal campaign by the Joint Chiefs of Staff (JCS), but the troops soon dubbed it "Operation Shoestring." The Marines, rushed into the commitment by the sudden appearance of an enemy bomber strip on Guadalcanal, hit the beach on 7 August 1942 with World War I–issue M1903 Springfield rifles and an assortment of early-vintage landing boats. The Japanese, taken completely by surprise, did not oppose the Guadalcanal landings, and the island's bomber strip was soon under new management.

But the Marines had *five* landings to conduct that day. The other four took place nineteen miles north across Sealark Channel (soon to be dubbed "Ironbottom Sound") at Tulagi and Gavutu-Tanambogo, adjacent to neighboring Florida Island. There, the *rikusentai* of the Kure 3d Special Naval Landing Force provided a grim foretaste of the war to come, defending their caves and spider holes to the last man and exacting a stiff casualty toll from the attacking Marines.

The Marines managed to get two M2A4 light tanks ashore under fire at Tanambogo, which was no small accomplishment, but their total lack of experience in working as a tank-infantry team led to disaster. One tank got too far ahead of its supporting infantry and, as the tank commander hesitated in the jungle growth, a *rikusentai* immobilized the vehicle by jamming an iron bar into the track. Other Japanese appeared out of the jungle and swarmed over the tank like fire ants. Using oil-soaked rags they set the vehicle afire, pulled the crewmen out, and hacked them to death. The driver was still fighting when the Marine infantry arrived and joined the brawl. Forty-two Japanese bodies lay sprawled around the disabled tank when the fighting ended.[9]

To their credit, the men of the 1st Marine Division learned from this incident. Increasingly, throughout their long commitments to Guadalcanal, Cape Gloucester, Peleliu, and Okinawa, the division became the premier tank-infantry force in the Pacific. None of the seven divisions fighting on Okinawa took better care of its tanks.

The division's M2/M3 light tanks performed well on Guadalcanal, essentially providing a mobile counterattack force for the threatened perimeter protecting Henderson Field. A tank platoon provided the coup de grace during the decisive Battle of the Tenaru on 21 August 1942, attacking frontally into the last pocket of resistance of the doomed Ichiki Detachment. Die-hard Japanese knocked out two Marine tanks, one with a slap-on magnetic antitank mine, but the survivors used shock action and devastating fire with 37mm canister rounds to terminate the vicious battle.[10]

In other battles around Henderson Field the light tanks had fewer soft targets to engage, and often had to have their firepower augmented by the 75mm half-tracks. Sometimes the tankers' own offensive spirit proved their undoing. On 14 September, after K Company, 3d Battalion, 1st Marines, repulsed a heavy attack by the Kuma Battalion, a platoon of six light tanks advanced unsupported down the ridge to the jungle's edge. Japanese troops unveiled 37mm antitank guns they had hauled by hand through the undergrowth and opened fire, destroying two tanks, disabling a third, and causing a fourth to tumble into the river, where it capsized, drowning the crew.[11]

The Marine LVT1 Alligators performed well at Guadalcanal but fulfilled a strictly logistical support role. The Alligators ensured the seamless delivery of combat cargo from ships offshore, through the surf, to inland supply dumps. When the ships of the amphibious task force disappeared following the disastrous Battle of Savo Island, the Marines used LVTs to distribute their limited supplies along the coastline and around the defensive perimeter. The Alligators could be relied upon to

LVT1 "Alligators" on their way to shore. *ARMOR* Magazine..

evacuate wounded through muddy swamps that would bog down all
other vehicles. Yet few Marines gave serious thought to transforming
the logistic support vehicles into assault vehicles. Their high silhouette
and thin (3/16-inch) armor made them vulnerable in an open fight. Afloat,
the underpowered engines were prone to flooding out in a choppy sea.
Bilge pumps worked only as long as the main engine ran. A dead engine at sea often resulted in a sunk vehicle—a fact that did not escape
the potential customers, the infantry.

The Japanese Seventeenth Army employed light and medium tanks
in its piecemeal attacks on the Marine positions around Henderson Field.
All came to grief. Marine infantrymen kept the Japanese crews buttoned
up with rifle and machine-gun fire until their own tanks or towed
antitank (AT) guns could find the range with high-velocity 37mm shells.
However, the Marines's AT weapon of choice at Guadalcanal remained
the 75mm gun on the half-tracks.[12]

The five-month battle for Guadalcanal caused enough casualties
from disease and enemy action to knock both the 1st and 2d Marine
Divisions out of action for nearly a year. Meanwhile, as fighting in the
Solomons proceeded "up the ladder" toward Rabaul in 1943, the light
tank platoons of the 9th, 10th, and 11th Marine Defense Battalions helped
spearhead Army advances. At Rendova, Munda, and Arundal,
Leatherneck tanks helped break a trail through the jungle for Army
troops, using 37mm canister rounds to clear underbrush, high-explosive shells to blow up Japanese log bunkers, and machine guns to cut
down escaping enemy survivors.[13]

Marines and Soldiers fighting the Japanese in the central Solomons began experimenting with mobilizing the cumbersome, man-packed flamethrower. At first they simply put the operator and his weapon on a combat vehicle, then they tried to adapt the flame dispenser to the ball mount of the bow machine gun in the Stuart light tank. Neither idea worked. The short range of the man-pack flame unit unduly exposed the operator and the tank, and vehicle vibration made the ball-mount modifications misfire. In one case, while a Marine light tank thus modified was supporting the 112th Cavalry on the Arawe Peninsula, the Leatherneck flame operator became so irate when the flamethower's fuel fired but failed to ignite that he risked his life by opening his hatch and tossing a thermite grenade at a thoroughly saturated bunker to spark the blaze.[14]

By the late spring of 1943, American forces in the Pacific had halted the Japanese offensive and taken the measure of the best of Imperial Japan's air, sea, and ground forces. But American offensive campaigns had been limited, bare-bones affairs, always (after Guadalcanal) undertaken within the protective umbrella of land-based air support. The Guadalcanal operation itself was best characterized as a strategic offensive coupled with a tactical defensive. Few subsequent amphibious landings were seriously opposed. The fledgling amphibious assault doctrine had not yet been validated. American tanks and amtracs had yet to participate in a true "Storm Landing" in the Pacific. As the JCS's Joint Planning Staff studied its maps of the maritime road to Tokyo, all of this was soon to change.

The Joint Planning Staff faced two unknowns as it evaluated the feasibility of opening a second front in the Central Pacific. The vast expanses of the Central Pacific would require each major campaign to operate beyond the range of land-based tactical aircraft. The advent of *Essex*-class fleet carriers and self-sustaining logistics forces might offset that disadvantage. But the problem of coral reefs surrounding the target islands in the region remained vexing. As good as the Higgins boats proved to be, none was able to traverse a barrier reef with less than a four-foot clearance. The staff asked the Marines to determine whether logistical LVTs could make it to shore through high surf and over a barrier reef.

Lt. Col. Victor H. Krulak was ordered to test four LVT1s in New Caledonia in May 1943. Field tests were hairy to execute but proved the vehicle's utility. Krulak reported that fully loaded LVTs could indeed negotiate a ten-foot plunging surf while crossing a coral reef. Encouraged by this news, the Joint Planning Staff issued a memo advocating an amphibious campaign to seize the Marshalls, suggesting that LVTs be used to assault coral atolls. The Joint Chiefs agreed, but decided to launch the campaign against the smaller, closer Gilbert Islands.[15]

In August the 2d Marine Division, still recuperating from Guadalcanal in New Zealand, received a warning order to prepare to assault heavily fortified Betio Island in the Tarawa Atoll in the Gilberts. Operation Galvanic, the first step in the Central Pacific campaign, would be the first major, opposed amphibious assault of the war.

Major Gen. Julian C. Smith commanded the 2d Marine Division and knew from the outset that he would need to convert his logistical LVTs into assault craft for the mission. However, the ravages of Guadalcanal had taken a toll on the division's Alligators. Only seventy-five were still operable. Many more would be needed to land sufficient assault troops on Betio. Smith directed his staff to make field modifications to the seventy-five functional vehicles. These included adding sheet-iron to protect the cabs, installing more machine guns, improving exhaust systems, and rigging stern-mounted grapnels to drag away barbed wire. Smith also submitted an urgent request for newer-model LVT2 Water Buffaloes known to be languishing on a pier in San Diego.

The Water Buffaloes offered an improved suspension system, a lower silhouette, and greater payload capacity, but their delivery to the Tarawa landing force could not occur any sooner than the morning of D day—if then. Smith's landing plan thus included two courses of action: one involving just the seventy-five LVT1s, and the other adding fifty LVT2s to the assault force. Meanwhile, Smith also requested a company of the new Sherman medium tanks and the means to get them ashore at Betio.[16]

The M4 Sherman medium tank had been in production since 1942, but the "Germany First" policy made them scarce in the Pacific until late in the spring of 1943. Everything about the Sherman appealed to the Marines—especially its 75mm gyro-stabilized main gun and improved armor protection—but nagging questions remained about its thirty-four-ton weight. Could Shermans operate in the jungle? Could they be expeditiously landed in amphibious operations characterized by urgent speed of execution? In May 1943, the Marine Corps adopted a new medium tank battalion TO&E that included four companies with fourteen Shermans each, but retained the light tank organizations until these questions could be answered in combat.[17]

Amphibious planners of all services knew that employing Sherman tanks in opposed amphibious assaults on Pacific islands would demand specialized landing ships. Thanks to the foresight of Andrew Higgins, the standard tank lighter, the LCM3, had been built big enough to carry both light and medium tanks, and there were plenty of LCM3s available. But until Tarawa there were precious few landing *ships* available in the Pacific—the ramped, flat-bottomed Landing Ship Tank (LST) or Landing Ship Medium (LSM)—or the revolutionary Landing Ship Dock (LSD), with its stern well gates and floodable well decks. These ugly, rough-riding auxiliaries would quickly change the nature of amphibious assault.

Until Tarawa, tanks were embarked on board the few available attack transports (APAs) or attack cargo ships (AKAs), usually stored in the square of the aftermost hatch. Since neither type ship possessed a ramp, the tanks had to be off-loaded one at a time by the ship's single, high-capacity, swinging boom. The process was hazardous and time-consuming. The ship's crew had to clear the hatch by launching the LCM3s, then bring the first LCM alongside, open the hatch, secure a four-point heavy cable to the lifting eyes of the tank, and then swing it out of the hatch, over the side and down into the heaving boat. It was a shoehorn fit into a moving target. The time it took to execute a single transfer—one tank into one boat—varied according to the skill of the winch crews and the sailors manning the steadying lines, the mechanical condition of the boom and winch, and the set and drift of the seas. A very good crew could do it in ten minutes. However, choppy seas always complicated the process. So did enemy fire. Launching a company of fourteen Sherman tanks in this fashion, even from a half-dozen ships, was subject to all of war's friction.[18]

The advent of the LSD changed all that. The same fourteen Shermans could literally "go to war" in their assigned LCM3s. The lighters would simply beach themselves at the embarkation point, the Shermans would back aboard, and the boats would swim into the flooded well deck of the LSD. The ship would then raise the stern gate, empty the water from the well deck, and boats and tanks would ride high and dry, griped down for the passage. In the objective area, the ship would reverse the process: flood the well deck, lower the stern gate, and launch the preloaded boats directly for the beach or reef. All fourteen Shermans could thus be launched in less time than it took to load one with a swinging boom.

Tank landing ships were even better in that they could beach themselves in relatively shallow water (three feet by the bow), open their bow doors, drop the ramp, and launch the tanks directly ashore, or at least into shallow water, thus doing away with boats altogether. But Japanese defenses in the Central Pacific rarely permitted this luxury on D day. Any LST seeking to approach a "hot" beach close to H hour at Tarawa, Saipan, or Peleliu would have been blown out of the water.

Marines favored launching their Shermans by the LSD method,

and the Navy's first LSD, the USS *Ashland*, was called on to deliver the first Shermans to Tarawa. Meanwhile, the Marines learned to use LSTs to transport and launch their LVTs seaward of the Line of Departure. Prior to Tarawa LVTs, like tanks, were launched from transports with a swinging boom. By contrast, the three LSTs employed at Tarawa launched their fifty LVT2s in less than five minutes. In a critical combat situation like this, with thin-skinned amphibious ships clustered close to an enemy island, lacking the relative safety of the open sea, the savings in launch times for tanks and LVTs often spelled the difference in surviving the inevitable Japanese aerial counterattacks.[19]

The contribution of landing ships to the employment of armor in amphibious operations is best illustrated by comparing the assaults on Bougainville and Tarawa in November 1943. Both were division-level operations, and both landing forces were augmented with freshly arrived Sherman medium tank companies. But at Bougainville, located in the upper Solomons, the amphibious task force consisted of a dozen APAs and AKAs. No major landing ships were available. Because of the proximity of the Cape Torokina beachhead to Japanese airfields, the landing force had a maximum of six hours to off-load from its ships on D day. Priority went to unloading 90mm antiaircraft guns. In the absence of landing ships, and aware of how dangerously long it would take to off-load his tanks the old way, the division commander had to leave his medium tanks back at Guadalcanal, out of action. Three companies of light tanks eventually landed by the tedious swinging-boom method, but none before the sixth day of the battle.[20]

The availability three weeks later of one LSD and three LSTs for the Tarawa operation made a world of difference. Launching the boated Shermans and the new LVT2 Water Buffaloes worked to perfection. The rampant confusion that characterized the pre–H hour activity in the transport area stemmed from having to launch the seventy-five Alligators and light tanks the old way, plus the difficulty encountered in marrying the infantry up with their assigned LVTs in the darkness. Later, given sufficient LSTs for the Saipan invasion, the landing force had the luxury of embarking assault troops with their LVTs on the landing ships, thus avoiding time-consuming, hazardous transfers between small craft in the open sea.

Tarawa proved to be an extremely tough nut to crack. A well-led, well-armed regiment of *rikusentai* defended tiny Betio Island with blazing tenacity. Many things went wrong for the Southern Attack Force. The heralded preliminary bombardment by ships and planes failed to destroy the island's five hundred pillboxes and heavy weapons emplace-

The detritus of war: LVT1 and LVT2 amphibious tractors and dead Marines beached at Tarawa. Naval Institute.

ments. A rare, seemingly inexplicable, apogean neap tide prevented passage over the reef by any boat during the first thirty hours. And no one knew if General Smith's field-converted cargo LVTs for the first three assault waves would work.

General Smith employed eighty-seven LVTs in the assault (the balance were spares, control craft, or simply too late in getting launched the old way). Surprisingly, despite a ten-mile, five-hour odyssey into the lagoon to attack Betio's least-protected shore—and despite an ill-advised cease-fire by the amphibious task force commander twenty minutes before touchdown—the LVTs crossed the barrier reef and hit the beaches, losing only eight vehicles to the stunned enemy gunners.

This was an undeniable turning point of the battle. Fifteen hundred Marine riflemen now clung to toeholds along Betio's north shore. They represented the thin margin of success for the next day and a half. All the rest of the assault troops and all supporting arms and reinforcements, embarked in Higgins boats blocked by the reef, would have to wade ashore from five hundred to seven hundred yards out. The Japanese met them with a hail of fire, inflicting grievous casualties on the attackers and thoroughly disrupting their formations. When General Smith tried to recycle the LVTs to shuttle the stalled Marines in from the reef, the Japanese recovered from their initial shock and cut loose with antiboat guns and horizontally fired antiaircraft guns at point-blank range. The 2d Amphibian Tractor Battalion lost its valiant Commander,

The M4 Sherman tank "Colorado" was damaged by shellfire at Tarawa, then supported the 8th Marines after being repaired. Marine Corps University Archives.

Maj. Henry C. Drewes, half its officers and men, and three-fourths of its combat vehicles. The few LVTs that survived the raking enemy fire soon ran out of gas, leaving the troops hugging the reef no alternative but to wade ashore.[21]

Marine tanks were sorely needed to suppress enemy fire and expand the beachhead. Thanks to the utilitarian LSD *Ashland*, all fourteen Shermans were boated and approaching the reef in timely fashion, eight on the left flank, six on the right. General Smith, having tested the reef-climbing abilities of the new medium tank by dispatching the first two Shermans to Fiji the previous month, felt confident that the vehicles could debark from their LCMs at the reef and trundle across. This they did, but the milky-green waters of the lagoon shoreward of the reef contained hidden hazards deadlier than the enemy fire.

The Shermans had arrived in the theater without fording kits. Not until the later landings in the Marianas in mid-1944 would Marine

Shermans wade ashore sporting the characteristic double stacks aft of the turret designed to protect their engine air intake and exhaust systems and allowing them to negotiate water six to seven feet deep. Absent such modifications at Tarawa, the tanks were in jeopardy in water over four feet deep. Hundreds of deep craters dotted the offshore approaches to Betio, a result of the inevitable near misses from the high-level, preparatory bombardment. Many of these holes were deep enough to swallow a tank, even a Sherman, and the turbidity of the water prevented the drivers from spotting the craters until it was too late. For this reason, the Marines decided to employ "human depth gauges," guides on foot who led the tanks ashore from the reef. If the guide suddenly disappeared, the driver would know a crater lay just ahead.

Unfortunately, the heavy fire and confusion on D day played havoc with the planned use of tank guides. On the left flank, the boat carrying the guides became separated from the tanks, ran aground on the reef, and suffered high casualties from machine-gun fire. The tanks pressed on unescorted, and one sank in an unseen crater, drowning the crew. On the right flank, Japanese gunners shot down guide after guide. Finally, Sgt. James R. Atkins volunteered to lead the column of six tanks under daunting fire. He died at the water's edge, his mission accomplished, but the passage the engineers had cleared for the tanks through the sea wall had become clogged with dead and dying Marines. The tank Commander chose not to grind over his fellow Leathernecks and turned the column around, looping back into the lagoon in search of a better access site. Without Sergeant Atkins or someone of his caliber to lead the way, four of the six tanks ran into bomb craters and sank. Ironically, no Shermans were lost to enemy fire in the risky run from the reef to the beach, but a third of the vehicles were lost in craters because of the lack of fording kits.[22]

In the fighting ashore, the Marines's inexperience with tank-infantry coordination led to a wasteful squandering of most of the nine surviving Shermans. Waved forward unescorted on the left flank, three tanks were knocked out by a hidden 75mm field gun, two fell prey to "friendly" fire from U.S. Navy carrier planes, and another, nearly blind without infantry, tumbled into a large crater and caught fire. On the right flank, a Sherman won its duel with a Japanese Type 95 light tank but at an embarrassing cost. The doomed Japanese tank fired one last round at its moment of destruction, a miracle shot that went squarely down the main-gun tube of the victorious Sherman. And in all corners of the battlefield, communications between tankers and riflemen were nonexistent because the tank radios could not match frequencies with the infantry's. Here was another irony. No Marine tank Commanders fell to enemy fire while in their armored turrets. The casualties in their ranks all occurred when the tankers had to dismount to coordinate with the infantry.[23]

From such an inauspicious beginning the 2d Marine Division learned how to use its few remaining Shermans to great avail. Enterprising maintenance crews worked throughout the nights to cannibalize the shot-up or sunken hulks for replacement parts—including a replacement 75mm gun for the tank disabled by the Type 95 tank's dying shot. Infantry began to cluster around each vehicle, banging on the hull with empty shell cases as crude signals. Using tanks, naval gunfire, field artillery, and carrier aviation with increased confidence, Maj. Michael P. Ryan led his patchwork battalion along the entire western coast on the second day, clearing the way finally for General Smith to land reinforcements with full unit integrity. Medium tanks helped withstand a ferocious Japanese counterattack on the third night and spearheaded the final sweep of the eastern tail of the island to end the savage seventy-six-hour battle.

The division's M2 and M3 light tanks, although valorously manned, contributed little to the final victory. They proved too thinly armored and too lightly armed against Japanese crew-served weapons and log and concrete bunkers. Nor did they get ashore any easier than the Shermans. Several fell victim to spectacular midrange sniping by a Japanese 127mm dual-purpose gun operating on the left flank on D day. The gunner fired single rounds at each of four LCMs laden with light tanks as they approached the reef and sank them all—with their tanks— in deep water.[24]

The 2d Marine Division suffered more than three thousand casualties in seizing Tarawa Atoll, but the experience validated the new doctrine of offensive amphibious operations and demonstrated the key role of armored units in such seaborne assaults. The candid assessments of battalion commanders after Tarawa shaped much of the way the amphibious war against Japan would be waged in the Central Pacific. LVTs would spearhead the assault waves, but many more of them would be required to support each division (up to three hundred), and not just the field-modified cargo versions, but new, factory-built assault craft with improved armor protection, gun shields, and a stern ramp. Assault waves would soon be preceded ashore by armored amphibians, the so-called amphibious tanks, with centerline turrets mounting 37mm guns—LVT(A)1s—or 75mm howitzers—LVT(A)4s.[25]

In the same critical assessment, the Sherman became the tank of choice for the Marines for the remainder of the war. The Stuart light tanks would persist, partly because of the valuable canister round avail-

An LVT(A)1 with 37mm gun. Naval Institute.

able only for their 37mm gun, and partly as experimental flamethrowers. Tarawa, if nothing else, had demonstrated the crucial need for mobile, protected flamethrowers to counter Japanese defensive skills and their fierce willingness to die in place. Finally, improved tank-infantry coordination would take top priority in amphibious training—including the provision of compatible radios and initial experiments with a bustle-mounted tank-infantry telephone.[26]

The generous sharing of the lessons learned so painfully at Tarawa by the 2d Marine Division enabled the new 4th Marine Division to prepare for the Marshall Islands campaign with confidence and focus. Operation Flintlock began barely two months after Tarawa ended, but the landing force had plenty of LVTs—including the first LVT(A)1s— and plenty of Shermans (plus the light tanks modified with experimental flamethrowers). It also had the landing ships needed to transport the entire lot. With greatly improved communications, naval gunfire, and amphibious underwater reconnaissance units available, Flintlock succeeded handsomely at extremely low costs. Tank-infantry coordination showed much improvement. Marine infantry and tankers easily dispatched three Japanese light tanks that sallied forth on Parry Island.[27]

An M3A1 light tank equipped with a flamethrower performs at New Caledonia in 1943. *ARMOR* magazine.

After seizing the Marshalls, the JCS decided to bypass heavily fortified Truk in the Carolines and strike instead at the distant Marianas, a full thousand miles closer to Tokyo and well within Japan's "Sphere of Absolute National Defense." Operation Forager was the campaign to seize Saipan, Tinian, and Guam. The islands were large and mountainous in comparison to the coral atolls of the Gilberts and Marshalls. For the first time, the Central Pacific Force would encounter major elements of the Imperial Army, including veteran and very proficient heavy artillery components.

The fact that Forager coincided in time and competed for resources with Operation Overlord, the epic landings in Normandy, France, is significant. Forager was a huge operation in its own right, employing two Army divisions, three Marine divisions, and a very large Marine brigade supported by a thousand ships. At no time before the summer of 1944 could the United States have launched simultaneous invasion forces of such size and scale. But the nation's wartime production and distribution systems had finally begun to function at full capacity. Neither theater would suffer for lack of essential resources.

At Saipan, for example, the 2d and 4th Marine Divisions would land on 15 June, carried ashore by nearly six hundred LVTs launched directly seaward of the Line of Departure from fifty LSTs. More than a hundred LVT(A)1s and LVT(A)4s led the assault waves ashore, firing their 37mm and 75mm guns directly at beach defenses and the nearest low hills. Many of the amphibian tractors were LVT4s, which featured a stern debarking ramp. The lessons learned from Tarawa continued to illuminate the ship-to-shore movement. There were sufficient LVTs for all of the assault waves, the time from launch to touchdown was reduced from five hours to twenty minutes, and the sturdier new vehicles were able to negotiate a wicked plunging surf and the barrier reef, and then survive a hellacious rain of Japanese artillery and mortar shells to deliver the troops ashore. All at a loss of less than two dozen LVTs.[28]

The medium tank battalions now organic to both Marine divisions rode to war in style, embarked on five LSDs. While bona fide fording kits would not appear until the Tinian landing, the Marines improvised sealants and extensions to protect the Shermans during the helter-skelter dash from the reef to the beach. The Shermans landed early and went right to work, already sorely needed. The original plans for LVT(A)s and LVTs to continue the assault inland to clear the beaches for the follow-on waves came a cropper when the Japanese began raking the invasion beaches with large-caliber weapons. No penetrations came easily at Saipan on D day, but those that did resulted from good tank-infantry coordination and provided breathing room on the congested beaches.

The 1st Battalion, 6th Marines, received the brunt of an armored counterattack by the Japanese 9th Tank Regiment on the second night of the invasion. Colonel Takashi Goto led forty-four tanks, mostly mediums equipped with 47mm main guns, against the Marine perimeter at 3 A.M. in one of the largest armored actions of the Pacific War. A company of Shermans moved up to meet the challenge, and the Marines added artillery, self-propelled 75mm guns, towed 37mm guns, and infantry weapons to the melee. Maj. James A. Donovan Jr., the battalion's executive officer, recalled: "The battle evolved itself into a madhouse of noise, tracers, and flashing lights. As tanks were hit and set afire, they silhouetted other tanks coming out of the flickering shadows to the front or already on top of the squads." The Marines sustained heavy casualties but held their ground as the Shermans, with their superior firepower, exacted a heavy toll on their less-well-armored opposite numbers. The Leatherneck tankers destroyed twenty-four enemy tanks in the three-hour battle.[29]

The Japanese commander on Tinian similarly squandered most of his tanks in a counterattack against the beachhead on the first night of the invasion there. However, the amphibious corps commander anticipated this assault and ensured that the 4th Marine Division not only landed sufficient organic artillery and tank units before dark, but augmented the perimeter defense with the 2d Division's tank and artillery units as well. The Japanese attack was well prepared and systematically executed—not at all like the wasteful banzai attack launched at the end of the three-week-long Saipan campaign—and the resulting combat has been rightfully described as "The battle of Tinian." Here, again, the combination of Marine supporting arms, nicely augmented by naval gunfire and star shells from the fleet, proved too much for the attackers. The failure of the Japanese attack broke the back of Tinian's defenders. For the next several days the Marines enjoyed the rarest of combat experiences in the Central Pacific: riding on tanks roaring pell-mell down country roads in pursuit of a fleeing enemy. Everything clicked at Tinian. The island fell in nine days with only light casualties suffered by the invaders.[30]

Guam, the biggest of Forager's three target islands, took longer to secure and cost more. There, the Army's 77th Infantry Division joined the 3d Marine Division and 1st Provisional Marine Brigade in an amphibious pincer movement that benefited from the longest preliminary naval bombardment of the war. The combination of surf and barrier reef made off-loading Shermans from LCM3s particularly hairy. Marine armored amphibians provided yeoman service under fire by anchoring the landing craft to the reef to facilitate off-loading. Other LVTs guided the tanks past submerged shell craters.

Japanese commanders on Guam deployed their limited tank assets more sensibly than in any other island battle in the Central Pacific. Resisting the temptation to squander the lot in one glorious charge, the Japanese instead employed their tanks in small units, usually at night, and always making maximum use of jungle cover and concealment. The Shermans prevailed in most one-on-one encounters (although one Sherman struck a thousand-pound bomb buried as a mine and literally disintegrated), but the GIs and Marines were fighting die-hard Japanese tanks until the last day of the three-week-long battle.[31]

Operation Forager demonstrated convincingly that the Americans had solved most of the problems of getting tanks and infantry ashore under heavy fire. After Forager, Gen. Douglas MacArthur's massive Philippines campaign would draw down the number of LSDs and other landing ships available for simultaneous operations in the Southwest and Central Pacific theaters, but experiments then underway with improved fording kits and flotation devices seemed encouraging. The

Marines also learned that while armored amphibians could never take the place of tanks ashore, they were instrumental in providing initial suppressive fires for the assault landing, followed by aggressive water-borne patrolling and close-in fire support along each coastal flank.

The major problem confronting armored forces in the western Pacific by late 1944 was the increasing lethality of Japanese defenses. Marine tankers could sheathe the sides of their Shermans with lumber to foil Japanese suicide crews armed with magnetic, slap-on mines, or festoon the entire topside with spare track blocks, sandbags, and sheet iron to reduce their vulnerability to the damnably accurate Japanese Model I (1941) 47mm antitank guns. But the greatest need was to develop a flamethrower that could be married to the Sherman turret. The M3A1s equipped with the "Satan" flame system had proven useful in the Marianas, but they were still light tanks.

Operation Stalemate, the campaign to seize Peleliu and Angaur in the southern Palaus, was a bitter pill, a reversion to the earlier, fumbling days of the war. Intended as a quick blitz to cover MacArthur's flank as he returned to the Philippines, the fighting on Peleliu stretched on for ten weeks, cost ten thousand casualties in the 1st Marine Division and the Army's 81st Infantry Division, and produced few strategic dividends.

The 1st Marine Division had executed the bare-bones landing at Guadalcanal, then returned to the South Pacific in late 1943 to assault Cape Gloucester. There the Marines made innovative use of their first Sherman tanks, including using them as "gunboats" during the long-range, shore-to-shore assault on the Willaumez Peninsula by positioning them atop stacks of dunnage so they could fire over the tops of the ramps on the LCMs. At Peleliu, however, there were insufficient LSDs, and the Marines had to leave a third of their Shermans behind, a costly penalty. The division's new LVT(A)4 armored amphibians did not arrive until the eve of embarkation. Lt. Col. Kimber H. Boyer, commanding the 3d Armored Amphibian Tractor Battalion, used factory blueprints of the new vehicles to train his crews.[32]

A reinforced Imperial Army infantry regiment defended Peleliu. The men were veterans of the Manchurian border fighting against the Russians, and quite likely the most formidable opponents any American landing force faced in the Pacific War. Their fighting skills, coupled with a sagacious use of the forbidding, cave-dotted Umurbrogol highlands, made Peleliu almost unassailable. Worse, the American preliminary naval bombardment, so brilliant at Guam, reverted to barely three days of ineffectual shelling. When lionhearted U.S. underwater demolition teams blew channels through minefields off the landing beaches, the Japanese commander dispatched equally brave swimmers of his own to replant hundreds of mines just before H hour. In their haste, however, most of the swimmers failed to arm the mines properly; otherwise, D day might have been twice as bloody.[33]

While the Japanese commander chose to lay low in the highland caves and fight a protracted battle of attrition, he did expend one infantry battalion and a tank company to bleed and disrupt the American landing. These doomed stalwarts performed brilliantly, knocking out close to sixty LVTs and amphibious trucks (DUKWs) and inflicting heavy casualties among the assault waves. The Marine Shermans fared better. Although struck repeatedly by Japanese antitank rounds on their run from the reef to the beach, the reinforced bow armor and rounded obliquity saved all but three of the Shermans. The tables were reversed later in the day when the Japanese tankers charged across an open airfield and provided the Marines with a welcome shooting gallery. The Japanese light tanks were shot to pieces.[34]

The Marines fighting on Peleliu experimented with the U.S. Navy Mark I flamethrower, a larger, longer-range weapon of greater duration and lethality. Developed for use in LVTs, the Mark I would have been spectacularly effective against Tarawa's defenses, which were so closely stacked along the beachfront. The system came along about a year too late. At Peleliu, the defenses were farther inland and less concentrated—plus, few LVTs in the initial waves got ashore unscathed. The Marines used the flame LVTs in the fighting ashore with limited success, but the amphibian tractors remained as vulnerable to the full panoply of enemy weapons as ever and thus required an inordinate amount of close support. Flame LVTs were not the solution to Japanese cave warfare waged inland from the sea. Nor were the small capacity, difficult-to-aim flamethrowers mounted in the bow machine-gun port of a Sherman tank.[35]

Despite Peleliu's high cost and the residual bitterness it engendered, the 1st Marine Division learned valuable lessons about tank-infantry attacks in rugged terrain that later served it well in the Okinawa campaign.

Meanwhile, the epic assault on the sulfuric fortress of Iwo Jima loomed, and the Marine Corps urgently sought to solve the deficiencies of mechanized flamethrower operations before embarking the landing force. The chief problem was how to mount a serviceable flame projector in place of the medium tank's main gun, thereby capitalizing on the mechanical traverse, elevation, and fire control features available in the turret. The concurrent problem was to avoid giving the flame tank a distinctive configuration that would attract fire from sharp-eyed enemy gunners before it could close the range. This meant the flame weapon would have to fire *through* the Sherman's 75mm tube. In Hawaii, an enterprising ad hoc committee of Marine tankers, Army Chemical Warfare Service officers, and Navy SeaBees assembled to tackle the problem. They quickly designed, tested, and adapted the POA-CWS-H1 flame system, which could fire a 6 percent napalm solution eighty yards through a salvaged 75mm gun tube. They then mounted the weapon on eight M4A3 Shermans. These special weapons subsequently proved invaluable when employed against the defenders of Iwo Jima's convoluted "Jungle of Stone." All would be hit, but none would be lost.[36]

Iwo Jima was the toughest and costliest amphibious assault of the Pacific War, thirty-six days of unremitting direct assault against a well-armed, well-sited enemy. Three Marine divisions (the 3d, 4th, and 5th) were assigned to seize the island, their ranks filled with veterans of the first three years' fighting. Their combined experience would be tasked to the limit to overcome Iwo Jima's steep beaches, powder-soft sand, and extensive minefields, tunnels, and caves. Japanese defenders would methodically inflict twenty-four thousand casualties on the landing force, the only battle in the Pacific War in which the Marines would sustain higher casualties than the island garrison they assaulted. In turn, the Marines would convincingly seize Japan's most heavily defended island in five weeks, a key outpost known as "the doorkeeper to the Imperial capital."

Marine LVTs—448 of them—delivered the eight thousand assault troops ashore in good order on D day, then kept the operation flowing by evacuating wounded and bringing in more ammo on the return run through the surf zone. Marine tanks, self-propelled guns, and armored bulldozers worked under the protective eye of riflemen, forward observers, and air spotters to scratch out cave after cave. All three tank battalions teamed up to storm across the airfield on the middle of the island against what seemed to be a solid sheet of antitank fire in the largest consolidated tank attack in Marine history. But Marine armor's greatest contribution came from the eight improvised "Zippo" flame tanks. According to Capt. Frank C. Caldwell, commander of a rifle company in the 26th Marines, "It was the flame tank more than any other supporting arm that won this battle."

Late in the campaign, as the 5th Marine Division cornered the last remnants of the Japanese 109th Division in "The Bloody Gorge," the 5th Tank Battalion expended napalm fuel at the rate of ten thousand gallons per day. The medium flame tank, integrated into an assault team

An M4 Sherman medium tank detonates mines on Okinawa with its main gun. *ARMOR* magazine.

of gun tanks, bulldozer tanks, riflemen, and combat engineers, provided the only way to bring the protracted struggle to an end.[37]

On Okinawa, 650 miles farther west, Lt. Gen. Mitsuru Ushijima followed the reports of the battle of Iwo Jima with keen interest. The commander of the Japanese Thirty-second Army knew that Okinawa would be the next major target for American amphibious forces. "The enemy's strength lies in his tanks," Ushijima concluded, and accorded antitank training a high priority for his hundred thousand troops. Ushijima also decided not to contest the American landings but to defend instead a concentric ring of interior defenses centered on the ancient Shuri castle complex. Once again Japanese soldiers began digging into the ground with hand tools. The broken terrain would become a killing ground of horrific proportions.

The U.S. Tenth Army, 182,000-strong, was created to assault Okinawa. Four Army and three Marine divisions (the veteran 1st and 2d, and the brand-new 6th) were Tenth Army's major components. Four divisions—two Army and two Marine—would land abreast on Okinawa's southwest beaches on 1 April 1945, April Fool's Day.

The Marines reorganized their standard division just before the Okinawa campaign. The M7 105mm self-propelled howitzer replaced the venerable 75mm half-track in the regimental weapons companies. The full-tracked M7 would provide enhanced mobility to Marine armored task forces and provide a welcome punch in direct-fire slugging matches against Okinawa's many "fire-port" caves. Tank battalions were now fully equipped with forty-six Sherman mediums, but the model varied according to combat preference. The 1st Tank Battalion opted to retain their diesel-engine M4A2s because they were leery of the fire hazard posed by the gasoline engine on the M4A3. The 6th Battalion considered the fire hazard an acceptable tradeoff in view of the improved power of the M4A3's five-hundred-horsepower Ford V-8 engine. But factory-built Sherman flame tanks had yet to reach the FMF, and the eight field-modified Zippo tanks used on Iwo Jima were too badly shot-up for use in another campaign.[38]

Into this gap rolled the Army's B Company, 713th Armored Flamethrower Battalion, newly outfitted with Sherman M4A1s mounting the modified Mark I flame system. One platoon would support the 6th Marine Division and two others were attached to the 1st Marine Division. It was a case of love at first sight. The Army tanks performed

An M4A1 "Zippo" flamethrower tank blasts a Japanese cave on Okinawa. *ARMOR* magazine.

brilliantly as the Marines provided the same smothering infantry and artillery protection as their predecessors had done at Iwo Jima. Each Army component won specific mention in the Presidential Unit Citations awarded the two Marine divisions.[39]

The Americans did not know in advance that General Ushijima was not defending Okinawa's beaches. Anticipating a major battle for the beachhead caused each division commander to insist on early delivery of his tanks. But the sheer size of the four-division landing complicated the problem of getting tanks ashore early in the assault. There were insufficient LSDs to accommodate the hundreds of tanks involved. Moreover, the newer model M4A3 Shermans would no longer fit in the standard LCM3. A newer, larger landing craft, the LCM6, appeared in time, but not in sufficient numbers. Marine tanks would have to hit the reef in some combination of landing craft or LSMs with a five-tank capacity, or risk employing the experimental flotation devices.

The Marines had heard about the disaster at Normandy, where the Direct Drive deepwater kits failed and twenty-six of twenty-nine Shermans in one battalion sank. They were more receptive to the experimental T-6 flotation device, which seemed more seaworthy and at least permitted the tank to fire all its weapons, even when wearing the ungainly long pontoons. The Marines adorned forty-five of their Shermans with these devices, despite the inordinate installation time (twenty-two hundred man-hours per tank). At least the T-6-equipped tanks could deploy to the battle on LSTs, which were in plentiful supply.[40]

An enormous armada delivered the landing force ashore under heavy protective fires on 1 April. The landings benefited from all the amphibious lessons learned during the Pacific War. More than fourteen hundred LVTs and LVT(A)s carrying GIs and Marines streamed into positions, crossed the reef without incident, and landed their troops over eight miles of beaches. Within an hour the Tenth Army had sixteen thousand assault troops ashore. Opposition was unbelievably light. The two valuable airfields at Yontan and Kadena fell the first day. Tanks indeed landed early, although there were horror stories aplenty. One LST skipper launched his Shermans in their T-6 flotation devices ten miles at sea, a bizarre standoff distance. Several sank after a five-hour odyssey. But most came ashore unscathed, blew off their cumbersome devices with built-in light explosives, and joined the advance across the island.[41]

For the Marines, the next several weeks were an idyllic reprieve as fast-moving mechanized forces swept the entire northern half of the island, pausing only to overrun a determined regiment of die-hard defenders on Mount Yae Take. The Army meanwhile discovered the true nature of the Japanese defenses and began battering against the formidable Shuri Line. When one division lost twenty-two of its thirty Shermans in a wild melee near Kakazu village, Lt. Gen. Simon Bolivar Buckner Jr., the Tenth Army Commander, asked for the 1st Tank Battalion as replacements. The Marines demurred, not wanting to dissolve the tight bonds between tankers and infantry the 1st Marine Division had forged so painfully at Peleliu. Instead, the entire division went into the line. Shortly afterward, both Marine divisions shouldered into the line—half of a four-division drive down Okinawa's narrow southern neck.

The battle for Okinawa became an eighty-day meat grinder. Tanks were invaluable but suffered heavy losses to Japanese antitank guns, artillery, and mines. By now the Japanese were well aware that the thinnest armor on a Sherman was its sides and rear. Many of the defensive complexes were designed to be mutually supporting, allowing hidden antitank gunners to take flanking shots at attacking Shermans with their high-velocity 47mm guns. Progress was slow and costly. For weeks the entire front advanced an average of fifty-five yards a day. In its first twenty days attacking the Shuri complex, the 1st Tank Battalion fired an average of twenty-five hundred rounds of 75mm ammunition and eleven hundred gallons of napalm-thickened flame fuel per day. Rifle squads were assigned to guard individual tanks, and did so zealously. In open country, where riflemen had zero cover, the Marines would forego the infantry and instead embark an artillery forward observer in one of the tanks to call down airburst munitions over the buttoned-up vehicles to discourage Japanese sapper attacks.[42]

The 6th Tank Battalion suffered its worst losses during the week-long battles for Sugar Loaf Hill and the Oroku Peninsula. In the latter operation, two Shermans were blown to pieces by direct hits from an eight-inch naval gun hidden in a nearby cave. At Kunishi Ridge, the 1st Tank Battalion provided the only possible relief to infantry forces clinging to the summit after a successful night attack but in dire need of reinforcements and with dozens of wounded to evacuate. The tankers trimmed their crews down to two men—commander and driver—squeezed in a half-dozen replacement riflemen, drove through the maelstrom to the ridge top, and delivered the reinforcements and retrieved the wounded via the driver's escape hatch, then carried the casualties to safety.

The Marines lost fifty-one Shermans and a tank retriever to enemy fire during the battle for Okinawa. Many more tanks were hit (one Sherman survived seven hits from a 47mm gun), but these were sal-

An LST discharging a LBT(A)4 with 75mm Howitzer. *ARMOR* magazine.

vaged and repaired by indefatigable maintenance crews who often raced Japanese sappers to reach a disabled tank. Both in the highly mobile sweeps through the north country, and the deadly, day-by-day, yard-by-yard advance in the south, Marine armor on Okinawa contributed mightily to the final victory. Maj. Gen. Lemuel C. Shepherd Jr., commanding the 6th Marine Division, stated after the battle, "if any one supporting arm can be singled out as having contributed more than any others during the campaign, the tank would certainly be selected."[43]

Marine armored forces had come a long way since the handful of four-ton Marmon-Herringtons first clanked across the training field at Quantico in 1937. The visionary Marines who first drafted the *Tentative Manual for Landing Operations* had foreseen both roles performed by Marine armor: first in direct support of the infantry, and second as a combat force in its own right, massed to exploit a breakthrough to the final objective.

As the Marines prepared for the inevitable invasion of the Japanese home islands, they voiced their opinions about the heralded new M26 Pershing heavy tank, with its 90mm gun and twenty-four-inch wide track, which had arrived too late for the fighting on Okinawa. Marine tankers agreed that the Sherman needed more armor protection, more mobility, and a wider track. But they liked the Sherman's high-velocity

75mm gun just fine, especially in view of the limited enemy armor threat they faced throughout the Pacific War. They also looked askance at the Pershing's forty-six-ton weight. The Marines had difficulty crossing many native bridges on Okinawa with the thirty-four-ton Sherman; surely the Japanese countryside would impose the same limitations. For that reason, the Marines suggested acquiring at least one company of the Army's new M24 Chaffee light tanks for each battalion. And—most importantly—legitimate, factory-built flame tanks of their own!

Similarly, the Marines had progressed steadily in development of tracked amphibians to lead the assaults across the Pacific. From the early days when Donald Roebling's prototype Alligator first eased into the ocean, the nation had manufactured 18,616 LVTs of all types. For the invasion of Japan, the Marines would have relied principally upon the Borg-Warner LVT3 (which arrived in theater out of sequence after the LVT4 but possessed an improved transmission and a greater payload) and the LVT(A)5, the newest armored amphibian, with a power traverse for the turret and a gyro-stabilizer for the 75mm howitzer.[44]

The atomic bombs mercifully precluded a protracted invasion of Japan. The Marines would enter the Cold War equipped with a proven mix of tanks and LVTs and a wealth of offensive combat experience—appropriate equity for the nation's premier amphibious assault force.

Notes

1. For a full account of this capstone battle, see Joseph H. Alexander, *Utmost Savagery: The Three Days of Tarawa* (Annapolis, Md.: Naval Institute Press, 1995).

2. 2d Tank Battalion, 2d Marine Division, Special Action Report, Tarawa, 14 Dec. 1943; and 2d Amphibian Tractor Battalion, 2d Marine Division, Special Action Report, Tarawa, 23 Dec. 1943, Archives Section, Marine Corps Historical Center, Washington, D.C. (Hereafter MCHC.)

3. Arthur E. Burns III, "The Origins and Development of U.S. Marine Corps Tank Units, 1923-1945," Marine Corps Command and Staff College, 2 May 1977, pp. 2-15.

4. HQMC, *Tentative Manual for Landing Operations,* 9 July 1935, pp. 225-26. Similar words appear in U.S. Navy, Fleet Training Publication 167, *Landing Operations Doctrine,* 1938, which codified much of the *Tentative Manual* and remained in effect throughout World War II.

5. Burns, "Origins," pp. 20-23. Marine tank battalions remained an integral component of the division from their creation in 1941 throughout the war.

6. The full account of the development and production of these vital landing craft is told in Jerry E. Strahan, *Andrew Jackson Higgins and the Boats that Won World War II* (Baton Rouge: Louisiana State University Press, 1994).

7. Victor J. Croizat, *Across the Reef: The Amphibious Tracked Vehicle at War* (Reprint, Quantico, Va.: Marine Corps Association, 1992), pp. 31-32; and Alfred Dunlop Bailey, *Alligators, Buffaloes, and Bushmasters: The History of the LVT Through World War II* (Washington: History and Museums Division, Headquarters, Marine Corps, 1986), pp. 34-4.

8. Burns, "Origins," p. 47. D-series Table of Organization, Marine Division, 1 July 1942, Reference Section, MCHC; and Brooke Nihart, "Armored Cars and the Marine Corps," *Fortitudine*, XXI (Fall 1991): pp. 12-15. Also see, interview with Col. Edward J. Driscoll, USMC (Ret.), veteran of division scout company operations in Guadalcanal and Tarawa, 13 Dec. 1993. The division scout company transitioned into the division reconnaissance battalion in 1944.

9. Frank O. Hough, Verle E. Ludwig, and Henry I. Shaw Jr., *Pearl Harbor to Guadalcanal: History of U.S. Marine Corps Operations in World War II,* vol. I (Washington: HQMC, 1957), p. 269.

10. Ibid., p. 291.

11. Ibid., p. 308. Richard B. Frank, *Guadalcanal: The Definitive Account of the Landmark Battle* (New York: Random House, 1990), p. 242; Interview with Col. Robert J. Putnam, USMC (Ret.) (commanding K Company, 3d Battalion, 1st Marines, throughout Guadalcanal), 9-10 Sept. 1990.

12. Frank, *Guadalcanal*, p. 350.

13. Charles D. Melson, *Condition Red: Marine Defense Battalions in World War II* (Washington: History and Museums Division, Headquarters Marine Corps, 1996), pp. 15-16, 28.

14. Leonard L. McKinney, "Mechanized Flame Thrower Operations in World War II," unpublished manuscript, Historical Office, Office of the Chief, Army Chemical Corps, 14 Feb. 1951, pp. 4-5.

15. Victor H. Krulak, "Tests of Amphibian Tractor under Surf and Coral Conditions," report to CG, I Marine Amphibious Corps, 3 May 1943, Archives Section, Marine Corps University, Quantico, Va. (hereafter MCU); Joint Planning Staff Report # 205, "Operations Against the Marshall Islands," ABC 384, RG 165, NA.

16. Alexander, *Utmost Savagery*, pp. 83-88.

17. Burns, "Origins," pp. 56-7.

18. The author served two years as Combat Cargo Officer on board a World War II–vintage attack transport in the western Pacific.

19. Alexander, *Utmost Savagery*, Chap. 5.

20. Commanding General, 3d Marine Division, "Combat Report of the Bougainville Operation," 21 Mar. 1944, Sections II and III, Archives Section, MCU.

21. 2d Amtrac Battalion Tarawa Report.

22. Interview with former Sgt. Hubert D. Crotts, USMC, 2d Tank Battalion, Tarawa, 6 Mar. 1997; 2d Tank Battalion Tarawa Report. Sergeant Atkins received a posthumous Navy Cross.

23. Ibid. See also Joseph H. Alexander, "Baptism by Fire: Sherman Tanks at Tarawa," *Leatherneck* 76 (Nov. 1993): pp. 34-37.

24. 2d Tank Battalion Tarawa Report, p. 6.

25. 2d Amphibian Tractor Battalion to CG, 2d Marine Division, "Amphibian Tractors, Recommended Changes in Equipment," 22 Dec. 1943, Archives Section, MCHC. The term "amphibious tank," while widely used to describe the armored amphibians or LVT(A)s of World War II, is a misnomer. Marine Corps amphibious vehicles must operate in the open ocean and surf zone. True tanks are too heavy, and armored amphibians, being light enough to float, are too thinly armored to qualify as a tank. Quarter-inch armor plating does little to deflect a heavy machine-gun bullet at zero obliquity. It is best to regard the LVT(A) as an assault gun, albeit an amphibious and damned useful one!

26. CG, 2d Marine Division, "Recommendations Based on Tarawa Operation, Number 4—Tanks," 2 Jan. 1944, Archives Section, MCHC.

27. Bailey, *Alligators, Buffaloes, and Bushmasters,* pp. 126-27, 142.

28. Croizat, *Across the Reef,* p. 243; Joseph H. Alexander, *Storm Landings: Epic Amphibious Battles of the Central Pacific* (Annapolis, Md.: Naval Institute Press, 1997), pp. 62-63, 71.

29. James A. Donovan, "Saipan Tank Battle," *Marine Corps Gazette* 32 (Oct. 1948): p. 26, and Henry I. Shaw Jr., Bernard C. Nalty, and Edwin T. Turnbladh, *Central Pacific Drive: History of U.S. Marine Corps Operations in World War II,* vol. III (Washington: HQMC, 1966), pp. 284-27.

30. Alexander, *Storm Landings,* pp. 83-84.

31. Shaw, et al., *Central Pacific Drive,* pp. 441, 562-67.

32. Larry L. Woodard, *Before the First Wave: The 3d Armored Amphibian Tractor Battalion, Peleliu and Okinawa* (Manhattan, Kans.: Sunflower University Press, 1994), pp. 86-90.

33. Underwater Demolition Team No. 6, "Anti-Invasion Mines on Peleliu," 1 Oct. 1944, Archives Section, MCU.

34. George W. Garand and Truman R. Strobridge, *Western Pacific Operations: History of U.S. Marine Operations in World War II,* vol. IV (Washington: HQMC, 1971), p. 110.

35. Ibid., p. 155; and McKinney, "Mechanized Flame Thrower Operations," pp. 83-90, 263-65.

36. Richard P. Hunnicutt, *Sherman: A History of the American Medium Tank* (Belmont, Calif.: Taurus Enterprises, 1978), pp. 405-406.

37. Joseph H. Alexander, *Closing In: Marines in the Seizure of Iwo Jima* (Washington: History and Museums Division, HQMC, 1995), p. 37.

38. Burns, "Origins," p. 92.

39. CG, Tenth Army Action Report, Ryukus Campaign, chap. 11, Staff Section Reports, sec. IX, Infantry and Tanks, pt. 2, Tanks, 6 July 1945, pp. 11-IX-5-12; and Benis M. Frank and Henry I. Shaw Jr., *Victory and Occupation: History of U.S. Marine Corps Operations in World War II,* vol. V (Washington: HQMC, 1968), pp. 885-86.

40. 6th Tank Battalion Special Action Report on Okinawa Operation, Phases I and II, vol. II, 15 June 1945, 5, 39, Archives Section, MCU; Hunnicutt, *Sherman,* pp. 422, 430; and Konrad F. Schreier Jr., *The Classic Sherman* (Canoga Park, Calif.: Grenadier Books, 1969), pp. 66-67.

41. Joseph H. Alexander, *The Final Campaign: Marines in the Victory on Okinawa* (Washington: History and Museums Division, HQMC, 1996), p. 34.

42. Burns, "Origins," pp. 93-94; and 1st Tank Battalion, "Notes Concerning Employment of Tanks With Other Arms, Period 6-21 May 1945 on Okinawa Southern Front," 23 May 1945, Archives Section, MCU, pp. 5-6. One illuminating conclusion therefrom: "The 75mm tank gun, using point fire into apertures and entrances and high-capacity tank flamethrowers have proved the most effective weapons for the wholesale destruction of deep caves and dug-in bunkers."

43. Frank and Shaw, *Victory and Occupation,* p. 386.

44. Bailey, *Alligators, Buffaloes, and Bushmasters,* p. 248; and Croizat, *Across the Reef,* p. 165.

7

Post–World War II and Korea: Paying for Unpreparedness

Philip L. Bolté

When World War II ended with Japan's surrender on 2 September 1945 there were only two superpowers in the world, the United States and the Soviet Union. America's leaders soon concluded they must contain what they perceived as a remorseless expansionist tendency in the policies of the Soviet Union.[1]

Meanwhile, in keeping with the traditional American view that armed forces are used to destroy occasional and intermittent threats, the American public clamored for demobilization.[2] Secretary of the Navy James Forrestal summed up what was happening when he observed that the country "was going back to bed at a frightening rate, which is the best way I know to be sure of the coming of World War III."[3] In spite of such concern among the nation's leaders, the public was going to get what it wanted. President Harry S. Truman proposed on 25 September 1945 to reduce the Army from 8 million to 1.95 million by the following June. Then, in March 1946, the War Department announced a proposed Army of 1.07 million by June 1947, with four hundred thousand of those to be transferred to the planned independent air force.

To further support such drastic reductions, the United States adopted a strategy of deterrence based on the atomic bomb monopoly the country then enjoyed.[4]

By 1948 the mounting tension between the Soviet Union and the United States was evident in such far-flung places as Greece, Iran, China, and Korea. Unfortunately, America's armed forces had reached a dangerous state of readiness. As Gen. George Marshall put it, "We are playing with fire and have nothing with which to put it out." Army strength had sunk to 552,000, with most of the troops deployed on occupation duty. The Army's contribution to the strategic reserve was a little over two divisions—a forty-thousand-man force that included only one armored combat command.[5] The Armored Force component in Europe was the U.S. Constabulary, consisting of three brigades. The basic building blocks for these units were squadrons composed of a light tank troop, a motorcycle platoon, and a horse platoon. The four infantry divisions on occupation duty in Japan each had only one company of light tanks.[6] It was the postwar low point of the active Army's Armored Force.

With the growing truculence of the Soviet Union and the 1947 establishment of the Truman Doctrine (the president's determination that America should help free peoples everywhere respond to communist aggression) American policy had reached a turning point.[7] One result was a modest increase in the strength of the armed forces. For the Armored Force, this meant strengthening the 2d Armored Division and activating the 3d Armored Cavalry Regiment in the United States, and converting the Constabulary brigades into three armored cavalry regiments in Europe.[8]

Meanwhile, the Army had been gleaning lessons from its World War II operations, in spite of postwar troop strength reductions and the emergence of the national strategy of nuclear deterrence. The result of this effort was reflected in both organizational changes and in the direction taken in weapons system development.

One of the earliest and most significant steps in this effort was made by the General Board that convened in Europe immediately after the war. One of the board's reports, "Organization, Equipment and Tactical Employment of the Armored Division," was important to the future shape of the Armored Force. The study confirmed the value of armored divisions and the tactics for employment developed during the war.[9] It also included a number of specific recommendations concerning organization and equipment. Some of these were ultimately adopted, while others were modified for later implementation.

Concerning organization, the General Board recommended that armored division strength be increased to that of the wartime heavy division. It favored a regimental structure that combined tank and infantry battalions, although the board members were unable to agree on the exact ratio to be established. The board recommended increasing divisional cavalry strength to a squadron, as well as increasing the strength of artillery and engineer units within the division. It also recommended that elements normally attached during operations be made organic to the division. Additional aircraft for liaison and command were recommended.[10]

Regarding future tank requirements, the thrust of recommendations was for three functional types of tanks: reconnaissance, exploitation, and infantry support. With the move toward higher velocity tank guns, the board felt that tank destroyers were no longer needed.[11]

The board also produced a study on "The Organization, Equipment and Tactical Employment of the Infantry Division" that affected the future Armored Force. The board recommended inclusion of an armored regiment of three tank battalions in the infantry division, while doing away with the infantry regimental antitank companies. The logic was that the best antitank weapon was a medium tank.[12]

Additional studies after the war resulted in an armored division that bore a striking resemblance to the 1940 organization, which had been divided into five functional components: command, reconnaissance, striking, support, and service. Two combat commands and a reserve command would be assigned battalions as the tactical situation dictated. Each of the three medium tank battalions had four medium tank companies instead of the earlier organization of three medium tank companies and one light tank company. This increased the division's medium tank strength from 168 to 216. A heavy tank battalion with three tank companies, each having four platoons, was added to the division. The division included a reconnaissance battalion and four armored infantry battalions with four rifle companies each, as well as three battalions of 105mm howitzers and one battalion of 155mm howitzers, all self-propelled.[13]

The Army also introduced a light armored cavalry regiment to replace the mechanized cavalry groups of World War II. The regiment included a headquarters company and three reconnaissance battalions. Each battalion consisted of a headquarters and service company, three reconnaissance companies, and a medium tank company. The regiment was designed to perform reconnaissance and security missions at the corps level, but was equipped and trained to engage in offensive and defensive operations, either mounted or dismounted.[14]

The other major organizational element in the Armored Force structure during the postwar period was the armored group. Consisting of a headquarters company and three tank battalions, the group was intended to be an administrative organization that provided its battalions for attachment to other units.[15]

Rather than accepting the board's recommendation of an armored regiment in each infantry division, the Army decided to add a tank company to each infantry regiment and a heavy tank battalion to the division structure.

Even before the war ended, Army planning to meet postwar weapons system requirements had started. In January 1945 the Army Ground Forces Equipment Review Board convened in Washington to consider requirements for all equipment to be used by ground troops and the Army Air Forces in direct support of land forces. The board recommended the development of light, medium, and heavy tanks weighing 25, 45, and 75 tons, respectively. Late in the year a new board headed by Gen. Joseph W. Stilwell reviewed and agreed with the findings. In addition, the Stilwell Board recommended the termination of all developmental efforts regarding tank destroyers and towed antitank guns. The board also recommended the development of special tank engines and improved running gear. Postwar tank development in the United States followed these recommendations until the appearance of the Army Equipment Development Guide in 1950.[16]

The last wartime light tank was the M24 Chaffee. It was lightly armored and armed with a 75mm gun. Although it was far superior to the M5 Stuart it replaced, the Army recognized that a new light tank was still required: a highly mobile, powerfully armed, reconnaissance vehicle. The postwar development program led to the production of the M41 series of light tanks, armed with the high velocity 76mm M32 cannon. Production, however, was not initiated until 1951.[17]

The last standard tank model fielded by the U.S. Army at the end of World War II was the M26 Pershing. It did not, however, reflect the latest design ideas. Its greatest deficiency was that it was underpowered. The five hundred horsepower engine used in the thirty-five-ton Sherman proved to be inadequate for the forty-five-ton Pershing. Although the Army recognized the need for a new tank, funding restrictions—coupled with the fact that there were two thousand M26s in the Army's inventory—resulted instead in the initiation of a product improvement program.[18] The result was the M46 Patton tank. However, although much improved, the M46 was still armed with only a 90mm gun, which disappointed proponents of a larger main gun. The fiscal year 1949 budget included funds for eight hundred new tanks. In addition, it was planned that 1,215 M26 tanks would be converted to M46s in fiscal year 1950.[19]

Although there was considerable experimental work with heavy tanks earlier in World War II, it was not until the middle of 1944 that combat reports from Europe concerning German heavy panzers stimulated a new interest in the Army. Procurement of 1,152 T29 heavy tanks was initiated in April 1945. Weighing over seventy tons when combat

M24 Chaffee light tank with 75mm gun. Patton Museum.

loaded, the T29 had seven inches of frontal armor, a 105mm gun, and was driven by a V-12 liquid-cooled engine rated at 770 horsepower. When the war ended the production contract was canceled.[20]

The 1945 requirement for a new postwar heavy tank specified a weight of about seventy-five tons, a five-man crew, and a gun not exceeding 90mm but capable of penetrating ten inches of armor at thirty degrees obliquity at a range of two thousand yards with special ammunition. While some experimental work was accomplished in the first years following the war, it was not until 1948 that a serious development effort was made. By then the Army and the Marine Corps had rejected the seventy-ton tank concept and began looking instead for a tank of approximately fifty-eight tons. Specifications called for a lightweight 120mm gun and the 810 horsepower Continental AV-1790 engine, possibly supercharged to 1,040 horsepower. This tank was designated the T43; however, the first prototype was not delivered until June 1951. Modifications based on testing resulted in the T43E1 tank, with a production schedule calling for delivery of three hundred tanks in 1953 and 1954. However the Army would not accept the tank until ninety-eight additional modifications were made. This ultimately resulted in the vehicle being type classified as the M103 in 1956.[21]

Clearly the Army's weapons system development and procurement programs, including that of tanks, were slowed in the postwar years because of the government's reliance on airpower and nuclear strategy. As stated by Gen. Omar N. Bradley, "The Army's role . . . was to support the strategic air offensive. Our main job would be to protect our air bases at home and abroad. . . . Much later, . . . the Army would occupy Western Europe and Russia in order to help restore law and order and stable government."[22] The Army would pay dearly as a result of that strategy.

On 25 June 1950 the Democratic People's Republic of Korea (DPRK) dealt the world's post–World War II complacency a shocking blow when it launched an invasion of the Republic of Korea. A stunned Ambassador John Muccio sent from Seoul, the South Korean capital, the following message: "It would appear from the nature of the attack and the manner in which it was launched that it constitutes an all-out offensive against the Republic of Korea."[23]

The division of that long-troubled land along the 38th parallel originally was intended to keep Soviet forces from advancing farther down the Korean Peninsula at the close of World War II by establishing zones of responsibility for the Russians and Americans to accept the surrender of Japanese forces in Korea.[24] However, the parallel, which neatly divides the peninsula in half, soon became the de facto boundary between two countries, with the North supported by the communist bloc and the South by the United States. It was one of a number of places in the world where the communist bloc and the free world shared a common boundary.

President Truman saw the invasion in the context of a threat to democracy and international peace. He wrote in his memoirs: "If the Communists were permitted to force their way into the Republic of Korea without opposition from the free world, no small nation would have the courage to resist threats and aggression by stronger Communist neighbors. If this was allowed to go unchallenged, it would mean a third world war."[25]

On 27 June the Republic of Korea (ROK) appealed to the United Nations for assistance. At the time, the Soviet delegation was boycotting the Security Council. Without the Soviets present to veto such action, the Council condemned the North Korean attack and called on UN members to furnish assistance to the South Koreans.[26]

Meanwhile, President Truman had authorized General MacArthur in Japan to use air and naval forces to support the South Koreans.[27] On 30 June, in response to an urgent request from MacArthur, the presi-

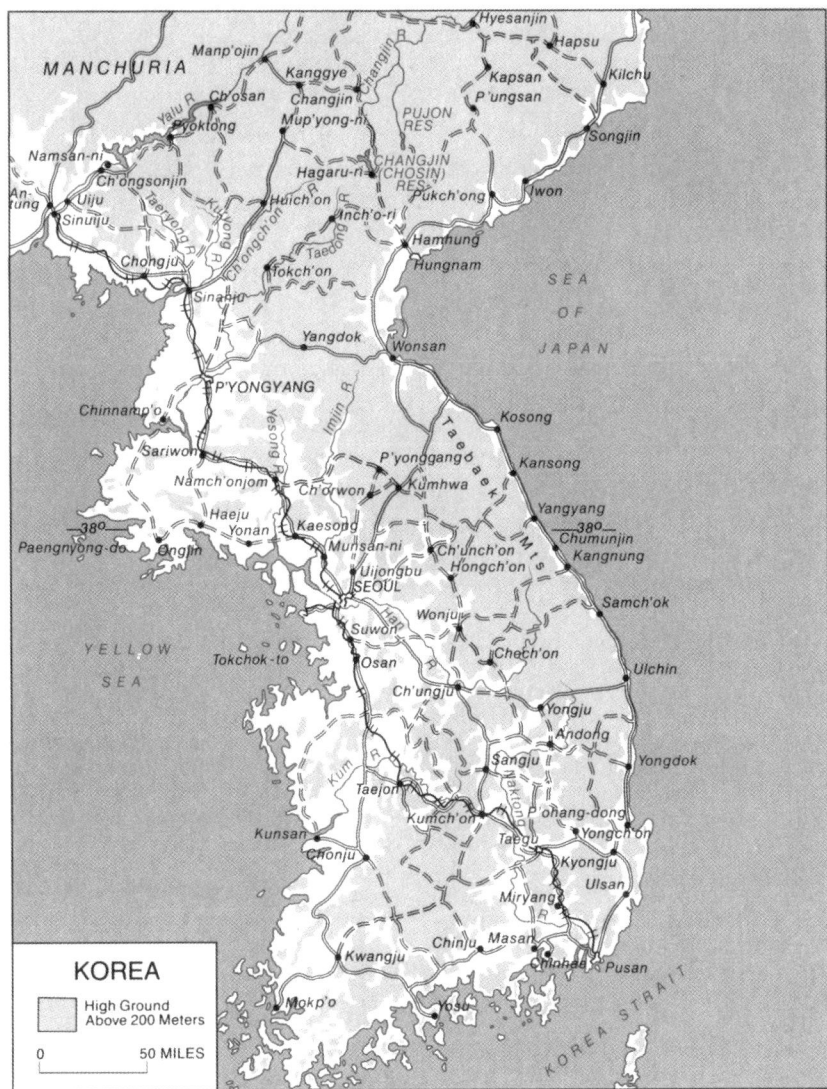

KOREA

High Ground
Above 200 Meters

0   50 MILES

dent authorized the commitment of ground forces.[28] Ready or not—
mostly not—the U.S. Army was again going to war.

By 1949, the Army that had fought World War II with eighty-nine
divisions could muster only fourteen divisions. Four of those were
organized solely for training purposes. Unwilling to go below ten combat divisions, the Army had reduced division strength by cutting one
battalion from each three-battalion infantry regiment and cutting one
battery from each three-battery artillery battalion. With the expiration

of the draft, the Army had to rely on short-term enlistments and low-ered mental standards to maintain its strength. Most of the Army's equipment was left over from World War II. As a result, much of it was antiquated.[29]

Four divisions were on occupation duty in Japan: the 7th, 24th, and 25th Infantry Divisions and the 1st Cavalry Division (also orga-nized as an infantry division), plus one separate regimental combat team. Each division had an authorized strength of 12,500, compared to its au-thorized wartime strength of 18,900.[30] Partly because of budgetary re-strictions—as well as the condition of Japanese roads—the divisions were especially deficient in tank strength. Instead of the standard comple-ment of one heavy tank battalion and three regimental medium tank companies, each division had just one company equipped with seven-teen M24 light tanks.[31]

In the U.S. Army in 1950 there were fewer than 900 serviceable and 2,557 unserviceable M24 light tanks, and 1,826 serviceable and 1,376 unserviceable M4A3E8 Sherman medium tanks, all of which were built during the war. There were only 319 new M46 medium tanks in addi-tion to a number of M26 tanks scheduled for conversion to the M46 configuration.[32]

While the U.S. Army's tank force, especially in Japan, was in poor shape to go to war, the ROK Army—which was being battered by a tank-led invasion force—was in an even more desperate situation. Earlier, in October 1949, in response to a South Korean request for 189 light tanks, the senior U.S. military representative had advised the Army chief of staff against fulfilling the request as "the rough terrain, poor roads, and primitive bridges militated against efficient tank opera-tions."[33] Nevertheless, the commander of the Korea Military Advisory Group (KMAG) told *Time* magazine the South Korean Army might be "the best damn Army outside the United States," except that it had no tanks, no medium artillery, no 4.2-inch mortars, and no recoilless rifles.[34] In short, the South Korean Army was virtually defenseless against tanks.

For the invasion of South Korea, the North Korean Army had ap-proximately 225 postwar thirty-five ton Soviet-built T-34/85 medium tanks. The major armored formation was the 105th Armored Brigade, composed of four regiments totaling 140 tanks. The 26th Armored Bri-gade was estimated to have forty tanks, and the 17th Armored Brigade another forty-five.[35]

Although the ROK Army fought hard in the initial battles north of the Han River, it had no means of stopping the T-34s. The battles were lost from the start. Disaster struck the defenders when the Han River bridges were blown prematurely, isolating most of the South Korean forces north of the river. By 28 June the ROK Army command could account for only twenty-two thousand of the ninety-eight thousand men it had carried on its rolls on 25 July. The ROK Army was virtually destroyed.[36] Once the North Korean Army crossed the Han, it could drive south almost unchecked. It was this situation that prompted General MacArthur to ask for authority to send U.S. ground forces to the peninsula.

As a result, the 24th Infantry Division was ordered to Korea. At the time, it was scattered throughout Japan and a shortage of shipping resulted in the division's piecemeal commitment. Task Force Smith, consisting of two understrength companies and a battery of 105mm howitzers, was the first force to deploy. When the North Koreans attacked with T-34s on the morning of 5 July, the force was quickly driven out of its blocking positions between Suwon and Osan.[37]

Meanwhile, General MacArthur requested that his divisions be brought up to authorized strength with both individual replacements and the restoration of missing units. Regarding armor, he requested trained and organized tank companies to bring the divisional heavy tank battalions up to strength. He also asked for three additional medium tank battalions.[38] The Army's response regarding armor units was to approve by 9 July the immediate shipment of three tank battalions.[39] The 6th Tank Battalion, equipped with M46 tanks, was sent from the 2d Armored Division at Fort Hood. Meanwhile, the 70th and 73d Tank Battalions were organized from school troops at Fort Knox and Fort Benning, respectively. These two battalions were a nightmare of equipment. The 70th Tank Battalion had two companies equipped with M4A3 Sherman tanks, which were shipped from Rock Island Arsenal to the port of embarkation. The other company was equipped with M26s gathered from around Fort Knox, where they had been placed as monuments. The 73d Tank Battalion was also a mixture of M4A3 and M26 companies.[40] One company of the 73d turned in its M24 light tanks at Fort Benning and drew M26s when it debarked at Pusan. After a brief period of orientation on the new tanks, the crews were committed to combat. Fortunately, the unit included a number of World War II Sherman tank veterans.[41]

However, until reinforcements arrived, the Far East Command had to make do with what it had. The 24th Infantry Division, followed shortly by the 25th Infantry Division and the 1st Cavalry Division, shipped out of Japan with its light tank and reconnaissance companies. The 24th Division's M24 light tanks were the first to engage the superior North Korean T-34/85 mediums. The mismatch problem was compounded by the fact that the Chaffees were almost worn out when they were committed to the battlefield.

The 24th Division's Company A, 78th Heavy Tank Battalion, was a heavy unit in name only when it arrived at Pusan on 4 July with fourteen M24s. It joined the 21st Infantry Regiment with thirteen tanks after one broke down in Pusan. Six days later the company was committed to combat and quickly lost two tanks to enemy tank fire and one to a problem with its 75mm gun. The next day, four tanks were lost—two due to mechanical problems, one when it was hit by artillery fire, and one that was disabled by small-arms fire and grenades. On 12 July two more tanks went down when one developed a transmission problem and the other became hopelessly mired in a rice paddy. Four days later one tank was disabled by antitank fire and a second was abandoned because of mechanical difficulties. A month later the last two M24 tanks in the company (one a replacement tank received on 3 August) were disabled by antitank fire and evacuated for repair.[42]

The 24th Reconnaissance Company, 24th Infantry Division, was also committed to combat in July with seven M24s. The company lost its first tank on 19 July and lost another three weeks later. One had to be destroyed and used for spare parts. The reconnaissance unit's tanks had already been rebuilt when the company deployed to Korea. The tanks that survived during the next three months averaged two thousand miles of travel.[43]

Perhaps the best summary of the lessons learned from committing light tanks intended for reconnaissance in the role of main battle tanks is an extract from an October 1950 report submitted by the 16th Reconnaissance Company, 1st Cavalry Division, which stated: "Many men recommend that the M24 light tank be declared obsolete and replaced with a tank that has more armor, more power, a larger gun, more space for the crew, and more turret vision."[44] In other words, they wanted a medium tank.

In Japan efforts were quickly initiated to provide medium tanks to the forces being deployed to Korea. On 28 June the Eighth Army Ordnance Chief found three M26 tanks at the Tokyo Ordnance Depot, all in need of extensive repairs. He immediately ordered them to be made serviceable. The repaired tanks finally arrived in Pusan on 16 July along with an officer and fourteen enlisted men who had been trained on M24s. The provisional tank platoon was shipped by rail to Chinju, arriving just before the 19th Infantry evacuated the town. Unfortunately, mechanical problems, a blown bridge, and a North Korean ambush resulted in the quick loss of all three tanks and half the men.[45]

In another attempt to provide a stop-gap measure for getting medium tanks to Korea, the Eighth Army activated the 8072d Provisional Tank Battalion on 17 July at Camp Drake, near Tokyo. The unit was equipped with rebuilt M4A3s. Seventy personnel were gathered from various units in Japan and another 155 were sent from Fort Hood. The battalion Commander, Lt. Col. Welborn G. Dolvin, was brought in from Fort Benning. At the end of July—only fourteen days after activation— Company A shipped out for Korea. Early in August the remainder of the battalion landed at Pusan, where it was redesignated the 89th Medium Tank Battalion.[46]

By this time, Company A had already been committed to combat in what was—discounting the disastrous brief career of the provisional M26 platoon—the first action involving American medium tanks in Korea. At 7 A.M. on 2 August, the 27th Infantry Regiment launched a counterattack with four tanks from the 2d Platoon in support. A platoon of infantry was mounted on the tanks. This mobile force—now several miles ahead—was followed by additional infantry in trucks. After advancing about twenty-four miles against light enemy resistance, the force was attacked by several antitank guns, losing two tanks. Although the battalion successfully neutralized the enemy, it was ordered to withdraw because the enemy had infiltrated the rear area. By the end of the day, the North Koreans had been driven out of the threatened rear.[47]

Four tanks from the 1st Platoon, Company A, supported an attack by the 19th Infantry in a similar operation on 4 August. Shortly after the attack began, the lead tank, operating with its hatches open, lost its entire crew when a mortar round struck the vehicle.[48]

More tank units began arriving in Pusan in early August. The three tank battalions shipped from the United States arrived on the seventh. Two were immediately assigned to divisions, the 6th Tank Battalion to the 24th and the 70th to the 1st Cavalry Division, and the 73d Tank Battalion's companies were parceled out to support various ground operations around the Pusan Perimeter. In mid-August the 2d Infantry Division arrived with its organic tank battalion, the 72d, and two organic regimental tank companies. The SS *Luxembourg Victory* departed San Francisco on 28 July with eighty more medium tanks for Eighth Army. Thus, there were more than five hundred medium tanks in Korea by the end of the third week of August.

Meanwhile, North Korean tank strength had significantly declined. By early September the ratio of U.S. to North Korean tanks was at least five to one. American tanks were about equally divided between M4A3 Shermans and M26 Pershings, with one battalion of M46 Pattons.[49]

With the assignment of the 89th Tank Battalion to the 25th Infantry Division, each of the four divisions had an organic tank battalion. There

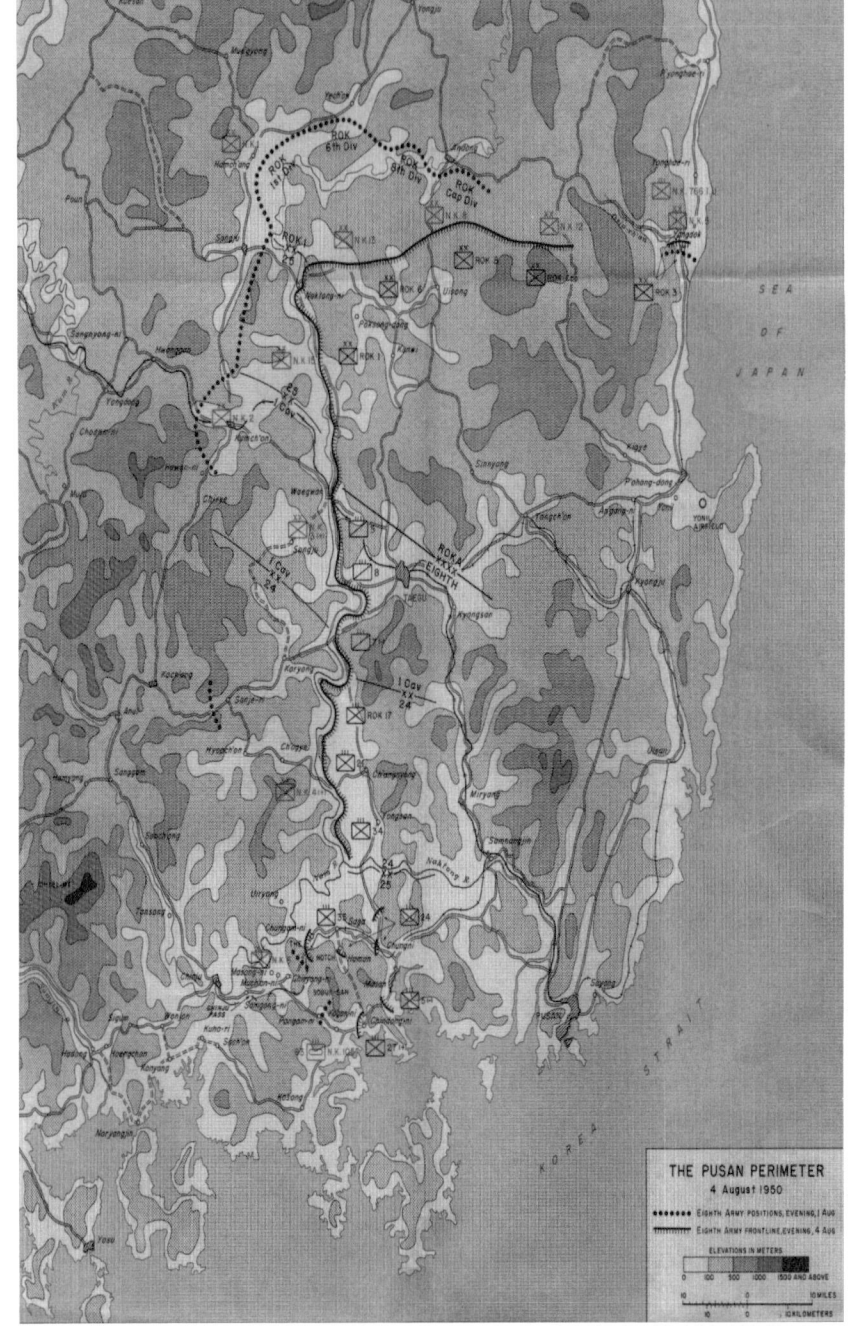

THE PUSAN PERIMETER
4 August 1950

●●●●●●● Eighth Army positions, evening, 1 Aug

━━━━━━━ Eighth Army frontline, evening, 4 Aug

ELEVATIONS IN METERS

0   100   500   1000   1500 AND ABOVE

10   0   10 MILES

10   0   10 KILOMETERS

were also a number of regimental tank companies equipped with medium tanks, while the division reconnaissance companies continued to operate with seven M24s.[50] With the stabilization of the Pusan Perimeter and the transition of armor units from light tank elements and ad hoc organizations, Eighth Army was ready to operate in a less desperate fashion.

By the end of August enemy attacks had been thrown back all along the front. It was clear to the North Korean high command that UN strength was growing. As a result, the North Korean People's Army planned a final, all-out effort to drive UN forces from Korea. The North Korean attacks, launched between 31 August and 2 September, were well coordinated and savagely executed. Eighth Army was hard-pressed to hold and, in some places, restore the Naktong River line. North Korean attacks against Masan and Taegu were repulsed with heavy casualties on both sides. By 12 September the NKPA offensive was spent. The Eighth Army and its ROK allies had held. The war was about to enter a new phase.[51]

Initially, U.S. ground troops had no effective means of countering North Korean T-34s. Enemy tanks usually advanced boldly, confident that the Americans could not stop them. They penetrated infantry positions in the front and flanks, overrunning command posts and artillery positions. As American airpower became more effective and 3.5-inch rocket launchers and medium tanks became available, North Korean armor losses quickly mounted. This forced the North Koreans to change their tactics. Enemy tanks now moved at night to forward positions, attacking at first light. T-34s moved cautiously, closely followed by infantry.[52]

During the early fighting U.S. tank elements were added to the infantry primarily to provide an antitank capability along with their mobile supporting firepower. For example, the war diary of Company A, 6th Tank Battalion, shows that on 8 September the company was "attached to the 19th Infantry, with the 3d Platoon attached to the 3d Battalion, 19th Infantry, in a blocking position."[53] In mid-August, a week after the 70th Tank Battalion arrived in Korea, Company A was committed. The tankers were attached to the 5th Cavalry Regiment, 1st Cavalry Division. When the enemy launched an attack against Hill 303 with infantry supported by tanks and artillery, a Company A tank destroyed a T-34 at a range of three hundred yards. Platoons from the company supported the 5th Cavalry in repelling repeated assaults, but by nightfall the enemy still held the hill. The following morning infantry supported by two platoons from Company A counterattacked, forcing the enemy to abandon the hill. The direct fire delivered by the tanks was critical in supporting the infantry.[54]

This "penny-packet" commitment of American armor, which initially was driven by the availability of tanks, continued throughout the war primarily because of the nature of the Korean terrain. Korea is predominantly a rough, mountainous country. The highest peaks are in the northeast part of the country, reaching nearly nine thousand feet in elevation. However, a nearly continuous mountain barrier extends southward from the mass of mountains in the north. These mountains closely border the east coast. Spur ranges extend southwestward from this main east coast range. Only a small part of Korea is lowland, and that small area is located mainly near the west coast between the spur ranges. The lowlands are intensively cultivated, with rice being the main crop. The rice paddies, which were flooded from June to October, generally proved impassable to tanks. The irrigation system of dikes, canals, and ditches proved to be a further hindrance to tank movement, even when the paddies were dry. The hill and mountain slopes are steep and eroded. They are either bare or covered with secondary growth, except in the extreme north, where there are extensive forests on the mountain slopes.

Consequently, with some exceptions, tanks were generally confined to roads or adjacent dry riverbeds. The roads were mostly narrow and unpaved. They were often built up from two to twenty feet above the adjacent ground, thus restricting tank movement. Many bridges were either destroyed or incapable of supporting tanks, although tanks could generally find a bypass.[55] Clearly, even after American tank forces were built up, there would be no armor operations of the magnitude experienced in Europe during World War II.

While his troops were fighting a desperate delaying action far to the south, General MacArthur had, by 20 July, decided on a counterattack. It strategically rested on an invasion of Inchon, a seaport city southwest of Seoul on Korea's west coast. Objections from the Navy and Marine Corps, plus a lack of enthusiasm within the JCS, did not discourage MacArthur. With persistence he gradually won the necessary support from the Joint Chiefs. To execute the landing, MacArthur activated the X Corps, and appointed his chief of staff, Lt. Gen. Edward M. "Ned" Almond, as commander. The corps consisted of the 7th Infantry Division—still in Japan—and the newly assembled and arrived 1st Marine Division. The 56th Amphibious Tank and Tractor Battalion, the Army's only such unit, was also assigned to the task force, but its few tanks and low readiness made it a token contribution. With this battalion's deac-

(*Above*) The lead T-34 tank in the attack on the 27th Infantry, supported by Company C, 73rd Tank Battalion, near Taegu, August 21, 1950. The tank approached the night position of Company C and was destroyed at a range of 125 yards by 90mm HVAP ammunition fired by an M26 tank. Ten enemy tanks were destroyed in the attack. Richard L. Coffman. (*Below*) An M4A3E8 Sherman advances northward with the 7th Infantry Division after the Inchon landings. Patton Museum.

tivation in 1954, the Army got out of the amphibious armored business for good.[56]

On 15 September, lead elements of the seventy-thousand-man corps, which included tanks from the 1st Marine Division's tank battalion in support of the 3d Battalion, 5th Marines, landed at Inchon.[57] Early the next morning, Marine aircraft attacked a column of six T-34s on the road from Seoul headed for Inchon. All were reported destroyed. However, when the 1st Marines and accompanying tanks saw movement in three of them, the Marine M26s finished them off. Later, on the same road, six tanks accompanying a North Korean infantry column were destroyed by Marine Pershings, rocket launchers, and recoilless rifles.[58]

As the 1st Marine Division fought its way along the Inchon–Seoul highway, the 7th Infantry Division, along with the 73d Tank Battalion, which had recently been withdrawn from the Pusan Perimeter, attacked. The tankers protected the division's right flank and engaged enemy units moving toward the battle area from the south. On the night of 21 September two T-34s engaged elements of the 73d Tank Battalion from Suwon. One American tank was destroyed with the enemy's first shot. The firing tank was then destroyed by other U.S. tanks; however, the second North Korean tank escaped. Before the night was over, another tank-versus-tank battle resulted in the destruction of two T-34s by M26s. Two others managed to escaped to the south.[59]

M26 tank in Korea. Col. Louis F. Dixon, USA (Ret.).

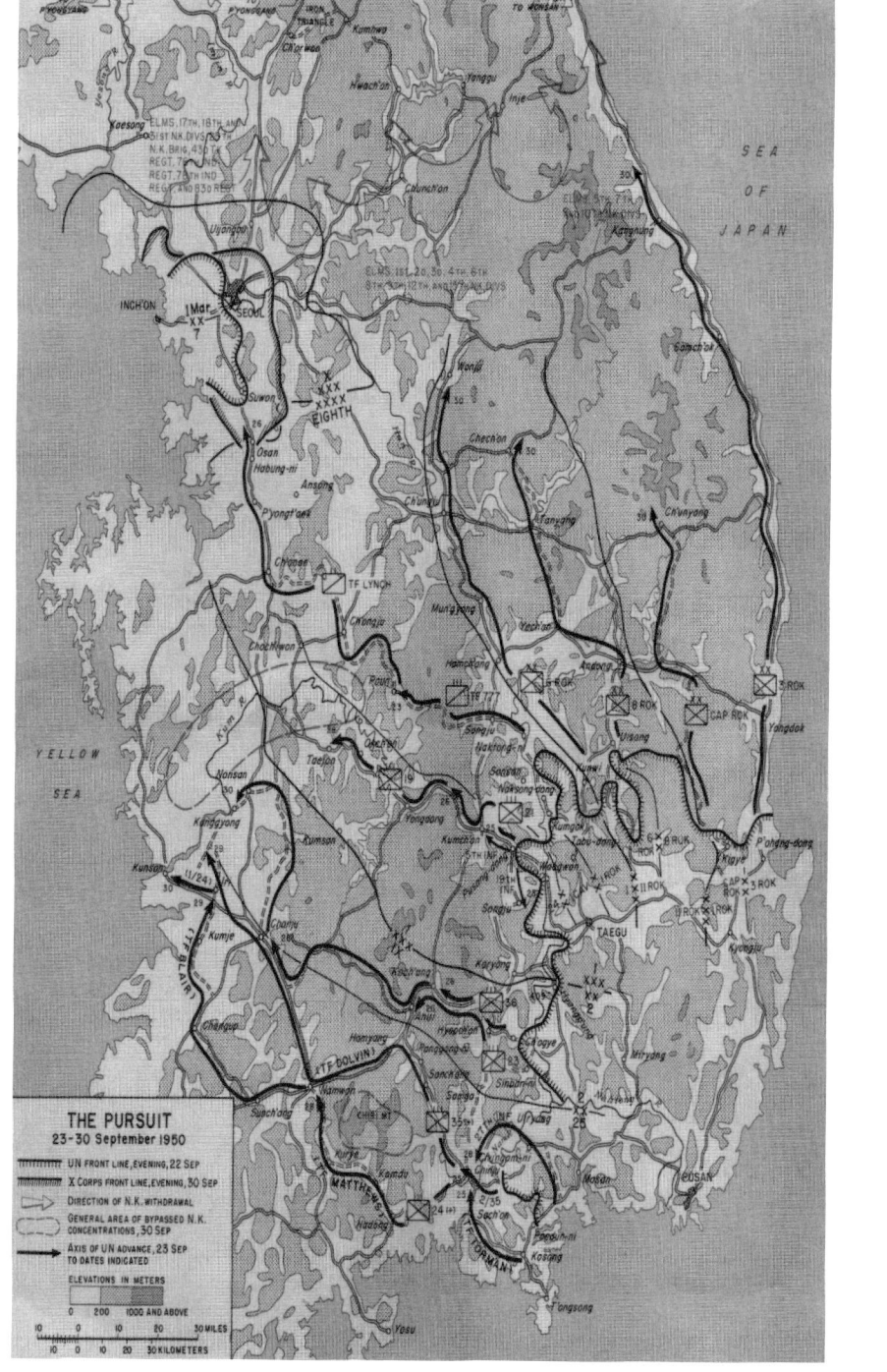

THE PURSUIT
23–30 September 1950

UN FRONT LINE, EVENING, 22 SEP

X CORPS FRONT LINE, EVENING, 30 SEP

DIRECTION OF N.K. WITHDRAWAL

GENERAL AREA OF BYPASSED N.K.
CONCENTRATIONS, 30 SEP

AXIS OF UN ADVANCE, 23 SEP
TO DATES INDICATED

ELEVATIONS IN METERS

0   200   1000 AND ABOVE

10   0   10   20   30 MILES

10   0   10   20   30 KILOMETERS

In spite of these and several other encounters with North Korean tanks, the Inchon invasion and the subsequent attack toward Seoul were primarily infantry battles. Although American tanks often supported the infantry—as in the early phase of the war—their absence sometimes affected the outcome of infantry engagements. Additional weapons in the hands of the foot troops, such as the 3.5-inch rocket launcher, made such situations less critical. During the capture of Seoul the Marines frequently sent two or three tanks against the city's numerous road-blocks, using their M26s to destroy antitank guns and automatic weap- ons and then breach the obstacles.[60]

With X Corps astride the NKPA line of communication, the stage was set for Eighth Army to break out of the Pusan Perimeter. Using X Corps as an anvil, Eighth Army prepared to hammer the NKPA forces fighting in the south.

MacArthur's plan was to launch a general attack along the Eighth Army front simultaneously with the Inchon landing. This kept the enemy from sending reinforcements toward Inchon from the Pusan front. If Eighth Army succeeded in breaking through, it was to drive north to effect a juncture with X Corps in the Seoul area. The battle line in the south was 180 air miles from the closest point to the landing in the enemy rear, although much farther by the winding mountain roads.

The Eighth Army operations order directed that the attack be launched one day after the Inchon landings. The main effort was to be directed along the Taegu–Kumchon–Taejon–Suwon axis with the mission of destroying the enemy along the line of advance and then linking up with X Corps.[61]

There was vicious fighting when the Eighth Army and ROK coun-terattack began. Except in the 2d Infantry Division's area of advance, few significant gains were made through 19 September. The previous night, 18 September, however, the enemy began to withdraw. It quickly became clear that the fighting in the Seoul area was influencing the actions of NKPA forces in the south. In most cases, enemy tanks were used in small contingents to support the infantry. As the North Korean forces retreated across the Naktong River under pressure, they lost heavily in their remaining tank strength. According to enemy sources, the 203d Regiment of the 105th Armored Division retreated to the west bank with only nine tanks, and the 107th Regiment with only fourteen. By 23 September the cordon around the Pusan Perimeter had been broken.[62]

The Eighth Army was now ready to sweep forward along the main axes of advance with motorized columns led by armored spearheads. On 22 September Lt. Gen. Walton H. Walker ordered Eighth Army to begin the pursuit. Elements were to "advance where necessary without regard to lateral security."[63] Walker, who commanded XX Corps under General Patton in Europe during World War II, was poised to launch a pursuit reminiscent of some of those halcyon operations, but on a smaller scale and without the armored strength of Third Army.

The mission of the 25th Infantry Division was to attack northwest in the direction of Chinju and then continue with unlimited objectives. Maj. Gen. William B. Kean organized two main task forces with armored support, Task Force Dolvin and Task Force Mathews.

Lieutenant Colonel Dolvin, Commander of the 89th Tank Battalion, led his task force in one of the few armor-led operations of the war. Task Force Dolvin consisted of Companies A and B, 89th Tank Battalion; Companies B and C, 35th Infantry Regiment; a platoon of heavy mortars, a medical detachment, and the task force trains. Two teams were composed of a tank company and an infantry company with, in each case, the tank company commander in command and the infantry company commander assisting him. On the morning of 26 September the task force moved out with the infantry riding on the back decks of the tanks. Shortly after moving out, the lead tank hit a mine, and the task force continued to come across minefields. It could only advance after the engineer platoon removed mines. Although several minefields were eliminated, enemy resistance stiffened, with sniper and small-arms firing covering the minefields. Suppressing fire from the tanks was required to allow the engineers to clear the mines.[64]

Once past the minefields, the task force encountered enemy mortar and small arms fire from ridges near the road. Timely tank fire and an air strike allowed the task force to break through and continue to advance. At one point, Dolvin sent tanks in a rapid advance to seize a bridge being prepared for demolition by the enemy. The task force then advanced at twenty miles per hour to its objective. After accomplishing its mission, the task force was dissolved at the end of September. The task force had captured or destroyed sixteen antitank guns, nineteen vehicles, sixty-five tons of ammunition, 250 mines, captured 750 enemy soldiers, and killed about 350 more, at a cost of three tanks and forty-six wounded.[65]

In the 24th Infantry Division tanks were used to support rapid advances once the pursuit began. A tank company from the 5th Regimental Combat Team (RCT) was attached to the 19th Infantry on 24 September and advanced with the 1st Battalion with the mission of exploiting the breakthrough and seizing Taejon. The enemy made only half-hearted attempts at delay, resulting in the task force advancing fifty miles in eighteen hours. The company lost only one tank while knocking out four T-34s at extremely close range by firing 76mm high-velocity armor-piercing ammo into the enemy tanks' turret rings.[66]

Tanks in the division continued to be employed by platoons in support of infantry battalions. For example, the war diary of Company A, 6th Tank Battalion, states that on 23 September, "1st Platoon attached to 1st Battalion, 19th Infantry, in an attack on Sonju from the north, 3d Platoon attached to the 3d Battalion, 19th Infantry, in an attack on Sonju from the northeast." The entry for the next day shows the 2d Platoon attached to the 2d Battalion, 19th Infantry, and on 25 September the 2d Platoon was attached to the 1st Battalion.[67]

Still east of the Naktong River, the 1st Cavalry Division launched Task Force 777, composed of the 7th Cavalry Regiment (less one battalion), the 70th Tank Battalion, and the 77th Field Artillery Battalion. Lt. Col. William A. Harris, the regimental Commander, commanded the task force. Harris assigned Lt. Col. James H. Lynch's 3d Battalion, with attachments, as the lead combat force. Task Force Lynch included the 3d Battalion, 7th Cavalry; Company B, 8th Engineer Battalion; two platoons of Company C, 70th Tank Battalion; the 77th Field Artillery Battalion (less one battery); 3d Platoon, Heavy Mortar Company; the regiment's Intelligence and Reconnaissance (I&R) Platoon; and a tactical air control party.

Fighter-bombers flew up and down the road attacking fleeing groups of North Koreans as Lynch's task force overcame scattered resistance, including some antitank guns, to reach the Naktong River that night and by morning secure the river crossing. The task force had covered thirty-six miles. The task force continued its advance the next day, crossing some tanks on a ferry constructed by the engineers. On the twenty-fourth, the engineers, assisted by Korean laborers, improved a North Korean underwater bridge. By noon all the tanks were across the river and quickly joined the infantry, which had pushed ten miles ahead on foot. Company K and a platoon of tanks advanced another thirty miles by dark. Stopped there initially by orders to hold in place, the division commander subsequently received authority to continue to a link-up with X Corps. With Task Force Lynch in the lead, the division started forward late in the morning on the twenty-sixth. The I&R Platoon and a platoon of tanks from Company C, 70th Tank Battalion, led the task force with orders to move at maximum tank speed and not to fire unless fired upon. The point element advanced so rapidly that it outran the main column, reaching the lines of the 31st Infantry, 7th Infantry Division, by 10:30 that evening. Lynch's point element covered 106 miles that day.

The remainder of the task force encountered T-34s that the point force had bypassed. Several enemy tanks were destroyed in a night engagement, and the task force stopped and waited for daylight. Finally, at 8:30 A.M. on the twenty-seventh, the task force's infantry linked up with elements of the 31st Infantry.[68]

By the end of September the NKPA had essentially ceased to exist. United Nations troops were everywhere no more than five miles below the 38th parallel. The NKPA suffered the loss of as much or more weapons and equipment as the ROK Army had in its retreat during the first week of the war. Although no reliable information is available concerning the number of damaged tanks the North Koreans were able to repair and return to action, the 239 tanks UN forces found destroyed or abandoned comes close to the total number used by the North Koreans in South Korea.[69]

On 27 September General MacArthur was given the authority to cross the 38th Parallel. This would allow him to fulfill his military objective of "the destruction of the North Korean armed forces." However, he was ordered not to cross over the Manchurian or Soviet borders with North Korea nor to use air or sea forces against targets in Manchurian or Soviet territory, which became a sanctuary for the rapidly increasing Chinese People's Liberation Army or Chinese Communist Forces (CCF).[70]

MacArthur's next plan called for Eighth Army to advance toward Pyongyang, the North Korean capital. Meanwhile, X Corps was to make an amphibious landing at Wonsan on North Korea's east coast and advance toward the west, thus encircling the capital.[71]

At noon on 7 October Eighth Army units crossed the 38th Parallel in the wake of ROK Army elements that had crossed several days earlier.[72] The 1st Cavalry Division led the advance with three regiments abreast. The division's progress was slowed in the center by enemy mines, and its armor spearhead had to halt repeatedly while engineers cleared the road.

On 12 October an enemy strong point with tanks and antitank guns stopped the 8th Cavalry's advance. Air strikes and artillery failed to dislodge the enemy.[73] On the morning of the thirteenth, the North Koreans were blasted with artillery and air strikes every thirty minutes. The enemy strongly resisted and then attempted counterattacks. In one of these actions enemy tanks rumbled out of the morning mist to strike a 70th Tank Battalion outpost. The defending tankers opened fire at a range of fifty yards. By the end of the day the 8th Cavalry had destroyed eight tanks, with Company B, 70th Tank Battalion, accounting for seven of them without a loss.[74]

Throughout the advance tank elements continued to come across minefields. On 8 October the 70th Tank Battalion had an M4A3 damaged while a tank platoon and an infantry company were probing to the north. Six days later another M4A3 was damaged when it hit a mine near a small concrete bridge. Likewise, an M26 was damaged when it hit a mine while bypassing a mined bridge.[75] Such casualties were typical as the road-bound tank-infantry teams advanced.

Despite stiff opposition from newly created NKPA forces, Eighth Army continued its rapid advance, particularly after 14 October. The 1st Cavalry Division, with the British 29th Infantry Brigade attached, and the 24th Infantry Division scrambled to see which division would reach Sariwon, the halfway point on the road to Pyongyang, first. Three days later the 1st Cavalry Division and the 1st ROK Division entered Pyongyang, and within a few days the North Korean capital was securely in UN hands.[76]

The same day UN troops entered Pyongyang, MacArthur issued orders for a general advance to a new objective some eighty to 130 miles north of the Pyongyang–Wonsan line, which would bring his forces within forty miles of the Manchurian border. The advance was to begin immediately following the capture of the North Korean capital.[77.]

The 1st Marine Division, part of X Corps's amphibious forces, began to load at Inchon. In the meantime, the 7th Infantry Division had moved by road to Pusan for embarkation. North Korean mining at Wonsan Harbor, however, delayed the landing. It was 28 October before combat elements of the 1st Marine Division were ashore. By that time the ROK I Corps had already captured Hungnam. The 7th Infantry Division, except for its tanks, was ashore in the Iwon area on 9 November, 150 miles north of Wonsan. With the capture of enemy territory west of Pyongyang, X Corps's mission was changed to an advance northwest toward the Manchurian border.[78]

Earlier, when Eighth Army began its advance, it was flanked by I Corps on the left and II ROK Corps on the right. When the units crossed the Chongchon River four days later MacArthur issued orders for all commanders to use the forces necessary to secure the remainder of North Korea. This order removed all restrictions on the use of non-ROK forces near the border. The advance, a continuation of the pursuit of the NKPA forces routed at Pyongyang, turned into a race by UN forces to reach the Yalu River. After some initial fighting, the ROK 6th Division advanced against no opposition, and one of its platoons reached the town of Chosan on the river on 25 October.

The next day, Chinese forces that had infiltrated into North Korea attacked the 6th ROK Division. Within three days the ROK II Corps—now badly crippled—was driven back to the Chongchon River. While

Chinese forces continued their efforts to outflank the ROK II Corps, General Walker became concerned about I Corps's security and assembled the 2d Infantry Division west of Sunchon for emergency use.

Just west of the ROK II Corps, I Corps pushed across the Chongchon on 25 October with the 24th Infantry Division and the attached ROK 1st Division on its right. The South Koreans were supported by Company C, 6th Tank Battalion. Designated Task Force Elephant, the company led the ROK 1st Division's advance.

Meanwhile, the 1st Cavalry Division was in Army reserve at Pyongyang. After the Chinese attacked the ROK 1st Division near Unsan, the 1st Cavalry Division's 8th Cavalry Regiment moved through the threatened South Korean division to form a semicircle around Unsan. However, before the relief was effected, a South Korean counterattack supported by elements of the 6th Tank Battalion made some progress. The 1st Battalion, 8th Cavalry, with a platoon of tanks from Company B, 70th Tank Battalion, launched an attack in support of the Koreans. Unfortunately, the company lost three tanks.[79]

Elsewhere, the westward pursuit initially went well. However, on 25 October the 27th Commonwealth Brigade, attached to the 24th Infantry Division, crossed the Taeryong River and encountered opposition on the west bank. Two days later the advancing brigade engaged the enemy in a bitter battle during which air strikes and artillery destroyed ten T-34s. The following day—now realizing that the advance was no longer a pursuit—the brigade held up just three miles from Chongju after moving fifteen miles. The next morning, an attached Australian battalion attacked in an effort to relieve pressure. The Australians, supported by tactical aircraft that smashed four enemy tanks, were able to engage and destroy an additional three T-34s with bazookas.

The night of 30 October, the 21st Infantry passed through the 27th Commonwealth Brigade and launched an attack. After midnight the North Koreans then tried to ambush the column with a force of five hundred troops and seven tanks. The nearest enemy tank opened fire at about three hundred yards. Although several tanks from the 6th Tank Battalion were hit, they were not destroyed. The American tankers immediately returned fire, aiming at the muzzle flashes of the enemy's 85mm guns. By dawn the North Koreans had abandoned their position, leaving five knocked-out T-34s. Air strikes later in the morning destroyed two additional T-34s while the infantry captured two that had been loaded on flatcars.

After the night battle, the 21st Infantry advanced against light resistance to within eighteen miles of the Yalu. During the midafternoon, seven T-34s and about five hundred enemy troops attacked the 1st

Battalion. Company A, 6th Tank Battalion, counterattacked and in a vicious half-hour tank battle destroyed the enemy's armor at a cost of two damaged tanks.

On 1 November, I Corps, responding to pressure farther to the east, ordered the 21st Infantry and the 5th RCT to its north to hold in place and prepare to defend in depth. Later that day the two regiments were ordered to return to the line of the Chongchon River, which they did that night.[80]

The situation had meanwhile turned critical to the east. Except for the ROK 15th Regiment, most of the ROK II Corps had given way, exposing I Corps's right flank. On 1 November Chinese forces cut the road south of Unsan and that night virtually surrounded the 8th Cavalry Regiment. Efforts by the 5th Cavalry to break through failed.

About 7:30 that evening the Chinese launched an attack on the 8th Cavalry's positions. The regiment's 1st Battalion was on the right flank. Tanks from Company B, 70th Tank Battalion, protected its right rear from positions overlooking the Samtan River. By midnight the ROK 15th Regiment had disintegrated and the cavalrymen reported large groups of enemy crossing the river. Heavily engaged and in danger of having its flank turned, the 8th Cavalry was ordered to withdraw. The tanks on the right flank had been forced back from the river but still held the road junction critical to the regiment's withdrawal. Meanwhile, other Company B tanks with the battalion withdrew through Unsan. As the last two tanks were withdrawing through the town, one slipped into a shell crater as the crew tried to bypass a burning truck. Chinese soldiers then killed the tank commander while he was attempting to get his tank unstuck. Other Chinese placed a satchel charge on the track of the other tank and disabled it. Of the ten crewmen on the two tanks, two were killed and five wounded.

While the 1st and 2d Battalions withdrew, four tanks from the 1st Platoon, Company B, 70th Tank Battalion, were placed at a critical road junction about a mile and a half south of Unsan. They remained there until two tanks from the 2d Platoon arrived, at which point they were ordered to move southeast and protect a crossing at the Kuryong River. However, the regimental column ran into a Chinese force that had effectively blocked the road. In the remaining hours of darkness—as the Chinese attacked the 1st and 2d Battalions—a few tanks with infantry were able to cross the river. Meanwhile, many crewmen scattered to the hills, leaving behind a number of tanks. Only four of Company B's tanks were able to assemble with the remains of the two battalions at Yongbyon.[81]

Supported by the 4th Platoon, Company B, 70th Tank Battalion, the 3d Battalion, 8th Cavalry, was southeast of the other two battalions, with the mission of protecting the regiment's rear. When it received the order to withdraw on the night of 1 November, the battalion had yet to be engaged by the enemy. At about 3 P.M. the Chinese launched their attack with a surprise assault from all sides on the battalion command post. A tank Commander at the command post fought enemy troops off the back deck of his tank with his pistol. Minutes later another tank was damaged by a satchel charge and exploded. The other tanks held off the swarming Chinese trying to cross the stream from the south. By morning, most of the battalion had sustained casualties. Meanwhile, three tanks had formed the nucleus of a defensive position. The following day air strikes temporarily contained the encircling Chinese forces. Efforts to relieve the battalion were unsuccessful. By dark the 1st Cavalry Division had abandoned any further attempts.

The battalion's reduced perimeter still included three tanks. That night they were all hit several times by mortar fire. One caught fire, killing a crewman who was attempting to extinguish the flames. With ammunition almost gone and gasoline low, the prospects for lasting through the night looked slim, causing the tanks to make a run for friendly lines. Three miles from the perimeter they had to be abandoned. After several desperate encounters, a few of the crewmen finally reached friendly positions.

After enduring another night of attacks, the remainder of the battalion—about two hundred men—broke out of the perimeter and attempted to make their way to friendly lines. Many were captured the next day. Unfortunately, about 250 wounded were left behind with several volunteers. The 3d Battalion had virtually ceased to exist.

When the last of the stragglers returned to friendly lines, it was apparent the regiment had lost about six hundred men during the action at Unsan. The 70th Tank Battalion lost about 25 percent of its strength.[82]

As the Eighth Army pulled back to the Chongchon River, General Walker decided to maintain a bridgehead on the north side. There the 24th Division, with the 27th Commonwealth Brigade, engaged in vicious fighting during the first few days of November. Farther east, the remnants of the ROK II Corps, supported by the 5th RCT, managed to hold off a Chinese attempt to get around the right flank. Then, on 6 November, just as suddenly they had materialized, the Chinese broke off their attacks and disappeared.[83]

Meanwhile, far to the east, General Almond's X Corps and the ROK I Corps were operating independent of Walker's Eighth Army. With the mission changed to seizing all of North Korea, the plan was for the ROK I Corps to advance along the coast to the Soviet border while the 7th Infantry Division pushed into the mountainous country to the north. The 1st Marine Division initially secured the X Corps rear area until the 3d Infantry Division, scheduled to arrive from the United States, landed. Thereupon the Marines would drive north, west of the 7th Division.

By late October the ROK I Corps had advanced along the coast, capturing a number of Chinese soldiers. The advance slowly continued until stopped by a Chinese force supported by North Korean T-34s. More Chinese prisoners were taken. In spite of increasing evidence of Chinese intervention, X Corps continued its advance to the north. The ROK Capital Division fought its way along the coast and captured Chongjin, sixty-five miles south of the Siberian border on 26 November. To the west of the ROK I Corps, the 7th Infantry Division had its first encounter with Chinese troops on 8 November. Continuing to attack north, the division reached the Yalu River on 21 November. With the arrival of the 3d Infantry Division, the 1st Marine Division advanced north to the west of the 7th Division. After overcoming a Chinese blocking force, the Marines advanced to Hagaru-ri just south of the Changjin (Chosin) Reservoir on 14 November. [84]

The Chinese, just as they had done in the west against the Eighth Army, broke contact after the first week in November and were not seen again for two weeks.[85]

Although northeast Korea's mountainous terrain essentially limited tank movement to roads, X Corps nevertheless included a number of tank elements. The 1st Marine Division's tank complement included the 1st Tank Battalion with four tank companies and a flamethrower platoon, plus a tank platoon with each of its three infantry regiments—the 1st, 5th, and 7th Marines. The 7th Infantry Division had the 73d Tank Battalion, with three tank companies, and each of its three infantry regiments—the 17th, 31st, and 32d—had an organic tank company. The newly arrived 3d Infantry Division included the 64th Tank Battalion with three M46 tank companies and a tank company with each of its infantry regiments—the 7th, 15th, and 65th. In addition, the two Army divisions each had a reconnaissance company with 7 M24 light tanks. Overall, X Corps had 393 medium tanks, 9 flamethrower tanks, and 14 light tanks. Thus when the 1st Marine and 7th Infantry Divisions attacked north they had approximately 100 and 150 medium tanks, respectively.[86]

In spite of the number of tanks in the two divisions, the number that went forward with the advancing infantry was limited by terrain considerations. Not only were the tanks generally limited to road movement, but a mechanical breakdown could create a serious obstacle to further movement. As a result, most of the tankers from the Marine tank battalion were used as infantry to keep the main supply route open between the coastal cities of Wonsan and Hungnam.[87]

With the temporary withdrawal of Chinese forces, X Corps renewed its offensive against light opposition. By the last week in November, the 5th and 7th Marines had advanced in subzero temperatures to Yudam-ni, west of the Changjin Reservoir. With no change in X Corps orders, preparations were made to continue the advance. However, it quickly became apparent that the Chinese were determined to block further progress. It also became evident from Chinese prisoners that their mission was to destroy the 1st Marine Division.[88] For the next several days, the two regiments were engaged in a desperate struggle for survival.

To lend further support to the planned advance of the force, the division Commander directed that the provisional tank platoon that had accompanied the Marines as far as Hagaru-ri be sent to Yudam-ni. Four M4A3 tanks made an attempt to join the infantry regiments, but the ice-coated road proved too great an obstacle. The tanks slid off the road about four miles from Hagaru-ri and one threw a track. Three were able to return to Hagaru-ri, but the fourth required recovery. A platoon of M26 tanks was then brought to Hagaru-ri and sent on the road to Yudam-ni with the hope that the heavier tanks might better negotiate the icy road. The platoon leader, Lt. Richard Primrose, and a driver started out on a one-tank test run. Traversing the narrow pass, the tank descended in about three hours to the perimeter at Yudam-ni. Primrose and the driver then returned by helicopter to bring up the remaining tanks. However, that night the enemy cut the road and no additional armor was able to get through. Staff Sgt. Russell Munsell volunteered to fly into the Yudam-ni perimeter to fight the tank in support of the infantry. During the next few days, he performed steadfast service as tank commander.[89]

With the 5th and 7th Marines fighting at Yudam-ni and the 1st Marines at Koto-ri, south of Hagaru-ri, the force at Hagaru-ri was a mixed bag of Army and Marine support personnel, two infantry companies, and a few tanks. The night of 29 November the perimeter came under heavy Chinese attack but survived the three-hour assault, partly through the efforts of supporting tanks. In the morning, 750 enemy dead were counted in front of the attacked positions.[90]

Meanwhile, on 28 November, Task Force Drysdale—made up of the newly arrived 41st Independent Commando, Royal Marines; a com-

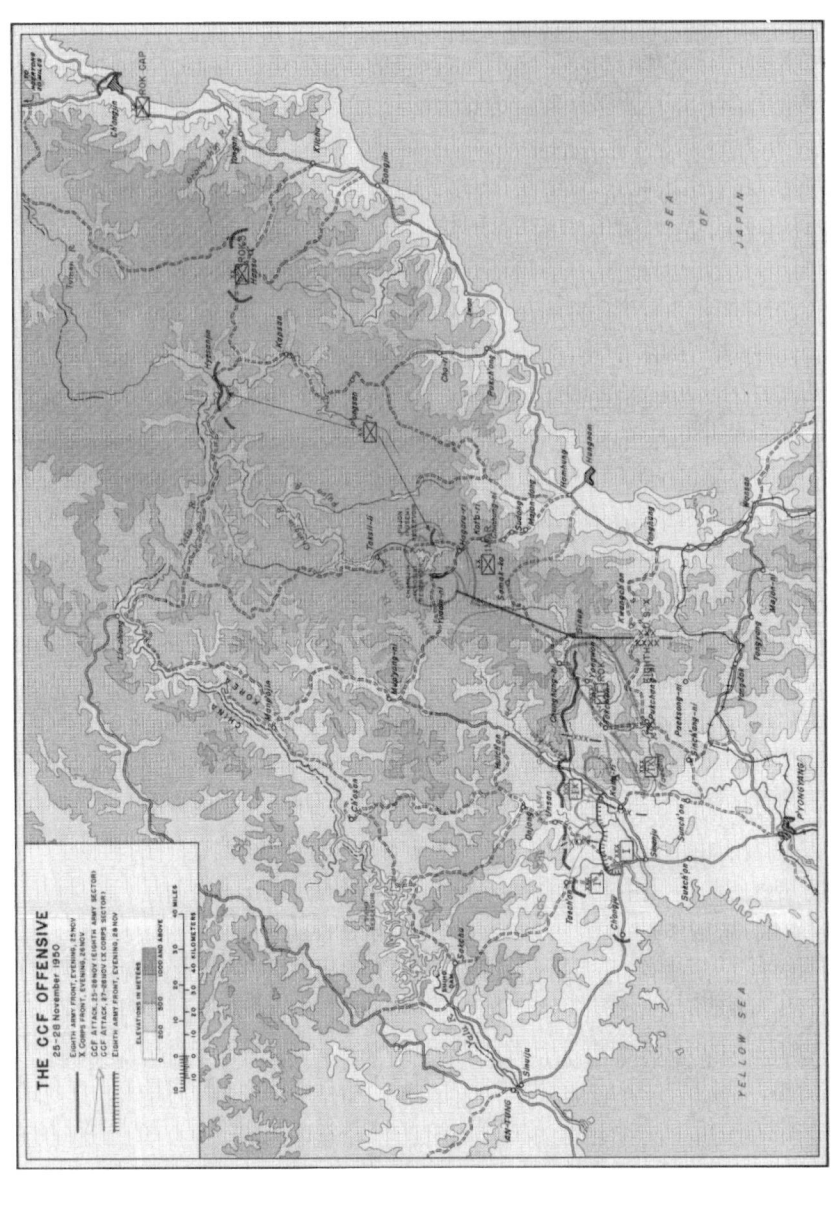

THE CCF OFFENSIVE
25-28 November 1950

pany of the 1st Marines; a company of the 31st Infantry, 7th Division; eight tanks; and a truck convoy—commanded by Lt. Col. Douglas S. Drysdale, the Royal Marine commander, was given the assignment of breaking through from Koto-ri to Hagaru-ri. Originally the tanks took the lead; however, after the column came under attack they were spread through the truck column. When the column was held up by enemy action, the tanks were able to restore movement by firing to the right and left. Although the part of the column with the tanks made it through to Hagaru-ri, the Commandos—cut off from the rest of the force by a roadblock—suffered 50 percent casualties. Some of the service troops in trucks at the rear of the column were also cut off and surrounded. They were forced to surrender.[91]

With the launching of the Chinese counteroffensive against Eighth Army, the X Corps commander, General Almond, decided to concentrate his forces in the Hamhung-Hungnam area. His first task was to extricate the badly mauled 1st Marine and 7th Infantry Divisions, which were under heavy attack in the Changjin Reservoir area. The 5th and 7th Marines were directed to return to Hagaru-ri, while the two-battalion task force from the 7th Division operating east of the reservoir was placed under the Marine division and directed to fight its way back to Hagaru-ri. By 3 December, despite heavy Chinese opposition and adverse weather, the two Marine regiments, along with Sergeant Munsell's tank, reached Hagaru-ri. However, the 7th Division task force was unable to break through the Chinese, and a rescue attempt by a small Army task force that included tanks failed to break through from Hagaru-ri.[92]

On the morning of 6 December the 1st Marine Division, with attached Army troops, began its move to Koto-ri. The 2d Battalion, 7th Marines, with a tank platoon attached, led the advance. The initial advance had made about two thousand yards when the column paused to allow the tank platoon to pass. The Chinese then took the column under fire. An infantry attack, supported by tanks and mortars, neutralized the Chinese and the column continued. The column was attacked repeatedly as it advanced, but after twenty-two difficult hours in extremely cold conditions, the lead elements negotiated the nine miles to Koto-ri. It was late in the evening of 7 December before the rear guard entered Koto-ri.[93]

Shortly thereafter, during a snowstorm, the 1st Marine Division and other Army and Marine elements in the Koto-ri perimeter began the final withdrawal to the coast. There were forty-six tanks spaced throughout the column. The 1st Marine Division's dismounted reconnaissance company acted as a rear guard behind a long column of wheeled vehicles. There was heavy fighting in the rear of the formation as the Chinese attacked the last few tanks and the supporting Marines. Several tanks were lost and the reconnaissance company suffered heavy casualties, but by the evening of 11 December the entire column had closed on Hamhung.[94]

Over on the Eighth Army front, with the sudden disappearance of Chinese forces on 6 November, General Walker gained renewed confidence as plans were revived for a final drive to the Yalu to end the war.[95] The Eighth Army attack was to be launched on 24 November. The plan called for a general advance to the Yalu, but Walker wanted this one to be a deliberate move rather than a pell-mell race to the river. The Chinese had earned his respect and concern as both forces fought in the bitter cold weather. Eighth Army's major commands, which included about 241,000 men, were—from west to east—I Corps, IX Corps, and the ROK II Corps.

Unbeknownst to Far East Command's intelligence staff, the Chinese were also ready on 24 November with six armies containing nineteen divisions with approximately 150,000 light infantry troops. In contrast to Eighth Army, the Chinese employed very few support troops.[96]

Unfortunately, Eighth Army's tank strength still reflected to some degree the reduced organization of the divisions that had been on occupation duty in Japan. The 2d Infantry Division, which had deployed from the United States, had its organic tank battalion, the 72d. It was equipped with one company of M26s and three companies of M4A3s. All three of its regiments also had tank companies equipped with M4A3s, making it the only division with a full complement of tanks. The 1st Cavalry Division had only the 70th Tank Battalion with one company of M26s and three of M4A3s, and no regimental tank companies. The 24th Infantry Division had the 6th Tank Battalion, equipped with M46s, and two of its regiments, the 5th and 21st, had regimental tank companies with M4A3s. The 25th Infantry Division had the 89th Tank Battalion consisting of one company of M26s and three companies of M4A3s, plus Company A, 79th Heavy Tank Battalion, with M24 tanks. There were no regimental tank companies. Each of the divisions had organic reconnaissance companies with the standard complement of seven M24s.[97] Except for the British, none of the other UN elements in Eighth Army had organic tank units. From time to time, U.S. tank units were called on to support UN forces operating without benefit of their own tanks.

Because the companies and platoons of divisional tank battalions were often parceled out, tank battalion reconnaissance platoons were usually left without normal employment. As a consequence, they found themselves sent on various support missions. Take for example Lt. Bill Ward's reconnaissance platoon from the 70th Tank Battalion. At various times, always independent of the battalion, the platoon supported the 1st Cavalry Division, the 2d and 3d Infantry Divisions, and the ROK 1st Division, as well as French, Greek, Thai, Belgian, and British units. Ward's platoon was "almost always on loan to somebody."[98]

The 25th Infantry Division, operating on I Corps's left flank, also advanced on 24 November with three elements abreast and one regiment in reserve. On the left was the 35th Infantry and on the right the 24th Infantry. Between them was Task Force Dolvin, led by Lieutenant Colonel Dolvin's 89th Tank Battalion. The task force consisted of a mix of tanks, infantry, combat engineers, reconnaissance units, and an assault gun platoon. In addition, two or three different 105mm howitzer battalions supported the task force. During the day, the task force advanced a few miles beyond Ipsok, driving off Chinese screening forces. That night the force temporarily halted. There was little enemy contact during the day within the two U.S. corps sectors.[99] On the ROK II Corps front, the most difficult terrain in the Eighth Army area, dug-in Chinese forces stopped the advance.[100]

By the end of the second day of the resumed offensive, I Corps had encountered few enemy. However, the attached ROK 1st Division intelligence staff reported heavy Chinese forces just ahead. Task Force Dolvin marked the farthest advance of the 25th Division. Farther east, the 2d Infantry Division advanced along the Chongchon River valley. However, a strong Chinese force stopped its 9th Infantry Regiment. The ROK II Corps made little progress on the twenty-fifth. Eighth Army troops would go no farther north; they had reached the Chinese concentrations waiting for them.[101]

On the night of 25 November the CCF launched its Second Phase Offensive. By the end of November and the first day of December, the Chinese had defeated Eighth Army all across the front. On the right flank, the enemy breakthrough in the ROK II Corps zone was decisive and complete. This reversal was rapidly followed by the defeat of American troops in the center of the UN line, primarily the 25th Division. Chinese forces then overwhelmed the 2d Division as it attempted to conduct holding actions.[102]

The next day the Chinese struck 9th Infantry Regiment's flank and overran the 1st and 3d Battalions and their supporting tank platoons. In both instances infantry were in positions on hills and the tanks were on the road. Apparently the inability to provide tank-infantry mutual support contributed to defeat of the American force. Tanks without infantry support, especially at night, were particularly vulnerable to attacking Chinese, who stealthily closed in on the doomed tanks, using satchel charges, antitank rockets, and recoilless rifles. The 2d Battalion, 9th Infantry, was not attacked that day, but was encircled the next night. With tanks in support the infantry was able to fight its way back across the river. Casualties were heavy.[103]

As they had before, tanks, in what was an infantryman's fight, played an important role in saving most of Eighth Army to carry on in spite of five days of defeats.

During the night of 25 November the headquarters of the 23d Infantry, one infantry battalion and supporting artillery, were attacked in positions along the banks of the Chongchon River. In spite of the enemy's penetration of the perimeter, six tanks from the regimental tank company temporarily assisted in holding the position. A company from the 72d Tank Battalion was sent to reinforce the beleaguered defenders and contributed to the defeat of the swarming enemy in an all-night battle.[104]

Also that night, Company L, 38th Infantry, supported by two tanks at two roadblocks, was driven from positions that formed a perimeter in the hills around the battalion command post. As the force gave way and withdrew with its many casualties, its salvation was three tanks that took the charging Chinese under fire.[105]

Meanwhile, Task Force Dolvin was also heavily attacked. The Ranger company had suffered such heavy casualties that its remnants were sent to the rear. The night of 26-27 November, Company E, 27th Infantry, suffered sixty casualties, including eight killed in action. The tanks and infantry, against almost overwhelming Chinese forces, had held. What was left of the depleted task force was withdrawn and then dissolved.[106]

For the next several days the 2d Infantry Division conducted a rear-guard action, withdrawing to the Kunu-ri area while holding the crossings of the Chongchon River for other retreating units. Next, the hard-pressed division acted as the rear guard for the entire retreating Eighth Army. On 30 November, with enormous effort, the division ran the CCF gauntlet from Kunu-ri to Sunchon.

From 30 November to 1 December the 2d Division withdrew through a narrow defile, passing under murderous fire from the high ground on both sides. The day before the division withdrew, a tank platoon from the 72d Tank Battalion had successfully passed through the oncoming Chinese forces. However, later in the day—after the division reconnaissance company was stopped—a stronger force, including an infantry company and another platoon of tanks to reinforce the reconnaissance company, failed to clear the roadblock. The Chinese forces were now in position. During the next two days—as the 2d Division ran the gauntlet—tanks again played a critical role. Interspersed in the long line of vehicles—sometimes with mounted infantry—the tankers engaged numerous Chinese troops attempting to mangle the columns. Almost exclusively restricted to movement on the road, the tankers' main function was to provide immediate fire support for infantry attacking the Chinese-infested hills. By the end of the withdrawal the 2d Infantry Division was down to the strength of a regimental combat team. The 72d Tank Battalion had thirty-three tanks left out of an authorized strength of seventy-six.[107]

As the units to its east withdrew, the right flank of the 24th Division became critical. Fortunately the 24th Reconnaissance Company and the 5th RCT protected the division's flank and covered the withdrawal as the 24th and ROK 1st Divisions withdrew under determined Chinese pressure.[108]

After 30 November Eighth Army continued to move south through Pyongyang, leaving behind massive destruction and abandoned supplies. It was not until Eighth Army reached the line of the Han and Imjin Rivers—twenty air miles north of Seoul and a hundred miles south of Pyongyang—that there was an attempt to stop and establish a defensive line. This was practically the original division between North and South Korea.[109]

During the early days of December the only contact with the enemy, which was generally light, was experienced by the 1st Cavalry Division and the attached 27th Commonwealth Brigade. These forces were guarding the Eighth Army's eastern flank. The tank casualties that occurred were almost exclusively the result of breakdowns and subsequent abandonment. Typically, when I Corps reported five M46 tanks needing repairs at Sinanju, three were repaired and two were destroyed.[110] There were some excellent efforts at saving major weapon systems, however. Eight tanks from the 25th Infantry Division reached a point where they could go no farther without new engines. The division ordnance company erected tents, built bonfires, and changed tank engines throughout the night. By morning, all eight tanks were up and running.[111] The 57th Ordnance Recovery Company responded in Pyongyang to a desperate request by the 6th Tank Battalion motor officer for help evacuating nine of the battalion's tanks that were limping along several miles north of the city. A number of the tanks were towed to Pyongyang and replaced with new tanks that arrived by rail. The towed tanks were then evacuated south on the same train.[112]

The Eighth Army had made plans to withdraw farther south and abandon Seoul if forced by the oncoming Chinese. When General Walker was killed in a jeep accident on 22 December and replaced by Gen.

Matthew B. Ridgway a few days later, Eighth Army was still waiting for a CCF attack. There were indications it was coming soon. General Ridgway—before arriving on the scene—had planned to launch an offensive. However, after examining the tactical situation, he reluctantly concluded it was out of the question. When the Chinese forces launched their Third Phase Offensive on the last day of December, he ordered a withdrawal to a Han River line. Ridgway was concerned about the complexities of fighting a withdrawal across the Han along with hundreds of fleeing government and foreign officials and thousands of refugees. With this in mind, he ordered Eighth Army to retire south of the river on 3 January.[113]

The next day, UN forces occupied a line across Korea about thirty-five miles south of Seoul, along the narrowest part of the Korean Peninsula. The line took on more strength when X Corps joined Eighth Army. Ridgway, meanwhile, had decided that for the time being his aim was to inflict casualties on the Chinese rather than acquire real estate. Thus, the basic strategy for fighting the war was changed.[114]

With little evidence of enemy forces to his front, Ridgway cautiously decided to begin an advance on a broad front on 25 January. However, Chinese resistance stiffened when the Eighth Army concentrated on a southward bulge in the center of the line. When a ROK unit was unable to contain the enemy, it broke, leaving an American artillery battalion unprotected. As a result, attacking Chinese destroyed the artillery battalion; 530 were men killed or captured, the most concentrated American loss of the war.

In mid-February the 23d Infantry and an attached French battalion were surrounded at Chipyong-ni. Three tanks were critical to a successful effort to plug a gap in the perimeter.[115] The 23d Infantry held and a relief column was able to get through. The relieving task force was formed from two tank companies, one from the 6th Tank Battalion and the other from the 70th Tank Battalion. The force also included a company of infantry from the 5th Cavalry and a detachment of engineers mounted on tanks. In spite of determined resistance, the task force broke through the Chinese encirclement. All the tanks made it, although three were damaged. Three tankers were killed and four wounded. Only twenty-three of the supporting infantrymen got through with the tanks.[116]

By the end of March, UN forces had formed a new line along the 38th parallel. Generally, plans were to limit subsequent advances to battalion-sized patrols to maintain enemy contact.[117] Although a major CCF counteroffensive was expected in April 1951, only senior officers were informed of a plan to conduct a fighting withdrawal to successive phase lines when the attack came.[118]

Then, on 22 April, the CCF launched their expected offensive. The main effort was directed at Seoul, with secondary efforts in the center of the UN line and in the mountains farther east. The Eighth Army held everywhere except in the center, where a ROK division broke when attacked. The Eighth Army was then forced to withdraw about twenty miles. Nonetheless, the 3d Infantry Division was able to stop the enemy north of Seoul. By early May the CCF had called off the offensive, having suffered horrendous casualties in a futile effort to capture the South Korean capital.[119]

During the UN forces' methodical advance from mid-February through March, in operations along the UN line established in late March, and in the April defensive operations, tanks were generally parceled out by platoons in support of infantry operations. Tanks were often dug in when deployed in defensive positions, and at times they were used as artillery by driving them onto ramps to provide greater gun elevation. This became the typical manner in which tanks were employed for the remainder of the war.

There were some exceptions, however. One was in an area northeast of Seoul called the Iron Triangle, bounded by Chorwon on the southwest, Kumhwa on the southeast, and Pyonggang on the northern corner. The Chorwon–Pyonggang axis formed an extension of the Uijongbu Corridor, one of the major approaches to Seoul. When UN forces resumed the offensive in early June 1951, the 3d Infantry Division had the responsibility for destroying Chinese forces in the Iron Triangle. The division plan was for the 7th Infantry to attack from the base of the triangle while the 64th Tank Battalion attacked up the left leg. In the meantime, the 25th Division's 89th Tank Battalion, with Company C, 64th Tank Battalion attached, was to clear the right leg.

North of Chorwon is perhaps the best tank country on the Korean Peninsula. The 64th Tank Battalion was able to attack up the valley with companies abreast while troops from the 7th Infantry faced tough opposition on the ridges within the triangle. Nevertheless, they were able in about a week to drive the enemy down the last slopes of the triangle into the waiting fire of two tank companies deployed on line. Many Chinese tried to escape through what appeared to be a one-kilometer gap in the line of tanks. The gap was deceptive, though, because tank platoons on either side of it could see all of the ground in between and thus were able to deliver destructive fire on the scattering Chinese. The two platoon leaders, Lts. Carmelo "Carm" Milia and Chuck Graham, both 1950 West Point graduates, conducted what one of them termed a "shooting gallery exercise." For the next two weeks, the 64th Tank Battalion aggressively patrolled the valleys in the Iron Triangle.[120]

By the spring of 1951 armistice talks had slowed the tempo of military operations. United Nations positions were stabilized essentially along the line occupied before the CCF offensive.[121] Nevertheless, the war was far from over. Authorization for the Eighth Army to make limited advances to reach the desired cease-fire line led to major battles in the "Punchbowl" in late August and early September. The valley itself was not important. However, the surrounding ridges, including "Bloody Ridge" and "Heartbreak Ridge," provided direct observation into UN positions. Infantry finally seized the high ground after heavy and exhausting fighting.[122] In spite of the terrain, tanks were used in support. Two regimental tank companies and a company from the division's tank battalion supported the 2d Infantry Division during the Bloody Ridge battles in August.[123]

Throughout the winter of 1951-52 and the following spring there was little activity on the battlefield other than patrolling. Both sides were dug in and both paid a hefty butcher's bill during offensive operations in 1952. Chinese attacks on "Old Baldy" and "Pork Chop Hill" cost both sides heavily and failed to alter the status quo.[124] Then, in the fall, a limited attack by two 7th Infantry Division battalions turned into a six-week-long debacle that resulted in nine thousand casualties and convinced the UN Command that such forays were not worth the cost.[125]

Mobile actions by armor remained the exception during this period. One such operation that succeeded was conducted by Task Force Gerhardt, assembled around the 187th Airborne RCT. The lead element, commanded by Maj. Charles A. Neuman, drove rapidly ahead with a tank platoon from Company B, 72d Tank Battalion. Neuman's tankers were supported by a platoon of engineers and a squad from the 187th RCT's I&R Platoon, which moved well ahead of the remainder of the task force. The aggressive fire from Neuman's force was devastating to successive Chinese forces encountered. After advancing several miles into enemy territory and reaching the objective, the force was reinforced. However, Lt. Col. Elbridge L. Brubaker, the 72d Tank Battalion commander, became a casualty. He was relieved on the spot by an impatient General Almond for making a tardy start.[126]

As the war settled into a stalemate and truce talks continued, tanks were generally deployed by platoons to augment infantry battalion positions. They provided direct firepower when needed. By 1952 a number of tanks were equipped with searchlights, which were useful during enemy night attacks and at times used to support friendly infantry night operations.

In one unfortunate case, when tank searchlights were used to support an infantry raid, the infantry unit suffered heavy casualties because the lights illuminated the attackers instead of the enemy defenders. Shortly after this incident occurred in a tank company, 65th Infantry, tank searchlights were removed. In the 3d Infantry Division half-tracks with quad-mounted .50-caliber machine guns from the division antiaircraft battalion were attached to the regimental tank companies and used along with tanks to provide fire support.[127]

This pattern continued throughout 1953. Tanks were mostly used in fixed positions, with occasional mobile operations, such as those conducted by the 3d Infantry Division in the Chorwon Valley near Kumwha in the spring of 1953.[128]

From the beginning of the Korean War there was suspicion in Washington that the North Korean attack might be part of a new Soviet global strategy. For the first time it appeared the Soviet Union was using forces of a satellite nation for external aggression. Was such action to become a worldwide pattern?[129] Thus, while the fighting continued in Korea, much thought was being given to maintaining a general reserve within the United States and building up a military force in Europe. Within the armor community, it was necessary not only to think about the tactical use of armor operations in places like Korea, but the possibility of major armor operations in the future, particularly in western Europe.

Even before the Korean War was over, the Army set about learning what lessons it could regarding tanks and armored warfare from its experiences in that conflict. Early in 1951 the Operations Research Office (ORO) of the Far East Command gathered data on armor operations in Korea and published a report that spring.[130] A number of its conclusions and recommendations are of considerable interest as they contributed to future tank development.

Considering the Korean experience, the study concluded that each infantry division could "usefully employ" about 150 tanks. In effect, this conclusion supported the division organization at the time: one divisional heavy tank battalion of 72 tanks and three regimental tank companies of 22 tanks each, for a total of 142 tanks. The battalion had 4 tanks in the headquarters and four companies of 17 tanks, each with three 5-tank platoons and 2 tanks in the headquarters. The regimental tank companies had four 5-tank platoons and 2 tanks in the headquarters.

The study supported a flexible organization in the tactical use of tanks, varying from one tank platoon per infantry company for a limited objective attack to one tank company per infantry company for capture of a distant objective or when heavy opposition was expected. Thus, the flexibility built into armor units from the start was validated.

The effectiveness of antitank mines was also noted. The study concluded that the North Koreans inflicted one tank casualty with every one hundred mines laid and that 34 percent of tank casualties were a direct result of the enemy's use of mines. It stated that tanks equipped with flails or some other antitank mine destroying device had reduced tank casualties. This, of course, was a lesson learned in World War II, but had either been forgotten or became the victim of budget restrictions.

Several other conclusions clearly contributed to future organization and tactical employment. The study concluded that tanks could be employed effectively at night if night vision devices were available and night training was conducted. It found that standard-issue winter clothing was impractical for tank crews. It also stated that light observation aircraft were invaluable to tank battalions.[131]

A number of meaningful recommendations were developed from conclusions reached by the study team.

Research and development (R&D) efforts were recommended in several areas. Efforts to improve reliability and maintainability, mobility and agility, and fuel consumption—as well as to reduce tank weight—were recommended. These, naturally, are the enduring goals of tank designers. A recommendation for an infantry support gun—such as a 105mm howitzer—mounted on a tank chassis harked back to the assault guns of World War II. Night vision and mine-clearing devices and improved radios were also recommended for R&D.

The study addressed the advantages of lighter tanks for general-purpose use, while still recognizing a need for both medium and heavy tanks. It recommended production of a thirty-five-ton tank while producing a light tank and a heavy tank "at a ratio of about one-tenth each of medium tank production."

An interesting recommendation, one that perhaps reflected the limited number of tank-versus-tank engagements during the Korean War, was to consider removing gyro-stabilizers, range finders, and cant correctors—all of which had proven to be of marginal utility in Korea.

The study's analysis of both friendly and enemy tank casualties resulted in recommendations for the development of antitank weapons and munitions.

Having concluded that using tanks as convoy escorts was inefficient, the study recommended development of a lightly armored convoy escort vehicle with multiple automatic weapons of "not less than cal .50 and not more than 40mm." This recommendation obviously reflected the effectiveness of the division self-propelled antiaircraft weapons used in a ground role.

There was only one specific recommendation concerning training, and it was a direct reflection of a weakness in the combat operations of the Eighth Army. The study suggested that both the Infantry and Armor Schools stress combined arms training, "emphasizing to each branch the capabilities and limitations of each member in the infantry-armor-air-artillery team." The authors observed that on too many occasions tanks drove off, leaving behind accompanying infantry that had dismounted. In addition, too many tanks were placed in night positions without infantry protection.[132]

The ORO, a Johns Hopkins University contract agency, produced a 1954 study entitled "Tank-versus-Tank Combat in Korea." The purpose of the study was twofold: first to present data to aid in evaluating the contribution of U.S. tank units to the destruction of the North Korean tank force, and, second, to assemble detailed quantitative data on tank-versus-tank engagements for use in the evaluation of proposed equipment changes. Because there were few tank-versus-tank actions after November 1950, the study covered only the period July–November 1950. Data was collected from 119 tank-versus-tank battles, using interviews, after-action reports, and other historical documents.[133] The study produced several significant conclusions.

American tanks were approximately three times as effective as North Korean tanks, destroying about 25 percent of the total enemy tank force. There were two major reasons. American tankers fired first more often (60 percent of the time), and had a higher first-round hit probability (0.64 versus 0.38).[134] This finding is logical given the number of North Korean tank crew prisoners who, in the previously mentioned ORO study, said that they had little training as tank crewmen.[135]

Among the observations of the study similar to those of the other ORO study was the need for more night training and better night vision equipment. Gunnery training emphasizing accurate first-round firing and opening fire at longer ranges, and equipment developments to support both, was also recommended.[136] This recommendation conflicts with the one in the earlier study calling for the elimination of such items as gyro-stabilizers, range finders, and cant correctors. These items of course contributed to long-range accuracy. The differing conclusions are likely traceable to the fact that the first ORO report was more historical, whereas the second, although based on historical data, focused on long-range improvement of armor's effectiveness in a tank-versus-tank environment.

The Korean War was a long and bloody affair. Yet, in the end, UN forces accomplished what they set out to do: maintain the independence of the Republic of Korea. The infantry carried the heaviest load, primarily because of the rugged terrain of the Korean Peninsula. Nevertheless, division tank battalions and reconnaissance companies and regimental tank companies played an important support role. Although outnumbered and outgunned at first, armor units did what they could, which was not much against the T-34s. As armor strength grew, tankers did much to strengthen and tactically support what were primarily infantry operations. Tanks were used in a variety of missions, not all of which resulted in their efficient use. However, once adequate tanks were available, they were generally used effectively.

Eighth Army was fortunate not to have been pitted against a sophisticated, modern tank force. Austere military budgets in the post–World War II years had left the U.S. Army—as in the post–World War I era—with mostly leftover weapons and equipment. The M4A3 tank developed early in World War II had serious problems dealing with the more heavily armored and armed German panzers. Toward the end of that war the M26 was available, but it was underpowered. Although the postwar M46 design represented an improvement over the M26, the slow rate of production meant that the U.S. Army received only a few of them in Korea.

That the tankers of the Korean War generation did as well as they did is a tribute to them. The U.S. Army went to war with a light tank company in the place of the heavy tank battalions authorized in each of the divisions stationed in Japan. World War II tanks retained as monuments on pedestals at Fort Knox had to be refurbished and issued for combat. Furthermore, ill-prepared crewmen had to meet their tanks just days before being committed to combat. Unfortunately, many of these untrained crews fought and died in tanks about which they knew little. This lack of preparedness should stand as an enduring message to the country concerning readiness.

While the war in Korea demanded much of the attention of America's leaders, it was also necessary for those leaders to see the war in the context of the overall world situation. The central question was whether the Chinese intervention in the war signaled the onset of World War III with the Soviet Union. This concern was serious enough that the JCS sent a warning message to all theater commanders and President Truman declared "a state of national emergency" on 15 December 1950.[137]

The start of the Korean War—as well as the prospect of eventual Soviet nuclear parity—also made it clear that the United States had to rethink its strategy of deterrence. It was apparent that America's military posture had not deterred limited communist aggression. The Korean experience suggested that forces-in-being and a demonstrable state of military readiness were required to deter acts of aggression. With Europe the focal point of the East-West confrontation, the United States began beefing up its NATO commitment. A ready ground defense force was the objective.[138] A new plan was formulated based on defending western Europe "as far to the East as possible," and the U.S. armed forces were gradually expanded to 3.5 million troops. The production of new weapons systems, including military vehicles, was to be increased by some 400 to 500 percent.[139] These decisions—as well as the need for modern weapons systems dramatically demonstrated during the Korean War—greatly increased activity in Armored Force structure and in tank procurement.

In early 1951 the 1st Armored Division was activated, joining the 2d Armored Division at Fort Hood, Texas. The next year the 2d Armored Division joined Seventh Army in Europe as part of a growing two-corps Army. Eight National Guard divisions were called into federal service; each had organic tank elements and a reconnaissance company. Two of these divisions went to Korea and two to Europe. Activation of the 11th Armored Cavalry Regiment brought the total of such regiments to five.[140]

Pre–Korean War combat vehicle development, which had been

M41 Walker Bulldog light tank. Patton Museum.

underfunded for years, was given new life. Production of the M41 light tank began in mid-1951, and by March 1952, more than nine hundred had been produced. The tank was nicknamed the Walker Bulldog in honor of the Eighth Army commander killed in Korea. The several thousand M41 and M41A1 tanks produced ultimately replaced the M24s in reconnaissance units. The chassis also served as the basis for self-propelled field and antiaircraft artillery weapons. However, the M41s arrived too late for use in the Korean War.[141]

Primarily because of budget constraints, the M41 tank was the only one fielded of the three tank types the Army needed—light, medium, and heavy. As noted earlier, a program was initiated to convert existing M26 Pershings to the product-improved M46 design.[142] Then, in 1947, military characteristics for a new medium tank were set forth and a program begun. By the time the Korean War started, development of prototypes of the T42 medium tank were underway. The urgent requirement to field a better tank than the M46 resulted in a decision to field an interim tank by combining the proven M46 chassis with the improved turret of the T42. That tank, designated the M47, was integrated into the Patton series. Although 8,576 were produced by 1953, none saw action in Korea. Furthermore, development of the T48 tank was well underway, so there was no additional need for the interim M47.[143]

With the end of the Korean War, America's military strategy concentrated almost solely on the Soviet threat in western Europe. Clearly, if there was to be a war there, the Armored Force would play a far more significant role than it had in Korea.

NOTES

1. Russell F. Weigley, *The American Way of War: A History of United States Military Strategy and Power* (New York: Macmillan, 1973), p. 366.

2. Ibid., p. 368.

3. Walter Millis, ed., *The Forrestal Diaries* (New York: Viking Press, 1951), p. 100.

4. Weigley, *American Way of War,* p. 365.

5. Millis, *Forrestal Diaries,* p. 373.

6. Mary Lee Stubbs and Stanley R. Connor, *Army Lineage Series, Armor-Cavalry,* pt. I, *Regular Army and Army Reserve* (Washington: CMH, 1984), pp. 74, 77.

7. Millis, *Forrestal Diaries,* p. 253.

8. Stubbs and Connor, *Armor-Cavalry,* p. 76.

9. The General Board, "Organization, Equipment and Tactical Employment of the Armored Division," U.S. Forces, European Theater, 1945, CMH Files, Washington, D.C.

10. Ibid.

11. Ibid.

12. The General Board, "Organization, Equipment and Tactical Employment of the Infantry Division," U.S. Forces, European Theater, 1945, CMH Files.

13. James I. King and Melvin A. Goers, "Modern Armored Cavalry Organization," *Armored Cavalry Journal* (July-Aug. 1948): pp. 47-50.

14. Ibid.

15. Ibid.

16. Richard P. Hunnicutt, *Patton: The History of the American Main Battle Tank,* vol. 1 (Novato, Calif.: Presidio Press, 1984), p. 9.

17. Richard P. Hunnicutt, *Sheridan: A History of the American Light Tank,* vol. 2 (Novato, Calif.: Presidio Press, 1995), pp. 6, 25, 53.

18. Hunnicutt, *Patton,* p. 10.

19. Ibid., p. 14.

20. Richard P. Hunnicutt, *Firepower: A History of the American Heavy Tank* (Novato: Presidio Press, 1988), p. 70.

21. Ibid., p. 111.

22. Clay Blair, *A General's Life* (New York: Simon and Schuster, 1983), p. 489.

23.. T.R. Fehrenbach, *This Kind of War* (New York: Macmillan, 1963). p. 65.

24. Ibid., p. 34.

25. Harry S. Truman, *Memoirs by Harry S. Truman: Years of Trial and Hope, 1946-1952,* vol. 2 (Garden City, N.Y.: Doubleday, 1956), p. 333.

26. J. Lawton Collins, *War in Peacetime* (Boston: Houghton Mifflin, 1969), p. 16.

27. Truman, *Memoirs,* p. 337.

28. Ibid., p. 343.

29. Billy C. Mossman, *Ebb and Flow, November 1950-July 1951* (Washington: CMH, 1990), p. 30; and Collins, *War in Peacetime,* p. 67.

30. James F. Schnabel, *Policy and Direction: The First Year,* U.S. Army in the Korean War (Washington: OCMH, 1972), p. 54.

31. Arthur W. Connor Jr., "The Armor Debacle in Korea, 1950: Implications for Today," *Parameters* (Spring 1992): p. 68.

32. Schnabel, *Policy and Direction,* p. 46. The "Easy Eight" was specifically the M4A3E8. This author is not sure whether all the M4A3 tanks in the inventory were M4A3E8s.

33. Ibid., p. 36.

34. Fehrenbach, *This Kind of War,* p. 17.

35. Vincent V. McRae and Alvin D. Coox, *Tank-vs-Tank Combat in Korea* (Washington: Operations Research Office, Johns Hopkins University, 1955), p. 10.

36. Fehrenbach, *This Kind of War,* p. 76.

37. Ibid., p. 98. Also see Eric C. Ludvigsen, "The Failed Bluff of Task

Force Smith: An 'Arrogant Display of Strength,'" *ARMY* (February 1992): pp. 36-40, 42-45.

38. Schnabel, *Policy and Direction*, p. 92.

39. Ibid., p. 94.

40. Connor, "Armor Debacle in Korea," p. 73.

41. Telephone interview with Col. Richard Coffman, U.S. Army (Retired), 30 Apr. 1997.

42. H.W. MacDonald, et al., *The Employment of Armor in Korea*, 2 vols., General Headquarters Operations Research Office, Far East Command, 1951, p. 2:186.

43. Ibid., p. 190.

44. Ibid., p. 188.

45. Roy E. Appleman, *South to the Naktong, North to the Yalu*, U.S. Army in the Korean War (Washington: OCMH, 1960), p. 231.

46. Connor, "Armor Debacle in Korea," p. 72.

47. MacDonald, et al., *Employment of Armor in Korea*, p. 2:172; and Appleman, *South to the Naktong, North to the Yalu*, p. 236.

48. Appleman, *South to the Naktong, North to the Yalu*, p. 240.

49. Ibid., p. 381.

50. MacDonald, et al., *Employment of Armor In Korea*, , p. 1:18.

51. Collins, *War in Peacetime*, p. 110.

52. MacDonald, et al., *Employment of Armor in Korea*, p. 1:22.

53. Ibid., p. 2:30.

54. Ibid., p. 2:200.

55. Ibid., p. 1:164.

56. Maj. Gen. William F. Ward Jr., U.S. Army (Retired), comment to author; and Stubbs and Connor, *Armor-Cavalry*, p. 394.

57. Fehrenbach, *This Kind of War*, p. 240.

58. Appleman, *South to the Naktong, North to the Yalu*, p. 509.

59. Ibid., p. 521.

60. Ibid., p. 535.

61. Ibid., p. 542.

62. Ibid., pp. 557, 572.

63. Ibid., p. 573.

64.. MacDonald, et al., *Employment of Armor in Korea*, p. 2:176.

65. Appleman, *South to the Naktong, North to the Yalu*, p. 578.

66. MacDonald, et al., *Employment of Armor in Korea*, p. 2:185.

67. Ibid., p. 2:30.

68. Appleman, *South to the Naktong, North to the Yalu*, p. 590.

69. Ibid., p. 600.

70. Collins, *War in Peacetime*, p. 147.

71. Appleman, *South to the Naktong, North to the Yalu*, p. 611.

72. Fehrenbach, *This Kind of War*, p. 274.

73. Appleman, *South to the Naktong, North to the Yalu*, p. 623.

74. Ibid., p. 628.

75. MacDonald, et al., *Employment of Armor in Korea*, pp. 2:147, 2:153.

76. Collins, *War in Peacetime*, p. 166.

77. Schnabel, *Policy and Direction*, p. 216.

78. Collins, *War in Peacetime*, p. 166.

79. Ibid., p. 180; and Appleman, *South to the Naktong, North to the Yalu*, p. 664.

80. Appleman, *South to the Naktong, North to the Yalu*, p. 681. On the battle of Chongchon and the entrance of Chinese forces, see Mossman, *Ebb and Flow*, pp. 61-83.

81. Appleman, *South to the Naktong, North to the Yalu*, p. 691.

82. Ibid., p. 700.

83. Collins, *War in Peacetime*, p. 189.

84. Ibid., p. 191.

85. Appleman, *South to the Naktong, North to the Yalu*, p. 756.

86. MacDonald, et al., *Employment of Armor in Korea*, p. 1:21.

87. Andrew Greer, *The New Breed: The Story of the U.S. Marines in Korea* (New York: Harper & Brothers, 1952), p. 248.

88. Ibid., p. 262.

89. Ibid., p. 243.

90. Ibid., p. 314.

91. Ibid., p. 317; and Mossman, *Ebb and Flow*, pp. 102-103.

92. Collins, *War in Peacetime*, p. 224.

93. Greer, *New Breed*, p. 349.

94. Ibid., p. 371.

95. Roy Appleman, *Disaster in Korea: The Chinese Confront MacArthur* (College Station: Texas A&M University Press, 1989), p. 26.

96. Ibid., p. 44.

97. MacDonald, et al., *Employment of Armor in Korea*, p. 1:19.

98. Maj. Gen. William F. Ward Jr., U.S. Army (Retired), to author, n.d., 1996.

99. Appleman, *Disaster in Korea*, p. 52; and Mossman, *Ebb and Flow*, pp. 63, 68.

100. Appleman, *Disaster in Korea*, p. 59.

101. Ibid., p. 76.

102. Ibid., p. 80.

103. Col. Charles W. Hayward, U.S. Army (Retired), to author, 21 Aug. 1996.

104. S.L.A. Marshall, *The River and the Gauntlet* (New York: William Morrow, 1953), p. 43.

105. Ibid., p. 152.

106. Ibid., p. 187.

107. Appleman, *Disaster in Korea*, p. 262.

108. Ibid., p. 166.

109. Ibid., p. 343.

110. Ibid., p. 354.

111. Ibid., p. 359.

112. Ibid., p. 383.

113. Collins, *War in Peacetime,* p. 241.

114. Joseph C. Goulden, *Korea: The Untold Story of the War* (New York: Times Books, 1982), p. 433.

115. Ibid., p. 449.

116. Collins, *War in Peacetime,* p. 260.

117. Ibid., p. 267.

118. Ibid., p. 268.

119. Ibid., p. 296; and Col. Carmelo P. Milia, U.S. Army (Retired), comment to author.

120. Col. Carmelo P. Milia to author, n.d., 1996.

121. Collins, *War in Peacetime,* p. 308.

122. Ibid., p. 309.

123. Russell A. Gugeler, *Combat Actions in Korea* (Washington: Combat Forces Press, 1954), p. 211.

124. Fehrenbach, *This Kind of War,* pp. 604, 626.

125. Collins, *War in Peacetime,* p. 322.

126. Mossman, *Ebb and Flow,* pp. 481-83, and Gugeler, *Combat Actions,* p. 193.

127. Col. Herman J. Vetort, U.S. Army (Retired), to author, January 1997.

128. Ibid.

129. Collins, *War in Peacetime,* p. 78.

130. MacDonald, et al., *The Employment of Armor in Korea,* vol. 1, Foreword.

131. Ibid., p. 1:2.

132. Ibid., p. 1:4; Gugeler, *Combat Actions,* p. 138; and Hayward to author.

133. McRae and Coox, *Tank-vs-Tank,* p. 1.

134. Ibid., p. 2.

135. MacDonald, et al., *Employment of Armor in Korea,* p. 1:83.

136. McRae and Coox, *Tank-vs-Tank,* p. 4.

137. Blair, *A General's Life,* p. 609.

138. Weigley, *American Way of War,* p. 396.

139. Blair, *A General's Life,* p. 610.

140. Stubbs and Connor, *Armor-Cavalry,* pp. 7, 228.

141. Hunnicutt, *Sheridan,* pp. 21, 199; and Mark A. Olinger, "Too Late for the War: The U.S. Industrial Base and Tank Production 1950-1953," *ARMOR* (May-June 1997): pp. 15-16.

142. Hunnicutt, *Patton,* p. 32, and Olinger, "Too Late for the War," p. 16.

143. Hunnicutt, *Patton,* p. 52, and Olinger, "Too Late for the War," 15-17.

8

The Marine Corps's Struggle with Armor Doctrine during the Cold War (1945-70)

Kenneth W. Estes

The Japanese surrender announcement found most of the Marine Corps, then some 458,000 strong, deployed in the western Pacific with the I, III and V Amphibious Corps, their six divisions and four aircraft wings in the Fleet Marine Force, Pacific (FMFPAC). Apart from demobilization concerns, their duties consisted of disarming Japanese forces and occupying parts of Japan and China. Postwar planning centered on a ready force of two divisions and two aircraft wings, plus corps troops, balanced between the Marine Corps's East and West Coast bases, for duty primarily with the Atlantic and Pacific Fleets. Marines quickly terminated their occupation duties in Japan, but deployments in China dragged on into early 1947. Nevertheless, by the end of 1946, barely fifteen thousand Marines remained on FMFPAC's rolls, and an even smaller number in the fledgling Fleet Marine Force, Atlantic (FMFLANT).[1]

The Pacific War saw the refinement of amphibious doctrine and the expansion of the FMF far beyond the scope imagined by the planners and dreamers of the interwar period. The amphibious tractor (amtrac) became the staple of the amphibious assault force, spearheaded by its howitzer-armed and better-armored cousin, the armored amphibian. By mid-1945, three armored amphibian battalions and nine amphibian tractor battalions stood ready for the final assaults on Japan. The demobilization left only a skeletal battalion with one amtrac company and an LVT(A) platoon in each division, but the 10th, 11th, and 12th Amtrac Battalions, each with an armored amphibian company included, were retained on paper with the Marine Corps Reserve.

More than eighteen thousand amtracs were produced during the war, but efforts to recover them from the western Pacific proved limited. Most were obsolete or worn out, and were left in Marianas or Hawaiian staging bases. Some three hundred LVTs were shipped from Guam to the depot at Barstow, California. Several hundred more made their way to the desert depot from Oahu. Most of the recovered vehicles were of the latest LVT3 and LVT(A)5 types, and by 1948 fifteen hundred amtracs had been mothballed at Barstow in various states of condition.[2]

Tank units in the Marine Corps had endured rapid changes during the war in terms of both equipment and tactical employment. The light tank generally failed in the infantry support role so often demanded of it, and the light tank battalions and mixed battalions eventually gave way to medium battalions, six of which stood ready in their parent Marine divisions for the final battles in 1945. The tank-infantry tactics that evolved from the Pacific island battles stressed close support of squads of riflemen by individual tanks, and vice versa. Tank platoons thus were organized with four tanks. By war's end, the flame tank platoons made up of three vehicles had been moved from the battalion structure and one was assigned to each line company, reflecting its essential part in the tactical array. The Shermans had proven unstoppable to all but the most desperate Japanese countermeasures and were widely credited by the Imperial Army with unhinging Japanese defenses. Although tied to the infantry support role, the tankers in the Corps still relished the opportunity to employ mobility and mass and did so on Roi-Namur, Guam, Okinawa, and in a massive attack on Iwo Jima's airfield by three battalions.[3]

Unlike the development and production of the amtracs, performed for the Marine Corps by the Navy's Bureau of Ships, the Corps came to depend exclusively upon the Army for the procurement of tanks. Army production schedules and logistics support policies began to dictate Marine Corps tank equipment decisions as early as 1944. The cherished twin-diesel M4A2 Shermans had to be replaced by gas-powered M4A3s rapidly, before the production line changed to the undesirable 76mm guns, simply to remain within Army logistics support and design configuration policies. Later, the Marine tank fleet was reequiped with M4A3E8s equipped with 105mm howitzers when the Army shifted to production of the M26 Pershing, keeping only the 105mm Sherman line open. Curiously, the advent of the M24 caused the Corps to again consider the employment of light tanks. Ten M24s were procured in 1945 for testing with a fording kit at Camp Pendleton. They remained in storage well after the Korean War.[4]

The recovery of tanks scattered throughout the western Pacific fared little better than that of the LVTs. An uneven flow of M4A2, M4A3, and M7B1 self-propelled artillery vehicles crossed Camp Pendleton. The 76mm M4A3E8 remained standard issue for the postwar battalions, but the 2d Tank Battalion on the East Coast had to make do with 75mm M4A3s drawn from local stores, as transportation funding shortfalls for the 105mm vehicles available on the West Coast delayed transfer until 1947. More urgent yet was the requirement to ship tanks to North China, where the 1st Marine Division needed its tank battalion to better handle the incipient combat situation there. The 1st Tank Battalion's platoons fought some minor skirmishes with M4A3E8 105mm tanks against Chinese forces well into 1946.[5]

The 75mm-armed Sherman flame tanks of the CWS-POA-H1 design were placed in storage, twenty-seven of them reconditioned for postwar requirements. By 1947, the Corps retained only two battalions of tanks—each with two companies equipped with seventeen M4A3 105mm tanks and six flame tanks. The Marine Corps Reserve retained the 10th and 11th Tank Battalions on paper.[6]

Training for the Corps's armored vehicles ground to a standstill after the war. The tank and LVT crewmen and mechanic courses were concentrated in a Tracked Vehicle School Battalion—later company—at Camp Pendleton. In 1947, the Commandant discontinued all courses except for LVT maintenance. Officer and NCO refresher courses conducted after war's end already had fallen prey to the budget ax. Many of the school personnel filled in deficiencies at the nearby 1st Amtrac Battalion, which boasted only three officers and fifty-six men trained on the LVT. The postwar budget cuts left little encouragement for the Corps at large, and the tracked vehicle units clearly bore their share of the impact.[7]

New equipment still concerned the Corps's leaders despite the clear lack of troops and operational funds. In a pattern to be duplicated almost every twenty-five years, the Corps obtained more modern tanks and LVTs and planned future organizations in the midst of severe budget limitations. The LVT3 fleet, designed in 1943 and first used at Okinawa, clearly required design improvements. Following the swamping of several vehicles during Camp Pendleton exercises, the commandant requested a redesign in 1948.[8] The Bureau of Ships added 860 pounds of side armor, cargo covers, escape hatches, a .30-caliber machine-gun cupola, and extended the bow to improve buoyancy, resulting in the LVT3C. The Long Beach Naval Shipyard began reconfiguring twelve hundred vehicles to this design in 1949.[9]

The tank modernization program proceeded from the twin pillars of dependency on Army initiatives and evident Marine Corps recognition that the tank would play as crucial a role on the battlefield of the future as it had in settling the issue against the Japanese garrisons in the Pacific. In March 1946 the director of Marine Corps Schools warned the commandant of the prevailing views: "In general, the tanks with which the Marine Divisions ended the war are now definitely obsolete. The tank for the future must be capable of withstanding greater punishment, be more mobile, and have improved hitting power. The present tanks are too slow and too vulnerable to antitank weapons."[10]

Actually, the commandant had already ordered the purchase of ninety-one M26 Pershings in late 1945 under the fiscal year (FY) 1948 budget. This heavy (later reclassified as a medium) tank mounting the excellent 90mm gun was slow in reaching the FMF because of frequent budget reductions, although this was partially rectified by supplementary appropriations and congressional forced transfers from excess Army stocks. Thanks to the wartime acquisition of the 90mm antiaircraft (AA) gun as standard, the Corps had plenty of ammunition for this tank, except for the M304 High Velocity Armor-Piercing (HVAP) round. The M26 tank entered service at this point as a mixed blessing, however. Its fording gear required much local welding and reworking to function, and the transmissions and controlled differential steering units gave headaches to Marine tankers accustomed to their more robust Shermans.[11]

In 1948 the Commandant authorized the M26 to be placed in service but limited it to use only in special exercises, experiments, and demonstrations. It was not to be used in amphibious operations, landing exercises, or maneuvers without his authorization. The commandant rescinded this stricture only in late 1949, after the staff was satisfied that the battalions had sufficient experience with the tank. Doubts about the M26 voiced by the Corps's leaders came in wake of an Army offer of additional M26s in late 1949. The staff noted that the "CMC desires [that the] Heavy Tank requirement be broached in connection with [the] FY51 budget and [that the] M26 cannot be classified as 'substitute heavy tank' at present."[12] Furthermore, the staff reported, "It is understood that [the] Marine Corps is only tentatively committed to the procurement of these additional tanks. It is requested that procurement not be finalized pending final determination of desirability of expenditure of funds for this equipment."[13]

Still, the commander of the 2d Tank Battalion had asked in late 1948 that he be allowed to swap more of his authorized M4A3E8s for

M26 tanks, to facilitate the rotation of the better tanks on the now-standard Mediterranean battalion landing team deployments. With only two line companies in each tank battalion, the five M26s per company permitted little slack in operations.

At the end of 1949, the Corps's 102 M26 tanks were mostly in the hands of depot personnel. Camp Lejeune had 37, 55 were in Barstow, and Quantico and Camp Pendleton had 5 each.[14]

In addition, the operating conditions for Marine Corps tankers seldom matched those of their Army counterparts. Not only did tanks doctrinally remain shackled to infantry support missions but the efforts of facilities engineers (and later environmentalists) also posed problems. In October 1948 the California Department of Public Works forbade the movement of both M26 and M4A3E8 tanks on tank transports between Camp Pendleton and San Diego.[15] And the last and most feared Quartermaster of the Marine Corps, Maj. Gen. W.P.T. Hill, ordered that tracked vehicles keep off of base roads, even in training areas: "It is the policy of the Marine Corps not to operate tanks on paved roads, road shoulders or drainage structures. Where necessary to follow the direction of a road, parallel tank trails shall be used and where necessary to cross paved roads, concrete tank crossings shall be constructed."[16]

Marine Corps doctrine for the two chief items of armor in use, the amtrac and the tank, continued to reflect the experience of the Pacific War. The Amphibious Operations series of manuals published during 1945-48 by the Marine Corps Schools were compendiums of experience and guidelines for operations. Most focused on specific functions or units, including the Employment of Tanks (Phib-18) and the LVT and LVT(A) (Phib-23). Each of these prescribed training, embarkation, ship-to-shore movement, and employment following the landing. The tanks were to support the infantry ashore, anchor the antimechanized defense and provide specialized vehicles, such as tank dozers and flame tanks to reduce field fortifications. A single page devoted to mechanized attack permitted the free use of tanks when the terrain and enemy situation permitted. The amtracs executed typical wave movements against an enemy beach, with armored amphibians leading and providing last-minute suppressive fires. After landing the assault waves, the amtracs reverted to ship-to-shore resupply tasks and the armored amphibians furnished direct-fire support as assault guns until the tanks landed, thereafter reverting to reinforcing artillery batteries for the remainder of the battle. These publications were converted with little change into the Landing Force Manual series of the 1950s and the FMF Manuals of the 1960s through the late 1970s. (LFM 17 and FMFM 9-2 governed amphibian vehicles, and LFM 14 and FMFM 9-1 covered tank employment, although the Army's FM 17 series of manuals was commended for tanks in land warfare). Not until 1981 would Marine Corps doctrinal publications outline task organizations for armored operations on a par with the Army's combined arms concepts of armored infantry and armor companies with appropriate attachments and supporting arms.

Major Gen. Lemuel Shepherd reported to the Commandant on the 2d Marine Division's amphibious landing exercise at Culebra Island in 1947. After watching the 8th Marines land followed by A Company, 2d Tank Battalion, an hour and a half later, he stated: "the time tanks should be brought ashore has always been subject to differences of opinion. I personally believe they should be landed as soon after the assault waves as possible, in order to provide direct support to the infantry during the most critical part of the landing." He also noted that the tank battalion was still using 75mm M4A3 tanks.[17]

Observing similar 2d Division exercises at Camp Lejeune in early 1950, amphibious expert Col. E.R. Smock wrote that tank units had only tank dozers for use in overcoming barriers. The need for more countermine and barrier equipment could not be overstated. However, the use of tanks as independent units earned no praise from him: "I am becoming increasingly concerned by the growing tendency of tank officers to seek means of breaking away from and discrediting the battle-tested and proved tank-infantry team and the employment of tanks in providing direct fires in support of infantry units." He also faulted the infantry for letting LVT commanders direct their vehicles in waterway crossings, roaming along looking for landing sites instead of the infantry commanders directing the vehicle movement and disembarkation on the other side.[18]

That the tank was considered an antitank weapon bore fruit in the Marine Corps in the form of the postwar infantry regiment's antitank company. The Corps had formed one antitank battalion during World War II, the 2d Antitank Battalion. It arrived at New Caledonia at the end of 1943 and was deactivated to provide antitank gun platoons to regimental weapons companies. Additional antitank gun batteries formed in the special weapons battalions of the wartime divisions. At the end of the war, and with the disappearance of the old defense battalions and their handy organic tank platoons, the postwar infantry regiment received an antitank company as compensation. Of the two platoons, only the antitank gun platoon was manned in peacetime, its incredibly obsolete 37mm guns giving way to 75mm recoilless rifles. The tank platoon of five medium tanks was carried on paper as a wartime organization. This structure remained in the division through 1957. Only during the Korean War did the regimental tank platoons enter service.

The sole attempt to rationalize the myriad forces affecting Marine Corps tank development in the immediate postwar period took form in 1949 in the report of the Armor Policy Board, convened by the Plans and Policy Division of the Marine Corps G3. It provided the following recommendations, which were affirmed in 1950 in the Marine Corps Equipment Policy:[19]

> The function of the tank in the amphibious operation or in base defense is the destruction of enemy forces by means of armor protected fire power, mobility and shock action.
> There continues to be an essential requirement for a *close support tank* [emphasis added] at division and regimental level.
> There is a requirement for a *destroyer type tank* [emphasis added] to destroy hostile heavily armored vehicles, assist in the reduction of heavy fortifications, and support the "close support tanks."
> There is no requirement for a reconnaissance vehicle of the light type.
> There is a requirement for a means of transporting tanks ashore in the ship-to-shore phase of the amphibious operation. This means of transportation must permit the tank gun to fire while enroute to the beach.
> There are urgent requirements for tank-mounted mine detectors and mine eradicating devices that can be operated from within the tank.
> All responsibility for developing is with the Army except for flotation devices.

The study is especially amazing in that it remained the single example of armor guidance in the Marine Corps through the 1980s, making it either a significant feat for the visionaries of 1949 or testimony of the Corps's reticence during the ensuing generation. No similar study or guidance ever again emerged from the Corps's leaders.

The force structure created to reflect this policy consisted of a divisional tank battalion of four medium tank companies with seventeen tanks each, together with the existing regimental tank platoons. The force (corps echelon) tank battalions would be formed with three companies each with seventeen heavy tanks. The Commandant's guidance for FY 1951 directed that the peacetime divisional battalions field thirty-five medium and five heavy tanks, with two force heavy tank companies with seventeen tanks each. However, the force companies were cut from succeeding revisions. Amtrac unit strength remained un-

changed, and two tank and three amtrac battalions remained in the Marine
Corps Reserve.[20] The medium tank would be the T42 design (later superseded by the T48 series), under development by the Army since 1946.[21]

On 22 May 1950 the inspector-instructor of the 10th Tank Battalion, Marine Corps Reserve, reported that he had on hand only four M4A3
tanks and one recovery vehicle, all in poor condition.[22] One month later,
North Korean forces attacked the Republic of Korea.

Marine Corps forces mobilized and improvised in 1950 from a weak
peacetime posture to fight the most varied five-month campaign in the
Corps's history. The 1st Marine Brigade took the few ready units to the
defense of the Pusan Perimeter. Then, the 1st Marine Division—hastily
amalgamated from the rest of the active forces and initial Marine Corps
Reserve mobilization—landed at Inchon and, reinforced by its third
infantry regiment, cleared most of Seoul. The division then shifted by
sea to Korea's east coast, landed administratively, and pushed north
toward the Yalu River—where it collided with the Chinese counteroffensive and was forced to withdraw ignominiously to the port of
Hungnam and reassemble near Pusan. At the end of 1950, the Korean
War changed direction, becoming first a slugfest to establish equilib-
rium on the former demarcation line between the two Koreas, and then
a wearying war of attrition to hold the line for an eventual peace.

The 1st Provisional Marine Brigade was formed hurriedly in July
1950 for immediate embarkation. The Barstow depot provided M26 tanks
to replace Company A, 1st Tank Battalion's M4A3s and new LVT3s for
the 1st Amtrac Company. A Company had one day on the range with
two tanks to familiarize its crews with the M26—enough time for each
gunner and loader team to fire two rounds of 90mm ammunition. The
choice of the M26 seemed obvious given the terror already generated
by the NKPA's T-34s. With the attached 5th Marines's infantry battalions operating with just two rifle companies per battalion, the tank
company's presence was insurance against disaster. That is not to say
that the opening moves were auspicious, however. The tank and amtrac
companies sailed aboard two Navy amphibious dock ships, one of which
accidentally flooded its well to a depth of five feet. Fourteen tanks and
three hundred rounds of scarce 90mm ammunition were damaged. Upon
return to port the tanks were repaired and the ship sortied again.

Arriving at Pusan on 2 August, the tanks were rushed into action
on the tenth. During the 5th Marines's attack on Paedun-ni and Kusong,
1st Platoon tanks crossed a river. One tank crashed through the bridge
and the second threw a track crossing the stream, stalling a long column of vehicles. The surviving section continued to support the rifle

North Korean T-34s destroyed in Naktong by Company A, 1st Tank Battalion, 1st Provisional Brigade. USMC History and Museums Division.

companies of the vanguard. Two days later, five tanks accompanied the 1st Battalion, 5th Marines, on an eleven-mile road march toward Sunchon. The NKPA attempted an ambuscade, but the tanks, roadbound but mobile forts, provided dominant fire support to neutralize and break the attackers.

This seasoning came none too soon, for the brigade was rushed into the first Battle of the Naktong when the NKPA 4th Division occupied a bridgehead with six battalions and at least four T-34 tanks. The 1st and 2d Battalions, 5th Marines, counterattacked on 17 August, each accompanied by a tank platoon. At 8 P.M., the 1st Battalion detected the approach of four T-34s. A flight of F4U Corsair fighter-bombers killed the trail tank and the rest hit the 1st Battalion's prepared positions, where they were greeted with a volley of 3.5-inch bazooka rockets and 75mm recoilless rifle rounds. The first tank was crippled, receiving the coup de grace from an M26. Two M26s combined to destroy the second enemy tank, and a combination of all weapons killed the last T-34. The first tank-versus-tank encounter for Marines since 1944 had gone perfectly. The M26 platoons worked with the infantry to clear the ridges and eject the NKPA from the brigade's zone of the Naktong Bulge.

In early September the sequence virtually repeated itself in the second Battle of the Naktong. Company A tankers caught three T-34s napping and destroyed them outright, going on to scatter accompanying infantry. Later, a fourth tank was knocked out in a thicket and a fifth, which had been abandoned, was captured.

In all these fights, the M26s remained practically road bound but untouchable to the 45mm antitank guns attached to the NKPA infantry. A few tank Commanders were wounded by small arms as they exposed themselves while searching out targets. The single setback came on 5 September when two T-34s and an armored personnel carrier (APC) accompanied a company of NKPA infantry moved against the 1st Battalion, 5th Marines, lines. With radios inoperable, the tank platoon rushed toward the action, the two lead M26s traversing their turrets in the wrong direction. The T-34s knocked out both, but the Marine tankers successfully bailed out. The infantry destroyed the enemy vehicles with a barrage of 3.5-inch rockets but too late. Of the eight hulks on that stretch of road, two were M26s.[23]

While the Naktong fighting peaked, the rest of the 1st Marine Division (less the reinforced 7th Marine Regiment, which was still assembling in California) straggled into Japanese ports and prepared for the amphibious landing aimed at the port of Inchon on the coast west of Seoul. Only 3,450 troops had remained at Camp Pendleton with the division cadre after the 1st Provisional Marine Brigade sailed. First to arrive as replacements were companies of the Organized Marine Corps Reserve: the 13th Infantry Company from Los Angeles, the 12th Amtrac Company from San Diego, and the 3d Engineer Company headquartered in Phoenix. While regulars were being shaken out of barracks, bases, and the 2d Marine Division and more reservists reported in, an order to keep troops under eighteen years of age out of the fighting resulted in the deliberate crippling of the 1st Armored Amphibian Battalion, which received five hundred of the division's minors. The bulk of the combat-ready Marines with tracked vehicle or weapons experience went to the 1st Tank Battalion, but many men without driving or gunnery experience had to be assigned as crewmen. Flame tank crews trained briefly on the vehicles at the Barstow depot as they were being issued. B Company was activated on 21 July and received the first eighteen M26s that arrived on 5 August. It was joined by A Company, 2d Tank Battalion, which was redesignated C Company, 1st Tank Battalion. The battalion began to embark for Japan on 16 August, but D Company did not form and catch up until late in September.[24]

The Inchon landing operation on 15 September 1950 remains shrouded in Marine Corps legend and lore. The truly legendary part should be the degree of improvisation resorted to by the Navy and Marine Corps planners, who could conveniently draw upon all the recent experience of the Pacific War against an opponent hardly reminiscent of that conflict. Thanks to the earlier disabling of the 1st Armored Amphibian Battalion, the Army's A Company, 56th Amphibian Tractor Battalion, had to be used for the LVT(A)5 spearhead. These eighteen armored amphibians led 164 LVT3C and eighty-five DUKW vehicles ashore from LSTs to land the 1st Marines southeast of the Inchon city limits. The 5th Marines, swinging around from Pusan, landed from landing craft directly in the port.

The key to success was the preliminary assault on Wolmi-do, an island guarding Inchon, to which it was connected by a causeway. Landing craft brought in the infantry, supported by a detachment of the now veteran A Company, 1st Tank Battalion: six M26s, one flame and two dozer Shermans, and a retriever. These quickly blasted and burned the NKPA troops from their caves and entrenchments as the infantry quickly overran the island. The causeway proved free of obstructions and the tanks quickly linked up with the rest of the landing force in the port.

The main landings took place with little resistance. The LVT waves of thirty vehicles each became disorganized after the first three touched down. Most LVTs stayed on the beach or bogged down trying to push inland. A floating supply dump of twenty-four LVTs had to wait until the next day after ten amtracs bogged down on the mud flats. The tanks landed over the next day and joined the two infantry regiments as they pushed east on the road to Seoul. Five miles ahead, Marine pilots spotted six T-34s moving without infantry support toward the town. Eight Corsairs hit them with rockets and napalm, knocking out two but claiming all six. That afternoon, the regiments settled in on their beachhead perimeter. A platoon from A Company, 1st Tank Battalion, spotted three intact T-34s at the site of the morning's air strike and destroyed them with twenty rounds fired by two M26s. Two hulks remained from the air strike, but T-34 number six was never seen.[25]

The advance continued and the 5th Marines halted the night of 16-17 September on the approaches to Ascom City. That morning, yet another incredibly inept NKPA detachment of six T-34s, this time with about two hundred infantry, blundered into the Marine position. This event provided vengeance for the platoon from A Company, 1st Tank Battalion, that lost two tanks on the Naktong position to T-34s. The M26s fired forty-five rounds of armor-piercing (AP) ammunition as the Marines from two adjacent battalions added bazooka and recoilless rifle fire. All of the NKPA were wiped out in the brief firefight, and tanks claimed five of the six tank kills. Certainly their hits had proved most

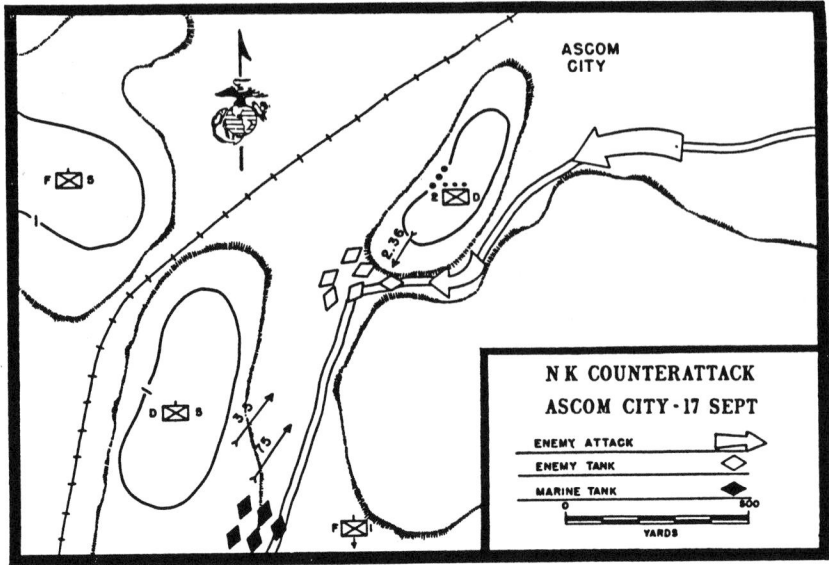

destructive, but all weapons hit all the tanks, save the first one, which was taken out by an outpost bazooka gunner. The combat had barely concluded when Gen. Douglas MacArthur, VAdm. Arthur D. Struble, and a dozen other flag officers drove up with X Corps staff officers and a gaggle of reporters. The made-to-order publicity scene lay before them as the Marines surveyed their handiwork. Kimpo Airfield fell to the tank-infantry teams that day, and the Han River presented the last obstacle to Seoul.[26]

Crossing the Han presented another opportunity for the Marine Corps to practice its amphibious art. The 5th Marines attempted the crossing early in the morning on 20 September, following on the heels of a reconnaissance party that had gone across in rubber rafts. The ubiquitous Army armored amphibians and 1st Amtrac Battalion LVT3Cs were in direct support. Tanks and artillery crossed later on bridge ferries. The result was a fracas as four LVTs stuck in the mud and the rest returned to the near shore. Greatly embarrassed, the Marines reorganized and crossed in daylight, the second battalion of infantry remaining mounted into the hill objectives, stopping only when swamps and light bridges barred further movement. The Army LVT(A)5 company led the 2d Battalion, ROK Marines, mounted in DUKWs in trace of the 5th Marines.

The 1st and 5th Marines closed on Seoul to the south and east on foot through rugged hills and growing NKPA resistance. The Amtrac Battalion carried the 1st Marines, the Army's 32d Infantry, and the ROK

An LVT(A)5 prepares to cross the Han River beach in September 1950. *ARMOR* magazine.

17th Infantry across the Han on 24-25 September, setting the stage for the clearing of the capital city.[27]

The Marine infantry narrowly avoided a setback as a platoon of T-34 tanks leisurely cruised around the industrial park at Yongdung-po the night of 21-22 September. A rifle company commanded by Capt. Robert A. Barrow, a future Commandant, remained dug into a berm, curiously immune from the fire of the 85mm guns. After volleys of 3.5-inch rockets finally convinced the tanks to withdraw, the company continued its mission.[28]

The 1st Marines spearheaded the clearing of Seoul with B Company, 1st Tank Battalion, in support. Tank-infantry teams netted many prisoners as they swept through the city. Several tanks were disabled by mines but recovered as the advance continued. A flame tank was lost to a surprise infantry assault, and tank-infantry coordination improved with that example as an incentive. Two SU-76 self-propelled guns fell to Marine tanks in the city center, and fighting ended on 27 September. The 1st Marine Division claimed a total of thirty-eight enemy tanks destroyed during the Inchon-Seoul operation, but aircraft probably accounted for a dozen of those. The Commander of the 1st Tank Battalion, Lt. Col. Harry T. Milne, cited his untrained and inexperienced crews, who he said were "able to operate efficiently and aggressively." He counted one flame, one dozer, and five M26 tanks lost to enemy action, but claimed his men destroyed thirteen tanks and fifty-six antitank or antiaircraft guns.[29]

During the next month, the 1st Marine Division moved to the east coast and joined the United Nations Command's drive toward the Yalu River and total victory. This campaign concluded with the massive riposte by Chinese Communist Forces (CCF) and the concomitant breakout of the division and neighboring Army units from the Changjin (Chosin) Reservoir sector. The 1st Amtrac Battalion and the 1st Tank Battalion's headquarters remained in the Wonsan and Hungnam rear areas. The 1st Tank Battalion fought with the division using its four medium tank companies as well as the newly activated tank platoons in the regimental antitank companies, which had been attached for administrative and maintenance support.[30]

In brief, the division was strung out in late November along a narrow, unimproved forty-five-mile-long road from the port of Hungnam to the front lines at Yudam-ni, where the 5th Marines prepared to continue the attack northwest through the 7th Marines. The division headquarters occupied the middle point at Hagaru-ri, with its auxiliary airstrip. The 1st Marines chafed further south at Koto-ri. Army troops and 76mm M4A3E8 tanks operated to the northeast, using the same main supply route (MSR). That this was not tank country was noted by author Eric Hammel, who described the arrival of a single platoon from D Company, 1st Tank Battalion, at Yudam-ni on the evening of 27 November as an "arduous ascent to the [Taktong] pass, teetering on the brink of bottomless chasms, scraping against rock outcroppings, squeezing by the abandoned [Army] Sherman tank."[31]

The Chinese offensive struck the forward regiments of the 1st Marine Division and 7th Infantry Division on the night of 28 November. The Marines held their positions against heavy attacks, but the Army's 31st Medium Tank Company failed in its attempt to relieve the Army vanguard east of the Chosin Reservoir. An attempt by the 1st Marines to push forward reinforcements to Hagaru-ri ended in a similar disaster, despite the use of eleven tanks from B Company, 1st Tank Battalion. The center of the long column, isolated from the tanks at either end, was devastated and routed by the CCF forces. On 2 December the general pullout began, starting with the 5th and 7th Marines and Army stragglers in the north. Tank D23 led the way, its engine conking out, batteries expiring, and tracks skidding into ditches, but somehow remaining in action and destroying roadblocks all the way to Hagaru-ri. Infantry battalions cleared the flanking hillsides and tank-infantry teams cleared the roadway as the regiments moved into Hagaru-ri, by then under siege from all sides. On 10 December the 1st Division moved toward Koto-ri and the 1st Marines. Bringing up the rear this time were the forty-six tanks of B and D Companies, 1st Tank Battalion, the battalion dozer tank platoon, the 31st Medium Tank Company, and the tank platoons of the three Marine infantry regiments. They remained at the end because of the doubtful condition of the vital bridge at the Changjiu Power Plant No. 1. In the end, all but the last eight—all M26s—made the crossing. A breakdown came at a tense moment when refugees and CCF troops pressed from the rear and men began to abandon their trapped vehicles. The bridge was blown and the column continued to fight its way to the port. One abandoned tank, B22, fought all night before the crew escaped, only to be captured two days later. Lieutenant Colonel Milne reported a total of fourteen tanks and ten trucks lost in the Chosin breakout, but surprisingly few casualties in the tank battalion (four missing, four wounded and thirty-one nonbattle casualties).[32]

The Marines evacuated Hungnam and reassembled at Masan, near Pusan, for rest and reequipping. The tankers made good use of the rest and maintenance time. Their shortage of twelve M26s and seven M4A3 dozer tanks was made up from shipments from the United States, and the support of the Army's 328th Ordnance Battalion proved exemplary. During this period the division performed counterguerrilla missions in the Pohang area, with the tank battalion assigned to a fifteen-by-fifteen mile quadrant south of Pohang. However, no actions resulted. The tank companies then split up to support the three Marine and one ROK Marine regiments in Operation Ripper, the first of a long series of short-legged counteroffensives designed to restore the prewar demarcation line. The advance continued in the summer into the Punchbowl zone, after which the lines remained static into 1952. The tank-infantry work produced no novelties and the routine support of tank platoons for infantry battalions and tank companies for regiments was hardly altered. Even the CCF spring offensives brought little change to tank tactics. The 90mm cannon dominated the battlefield wherever it could be brought to bear. Bunkers provided the bulk of targets, and some tank-infantry raids occurred in the fall defensive period. Attempts to dragoon the tankers into reinforcing the artillery with indirect fire were eventually thwarted by the prospect of wearing out the 90mm gun tubes with no replacements on hand. The tank battalion, almost a thousand Marines in number, operated and maintained its tanks and those of the regimental antitank companies. Mines proved to be the greatest problem for tanks, but few became total losses. Ammunition expenditures averaged over three thousand rounds per month for the main gun (over

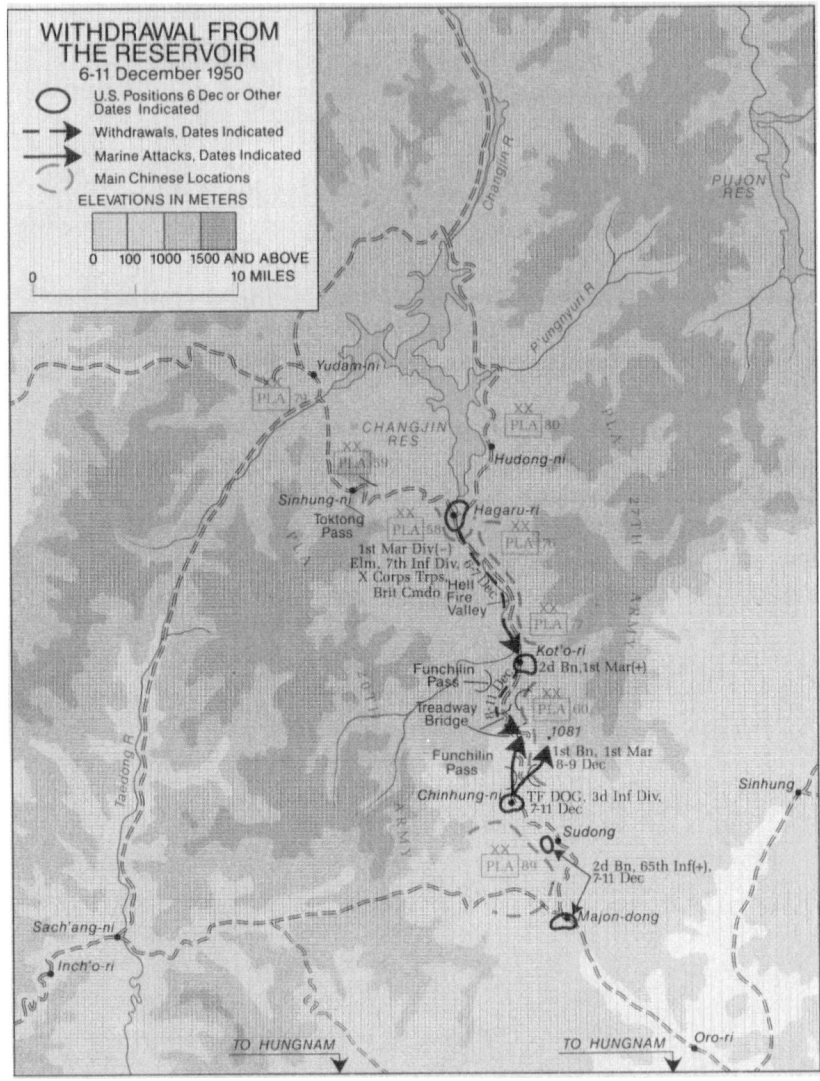

thirty rounds per tank, considering the ninety-seven authorized in the division, not counting the nine flame tanks). During the July–November period the tank units received the M46 tank as replacements for their venerable M26s, a great improvement in mobility, but too late for the more mobile phases of the war.[33]

In March 1952 the 1st Marine Division made its last shift on the front, taking responsibility for the west coast and the frontage extending past the Panmunjom Peace Corridor to the left boundary of the British Commonwealth Division. The division thus recovered close proximity to its amtrac and armored amphibian battalions. The latter had remained since 1950 on the Kimpo Peninsula attached to the I Corps Kimpo Provisional Regiment. The Amtrac Battalion spent the post-Masan period in the Ascom City zone and the Kimpo Peninsula, with B Company detached to Pohang to off-load supplies for Marine Aircraft Group 33 there. D Company was activated at Camp Pendleton in late 1950, but remained there throughout the war.

The long and tortuous front occupied in 1952 remained vexing to the Marine Corps's leaders. The Kimpo Provisional Regiment remained much the same under 1st Marine Division control. The 1st Amtrac Battalion was placed into the line to the east of the Kimpo Provisional Regiment and west of the ROK Marine regiment. The three Marine infantry regiments rotated in and out of the line for the rest of the front. The 1st Tank Battalion remained in division reserve, with two companies in support of frontline regiments and the other two companies in reserve assembly areas. The amtrackers responded to their infantry duties with aplomb, initiating a tradition and doctrine of readiness to serve as an infantry battalion as a secondary mission. Some thirty positions on a fourteen-thousand-yard frontage were manned by three LVT platoons and the divisional reconnaissance company. Weapons included 106 heavy and light machine guns, four 60mm and six 81mm mortars, and fourteen 3.5-inch bazookas. Company A supported the Kimpo Provisional Regiment, generally with waterborne patrols on the river and estuary.

The static positions created no other new missions for Marine armored fighting vehicles. The 105mm guns of the flame platoon and 90mm guns of D Company, 1st Tank Battalion, were pressed into artillery reinforcement in April 1952. A ROK Marine tank company was trained in the use of its new 76mm M4A3E8s and attached to the 1st Tank Battalion. Tank raids, now including the much-feared flame tanks, broke the monotony of occasional sniping and bunker shoots on the main line during the tank companies' rotations. A number of M39 armored personnel carriers were received from the Army and operated by the tankers. They primarily served as part of a rescue element kept ready during the Panmunjom talks. They also formed an "armored utility vehicle" platoon for general use in the battalion, including rearming the tanks on the front lines. Other improvisations occurred, such as the borrowing of a 76mm M4A3E8 platoon and some 90mm AA guns by the 1st Armored Amphibian Battalion and, on receipt, manning them with battalion headquarters personnel in support of the Kimpo Provisional Regiment. At least one of the regimental antitank companies also "borrowed" 90mm guns for use in the front lines.[34]

As the truce talks proceeded, the UN troops grew to mourn the earlier monotony, especially when several fierce fights broke out in the late summer and fall of 1952 along the outpost line. Thanks to the new eighteen-inch (incandescent) searchlights mounted on the M46 tank, the tankers could provide firepower to support many of the outposts, but not all could be saved by fire alone. Infantry casualties caused the division to send replacements from the Amtrac Battalion and place thirty-five LVTs out of service in Ascom City.

From March through May 1953 the division rotated out of line and into corps reserve. However, tank and artillery units remained in place and supported the Turkish Brigade and the 25th Infantry Division in more outpost fights. When the division returned to the front in June, a platoon of M46s was used to reinforce a key outpost named Boulder City. It held against a fierce CCF attack on 24-25 July. Marines now considered the "Army method" of occupying firing positions full time to be better than the earlier notion of occupying firing positions only after an action had begun. Overall, the defense of the outpost positions during 24-27 July had involved over thirty tanks, which fired approximately thirteen hundred shells and fifty-five thousand bullets. Ironically, the armistice took effect on 27 July after one of the most costly months of fighting endured by the division.[35]

The fighting over, the 1st Marine Division remained in Korea until its 1955 redeployment to Camp Pendleton. A new main line of resistance and fallback positions demanded the construction of some two hundred tank slots. The 1st Tank Battalion turned over its M46s and M39s to the Army's 6th Tank Battalion on 17 March (armistice terms froze equipment in Korea for a specified period), and embarked at Inchon for Pendleton and a new issue of M48A1 tanks. The 1st Amtrac Battalion remained in Japan and Okinawa and served with the 3d Marine Division.

The impact of the Korean War upon armor policy in the Marine Corps can be minimized. The Corps's leaders had already determined a modernization policy for tanks and amtracs in the face of appalling budget reductions, and this paid off handsomely during the war from a materiel standpoint. The M26, regarded with some suspicion and purchased with evident caution, had provided a necessary overmatch against the NKPA armor. Since the CCF used no armor and had no artillery to speak of before the final static phase of the war, the Marine Corps tank policy may be said to have correctly gauged the requirements. The LVT3C had performed well in the hands of inexperienced personnel, and no continuing amphibious campaign ensued to demonstrate Corps proficiency in the art. Tactics had remained those of the

Pacific War, in keeping with the character of the Corps itself. Innovations were few, more in line with mere equipment upgrades, such as improved vehicles, tank searchlights, and command vehicles. An innovation quickly forgotten was the so-called porcupine tank, which featured extra radios and antenna for directing air and artillery support from a tank. Both M46 and Sherman models were developed in depot and in the field, yet no command or fire-support control tanks emerged in the aftermath of the war.[36]

Operations in Korea also witnessed the introduction of the helicopter as a source of tactical mobility for the infantry formations. It seemed to offer the infantry complete freedom from the tyranny of terrain restrictions. This factor proved to be of far more importance to the Corps's armored fighting vehicle policy and doctrine than did any accomplishments or omissions by the tank and amtrac troops. However, it would be a decade before the impact of the Corps's fascination with the helicopter would truly be felt.

The Korean War and the mobilization of national resources it spawned allowed the Marine Corps to achieve many of its program objectives otherwise unattainable in the 1940s. An entirely new family of amtracs was designed and produced, and the tank program was fleshed out to the approximate levels that the 1949 Armor Policy Study had recommended.

As the 1st Marine Division prepared to embark for the Far East, the Corps ordered its first eighty M46s, essentially M26 tanks fitted with the Continental twelve-cylinder engine coupled to a modern cross-drive transmission. Further purchases of the M46 gave way in October 1951 to the Army's new M47 production series, which was based on the M46 hull married to the T42 turret. M47s equipped the new 3d Tank Battalion, which deployed to Japan in 1953 with its parent division, as well as the rebuilt 2d Tank Battalion. The M47s also outfitted the training pools of the two reserve battalions (now designated 1st and 2d Tank Battalion, USMCR). In addition, the Corps formed two force tank battalions, the 7th and 8th, on the West and East Coasts, respectively. All tank school training converted to the M47 in October 1952, despite the retention of the M46 in the 1st Marine Division for the rest of its service in Korea.[37] The M47 introduced an improved 90mm cannon and the first range finder to the Corps, but otherwise had little effect, for the M48 series replaced it in short order. The 3d Tank Battalion kept it the longest (until 1959), owing to its isolation in the western Pacific, whence that battalion began its tradition of withering at the extreme limits of the limited Marine Corps logistical support chain.[38]

The M48-series medium tank, derived from the T42 through T48 prototype series, began to enter Marine Corps service in 1954. It remained in service longer than any other tank in the Corps. Rushed to production as part of the Korean War rearmament program, it suffered an inordinate amount of teething problems and production delays, exceeded only by its heavy tank stable mate. Marine Corps medium companies of both divisional and force battalions operated it in the now-standard seventeen-tank company, one of the vehicles receiving the dozer kit (reduced from the three per company of the Korean War in 1957). Outfitting the active and reserve battalions was not completed until 1959. The initial order of M48s in 1952 also included twenty-eight of the eventual seventy-four M67 (T66) flame tank variants, which substituted a flame projector for the main gun, using a mock 90mm gun tube to conceal its special nature. With the provision of M48 dozer kits and flame tanks, the Sherman tank finally disappeared from the Corps battalions in 1959.[39]

The unique story of the heavy tank in the postwar Marine Corps followed a much more confusing course of events. The last of the T41-43 family of light, medium, and heavy tanks developed from 1946 by the Army, the sole postwar heavy tank to reach production owed much

M67A1 flamethrower tank. USMC History and Museums Division.

of its existence to the Marine Corps. The Armor Policy Board of 1949 and the Marine Corps Equipment Board of 1950 had recommended that all force (corps-level) tank battalions be heavy. This concept was reversed by the Basic Organization Structure Board of 1952, which held that the Corps had no need for a heavy tank, but that the Corps would still accept the T43 in production. In a curious act of doublespeak, the commandant decided that heavy tanks be issued to the 8th Tank Battalion, but that the force battalions eventually receive the M48s.[40]

The T43 design represented the pinnacle of U.S. tank engineering in the late 1940s, with its cast elliptical hull and turret, Continental AV-1790 engine, cross-drive transmission, torsion bar suspension, and range-finder directed 120mm main gun in an electro-hydraulic turret, among other features. In 1950, it existed only as a full-scale mockup. The outbreak of war brought a rush order in December 1950, which led to a complete production run of three hundred vehicles—considered sufficient for Army and Marine Corps requirements.

The production T43E1 heavy tank featured the following characteristics:

Weight: 62.5 tons loaded, 58.5 light
Crew: 5 (two loaders)
Radius of action: 80-100 miles on 280 gallons of fuel
Speed: 21 mph
Armor: front, 4 inches at 53 degrees; side, 3 inches; and turret
 front 7 inches/side 3 inches, all at 0 degrees
Armament: 120mm T123E1 with 34 rounds (later 38), .50-cal.
 M2 with 1,000 rounds, and .30-cal. M37 with 5,250
 rounds
Suspension: torsion bar, seven road wheels, six return rollers
 each side.

The 120mm gun proved exceptionally powerful, firing a fifty-pound shot at a muzzle velocity of thirty-three hundred feet per second with 48,000 pounds per square inch chamber pressure. Penetration performance against 30 degree sloped armor compared very favorably to the 90mm T119 gun:

120mm AP221mm at 1000 yards/196mm at 2000 yards
90mm AP119mm at 500 yards/117mm at 1000 yards
90mm HVAP221mm at 500 yards/199mm at 1000 yards

Also available for the 120mm gun were high explosive (HE), high

M103A2 heavy tank with 120mm gun. USMC History and Museums Division.

explosive antitank (HEAT), white phosphorous (WP), and training projectiles.

As might have been suspected from the rush to production, the T43E1 failed its initial trials at Fort Knox, mostly because of erratic gun controls and poor ballistic performance of the projectiles. A modification program (correcting more than 110 discrepancies) resulted in the standardization of the T43E1 as the 120mm-equipped M103 combat tank in 1956. Throughout all these perturbations, the Marine Corps maintained its interest in the tank and applied its program funds to the appropriate modifications as well as storage costs pending the reworking of the 218 tanks it took from the program. The Corps alone participated in the M103A1 upgrade of turret controls, sights, range finder, and ballistic computer. The new heavies finally reached the troops in 1958-59. So impressive was the M103A1 that the Army took seventy-two of them on temporary loan from the Corps to outfit its sole heavy tank battalion in Europe during 1959-62. The heavy tank proved fairly popular with the troops, who above all respected the powerful armament it carried. Many challenges to the crewmen, such as the job of the second loader to hand chamber both the projectile and the propellant cartridge in a single movement within the confines of a narrow turret, were taken on with a sense of pride. It shared all the teething problems of the M48 series, exacerbated by its heavier weight and even shorter radius of action.[41]

Development of the amphibian tractor enjoyed similar impetus in the post-Korea period. Not only was a new design possible, but a complete family of carrier and specialized support vehicles came into service through the efforts of the Navy's Bureau of Ships with Marine Corps collaboration. The LVT Continuing Board, which had been functioning since 1943, was dominated by the Marine Corps in the 1950s—a process that began with the switch of the senior member assignment from the Bureau of Ships to the president of the Marine Corps Equipment Board.

In December 1950 the Bureau of Ships contracted the Borg-Warner Corporation's Ingersoll Division to design a family of amtracs by drawing upon engineering developments that had continued since 1946. The LVTP5 personnel carrier variant would be accompanied by similar developments of an armored amphibian (LVTH6), command (LVTC5), air defense (LVTAA1) recovery (LVTR1) and combat engineer/breaching (LVTE1) vehicles. A competing design from the FMC Corporation emerged in 1951, based upon the Army's M59 APC, and development proceeded on this series as the LVTP6. The Borg-Warner LVTP5, however, proved superior in seaway and surf handling and entered full production in 1952. A total of 1,112 LVTP5s and 210 LVTH6s (armed with a fully stabilized 105mm howitzer) were built by 1957. Although the AA version failed surf testing, some fifty-eight command vehicles were modified, sixty-five LVTR1s built, and fifty-five LVTE1s modified. As with the Army's T43 and T48 series tanks, the rush to production of the new LVT family resulted in numerous defects requiring extensive modification by the factory, provided in the end by the FMC Riverside, California, plant. The resulting delays permitted extended troop tests of the initial vehicles, compounding the requests for changes to the design and modification kits. Headquarters ordered the rebuilding of five hundred LVT3Cs in April 1953 to cover the gap until the LVTP5 family finally entered full service in 1956.[42]

For all its mechanical woes, the LVTP5 family provided the Marine Corps with a superior mechanized amphibious assault capability. With the same engine and cross-drive transmission employed by the M48 tank series, the LVTP5 could carry its troops and cargo at speeds of 5.6 knots on water and thirty miles per hour on land. Armor protection resisted small arms and artillery fragments, but not heavy machine gun AP rounds. A cupola-mounted .30-caliber machine gun provided close-in protection. Its cavernous covered troop compartment and large bow ramp permitted carrying a large troop complement (rated at thirty-six men, but not achieved with infantrymen carrying typical field equip-

LVTH6 with 105mm howitzer. Naval Institute.

ment loads) or five tons of cargo, including a jeep or a partly disas-
sembled 105mm howitzer. Although designed mainly for waterborne
operations, its torsilastic suspension provided a better ride for occupants than the earlier generation LVTs. However, despite initial enthusiasm for its potential as an APC, extended land use of the new LVT
only increased the difficulties of maintenance. Engine or transmission
replacement remained a day-long task, and the crew of three usually
had its hands full keeping this large vehicle operational.[43]

The final and perhaps most unusual armored development by the
Marine Corps in the wake of the Korean War was the development of
a mobile antitank vehicle for the division. The regimental tank platoons
were not manned in peacetime, and the tank required specialized support barely adequate in Marine Corps tank battalions. Moreover, the
weight and size of the medium tank had rendered it road bound for too
much of the Korean War campaign. Although the Army dropped the
tank destroyer concept after World War II, the Marine Corps's development of the M50 Ontos (also known as the "Thing") must be considered
a reversion to that doctrine.

In 1951, the Corps and the Allis-Chalmers Corporation began development of a light vehicle employing the new and more powerful

106mm recoilless rifle and using highly mobile "shoot and scoot" tactics to reinforce the division's infantry, which used the same weapon on a fixed mount at battalion level. The result was a fully tracked, three-man vehicle, weighing in at nine tons, and capable of firing its six rifles in sequence, pairs, or together in a single devastating volley directed by simple telescopic sights and .50-caliber spotting rifles. A 145 horsepower gasoline engine drove it at thirty miles per hour. It carried a basic load of eighteen rounds of ammunition. Forty-five Ontos vehicles were allocated to each of the Corps's new divisional antitank battalions. Manned by infantry specialists, the antitank battalions supplanted the antitank companies in the infantry regiments, entering service in 1958. Production of the M50 commenced in September 1956. The installation of a Chrysler 361B engine developing 180 horsepower and new track during 1963-64 resulted in the M50A1, which had sufficient automotive improvements to enable it to reach speeds for which it was originally designed.[44]

Operation of the Ontos was easy due to simplified steering. Sighting and spotting were designed to minimize the training demands usu-

The M50 Ontos self-propelled antitank system with its six 106mm recoilless rifles. USMC History and Museums Division.

ally associated with armored fighting vehicles. Indeed, the night-fighting capability installed in the M50A1 consisted of a quadrant and azimuth indicator to permit range-card firing unobserved! The spotting rifle tracers burned out at fifteen hundred yards, thus extended range firing was conducted using the burst-on-target technique. The weapons' limited traverse (40 degrees to each side, elevation –10/+20 degrees) and back blast of the rifles, coupled with the vehicle's light armor, forced Ontos crews to train to fight from the ambuscade. The five-vehicle platoons typically were attached to individual infantry battalions for deployments and the fifteen-vehicle company usually operated in direct support of the regiment, although the division commander could retain the battalion as an antitank reserve.[45]

The inclusion of antitank battalions in the Marine division TO&E at least had the favorable effect of relieving the tankers from the responsibility of providing close-in antitank support to the infantry, freeing them for more conventional armored missions. However, doctrinal change in the Corps made the issue moot. The helicopter had sufficiently matured by the early 1960s that Marine Corps planners looked to it to solve most of the traditional mobility concerns of the FMF. Study groups at Quantico had, since the late forties, forecast great potential for the helicopter in the landing force, and by the mid-50s they were ready to assume that new models would lift artillery and prime movers ashore. Neither the performance nor the numbers of helicopters needed to conduct a vertical assault ever materialized, but the seeds were sown. The Commandant at the time, Gen. Randolph McC. Pate, instructed the 1956 FMF Organization and Composition Board, also known as the Hogaboom Board, that: "The helicopter will become the primary means of achieving tactical surprise and flexibility. However, surface landing craft and land vehicles will continue to be the primary means of mobility at the objective until sufficient helicopters of improved capability . . . are available to permit the landing, tactical maneuver and logistic support of all assault elements of a Marine Division. As the helicopter capability increases, the need for surface landing craft and land vehicles will decrease."[46]

The board responded by recommending reducing the size of the division and the weight of equipment in an effort to make it fully air transportable, thus enhancing the potential for vertical assault doctrine. In 1957 the commandant ordered the adoption of the "M-series" TO&E. The change meant placing all tracked vehicles (except the Ontos), the heavy artillery, and much of the transportation and engineering assets of the old World War II–pattern Marine division into the force troops. It also meant replacing most of the remaining artillery howitzers with heavy mortars and reducing the strength of the infantry regiments. The divisions conducted early trials of the reorganization division. The addition of a fourth rifle company and a foot-mobile reconnaissance battalion scored well, but artillery remained weak in terms of counterbattery capability, and antitank firepower was improved only for "the initial encounter." In addition, "the mortality rate of the Ontos is still an unknown and highly critical factor. The armored shock power of the division is negligible."[47]

The new organization remained in force well into the 1980s, modified only by the return of conventional artillery and tank units to the division in the 1970s as a response to the increased orientation on the Soviet-pattern ground threat. Regardless of what happened with helicopter development in the aftermath of the Hogaboom Board, the direct effect upon Marine Corps armored fighting vehicle concepts was to reduce their importance both tactically and in the budget program.

As the Corps edged toward its seven-year campaign in Vietnam, the only significant change in the armored vehicle program was the adoption of the diesel engine for the tank fleet. As in World War II, the move to the new AVDS-1790-2A engine was driven by Army decisions rather than Corps concerns for the efficacy of the tank fleet. However, the Corps did pursue a series of developments in hybrid high-speed amphibious landing vehicles, employing various concepts such as planing hulls and hydrofoils combined with wheels or tracks. These did not progress to the stage of fighting vehicles but rather were projected as a high-speed cargo carrier designed to replace the DUKW. The goal was to produce a vehicle that would "transit non-mobile loaded cargo directly from ship to inland beach dumps without the laborious landing craft–crane–truck transfer operation presently required at the shoreline." Corps amtrackers fervently hoped for a new series of LVTs emerging from such testing, but none of the designs attempted during the 1960s proved successful, although much engineering knowledge was obtained in hull forms and water-jet propulsion systems.[48]

Just as doctrine remained static for land vehicles in the Corps, training also became stultified. Although some forty students attended Army schools at Fort Knox in FY 1955, the number declined to twenty-three in FY 1958. The Marine Corps Schools Catalog of 1958 failed to list the Fort Knox armor basic course for officers and set requirements for advanced course attendance by captains at zero. Six officers were slated for the abbreviated fifteen-week course for company grade officers. More and more, the training of armor officers and crews was relegated to on-the-job training. The headquarters training staff disapproved a request to extend officer training in the basic tank/amtrac course from four to six weeks. While recognizing that the proposal would produce a better-trained officer, the staff asserted that the purpose of the course was to orient the officer to his anticipated duties in order to continue his military education on the job. "The temptation to use these and other courses to qualify officers in an MOS [military occupational specialty] cannot be followed, however, without undue loss of service in the operating forces and increase in students and overhead in the training base."[49]

The modernization of the tank fleet to diesel engines marked final decisions on the organization and fielding of tank units. The Corps's 421 medium M48A1 tanks entered the M48A3 overhaul program in December 1962 at a rate of twenty-five vehicles per month, and the M67A1 to M67A2 overhaul program proceeded at a rate of five per month. The 218 M103 heavies were cycled through beginning in August 1963 at a rate of twenty-five per month. The M50 Ontos vehicles in the inventory were brought up to M50A1 standards at the same time as the M103 upgrades.[50]

While the three amtrac and three antitank battalions remained unchanged from the mid-1950s through the Vietnam War, a final cutback in tank strength occurred as the vehicles received their final modernization. The last force battalions disappeared, and the heavy tanks were assigned to the third companies in the 1st and 2d Tank Battalions. The Marine Corps Reserve tank battalions were designated the 4th and 8th on the West and East coasts, respectively, with the 8th Tank Battalion and C Company, 4th Tank Battalion, retaining the rest of the heavy tanks. A force tank company remained on paper, with the same "destroyer tank" mission of 1949, to be activated from the reserves when necessary. The tank battalions thus fielded fifty-three medium or heavy tanks, nine flame tanks, and four retrievers. The amtrac battalions had 190 LVTP5, including ten command, six retriever, and eight engineer variants. The armored amphibian company was in depot storage, with a paper strength of eighteen LVTH6s, two LVTP5s, and a single LVTR1. The antitank battalion continued unchanged with its forty-five Ontos vehicles.[51]

Increases in military spending under the Kennedy Administration improved readiness in the Marine Corps across the board—manpower, training, and materiel in particular. One demonstration of the renewed verve in Marine Corps operations was the execution of two major amphibious exercises at almost the same time in the Atlantic and Pacific. Operation Steel Pike (October–November 1964) saw the II Marine Expeditionary Force (MEF) embark over twenty-one thousand troops for a landing near Huelva, Spain. Not only were three full infantry regiments landed (eight U.S. Marine and one Spanish naval infantry battalions), but also put ashore were forty-five tanks, thirty-five Ontos, and fifty LVTs. A mere three months later, Operation Silver Lance (February–March 1965) saw III MEF land fifteen thousand troops, thirty-six tanks, and twenty-four LVTs at Camp Pendleton. The new Navy LCU and LCM8 landing craft also took part in this landing. The crafts' sides and ramps were specially constructed to permit tanks to fire in the forward arc while afloat, one of the last elements of the 1949-50 armor and equipment policy concepts to be accomplished.[52]

Nothing could prepare the Corps fully for Vietnam—its longest war and the greatest challenge ever to its institutional fiber. There would be fewer shocks than in Korea, but once again the Corps undertook tremendous expansion while continuing to operate in strength in other parts of the globe. The Corps's strength peaked in 1969 at over 314,000 (from over 190,000), including more than 80,000 on duty with the III Marine Amphibious Force (MAF) in Vietnam. Another division, the 5th, was activated, with its constituent tank, amtrac, and antitank battalions. Two and a third divisions engaged in combat in Vietnam, fighting a combination campaign of pacification within the northern provinces of the Republic of Vietnam, and a more conventional border conflict against the Army of the People's Republic of Vietnam (NVA).[53]

Tank, Ontos, and amtrac companies landed with the 9th Marine Expeditionary Brigade (MEB) at Da Nang to begin the formal campaign in March 1965. Within a year, the tank, amtrac, and antitank battalions of the 1st and 3d Marine Divisions were also in country. The 5th Marine Division's armored units were activated in July 1966, but equipment and personnel shortages plagued them. Each battalion furnished a company to deploy with the 27th Marines to Vietnam in the summer of 1967. Except for the M103 heavy tank, all the equipment procured in the preceding decade received a rigorous trial by fire. Unlike the Korean War, the enemy infantry in Vietnam had capable antitank weapons in the form of rocket-propelled grenades (RPGs) and recoilless rifles. Full of fight and operating in infantry-favorable terrain, the NVA and their local Vietcong (VC) henchmen proved their mettle on more than one occasion against Marine tracked vehicles and their crews.[54]

Tactics remained the same as in the previous conflicts. A few company sweeps occurred with tanks and LVTP5s, the latter seeing considerable use out of their element in the role of APCs. The Ontos proved extremely vulnerable to the principal threat of mines and spent most of the war in static positions, guarding bridges and other vital points. Tanks made road sweeps, operated in tank-infantry operations with companies in direct support of infantry regiments, defended during the monsoon season, and fired much-despised indirect fire missions to reinforce the artillery. In the absence of enemy armor, the tanks dominated any battlefield that they entered, and the 90mm gun's superb mix of ammunition made it ideally suited for a wide variety of missions. Collateral missions for armor quickly mounted. The 1st Tank Battalion was tasked late in the war with operating the Southern Sector Defense Command, having to cadre one company to field two full companies in support of the three U.S. Marine regiments and the 2d Brigade of the ROK Marine Corps. In similar fashion, the 3d Tank Battalion found itself charged with coordinating the Combined Action Platoon (CAP) program in the 3d Division's sector. Meanwhile, the Marines of the neighboring 1st and 3d Amtrac Battalions, when not operating their vehicles, served as temporary infantrymen patrolling the "Rocket Belt" cordon around the Da Nang air base. The 1st Amtrac Battalion essentially became a full-time infantry battalion, reprising its role from the Korean War, while the 3d Amtrac Battalion pulled more troop resupply, convoy, perimeter, and landing zone security duty.[55]

The LVTP5, with its belly fuel cells, proved especially vulnerable to mines, so its use as an APC was strictly limited. Frequently, the infantry rode on the roof, with crew-served weapons up to 106mm in size sandbagged and secured to the vehicles. In the Marble Mountain sector, an average of 2.7 LVTs were damaged each month by mines. One LVTP5 suffered the ignominious fate of being sunk in a river in 1967 after being hit by a volley of RPG-2 rockets.

The armored amphibian returned to war with the activation of the 1st Armored Amphibian Company in June 1966. One platoon had preceded it in 1965 and the LVTH6s operated as assault guns and reinforcing artillery until their final, definitive decommissioning in 1970.[56]

In all, some three hundred LVTs were lost during the Vietnam War, reflecting a high rate of use. Most of the sixty-two landings by Corps units of III MAF and the Seventh Fleet were conducted with LVTP5s, but only a few, such as Operation Starlight, attained any scale. Mines were the cause of most of the LVT losses. There were far fewer tank and Ontos losses, and their causes were not reliably recorded. Many of the damaged tanks and Ontos were returned to action by support or depot repair. Logistical support proved inadequate, coming as no surprise to armor units long frustrated with the peacetime procurement of spares. The tanks fared best because of the Army's use of the M48A3 in the war and the activities of tank depots on Okinawa and in Japan. The 1st and 3d Antitank Battalions were decommissioned in Vietnam in December 1967, their A Companies being attached to the constituent tank battalions.[57]

Marine Corps forces began to redeploy from Vietnam in 1969—a withdrawal that was finally completed two years later. The tank and amtrac battalions rotated early, since fighting had ebbed considerably in the III MAF zone. By mid-1970, all had relocated to their former garrison bases, and tracked vehicle units picked up a new direction for training and operations as the Corps began to prepare for Europe and NATO commitments. Although the Marine Corps received little in the way of additional funding in the 1970s, the need to prepare for conflict involving large mechanized forces blew new life into the tank and amtrac units.

NOTES

1. Allan R. Millett, *Semper Fidelis: The History of the United States Marine Corps* (New York: Macmillan, 1980), pp. 438-47.

2. James Ralph Davis, "From the Sea: The Tactical Development of Marine Corps Tracked Amphibian Vehicles" (master's thesis, University of San Diego, 1995), pp. 396-422. Davis says the Barstow inventory of 1 Sept. 1948 included 1,273 LVT3, 59 LVT4, 14 LVT4(A), 80 LVT(A)5 "new", 11 LVT3, 22 LVT4 and 52 LVT(A)4 "used." Subsequent experience showed that maintenance in storage of such materiel remained beyond the capabilities of the depot system in this period. See Commandant, USMC (CMC), to Quartermaster, USMC (QMMC), 16 Aug. 1950, Headquarters Marine Corps Correspondence, 63A-2000, Box 191, NA, Suitland, Md. (Hereafter NARS/Suitland/63A-2000/191).

3. Arthur E. Burns III, "The Origin and Development of U.S. Marine Corps Tank Units: 1923-1945," student paper, Marine Corps Command and Staff College, Quantico, Va., 1977, pp. 105-107. On the Imperial Army's inability to cope with tank attacks, see Meirion and Susie Harries, *Soldiers of the Sun* (New York: Random House, 1991), p. 353.

4. Correspondence of the CMC, 1940-50, Washington D.C., Box 2148, RG 127/18, NA (hereafter NARS/RG 127/18/2148) contains files of the Chief of Staff, Plans and Programs, related to tanks and spares.

5. Ordnance Supply Division to Plans-Policy Division, 18 Oct. 1946. The Commanding General (CG), FMFPAC, in a message dated 9 Dec. 1945, ordered fifty M4A3E8s to China for 1st Marine Division; seventeen were lost in a typhoon. NARS/RG-127/18/2148.

6. CMC to QMMC, 6 May 1946, subject: flame tanks; and CMC to QMMC, 11 June 1947, subject: postwar allowance, NARS/RG 127/17/2148. The CWS-POA-H5 version with the 105mm gun served in Korea.

7. Letter, CMC, 16 Dec. 1948, where he continues the policy of his 4 Mar. 1947 decision cited in Plans-Policy memo, 17 Apr. 1947, NARS/RG 127/18/497.

8. Davis, "From the Sea," p. 420.

9. The LVT(A)5 also received a turret cover and bow extension. Good sources on LVT evolution are Steve Zaloga, *U.S. Amphibious Assault Vehicles* (London: Osprey, 1987), Robert J. Icks, *AFV Profile 16: Landing Vehicles Tracked*

(Windsor: Profile Publications, n.d.). A lucid essay by an LVT pioneer is Victor J. Croizat, "Fifty Years of Amphibian Tractors," *Marine Corps Gazette* (Mar. 1986): pp. 69-76.

10. Director, Marine Corps Schools to CMC, 22 Mar. 1946, NARS/RG 127/18/487.

11. QMMC to CMC, 29 Sept. 1947; CMC to QMMC, 23 Oct. 1947; and Navy Bureau of Ordnance to Army Chief of Ordnance, 23 Oct. 1947; NARS/RG 127/18/2148.

12. G4 annotation on CG, 2d Division message, 052220Z July 1949, NARS/RG 127/18/2148.

13. Plans-Policy Division to Ordnance Supply Section, 27 Oct. 1949, NARS/RG 127/18/2148.

14. Ibid.

15. California Department of Public Works, 1 Oct. 1948, NARS/RG 127/18/2148.

16. QMMC to Director, Marine Corps Schools, 13 Sept. 1951, on approving request for tank platoon for Quantico Schools Battalion. NARS/Suitland/63A-2000/191-93.

17. Maj. Gen. Lemuel Shepherd to CMC, 20 Mar. 1947, NARS/Suitland/63A-2000/1118.

18. Col. E.R. Smock to Plans-Policy Division, 10 May 1950, NARS/Suitland/63A-2000/1099.

19. Marine Corps Equipment Policy quoted in J.R. Mundy, "The Future of Tanks in the Marine Corps," Marine Corps Development and Education Center study, 1953; Marine Corps University Archives, Quantico, Va.

20. Letter, CMC, 23 June 1949, on FY51 tank requirements, citing Armor Policy Board Report (S) 003D10449 of 1 Apr. 1949; NARS/RG 127/18/2148.

21. CMC to QMMC, 14 Apr. 1950, FY52 Budget Plan, directed buy of maximum quantity of T42 for $2.45 million for war reserve; NARS/RG 127/18/685.

22. NARS/RG 127/18/2148.

23. For the story of the 1st Marine Brigade, see Lynn Montross and Nicholas A. Canzona, *The Pusan Perimeter*, vol. I, U.S. Marine Operations in Korea, 1950-53 (Washington: MCHC, 1954).

24. Montross and Canzona, *The Inchon Seoul Operation*, vol. II, U.S. Marine Operations in Korea, 1950-53 (Washington: MCHC, 1954), pp. 24, 76, 131; 1st Tank Battalion War Diary for Aug. and Sept., NARS/Suitland/65-A5099/81. Records for the Korean War have since been transferred to NARS/Washington.

25. Montross and Canzona, *Inchon Seoul Operation*, pp. 90-139.

26. Ibid., pp. 147-56.

27. Ibid., pp. 188-251.

28. Ibid., p. 229. This incident may well form the basis for Barrow's refusal as CMC to modernize the Marine Corps tank fleet.

29. Ibid., pp. 259-97. 1st Tank Battalion War Diary for Sept., NARS/Suitland/65-A5099/81.

30. 1st Tank Battalion War Diary for Nov.-Dec. 1950.

31. Eric Hammel, *Chosin: Heroic Ordeal of the Korean War* (New York: Vanguard, 1981), p. 50.

32. Ibid., 273-80, 409-18. 1st Tank Battalion War Diary for Dec. 1950.

33. Lynn Montross, Hubard D. Kuokka, and Norman W. Hicks, *The East-Central Front,* vol. IV, U.S. Marine Operations in Korea, 1950-53 (Washington: MCHC, 1962), pp. 13-216, passim. 1st Tank Battalion January-December 1951, perhaps as a sign of improved morale, the commander reported 20 cases of gonorrhea at Masan the first week of January.

34. Pat Meid and James M. Yingling, *Operations in West Korea,* vol. V, U.S. Marine Operations in Korea, 1950-53 (Washington: Marine Corps Historical Center, 1972) pp. 17-98, passim; Command Diary, Antitank Company, 1st Marines, Apr. 1953 NARS/ Suitland/65-A5099/70; Command Diary, 1st Armored Amphibian Battalion, Dec. 1952, Box 72; and Command Diary, 1st Amtrac Battalion, Box 71.

35. Meid and Yingling, *Operations in West Korea,* pp. 114-392, passim; and 1st Tank Battalion War Diary, Apr.-July 1953.

36. 1st Tank Battalion War Diary, June 1951; and Supply Dept., QMMC to CMC 27 Mar. 1951, NARS/Suitland/63A-2000/191. For an evaluation of the Russian 45mm AT gun vs. the M46, see Special Report type "C," Antitank Company, 1st Marines, Box 70: eleven hits at fourteen hundred yards, summarizing superficial damage.

37. Ordnance Section to G4, HQMC, 11 Aug. 1950; CMC to QMMC, 2 Oct. 1951, subject: decision on M47; QMMC to I&I, Syracuse, 13 Nov. 1952, subject: issue to 2d Reserve Battalion; and CG, Camp Pendleton to CMC 30 Dec. 1952, subject: M47 in schools; NARS/Suitland/63A-2000/191. No interest remained in a light tank in the Marine Corps, yet the M24 tanks acquired in 1945 for amphibious testing still (1950) languished in storage at Barstow, albeit reduced by one to nine. See QMMC to Plans-Policy Division, NARS/RG 127/18/2148.

38. Command Diaries, 3d Tank Battalion, Jan.-Dec. 1955, showing 35-55 percent deadline rates. NARS/Suitland/65-A5099/94; QMMC to G4, HQMC, 1 Dec. 1958, notes shipping of M48 to 3d Marine Division, with M47s to be disposed of "in best interests of government," NARS/Suitland/63A-2000/191.

39. G4 to Ordnance Branch 19 July 1952, on T48 deficiencies; CG, 2d Marine Division to CMC, 28 Dec 1954, on modifications to M48; QMMC to CMC, 2 Dec. 1955, 15 July 1955, notes CMC intention to reequip 8th Tank Battalion with M48 until heavy tank ready; CMC to CG 2d Marine Division, 7 Jan. 1956, citing M48 dozer kit in production Sept. 1956. Marine Corps M48s became M48A1 tanks through the normal depot overhaul cycle. NARS/Suitland/63A-2000/191. On the M67/T66, see CMC to CG, Army Chemical Corps, 18 June 1952, NARS/Suitland/63A-2000/191; and Rich-

ard P. Hunnicutt, *Patton: A History of the American Main Battle Tank* (Novato, Calif.: Presidio, 1984), p. 252.

40. CMC to Director, Educational Center, 3 Dec. 1952, NARS/Suitland/63A-2000/191.

41. Richard P. Hunnicutt, *Firepower: A History of the American Heavy Tank* (Novato, Calif.: Presidio, 1988), pp. 113-28. The initial M103A1 tank issues went to 1st and 8th Tank Battalion, with later distribution to the reserve battalions. 1st Tank Battalion turned in forty vehicles in 1961, apparently to revert to a mixed battalion. With the demise of 8th Tank Battalion as an active battalion, most heavies remained in the reserve structure and depot storage. See QMMC to CMC, 2 Dec. 1955; QMMC to G4, 7 Mar. 1961; and Army Deputy Chief of Staff, Logistics, to G4, HQMC, 15 Sept. 1959, NARS/Suitland/63A-2000/193. The storage period produced serious deterioration of the oil coolers, which suffered high failure rates. However, funds existed only to rectify under the later depot overhaul cycle. See CMC to CG, Barstow, 6 Feb. 1959, NARS/Suitland/66A4849/17. See also QMMC to CMC, 19 July 1958, announcing first shipments in Oct. 1958; Ordnance Section, 26 July 1956, with shipping instruction for 104 M51 retrievers: and CMC to Chief of Staff, Army, message 282002 Mar. 1956, requesting modification of Marine heavy tanks as soon as possible, with no requirement for dozer kits, NARS/Suitland/63A-2000/193.

42. Bureau of Ships (BuShips) memo for file, 18 July 1951, on LVTP5/H6 production; CMC to FMF commanders, 14 Apr. 1953 on LVT3C rebuild; HQMC chief of staff to G4, 3 Jan. 1953, on initial deliveries; Advance logistic data for LVTE1, 31 July 1963, the last of the production; NARS/Suitland/63A-2000/191. G4 to CMC, 9 Sept. 1963, Staff Study 4-63, Annex A: "A History of LVT Management 1940 to 1962, NARS/Suitland/63A-7645/19. The archives of the Amphibious Vehicle Test Branch, functioning since 1948 at Camp Pendleton, contain the details of the engineering development and testing of all amtrac and related amphibians considered by the Corps since then.

43. See also Zaloga, *U.S. Amphibious Assault Vehicles.* After-Action Report: B Company, 3d Amtrac Battalion, 7 Apr. 1955, on landing Army 38th RCT in exercise near San Simeon, using fifty-six LVT3C and four LVTP5 during Feb.-Mar. 1955, NARS/Suitland/65-A5099/94. The rival LVT6 program was canceled on 10 Oct. 1956 by CMC. The design was retained as an easily mass-produced alternate LVT in general war. CMC reported to the BuShips in a letter dated 28 Dec. 1956 that the Corps was "satisfied" with the integral armor of the P5 and H6, and had no interest in pin-on armor kits, preferring welded kits in the event of future requirements. NARS/Suitland/63A-2000/191.

44. William B. Allmon, "The Ontos," *Vietnam* (Aug. 1994): pp. 12-16. E.L. Bale Jr., "Ontos," *Marine Corps Gazette* (Oct. 1957): pp. 48-50. Regimental antitank companies disbanded 20 Dec. 1957. See press releases and unit

histories from Ontos file, Reference Section, MCHC; QMMC to G4, 15 June 1956 on Ontos production, QMMC to G4, 15 Oct. 1961, on M50A1, NARS/Suitland/63A-2000/191. Also see advance logistic data on M50A1, 16 Oct. 1963, NARS/Suitland/69A-7645/19.

45. LFB-16, *Employment of the Rifle, Multiple, 106mm, Self-propelled M50(T165E2) (Ontos),* 5 Nov. 1955, and LFB-23, *Employment of the Antitank Battalion,* 23 Dec. 1959, gave way in 1965 to FMF Manual 9-3, *Antimechanized Operations.*

46. Davis, "From the Sea," p. 502, and Millett, *Semper Fidelis,* pp. 524-27.

47. 1st Marine Division, Final Report of Reorganization Test, 4 June 1957, NARS/Suitland/65A-5099/4770.

48. CMC to various CGs on tracked vehicle modernization, 17 May 1962, NARS/Suitland/68A-4244/18; and CMC to CNO, 16 June 1965 on LCA vehicle, NARS/Suitland/70A-5214/20.

49. CMC to Army Chief of Staff, 29 Oct. 1956, subject: Formal Schools Training Program, 22 May 1958; and CMC to CG, Camp Pendleton, 17 Apr. 1957, NARS/Suitland/63A-2000/128. Millett, *Semper Fidelis,* p. 534, notes overall decline in the training establishment from 23.5 to 17.5 percent of total personnel strength during 1957-61.

50. CMC to G4, 7 Aug. 1961 and 15 Nov. 1961, NARS/Suitland/67A-6485/29. Advance Logistic Data for M103A2, 15 May 1964, NARS/Suitland/70A-5214/20.

51. Numbered battalions corresponded to parent divisions, except for the 1st Amtrac Battalion. Stranded in the Far East after the Korean War, it remained thereafter attached to the 3d Marine Division. One company each of the 1st Amtrac and 3d Antitank Battalion served in Hawaii with the 1st Marine Brigade.

52. Steel Pike folder, Reference Section, MCHC.

53. Millett, *Semper Fidelis,* pp. 559-60, 577.

54. Command Chronologies (CC), 5th Tank, Amtrac, and Antitank Battalions, Operational Archives, MCHC. Unit history sketches and the personnel diaries are held by the Reference Section. The heavy tank D Company, 1st Tank Battalion, remained in Camp Pendleton, attached to the 5th Tank Battalion. The 5th Tank Battalion received its last required M48A3 in Feb. 1969 and was deactivated at the end of the same year. The 5th Amtrac Battalion experienced a 50 percent administrative deadline rate (no crews) in mid 1968. But the 5th Antitank Battalion busied itself, firing 3,200 and 7,750 rounds of 106mm ammunition the last semester of 1968 and first of 1969.

55. CC, 1st Tank Battalion, Sept. 1968, Dec. 1968, Apr. 1969, Dec. 1969; 3d Tank Battalion, Jan.-Feb. 1966, Mar.-Apr. 1966, Aug.-Sept. 1967, Jan. 1968; Operational Archives, MCHC; and Davis, "From the Sea," pp. 545-66.

56. Davis, "From the Sea," p. 536ff.

57. Ibid., pp. 568-69, 576; and Allmon, "Ontos," p. 16.

9

The Patton Tanks

The Cold War
Learning Series

Oscar C. Decker

This chapter considers the various stages of tank development and acquisition during the Cold War era. It is a history of the never-ending struggle to balance firepower, protection (survivability), mobility, and, in later years, fightability, in the best way to support armor soldiers by providing them the materiel means to decisively defeat the enemy. It is noteworthy that, although numerous programs were initiated in an effort to develop and produce the "ultimate" tank, Cold War emergencies and funding constraints repeatedly overtook those programs, leading to the production of interim tank models. The reader should also note that while some of the excursions taken in the pursuit of improvements cannot be covered because of the space that would be required, many of them influenced the tank models that were produced. Some steps were small but others were visionary. The discussion that follows focuses on those developments that appear to have had the most impact, either at the time or in retrospect, as the Patton tanks evolved.

Between 1940 and 1945 there was a significant change in how the Army viewed the tank. It was no longer just an armored weapons platform designed to support the infantry. The tank had become a decisive weapon in its right, capable of moving rapidly in the face of combat action that no other ground weapon system could withstand. In spite of this acceptance, the war years had shown that U.S. tanks left much to be desired.

Combat experience had shown that U.S. tanks needed to be better protected and armed than their adversaries. There was also a realization that tanks had to be mechanically reliable and easy to operate, maintain, repair, and replace. This showed the necessity for standard components and interchangeable parts. These characteristics moved into the forefront as planners began to consider recommendations for the type of military equipment that should be developed after the war.[1]

The Army Ground Forces (AGF) Equipment Review Board convened in Washington, D.C., in January 1945 to consider the service's postwar requirements and produced several recommendations concerning tanks in June 1945. The board recommended the development of three classes of tanks—light (25 tons), medium (45 tons), and heavy (75 tons)—and an experimental 150-ton super heavy tank. In spite of their views on high end weights, the board showed significant foresight in specifying stabilization in both azimuth and elevation, a radar range finder with the ability to automatically identify friendly and enemy vehicles, and an automatic loader. They also noted that special power plants should be developed for armored vehicles, with consideration given to both multifuel and gas engines.

This report was followed closely by one submitted by the War Department Equipment Review Board (also known as the Stilwell Board because it was presided over by Gen. Joseph W. Stilwell). The Stilwell Board's January 1946 report mirrored that of the AGF Board except that it dropped the requirement for a 150-ton tank study. It also emphasized the importance of developing components specifically for tanks.[2] Even then there was a realization that the tank is a special vehicle whose design does not lend itself to the use of standard automotive components.

It is interesting to note that at almost the same time as the Stilwell Board was meeting, a special industry committee of automakers and suppliers, organized by Chrysler's K.T. Keller, convened in response to a request from Maj. Gen. Levin Campbell, the chief of ordnance. Keller had been instrumental in 1940 and during the war years in mobilizing the automotive talent necessary to build the famous Detroit tank plant. The Keller Committee's report, issued in April 1946, emphasized that preparedness also meant continued research because "It is obvious that technological developments will continue to be dominant factors in determining the outcome of wars between nations." Furthermore, the committee members believed it was "imperative for our nation, in peace, to pursue an intensive research and development program which will place our nation technologically in advance of other powers of the world." The Keller Report went on to state that "the continued development of tank-automotive equipment has increased in importance."[3]

These committee recommendations were very perceptive. It can be readily concluded that at that moment in 1945 and early 1946 there was close alignment between the military and industry concerning what needed to be done for armored equipment after the war. Ironically, within a matter of months the activity both the soldiers and industrialists viewed as essential to the army's survival on future battlefields would be virtually shut down.

Within the constraints of reduced postwar funding, the recommendations of the Stilwell Board continued to be followed until December 1950, when the Army Equipment Development Guide appeared. This document defined the mission of armor as "the destruction of enemy forces by means of armor-protected firepower, mobility, and shock action," and stated that "the tank is the principal weapon of armor." Firepower, mobility, and maneuverability were the characteristics most desired. Armor protection was relegated to a secondary requirement (contrary to the M1 Abrams requirements developed in 1972, which rated survivability over mobility), although relative immunity to frontal attack was considered essential. Tank guns were to have high muzzle velocity and provide a high probability of first-round kills at ranges up to two thousand yards. Special attention was also directed to ruggedness, simplicity of operation, mechanical reliability, durability, ease of maintenance, and fuel economy. Size constraints were also specified so as not to exceed the transportation capabilities of ships, roads, bridges, and tunnels in the United States and abroad.[4]

Another important action of the 1950 Army Equipment Development Board was to introduce a new classification for tanks based on the size of the gun the vehicle mounted rather than its weight. The new classes were light gun, medium gun, and heavy gun tanks, with 76mm, 90mm, and 120mm guns associated with the respective sizes. The importance of this action was to recognize that tanks should be built around new guns that were specifically designed as tank weapons and not modified artillery or antiaircraft weapons.[5]

Unfortunately, America was in no mood for preparedness immediately following the war. By April 1946, Detroit Arsenal activity had come to an abrupt halt and the arsenal was turned over to a small maintenance crew. The Keller Report had also recommended that the United States establish a permanent tank-automotive components laboratory at the arsenal with the hope that such a laboratory would provide a common meeting place for industry engineers and Army ordnance personnel. However, funds to initiate this were minimal.[6]

While there were no new tanks being produced from the end of 1945 to 1951, when the M47 went into production, there were important automotive component developments.

The M26 Pershing, fielded in the waning months of World War II, was grossly underpowered. Its five hundred horsepower Ford engine, which propelled the thirty-five-ton M4 Sherman tanks, struggled to move the Pershing's forty-five tons. Two major developments that would affect the M26's mobility were the 810 gross horsepower air-cooled Continental AV-1790-1 engine and the CD-850-1 cross-drive transmission. Maneuverability suddenly had the potential to become measurably better. Not only was significantly more horsepower available, but the automatic, cross-drive transmission would allow the vehicle to pivot rapidly in its length. Marrying this transmission and engine with a torsion bar suspension instead of volute springs also promised improved cross-country mobility. These developments would have a major influence on the tactical employment of the tank, changing it from a heavy, somewhat lumbering armored vehicle with a sixty-foot turning radius to one with agility, quick response, and better cross-country mobility. Driving a tank changed from a feat of strength using two lateral levers for steering control to a hydraulic system tied to an easily controlled "wobble stick." This new power package was tested in the M26E2 at Aberdeen Proving Ground, Maryland, in 1948.

The new M26 model evolved as part of the T20 medium tank series. Design work began in 1942 after British forces met the Afrika Korps in the Battle of Gazala in May. American armor officers were concerned by the fact the German Pzkw III series tanks encountered there were clearly superior to the American M3 medium tanks on loan to England. The Pzkw IVF2, which mounted a much more powerful weapon than the 75mm on the M4 Sherman tank, soon followed the Pzkw III. This caused Maj. Gen. Gladeon M. Barnes, chief of ordnance R&D, and Col. Joseph Colby, chief of the Development and Engineering Department at the Detroit Arsenal, to propose what was considered a radical change in tank design. The hull was to be simplified, making it into a box by eliminating the sponsons that had to be protected by armor on the Sherman tank. The result was less volume to protect, which allowed thicker armor with approximately the same weight.[7]

A number of models employing various components were tested, including at least seventeen within the T20 series. These included vehicles mounting 75mm, 76mm, and 90mm guns, and even a 105mm howitzer on one model (the T26E2). In the automotive area, the Ford engine with a torqmatic transmission was used in most models except where it was paired with an electric drive in the T23 models, the T25, Pilot No. 2, and the basic T26. The major breakthrough in mobility came with the M26E2. While stabilization in elevation had been tried on several of the models, it was not included on the M26 (T26E3) Pershings fielded in Europe in 1945.[8]

In spite of the sharp reduction in funds for vehicle research and development after V-J Day, some progress was made—as evidenced in the M26E2 with its more powerful air-cooled engine and cross-drive transmission. When the Korean War began in mid-1950, some of the M26s had been modernized and redesignated as M46 medium tanks, the T41 light tank was under development, and projects for the development of an improved medium tank, the M47, and a heavy tank, the T43, were underway. Fire control equipment research had progressed in the pursuit of reducing the time required for target acquisition and increasing the probability of first-round hits. Armor research also continued, with the focus on lighter metals such as aluminum and titanium.

In evaluating the M26E2 program, designers decided to incorporate the M3A1 90mm gun with bore evacuator and a single-baffle muzzle brake in the next model, the medium tank T40. The fire control system was improved with the M83 telescope, which featured two moveable reticules in a mount that corrected for cant. Ten T40 tanks were authorized using FY 1948 funds. They were powered by the AV-1790-3 engine with 810 gross horsepower, and used the CD-850-1 or -2 transmission. Hull design and top armor changes to the M26 had to be

M46 Patton medium tank. Patton Museum.

made for cooling air, but frontal armor remained at four inches at forty-six degrees from the vertical on the upper hull and three inches at fifty-three degrees from the vertical on the lower hull. The suspension system was modified to include a small track tension idler to assist in maintaining track tension during turns and while maneuvering over rough terrain. This vehicle was type classified as the M46 in July 1948, and nicknamed the Patton in honor of Gen. George S. Patton Jr. An initial production run of eight hundred M46s was authorized in the 1949 budget. The production M46s used the AV-1790-5 engine and the CD 850-3 transmission. A total of 1,215 were expected to be available for the 1950 budget. However, events in Korea changed all of this.[9]

Suddenly more tanks were needed than could be provided by modifying the M26s. In addition, deploying many of the M26s to Korea had caused a shortfall in those available for modification, so there were only 319 of the new standard M46s.[10] Production of some vehicle was a necessity, but the T42 program was not considered far enough along to go into production. Work on the T42 began in December 1948 with the release of approved specifications for a new medium tank weighing thirty-six tons, with better armor protection than the medium M46, and equivalent armament. The 90mm gun was to be stabilized in elevation and the turret in azimuth, with provision for an automatic loader and a stereoscopic range finder. Turret space was to be saved by using a

M47 Patton medium tank. Patton Museum.

concentric hydro-spring recoil system for the cannon rather than the usual recoil cylinders. The 90mm M119 gun was required to penetrate over eleven inches of homogeneous armor at a thousand yards using armor-piercing discarding sabot (APDS) rounds. A light tank power package was retained with a five hundred horsepower AOS-895 engine and CD-500 transmission. However, this low power-to-weight ratio caused trouble as the program progressed, and the T42 tank was dropped for production. It is noteworthy that the Ordnance Department continued the development program for some time in the hope that it would result in a lighter, more economical medium tank.[11]

In response to the requirement to produce more tanks quickly, the decision was made in September 1950 to produce an interim tank by installing a T42 turret on an M46 hull modified to provide increased armor protection and improved mobility with the 810 horsepower Continental AV-1790-5B engine and CD-850-4 transmission. Hull frontal armor angles were changed to provide four inches of armor at sixty degrees from the vertical on the upper glacis and three to 3.5 inches at fifty-three degrees on the lower glacis. The T42 turret had significantly improved armor protection over the M46. In November the new tank was designated the M47 even though it had not been standardized. The first pilot model was shipped to Aberdeen Proving Ground for testing in March 1951. Production started in June at Detroit Arsenal even before testing was completed, and at the American Locomotive Company some time later. Tests of production vehicles began in August 1951 and continued for a year. A number of the deficiencies that had to be corrected before issuing the tank to the field were related to the elimination of the IBM stabilization system and related components that had not been tested when stabilization was dropped. The first M47s were shipped with flat plates installed over the ports for the delayed production range finder blisters. Besides stabilization not being possible, there was no automatic loader as required in earlier specifications. The M47 was standardized in April 1952, and later named the Patton. A total of 8,576 M47s were built before production halted in November 1953.[12]

While the decision was being made to produce the M47 as an interim tank, plans to replace it were already being made in October 1950. Among other things, an elliptical hull and turret for improved armor protection were being favorably considered. The diameter of the turret ring grew again, this time to 85 inches (as in the heavy T43) versus the 73 inches of the M47 and the 69 inches of the Sherman. The main armament was to be the high velocity T139 90mm gun, which featured a quick-change tube. The tracks were to be wider than those on the M47 for better flotation, and the crew size, as in the T42, was reduced from five to four by eliminating the bow gunner. The fire control system was to incorporate a range finder, ballistic computer, and ballistic drive. The power package, however, was to be basically the same as the M47, with an 810 horsepower AV-1790-5B (which later moved through the -7 series with improvements) air-cooled gasoline engine and a CD-850-4 cross-drive transmission. Combat weight was to be forty-five tons and the cruising range one hundred miles.[13]

Because of the critical world situation, production was again begun without thorough testing, which predictably led to problems. To obtain the most modern weapons at the earliest possible date, Chrysler Corporation was given a letter order in December 1950 to build six engineering pilot models and 542 production vehicles. Quantity production of the T48 began in Chrysler's Newark, Delaware, plant in March 1952. Two other manufacturers had also been brought into the program: the Ford Motor Company and the Fisher Body Division of General Motors Corporation. In April 1953 the T48 was standardized as the M48 Patton tank. By that time, almost nine hundred tanks had been built. There were two distinct versions. One had a small driver's hatch and a Chrysler-designed Commander's cupola. The other had a large driver's hatch and a Model 30 Aircraft Armaments–designed cupola. Both cupolas permitted the .50-caliber machine gun to be fired from inside the vehicle. In the Chrysler version of the cupola, however, the Commander

M48 Patton medium tank with 90mm gun. Patton Museum.

had to expose himself to reload the machine gun. The Chrysler version eventually became the M48, and the Aircraft Armaments version became the M48A1.[14]

Several fire control systems were also installed in the M48. Four phases of installation were determined as problems with the T30 ballistic computer delayed the installation of the complete fire control system. The Phase I system was relatively simple. The gunner used the T35 periscope sight that later became the M20. The tank commander's T161 telescope was installed in the right range finder blister and was used for sighting and target designation. The Phase II and III systems replaced the T161 telescope with the stereoscopic T46E1 range finder at the tank commander's station. The T24E1 ballistic drive connected the gun elevation mechanism with the M20 sight and the T25 range drive. The Phase IV system replaced the T25 range drive with the T30 ballistic computer. The T24E2 ballistic drive directly linked the range finder to the ballistic computer, thus tying the whole system together. Normal operation required the tank Commander to announce the type of ammunition and the gunner to index it into the computer. When the tank Commander lined on the target, the range was transmitted mechanically from the range finder to the computer. The gunner then matched the indices on the computer dial, which in turn transmitted super elevation through the ballistic drive to the sight and range finder, elevated the gun to place the crosshairs on the target, and fired.[15]

In the continuing quest for greater operating range, Continental Army Command (CONARC) established a requirement for a medium tank that would use less fuel and have greater range than the M48 or M48A1. The first research project with this goal, the T48E1, was abandoned. This was due to requirements that would cause significant modifications, resulting in a major rebuild. However, a second project, the T48E2 was initiated in 1953 to improve the fighting effectiveness of the M48 Patton by improving its power package. The AV-1790 air-cooled engine was modified to be fuel injected and became the AVI-1790-6 with the CD-850-4B as the modified transmission. It was a significant operational change because it extended the tank's operating range to 170 miles. The new power package required redesign of the top deck to provide adequate cooling air for the engine.[16] Another important change was the selection of the newly developed Cadillac Gage constant-pressure hydraulic gun elevation and turret control system, replacing the electric hydraulic system used earlier. Minneapolis-Honeywell designed an electric amplidyne system that also provided better performance, but the Cadillac Gage system had the advantage of taking up less space and generating less heat. This tank was type classified as the M48A2 in

October 1955 and approved for production in December. Additional modifications were made to the M48A2 to make it the M48A2C. These included replacing the stereoscopic range finder with the M17 coincidence range finder and including a temperature-controlled linkage for the M5A2 ballistic drive. Eventually 1,344 M48A2 tanks of the 2,328 previously produced were modified to become C models.[17]

The Army Equipment Development Guide of 1950 had recommended that a 105mm tank gun be developed. One of the reasons for the eighty-five-inch turret ring in the M48 was to provide room for a larger weapon if required. In July 1951 a program was initiated to design two vehicles with 105mm guns. These vehicles were designated the T54 and T54E1. Both used the T48 chassis, but the T54 had a conventional turret and the T54E1 featured an oscillating turret. Muzzle velocity was increased to thirty-five hundred feet per second for an armor piercing round. Both systems were to have automatic loaders, so the gun was inverted, with the breechblock moving down to close and up to open. Rheem Manufacturing Company received the contract for the T54, and the United Shoe Machinery Corporation was awarded the T54E1 contract. Two separate designs were followed for the automatic loaders. The oscillating turret in the T54E1 simplified the loading problem by maintaining the loader in a fixed position. Work progressed on both models, but design problems and delays in obtaining government-furnished equipment caused it to be overtaken by tanks with more powerful guns and lighter chassis, such as the T95 series.[18] This was the first of several attempts to develop an automatic loader, and, although it died in the developmental process, it provided information for the next generation in the T54 series.

In May 1953 two pilots of the T54E2 were authorized, but without an automatic loader. The effects of the shock from a hit and of thermal effects on the turret and gun accuracy began to be studied. The primary fire control equipment on the T54E2 was isolated from the turret armor so that the Commander's T46E3 range finder, the T32 ballistic computer, and the T37 ballistic drive were all installed on a shock mount. Pilots were tested at Fort Knox and Aberdeen, but development action was concluded officially in January 1957 with the turrets being assigned to the T95 program.[19]

While Rheem Manufacturing Company was working on the T54, it received a contract to produce two pilot models of the T57 120mm gun tank incorporating an oscillating turret. The turret was first mounted on the hull of the T43E1 heavy tank. However, it appeared that it might be feasible to build the turret with lighter armor and mount it on the T48 chassis. The project was initiated in May 1953, and was designated the T77 120mm gun tank. Including an automatic loader was also a requirement. Slow development led to the T57 project's cancellation in 1957.[20]

In response to limitations of the M48 series tanks, several new programs were initiated. The first was the T87, which was proposed as a lighter, more efficient medium tank. As a result of the Questionmark III Conference in June 1954,[21] a new program, the T95, was launched in January 1955 to develop a lighter, more powerfully armed system. It was an effort to satisfy the never-ending quest for a tank that was lighter in weight, smaller, lower in cost, easier to maintain, simpler and more economical to operate, and would provide greater tactical and strategic mobility. The design included a number of what were then considered unconventional features. Its 90mm hypervelocity gun was to be smooth-bore and rigidly mounted in the turret. The suspension was to be flat track. The tank was to use the Optar range finder, which was then under development.[22] Gross weight was to be forty-one tons—8.5 tons less than the M48A2.[23]

When measured against today's programs, there was significant tank development work in progress when the T95 pilots were authorized in 1954—the T71 and T92 76mm gun light tanks, the T96 105mm gun heavy tank, and the T110 120mm gun assault tank. At a meeting in Detroit in July 1956 to review the T95 full-scale mockup and the T95 and T96 programs, several conclusions were reached. The T96 hull and chassis no longer represented the optimum solution to the heavy gun tank problem because of weight constraints. Its automotive performance was more than required for a heavy tank. At the same time, the reviewers found that the T95 would have automotive performance superior to what had been originally anticipated. They concluded that the T95's chassis could be used in conjunction with a number of different turrets to satisfy both the T95 and T96 program requirements. The conference recommended that four tanks be developed using the T95 chassis with better-armored turrets from the M48A2 and T96, the T54E2 turret, and a turret incorporating the British 120mm gun. These recommendations were approved in October 1956.[24]

In January 1957 the Joint Coordinating Committee on Ordnance requested that the assistant secretary of the army for R&D establish a panel to review the status of tank development. The following month, the army chief of staff established the Ad Hoc Group on Armament for Future Tanks or Similar Combat Vehicles (ARCOVE) to study tank armament requirements for the period after 1965, with special consideration to be given to the effects of atomic weapons. ARCOVE submitted its recommendations in May 1957, proposing that a maximum effort be made to equip tanks with a guided missile by 1965. Unfortunately, ARCOVE also recommended that work on conventional weapons be sharply curtailed. In August the Chief of Staff approved a new tank development program that would have only two tank type vehicles in the future. One was to be an armored reconnaissance/airborne assault vehicle (a light tank that later became the M551 Sheridan) and the other a main battle tank. The main battle tank was expected to be one of the T95 series, with a hypervelocity smoothbore weapon and the new compression ignition version of the AV-1790 engine. Tests at Yuma Proving Ground during the summer of 1957 showed a 60 percent improvement in fuel economy with this engine over the gasoline engine.[25]

There was an unusual twist during the first half of 1958 with the Bureau of the Budget (BOB) pressing the Army, rather than the reverse, because the tank modernization program was not progressing rapidly enough. The BOB wanted the Army to pursue all means to replace the M48A2 tank. In the ensuing debate the T95 lost out to those who believed that the two most important improvements—improved fuel economy and greater firepower—could be achieved at less cost by installing a more powerful gun and a compression ignition engine in the M48A2. In that way the T95, as a result of M48A2 product improvements, became just another interim design.

In the meantime, the long-term goal became the development of a main battle tank equipped with a guided missile launcher and radiological protection.[26] This awareness of the potential impact of nuclear war on armored vehicles—as well as their potential value in nuclear warfare—was beginning to affect thinking in both the R&D and armor communities.

When the BOB's decision prohibiting procurement of the M48A2 was announced in May 1958, the Tank Automotive Command was already at work formulating a new program: the XM60. The planners settled on the M48A2 chassis and the AVDS-1790 engine. Siliceous cored armor was proposed for both the hull and turret. The main gun was to be selected after comparative firing tests at Aberdeen in October. For comparison purposes, tests included two 90mm versions—one smoothbore and one conventional—two 105mm versions (one was the British X15E8), and two 120mm versions. Although Ordnance preferred the 120mm gun, it was not a contender, possibly because of its slow rate of fire due to using separate loaded ammunition. The British gun ranked first overall by a slight margin. The American version of the British gun, the T254, was selected and the British tube was specified until American-manufactured tubes could provide comparable accuracy. In December 1958 there were still unanswered questions concerning major components. Siliceous cored armor was not available both because of cost and availability of production facilities. The T95E7 long-nosed turret, which provided improved ballistic protection, was preferred but could not be made available for production by 1960. As a result, the M48A2 turret was specified for initial production, and the long-nosed turret was introduced on the M60E1, which was type classified as the M60A1 in October 1961.[27]

While the last production M48 series tank was the M48A2C in 1959, it was not the end of improvements for the series. Three conversion programs followed the M48A2C upgrade. The first, in 1963, converted the M48A1 to the M48A3, the variant that was used in combat in Vietnam. The most significant changes were the installation of a new diesel engine and transmission (the AVDS-1790-2A and CD-850-6A), changing the turret controls to the Cadillac Gage system as in the M48A2, adding a xenon searchlight to improve the tank's night-fighting capability, and incorporating the new M17B1C coincidence range finder. The initial conversions were accomplished at the Anniston and Red River Army Depots. Conversion of M48A1s to M48A3 (MOD B) 90mm tanks began in July 1967. These changes were primarily for ease of operation and maintenance except for the addition of the M2A2 chemical, biological, radiological (CBR) system for crew safety. The MOD B conversion provided infrared fire control equipment for the gunner and changed the Commander's cupola to provide more space and improved vision. A program that converted M48A1s and M48A3s to M48A5s began in 1974. The M48A5 was armed with a 105mm gun, used the new T142 track, and had a 2.2 kilowatt searchlight. It used as many of the M60A1's components as possible. A later product-improved version also incorporated the Israeli Defense Forces low-profile cupola and a loader's machine gun.[28]

The XM60 was ordered into production in December 1958 because of its improvements in firepower, protection, and cruising range. The first 360 tanks were built at Chrysler's at Newark, Delaware, plant. Production then switched to the Detroit Tank Plant, where it continued for a total run of 2,205 tanks. The M60 was replaced by the M60A1 in October 1962. The production version of the Continental diesel AVDS-1790-2 developed 750 horsepower and was mated with the CD-850-6 transmission. M60A1 frontal armor was increased in the upper plate from 3.67 inches to 4.29 inches at sixty-five degrees from the vertical. Armor on the sides of the crew compartment was also increased an inch, to 2.9 inches at zero degrees.[29] This interim tank, which mounted a 105mm gun, continued in production for more than twenty years with numerous improvements prolonging its life. It thus spanned the careers of many U.S. armor soldiers and leaders.

The ARCOVE Report issued in January 1958 had recommended a major effort be made to develop a guided missile system for a tank. Many of the army's leaders believed that tank gun development "appears to be reaching the point of diminishing return."[30] The solution, they reasoned, was an armored vehicle-mounted guided missile system. As a result, Gen. Maxwell D. Taylor, then serving as Army Chief of Staff, directed two tank programs for the future, both to be armed with a missile system. In June 1959, a program was initiated with the Aeroneutronics Division of Ford Motor Company to develop the XM13 Shillelagh Combat Vehicle Weapon System. This was incorporated in the XM81 152mm gun-launcher, which could fire a conventional projectile or launch the Shillelagh missile. The gun-launcher was approximately half the weight and length of the 105mm gun, which made it very attractive for combat vehicles. The 152mm conventional rounds were spin stabilized, and the missile rode on top of the rifling lands, engaging a keyway to prevent rotation.

One of the two tanks factored into the Army's modernization program was the M551 Sheridan armored reconnaissance/airborne assault vehicle. Its development began on 17 September 1959 and hinged on the ability to develop two new and untried features: 152mm conventional antitank combustible-case ammunition and an antitank guided missile. The Sheridan dual gun-launcher system became the foundation of other tank improvements, such as the standard M60 and the newly built MBT70/XM803. Slippage in development occurred almost immediately, causing Congress to launch an extensive review of the program on 13 March 1969. The subcommittee charged with the investigation accused the army—especially the developer, the Army Materiel Command's (AMC) R&D chief and the Munitions Command at Picatinny Arsenal, New Jersey—of supporting Standard A classification of the system in May 1966 before all designs and modifications were verified and corrected. The developer drove what became a degraded project for fear of losing program funds. Opposing Standard A classification for the M551 were user representatives from the CONARC, the Combat Development Command, and the Assistant Chief of Staff for Force Development (ACSFOR).[31]

Another user's concern surfaced in Panama in June 1967 when the Test and Evaluation Command—headquartered at Aberdeen Proving Ground—again found deficiencies in the conventional 152mm caseless ammunition. A year later, in the fall of 1968, an abrupt decision was made to supply cavalry units in Vietnam with the Sheridan because it offered greater mobility and firepower than the M113 armored cavalry assault vehicle (ACAV). This decision was made before all the Sheridan fixes were resolved. The major problem was the smoldering residue left by the caseless cartridge and the inability to design a satisfactory breech scavenger. However, there was an urgency to get the Sheridans in the hands of troops before the vehicle achieved its specific design goals. The troops thus were forced to accept a vehicle that was not fully functional. Another problem with the vehicle was its welded aluminum hull, which made it vulnerable to mines and antitank missiles. Nevertheless, the 152mm canister round proved devastating, thus providing the cavalry with considerable firepower and enhanced vehicle mobility.[32] In the end, the Sheridan gun-launcher program was driven by project managers who became captive to the system they were supposed to be managing.

Meanwhile, studies were begun to determine the feasibility of upgrading the existing tank fleet with the new weapon system. In August 1961 it was decided to mount a gun-launcher system in an M60. During the next three years, a number of turret concepts and designs were considered. In 1964, four concepts were reviewed and a compact version was selected for further development. In February 1965, the M60A1E1 project, conceived as a fairly low-risk program, was officially recognized. Prototypes were delivered for tests in late 1965 and early 1966. Tests started, and so did the problems—which reaffirms the lesson that "low-risk" programs involving a radical change require very thorough analysis before they are declared low risk. Troop confidence in the new ammunition plummeted with each delay. However, while the problems were still being solved, procurement of 243 turrets was authorized with FY 1966 funds and three hundred vehicles with FY 1967 funds. A new scavenging system had to be designed to remove smoldering residue from the gun before opening the breech. In addition, the fire control system did not include a range finder, and night vision was limited. The fire control problem was very serious because it restricted use of the missile to daylight or with a searchlight. Since the maximum range of the searchlight was only a thousand meters, and the missile was basically ineffective at ranges closer than that distance, it was a less than desirable situation. Nevertheless, the tanks were issued to troops in 1974. But the M60A2 remained an unpopular tank. It was taken out of service as interest shifted away from missiles to significantly improved high-performance kinetic energy rounds.[33]

Concurrent with the product improvement programs for the M60 tank, the MBT70 (Main Battle Tank-1970) program, a joint U.S. and

German effort, was initiated in 1963. This program is discussed in detail later in another chapter. As it was initially contemplated, the MBT70 was to have a three-man crew and a gun-missile combination, with the missile designed to meet long-range requirements and the gun to fulfill close-in high explosive and antipersonnel mission requirements. Inherent in the 152mm gun design was the potential for future nuclear warhead use.[34] As the MBT70 program developed, the nation's goals and requirements became problematic. Some of the expressed tank program goals had their roots in NATO commonality and the strengthening of international alliances. There were, however, probably other motives. Each country wanted something. The Americans wanted a new tank as soon as possible whereas the Germans seemed unhurried—perhaps because their Leopard had not completed a full production run. In addition, the potential operating environments for use of the tank caused national requirements to differ somewhat.[35]

While the MBT70 and M60A1E1 gun/missile tanks were being developed, other studies were being undertaken to improve the M60.

M60A2 Patton medium tank with 152mm gun-missile launcher. Patton Museum.

One of these was the Chrysler-proposed "K tank." It featured a modified MBT70 152mm gun-launcher with a turret made from rolled armor plate (for ease of production) with highly sloped sides and front. It was also to have a version of an automatic loader, and mobility improvement would be gained by using what was then the new tube-over-bar suspension system, which in effect doubled the length of the torsion spring action. Due to limited resources for projects of this type, probably because of financing the war in Vietnam, this concept was not supported.[36]

As costs for the new main battle tank continued to mount, it became obvious that the M60A1 would require a longer life cycle than had originally been intended. The Senior Officers Materiel Review Board meeting in December 1969 recommended the M60A1 move into another phase of product improvement to eliminate several problems. Dirt and dust ingestion that caused significantly shorter engine life had been a problem with the M60 series. The top-loading air cleaner (TLAC) was developed in response to this problem. It was the first of the new components to be incorporated into the product-improved M60A1 in 1971. This was followed by add-on stabilization (AOS) in 1972, marking another significant change in the tank's operational capability by providing a 50 percent probability of hitting targets on the move against essentially a zero probability without it. The next change was to replace the T97 track with twenty-eight-inch wide T142 steel track with replaceable track pads. With add-on stabilization, the top-loading air cleaner, and T142 track, the tank was type designated as the M60A1(AOS).[37]

As these improvements were being made, the outbreak of the October 1973 Arab-Israeli war placed a sudden demand on new tank production. In spite of regular industrial mobilization base studies that had been made over the years to ensure that tank production could be rapidly increased if needed, the lack of real tank industrial base mobilization capability became apparent as a result of the Yom Kippur War. Requirements for tanks outnumbered the supply. Some tanks sent to Israel were removed from reserve stocks with a promise to replace them later. Since the United States was not at war, there was a reluctance to use the wartime priority system to supersede civilian product requirements. The most important of the requirements for production was the foundry capability to cast hulls and turrets, although significant second sources were required for gun mounts and fire control. Even though mobilization plans showed foundry capacity would be available if needed in a crisis, in reality it was not.

Suddenly, when tank production had been at a rate of under forty units per month and on the decline, active discussions were underway to determine how best to go about building twelve hundred new tanks, plus those required to pay back the tanks being taken from reserve stocks. All options were being considered. High-level meetings were held between Gen. Henry Miley, the Army Materiel Command (AMC) commander, and the president of Chrysler, the U.S. tank producer. General Motors was invited as a possible second source for tank production. American companies and several companies overseas were contacted concerning the possible casting of hulls and turrets. Congress wanted to know how such apparent shortfalls in planning had occurred. These questions came from the same body that earlier had ordered reduction of the hull and turret foundry base from three producers to one even though the cost of maintaining the extra producers was marginal. Intensive management and scrutiny of the tank program by the undersecretary of the army became the daily fare. Finally, after almost two years of painful deliberation, a second foundry source was contracted.

In December 1974, Maj. Gen. Chester McKeen was appointed to serve as a special "Czar for Tanks" with the authority to direct the army staff as well as lower headquarters in matters of improving production and modernization of the tank fleet. This centralization of authority served to focus the attention of all concerned on increasing M60 production while improving the M48 tank.

The M60 production crisis of 1973-76 resulted in a number of lessons. One, the subject of annual discussions in the acquisition community, is that the industrial production base for specialty equipment such as tanks cannot be turned off or reduced without serious repercussions later, if and when the capability and capacity are again required. Tanks cannot be produced in automobile factories, and the expertise required to design and build the "ultimate" tank is not an "on demand" skill.

A second lesson is that true mobilization planning requires regular funding. For many years, mobilization base planning had depended on the goodwill of various companies involved in filling out certain industrial preparedness forms and stating their readiness to produce required items in the event of mobilization. There normally was no government funding involved to compensate for their planning. Therefore, planning was minimal or nonexistent.

A third lesson is that the military industrial priority system is difficult to invoke when the country is not at war. This perhaps is more important today than it has ever been in the past for several reasons. There are fewer depot stocks and more dependence on premium transportation and just-in-time supplies; at the same time there are fewer producers in the defense business. In addition, today's troop deployments are more for police actions and peacekeeping than in the past. For the last reason particularly, invoking the priority system that is normally associated with wartime can be expected to be difficult and open to argument because of its impact on civilian products.

Another lesson is the requirement for system control of weapons system design. When the M60 production crisis developed, responsibilities were split across several AMC subordinate commands as well as between development and readiness staffs. This contributed to some of the initial problems that arose.

The next phase of the program to improve the M60A1 resulted in the M60A1(RISE) variant, which was introduced in 1975. Detailed engine studies had identified areas that could be improved. The resulting engine package went five thousand miles without replacement. This improved life was a combination of the top-loading air cleaner, an improved starter, fuel injection lines and nozzles, stronger cylinders, and improved turbochargers. Other additions included a new electrical system with a 650-ampere oil-cooled alternator, a solid-state regulator, and a wiring harness with quick disconnects. The M60A1(RISE) (PASSIVE), which followed the (RISE) version, resulted from adding a smaller searchlight, the AV/VSS-3A, which produced both white and infrared

M60A3 Patton medium tank with 105mm gun, thermal sights, and appliqué armor. Patton Museum.

light. The gunner and commander were both provided with passive night sights and the gun tube was enclosed in a thermal shield to minimize tube distortion resulting from heating and cooling.[38] The operational desirability and capability of night sights was beginning to be recognized.

Many of the same components were incorporated in the M60A1E3 in 1974. The major product improvements incorporated into this model were a laser range finder, solid-state computer, and tube-over-bar suspension. Although the tube-over-bar suspension system significantly improved cross-country mobility, it was not released for production. The hydro-pneumatic and high-strength torsion bar suspension systems became competitors. While favorable tests were conducted on both systems at Aberdeen and Fort Knox, neither was released for production.[39]

At the annual American Defense Preparedness Association Combat Vehicle Conference in June 1976, there was an excellent summary of the periodic calls for action regarding combat vehicle requirements, as well as an outline of what the user was trying to accomplish. Some were almost traditional, but many were visionary. Gen. William DePuy, then the Training and Doctrine Command commander (TRADOC), led the parade of speakers as he discussed how the United States had to equip and train for NATO to "fight outnumbered and win." His simple formula was "Weapons × Crew Proficiency × Tactics = Effectiveness" in active defense doctrine to be employed against massive armored attacks. His solutions included mobility, weapon effectiveness, increasing the firing rate, and reducing losses. Brig. Gen. William B. Burdeshaw, a member of the TRADOC staff, went on to expand on the mobility theme with an example of collecting troops and vehicles from forty kilometers to concentrate at a single point. He called for artillery and armored infantry to have the same protection as a tank, and emphasized that staying power translated into increased fuel and ammunition payloads. Concluding, he called for day-night capabilities and questioned why training simulators could not be built into the equipment.

Brig. Gen. David K. Doyle also expanded on mobility by asking for smaller power plants repairable in the field, land navigation aids, agility, and survivability improvements desired with advanced armor, fire reduction, and signature reduction. The ability to identify friend from foe and provide chemical, biological, and radiological (CBR) protection also ranked high on his requirement list, as did the ability to live in the tank for twenty-four hours a day.[40]

As items were released for production, the M60A3 took shape. It included the ruby laser range finder and a solid-state ballistic computer.

The ballistic computer—based on information from the range finder and wind sensor—provided compensation for drift, crosswind, horizontal target motion, altitude, gun tube wear, cant, sight parallax, and gun jump. Fire control solutions became much more sophisticated with the entry of the M60A3. It was standardized in May 1979, although production began in 1978.

The next major improvement was the Tank Thermal Sight (TTS) for the gunner, with a light pipe providing the commander a view of the gunner's screen. Using heat from the target to provide an image, the TTS could sense targets through smoke, fog, or rain as well as at night. Suddenly, the potential for control of the night was at hand, with a major influence on tactics. The systems that received the TTS were designated M60A3(TTS).[41]

Numerous other studies were made to improve the M60 series tank. These included several different engines and transmissions, the Teledyne variable compression ratio engine with a different transmission, later proposed as the M1 power pack. A hydro-pneumatic suspension was also tested through the private efforts of Teledyne Continental. None of these were taken into production, because of the competition for funds with the M1. However, they did help recognize the potential for upgrading the M60 and the M48 series. In 1977 the army vice chief of staff directed that a task force meet to determine the best tank fleet balanced between M60 product-improved systems and the M1. After considering funding, the task force concluded that the most effective combination would be a force consisting of seven thousand M1s and thirty-six hundred M60A3s. The added improvements approved for the M60A3 were the TTS, a muzzle reference system, a Halon fire extinguisher system, and hardware to allow mounting auxiliary devices such as a chemical alarm.[42]

Many factors influence tank programs. Some are external to the immediate U.S. armor user, and others are internally driven. External influences include the pressures of allied countries and the perceived threat. Since 1951 and the proceedings of the Tripartite Conference on Armor and Bridging and other proceedings since then, there has been an underlying belief that there are opportunities for international cooperation in the development of armored equipment, although this belief has not yet resulted in a common tank.

Although the U.S. has been involved in efforts with several allied nations to cooperate on tank programs since the end of World War II, and regular study groups have convened under the aegis of NATO, few major successes in common tank development resulted. Beginning with the 1951 Tripartite Conference, representatives of the United Kingdom, Canada, and the United States have sought agreement on the part each nation should take in developing tanks for common use, the design and performance characteristics most desirable, and the time-phasing for the development program. The Tripartite Conference classified equipment as "short-term" if it could be made available by 1956, as "interim" if it could be made available between 1956 and 1960, and "long-term" if development would extend beyond 1960. The conferees also adopted characteristics for a new light gun tank with the United States to develop the weapon. For the medium gun tank, they agreed that the American M48 with a 90mm gun was acceptable for combined use, as was the British Centurion tank, which carried a twenty-pounder. They further agreed that the United States and the United Kingdom should each develop a 105mm gun tank to weigh no more than ninety thousand pounds, the winning design to be selected by competitive testing. They also agreed to work on a heavy gun tank. The United States was to first develop a 120mm gun for immediate use. The United States was also to develop a 155mm tank gun, and the United Kingdom a 180mm gun. These would be competitively tested to determine whether either should be adopted as a replacement for the 120mm gun tank approved for combined use.[43]

While the influence of this tripartite group was felt for a long time, things had changed sufficiently so that in 1958 there was a tank gun shoot-off between two 105mm guns. Although, as indicated earlier, the British gun won by a slight overall margin, the Americans chose to use an American-made tube.

The next major cooperative effort was with Germany on the MBT70. That effort was followed closely by the tripartite tank gun competition. The United Kingdom, Germany, and the United States began a program in 1973 to competitively test three different tank guns. The U.S. 105mm (originally a British gun tube) won based on the criteria established before the shoot-off. While the 105mm gun was selected, recommendations resulting from this effort included the continuation of work on the 120mm gun as well as on 105mm ammunition. This effort and testing was, in effect, rerun several years later with a decision to pursue the 120mm gun. A significant outgrowth of these efforts was the ability for several countries to use standardized ammunition, one of the largest contributors to the logistical support of tank warfare. It allowed the allies to have two significant common items of logistical support: ammunition and fuel.

The driving factor behind most actions taken by U.S. armor leaders and tank developers during the Cold War was the requirement to counter massed Warsaw Pact armored attacks in Europe. Although the capability to operate in any part of the world remained, equipment capability and soldier training were oriented toward Europe. If an attack came, it was expected that it would be through the Fulda Gap or across the north German plain, both affording access to the best tank country for exploitation with massed armored forces—a cornerstone of Soviet armor doctrine. The mission of NATO forces was to quickly defeat such an attack and delay, disrupt, and defeat Soviet follow-on forces—all in the limited maneuver area available in western Europe.

With a force ratio of more than five-to-one in favor of the Soviets in 1976,[44] there was a realization that the U.S. and its allies could not meet these potential attacks man-for-man or vehicle-for-vehicle. This reality brought recognition that sound operational doctrine, good equipment, effective organizations, and superior training would have to provide the answers.While the potential for large-scale enemy armor attacks like those visualized during the Cold War is not as great a concern today, the power of the massed armor attack executed by the U.S.-led coalition during the 1991 Gulf War demonstrated the effectiveness of a combined arms force built around the tank. Technology and the superb training of U.S. soldiers, along with a swift, devastating attack, made Desert Storm a short operation. However, it must be emphasized that technology development has to proceed on a continuing basis if America is to retain that cutting edge. That, along with a strong dedication to training armor soldiers, is key to the effectiveness of battlefield technology. Critics of the armored force competing for technology funding periodically claim that the tank has outlived its usefulness. It has not. It remains the centerpiece of the combined arms team.

The history of U.S. tanks shows that America has moved through several cycles in the development and production of tanks since the beginning of World War II—a period of rapid development and mass production of tanks. After the war there was little budget support for new tank development. Neither was there much industry interest in manufacturing tanks when weighed against the high potential for civilian business as a result of pent-up demand for consumer products. Although no new tanks were manufactured during the immediate postwar years, a few tanks were reconstructed or modified, and the M46 (the first tank in the Patton series) program began as a conversion program from the M26 in 1948. However, the outbreak of the Korean War rekindled the country's interest in producing new tanks, leading to a rush to produce smaller numbers of several interim Patton tank models: the M47, M48, M60, and then a relatively long production run of product-improved M60 models. Finally, the M1 Abrams was allowed to follow its development course and resulted in what was then the "ultimate" tank. The Abrams represented advanced armor design technology and gun-ammunition development. Lessons from two decades of experience with the Patton series were carefully woven into requirement documents for the M1, laid down at Fort Knox in the early 1970s and amended to reflect lessons of the 1973 Yom Kippur War.

Unfortunately, history repeats itself with no new tank production. The May 1997 *Armed Forces Journal* article, "Dooming the Behemoths," is one of the more recent attempts to speculate on the future of tank configurations.

America has come a long way since the 1940s. At one point, William Knudsen, who, while president of General Motors, was called to Washington to be the industrial specialist on the National Defense Advisory Commission, called K.T. Keller, president of the Chrysler Corporation, and asked if he would make tanks.

"K.T., will you make tanks?" Knudsen allegedly asked.

"Yes, Bill," Keller responded. "Where can I see one?" He then sent some of his engineers to Rock Island Arsenal to get whatever blueprints there were—about 180 pounds of paper.[45]

NOTES

This chapter was made possible by the cooperation of the following:

Historian's Office, U.S. Army Tank Automotive and Armaments Command (TACOM), Warren, Michigan.

Richard P. Hunnicutt, author of the following series of outstanding books tracing in significant detail the history of U. S. Army tanks and armored vehicles: *Pershing: A History of the Medium Tank T20 Series* (Berkeley, Calif.: Feist Publications, 1971); *Sherman: A History of the American Medium Tank* (Belmont, Calif.: Taurus Enterprises, 1978); *Patton: A History of the American Main Battle Tank,* vol. 1 (Novato, Calif.: Presidio Press, 1984); *Firepower: A History of the American Heavy Tank* (Novato, Calif.: Presidio Press, 1988); *Abrams: A History of the American Main Battle Tank,* vol. 2 (Novato, Calif.: Presidio Press, 1990); *Stuart: A History of the American Light Tank,* vol. 1 (Novato, Calif.: Presidio Press, 1992); and *Sheridan: A History of the American Light Tank,* vol. 2 (Novato, Calif.: Presidio Press, 1995). The Hunnicutt books should be a part of any serious U.S. armor vehicle library. They will be referred to extensively in this chapter. It is suggested the reader review *Patton: A History of the American Main Battle Tank,* which features extensive photographic coverage of the technology discussed in this chapter.

1. R.C. Engelman, "A History of the Tank Development Program, 1945-

1956," Ordnance Tank Automotive Command, 1957, on file at the History Office, TACOM, pp. 1-2.

2. Hunnicutt, *Patton*, p. 9.

3. Kevin Thornton and Dale Prentiss, *Tanks and Industry, The Detroit Arsenal, 1940-1954* (Warren, Mich.: History Office, TACOM, 1995), p. 54.

4. Engelman, "History," pp. 3-5, as taken by the author from the 1950 Army Equipment Development Guide, chap. 2, sec. I and II, pp. 25-28.

5. Ibid., pp. 5-6; Ordnance Committee Minutes (OCM) 33476, 9 Nov. 1950, author's files.

6. Thornton and Prentiss, *Tanks and Industry*, pp. 55-56.

7. Hunnicutt, *Pershing*, pp. 49-50.

8. Ibid., pp. 94, 217. *Pershing* contains an excellent presentation of the many steps and models that the T20 series tanks went through en route to becoming the M26E2. It includes decisions affecting these models and the specifications for each.

9. Hunnicutt, *Patton*, pp. 12-16.

10. Thornton and Prentiss, *Tanks and Industry*, p. 60.

11. Hunnicutt, *Patton*, pp. 32-35.

12. Ibid., pp. 55, 58-59.

13. Engelman, "History," pp. 54-55; and Hunnicutt, *Patton*, p. 100.

14. Engelman, "History," pp. 56-58.

15. Hunnicutt, *Patton*, p. 94.

16. Engelman, "History," pp. 60-61.

17. Hunnicutt, *Patton*, pp. 114, 119.

18. Ibid., pp. 126-31.

19. Ibid., pp. 134-41.

20. Ibid., pp. 143-44.

21. The Questionmark conferences were held with high-level armor leaders, Department of the Army representatives, and Tank Automotive and Weapons Command key personnel to present concept alternatives. They were decision-making conferences that did much to expedite the total process of determining development actions and objectives. To the author, who was present at several of these conferences, they appeared to be relatively rapid ways to reach consensus and decisions.

22. Flat track suspension used the road wheels to return the track, whereas the conventional system of the time used support rollers for that purpose. Flat track suspension saves weight. The Optar range finder used pulsed light for ranging and was thought by Tank Automotive engineers then to be the wave of the future. It measured the time that it took for the light pulse to get to the target and back.

23. Engelman, "History," pp. 76-79.

24. Ibid., pp. 79-81.

25. Hunnicutt, *Patton*, pp. 149-50.

26. Ibid., p. 152.

27. Ibid., pp. 152-71.

28. Ibid., pp. 219-39; and "48 Medium Battle Tank: Handoff Notebook," pp. 20-29, on file in the Historian's Office, TACOM.

29. Hunnicutt, *Patton*, pp. 164-74.

30. House Committee on Armed Services, *Review of Army Tank Program*, Report: Armed Services Investigating Subcommittee, 91st Cong., 1st sess., 9 July 1969, (Washington: GPO, 1969), p. 1.

31. Ibid., pp. 12-13.

32. Ibid., pp. 28-24. Also see Hunnicutt, *Sheridan*, pp. 101-18.

33. Hunnicutt, *Patton*, pp. 178-93.

34. Lt. Gen. Welborn G. Dolvin, oral history interview, TACOM, Warren, Mich., pp. 21, 25.

35. Brig. Gen. Bernard Luczak, oral history interview, TACOM, Warren, Mich., pp. 24, 26.

36. Hunnicutt, *Patton*, pp. 196-98.

37. Ibid., pp. 199-200.

38. Ibid., pp. 200-202.

39. Ibid., pp. 201-207.

40. O.C. Decker, personal notes.

41. Hunnicutt, *Patton*, pp. 208-10.

42. Ibid., pp. 210-15.

43. Engelman, "History," pp. 6-7.

44. Arthur J. Alexander, *Armor Development in the Soviet Union and the United States* (Santa Monica, Calif.: Rand Cooperation, 1976), p. 115.

45. Thornton and Prentiss, *Tanks and Industry*, p. 15.

10

Adaptation and Impact

Mounted Combat in Vietnam

Lewis Sorley

The definition of what constitutes armor has from at least the close of World War II been complicated by the fact that there is a branch called "Armor" composed of some, but only some, of those elements that in the recent war had made up the armored force. Maj. Gen. Adna R. Chaffee Jr. described that force as "a balanced team of combat arms and services of equal importance and equal prestige." Tanks, armored infantry mounted in half-tracks, armored field artillery, tank destroyer elements, and the whole range of what are now known as combat support and combat service support components were thus assigned or attached to the World War II–era armored divisions and armored cavalry groups.

The evolution of the armament and equipment of various elements of the armored force over the past half century, and of the doctrine for its employment—especially in terms of mechanized infantry and helicopters—has further expanded the envelope of what might be considered the armored force, while at the same time taking it perhaps even a little farther beyond what falls unequivocally within the narrower purview of the branch designated Armor.

For the purposes of this essay an inclusionary approach has been chosen. Consideration will thus be given to a wide range of units and their equipment and operations, to include tank and armored cavalry, air cavalry, mechanized infantry, and a number of specialized systems. It goes without saying that, as proof of the enduring validity of General Chaffee's observation, all the participants in armored operations in Vietnam were critically dependent on every type of support, from logistics and medical services through quartermaster, transportation, maintenance, and more, all "of equal importance and equal prestige."

The Armored Force monument near Arlington Cemetery in Washington provides further perspective on the categorization of units, memorializing the Vietnam service of a U.S. Army armored cavalry regiment, three tank battalions and a separate tank company, six armored cavalry squadrons, ten mechanized infantry battalions, twenty-two armored artillery battalions, and four armored cavalry troops, along with two Marine Corps tank battalions, two amphibian tractor battalions, and an armored amphibian company.[1] The armored cavalry regiment and the divisional armored cavalry squadrons also had organic air cavalry elements, while ground armor and cavalry elements habitually operated with aviation elements under their operational control or in support.

Initially it appeared that armor would have little part in the fighting in Vietnam. Gen. William C. Westmoreland, commander of American forces there from June 1964 until the summer of 1968, was skeptical of armor's usefulness and ability to operate in the combat environment as he assessed it. An artilleryman himself, Westmoreland's service had been entirely in infantry and airborne units. Nor had he any theoretical knowledge of armor, for the Army's famous school system had almost entirely passed him by. So limited was Westmoreland's military education, in fact, that according to his biographer "the only service school he ever undertook in his Army career" was "the cooks and bakers school at Schofield" Barracks, Hawaii.[2]

General Westmoreland's duties in Vietnam included being senior adviser to the South Vietnamese military forces. Indeed, the designation of his headquarters—U.S. Military Assistance Command, Vietnam (MACV)—emphasized that role. Had Westmoreland been more observant in terms of the armored elements in the Army of the Republic of Vietnam (ARVN) he was charged with advising, he would have seen that they were able to function effectively in many parts of the country, terrain and weather notwithstanding.

But, lacking practical experience, theoretical knowledge, or observation of his ally, Westmoreland concluded that armored operations were not feasible in Vietnam. In a July 1965 message to Army Chief of Staff

Gen. Harold K. Johnson—at a time when the Army's massive buildup of ground forces in Vietnam was just getting underway—Westmoreland asserted that "except for a few coastal areas, most notably in the I Corps area, Vietnam is no place for either tank or mechanized infantry units."[3] That outlook was probably reinforced by the preponderance of airborne infantry officers with whom Westmoreland populated his MACV staff. It was also reflected at Department of the Army, where General Johnson, an infantryman by background and experience, shared Westmoreland's views on the unsuitability of armor for the Vietnam battlefield.

The 1st Cavalry Division (Airmobile) was the first full division shipped to Vietnam. Given its specialized air cavalry configuration, the issue of ground armor elements did not arise, and the division's air cavalry squadron—the 1st Squadron, 9th Cavalry—was an integral part of its newly developed airmobile capability. But the next large outfit shipped, the 1st Infantry Division, had two tank battalions and two mechanized infantry battalions, along with a divisional cavalry squadron. Department of the Army stripped the deploying division of its tank battalions and dismounted the mechanized infantry, allowing only the cavalry squadron to retain its tanks and APCs. This was, General Johnson explained, a decision based on Korean War experience with mine warfare, on a lack of information concerning the use of armor by the South Vietnamese, and on a concern that "the presence of tank formations tends to create a psychological atmosphere of conventional combat."[4] Such an atmosphere was destined to become far more than just psychological in the very near future, but in the meantime early deployment of armor was severely restricted.

Once on the ground, the 1st Infantry Division found even more inhibitions placed on use of its arbitrarily limited armor assets. There existed in MACV at that time a "no tanks in the jungle" attitude. "Because General Westmoreland saw no use for tanks," observed Gen. Donn Starry, the M48A3 tanks of the divisional cavalry squadron were withdrawn from the line cavalry troops and held at Phu Loi base camp. It took six months to convince Westmoreland that tanks could contribute to the division's combat operations.[5]

Once it became clear what armor could do, the accelerating buildup of U.S. ground forces was richly augmented with tank, mechanized infantry, and armored cavalry units. By the end of 1965 MACV had requested deployment of both the 25th Infantry Division and the 11th Armored Cavalry Regiment. Maj. Gen. Frederick C. Weyand, then commanding the 25th Division, overcame staff resistance and brought with him the division's tank battalion, a mechanized infantry battalion, and its armored cavalry squadron.[6] As for the armored cavalry regiment,

Gen. Michael S. Davison, then serving as deputy to the Assistant Chief of Staff for Force Development, recalled, "we had a hell of a time selling that one, not only had difficulty in selling it, but difficulty in selling the use of tanks at all in any form in Vietnam, because General Westmoreland and Bill DePuy, who was his J3 in this period, couldn't conceive of tanks or armored cavalry being able to do anything in Vietnam." He further explained that "this is a factor of, really, their own lack of experience with armor."[7]

Once these and similar units in succeeding increments of the buildup got into combat, they quickly proved their worth. The assertion "you can't use armor in Vietnam" was quoted in an early segment of the Army's videotaped "Vietnam Training Report" series, along with the answer pounded out on the battlefield: "Like hell we can't!"[8]

The stubborn opposition to the deployment of armored force elements to Vietnam apparently stemmed from an overall impression that the operational environment was inhospitable to such forces. That outlook failed to consider the wide range of terrain conditions across the length and breadth of the country, and that seasonal weather patterns further affected the key consideration of trafficability at given times and places. Recalling the classic observation that there are only two kinds of terrain, good and bad, and that good terrain is tank country and bad terrain is not, it turned out that in Vietnam there were—depending on season and locale—many areas where tanks could operate and even more where the workhorse armored personnel carrier could make its way. It was never easy, but tankers and cavalry troopers used ingenuity, aggressiveness, and an enormous amount of labor to make it happen.

Two environmental factors, weather and terrain, were the primary determinants of when and where tracked vehicles could operate effectively in Vietnam. The weather in that part of the world is marked by very distinct seasonal variations. These in turn are a function of the prevailing monsoon winds. The southwest monsoon prevails during the summer months, typically bringing heavy rains to the lower part of the country, while in winter the northeast monsoon similarly drenches the northern sector. As might be expected, these alternating weather patterns had a dramatic effect on trafficability.

A detailed study conducted in the spring of 1967 determined typical conditions for the wet and dry seasons in each of the four corps areas into which South Vietnam was administratively and tactically divided. Results were reported separately for tanks and for APCs. The findings were in some respects surprising. In the IV Corps area, which included the Mekong Delta, APCs were found to have an 87 percent "go-trafficability," regardless of the season—and this despite the preponderance of waterways and inundated tracts in that region. In fact, APCs were found to be more mobile than foot soldiers during the wet season there. The APC's flotation and swim capability was obviously an important advantage in such conditions. By contrast, the tank could manage a high of only 61 percent in the dry season in IV Corps, and that plummeted to *zero* in the wet season.[9]

The tank fared far better elsewhere. Its "go-trafficability" rating ranged from 92 percent in III Corps, the group of provinces surrounding Saigon, in the dry season—reduced to a still robust 73 percent in the wet season—to 44 percent dry and 36 percent wet in I Corps, which encompassed the northernmost group of provinces. In the Central Highlands region of II Corps, the tank surprisingly did virtually as well in the wet season, at 54 percent, as it did in the dry, when it could traverse 55 percent of the terrain. APCs were more mobile than tanks in all regions of the country. In III Corps APCs managed a 93 percent rating regardless of the season, and in I and II Corps 44 and 55 percent, also wet or dry. Overall it was calculated that tanks could traverse 60 percent of the terrain in the dry season, reduced to 45 percent when the monsoon rains descended, whereas APCs carriers could negotiate 65 percent regardless of season.[10]

These results, it should be emphasized, were achieved through the ingenuity, effort, and experience of the tracked vehicle crews. Much of the relevant experience was amassed by South Vietnamese crews operating M41A3 tanks and M113 APCs. As early as March 1966, the ARVN had distributed the draft version of its Field Manual 3-1, "Armor Operations in Vietnam," containing a wealth of useful information. Developed in cooperation with U.S. advisers, then headed by Lt. Col. Raymond R. Battreall Jr., the draft contained much of value on how tracked vehicles could maximize their mobility in Vietnam. This included detailed descriptions of self-recovery techniques such as the use of capstan kits, a block and tackle, or long tow cables. Practical suggestions—obviously derived from hard experience—abounded, including the suggestion that vehicles move with tow cables already attached instead of trying to get them hooked up after a vehicle had sunk into the mire. American armor crewmen soon developed their own rules of thumb. "We were able to maneuver fairly well in this mountainous, overgrown terrain, providing we observed certain rules," Sgt. Ralph Zumbro recalled of his service with a tank battalion in the Central Highlands. "The most important was not to tackle the steep slopes with tanks."[11]

Much useful material on driving techniques had also been amassed by the South Vietnamese. "Momentum is vital to movement through soft ground," the draft ARVN armor manual observed. It also paid to look before leaping, since "areas of good trafficability are paddies with clear water and green reeds." However, "inundated areas where reeds are yellowish and water is cloudy usually have soft mud bottoms in which the armored vehicles will be unable to move." And there were yet other indicators that could tip off an observant driver to good trafficability, including "inundated areas where water buffalo are feeding (buffalo will not remain static long enough to graze if bottoms are soft)." Some young tankers may have been surprised to find they could learn something useful from the water buffalo, but ARVN soldiers had been doing so for a long time.[12]

American advisers with Vietnamese armor units took the initiative in passing this experience to deploying U.S. units by preparing an information packet covering terrain, tactics, and equipment. The latter point was particularly important, for it described modifications to the M113 devised by the South Vietnamese that made it a far more versatile and effective combat vehicle.[13]

In addition to weather and terrain, always the two primary factors affecting mobility, the load-bearing capacity, width, and condition of numerous bridges could be constraining factors. Finally, and of great significance, there was the unique combat environment. Vietnam was, as is widely recognized, a war without fronts. There were no established front lines, and thus no reliably secure rear areas. Terrain, with the exception of the base camps and certain major population centers, was seldom seized and held. Thus lines of communications were routinely insecure. That reality necessitated they be repetitively cleared, a time-consuming and dangerous job, and one frequently assigned to armored units.

Three U.S. Army tank battalions were eventually deployed to Vietnam, all equipped with the M48A3 Patton medium tank.[14] A good, solid tank mounting a 90mm cannon for which it carried high-explosive (HE), high-explosive antitank (HEAT), white phosphorous (WP), canister, and beehive rounds, the M48A3 was also armed with a coaxial-mounted 7.62mm machine gun and a cupola-mounted .50-caliber machine gun. Powered by a 750 horsepower diesel engine, the 49.5-ton tank had a stated cruising range of three hundred miles. A very important accessory was the xenon searchlight, which could be employed in a white light or infrared mode. Selected tanks also mounted a useful dozer blade.[15]

Often used as mobile battering rams in "jungle busting" operations, these tanks took a beating in Vietnam, especially their suspension

The three tank battalions deployed to Vietnam were all equipped with 90mm M48A3s like these. They were also assigned to divisional cavalry squadrons and the 11th Armored Cavalry Regiment. National Archives.

systems, which also suffered frequent damage due to the enemy's extensive and effective use of mine warfare. One squadron executive officer, frustrated by trying to "piece together our fifteen-year-old tanks as best we could from the repair parts that we managed to 'expedite,' wondered at a policy that sent new tanks to Europe and old tanks with only inadequate repair parts available to the combat zone."[16]

However, the workhorse of armor in Vietnam was the M113 APC, and later the diesel-powered M113A1, to which the fleet converted by about mid-1968. The M113 performed a myriad of roles reliably and effectively, and, modified in ways pioneered by the South Vietnamese, it transformed armor doctrine governing the employment of armored personnel carriers. The subsequent developmental effort for such systems was redirected as a result of the Vietnam experience.

The M113/M113A1 was an aluminum-hulled tracked vehicle weighing 11.3 tons. Powered by a 215 horsepower engine, it had a cruising range of two hundred miles (three hundred miles for the diesel version) and possessed an amphibious capability. "The M113 APC," stated an Armor School publication at the height of American deploy-

M113 Armored Cavalry Assault Vehicles like these were the workhorse in U.S. Army armored cavalry squadrons and mechanized infantry battalions. National Archives.

ments, "enjoys the highest degree of mobility of any vehicle presently employed in Vietnam."[17]

As originally configured, the M113 mounted one .50-caliber machine gun at the vehicle commander's position. The significant innovation introduced by the South Vietnamese was the addition of an armored shield for the .50-caliber gunner (usually the vehicle commander) and two side-mounted 7.62mm M60 machine guns, also shielded. Thus reconfigured, the M113 was referred to as an armored cavalry assault vehicle (ACAV). So altered, observed John Albright, "the vehicle took on some of the characteristics of a light tank."[18] Subsequent modifications provided thicker belly armor to protect crews from mine explosions, the relocation and strengthening of the fuel line to lessen the danger of fire, and stand-off side shielding designed to cause the premature detonation of the enemy's lethal rocket-propelled granades (RPGs).[19]

Simon Dunstan has maintained that "undoubtedly the most significant innovation in the employment of armor in Vietnam was the use of the M113 as a fighting vehicle."[20] The modifications in armament, plus the provision of armor protection for crew members at their firing stations, made that possible. The more robust capabilities of the vehicle led in turn to revolutionary changes in its employment, transforming what had been simply a protected means of transporting infantry into combat, where doctrine held that they were then to dismount and join the battle on foot, to a formidable armored fighting vehicle. No longer was the unlucky foot soldier obliged to abandon his armored protection at the point of greatest peril. Instead he could remain mounted, contributing significantly more to the fight with his new-found firepower, tracked mobility, and robust communications capability. In time, the wisdom of this approach having been conclusively demonstrated on the battlefield, doctrine was revised accordingly and reflected in the subsequent development of such systems as the Bradley Infantry Fighting Vehicle. Except for the helicopter, which of course came into its own during the Vietnam War, no vehicle underwent more of a combat metamorphosis than the humble armored personnel carrier.

Solid evidence of how well mechanized infantry units were performing soon led to a decision that a number of dismounted infantry units should be designated for in-country conversion to mechanized infantry, including the two formerly mechanized battalions of the 1st Infantry Division that had been dismounted before deploying to Vietnam. In this way, the armored force in Vietnam was dramatically augmented on the ground. At one point, tankers from the 1st Battalion, 69th Armor, provided drivers and mechanics to help soldiers of another divisional unit, the 2d Battalion, 8th Infantry, learn to drive and maintain their newly acquired APCs. That reconfigured outfit, said an armor officer who was one of its Commanders, "was almost like a cavalry squadron today because it had a self-propelled 155mm artillery battery, a tank company, and three mechanized infantry companies."[21]

Meanwhile, the versatile M113/M113A1 performed many, many other useful roles configured as a command post vehicle, mortar carrier, ambulance, cargo hauler, bridge launcher, flamethrower, and tube-launched, optically sighted, wire-guided (TOW) missile platform. By one estimate, more than forty thousand APCs saw service in Vietnam in some sixty variants.[22]

In March 1967 the Army's "Evaluation of U.S. Mechanized and Armor Combat Operations in Vietnam" recognized that the M551 Sheridan armored system was about to become available and noted that there was a requirement for a light tank with the "going" characteristics of the M113. However, the "Evaluation" stated that the M551 in its present state had several significant deficiencies that precluded its use in Vietnam, such as a lack of a suitable antipersonnel round for the main gun, and the absence of night-fighting capabilities, a bulldozing kit, and additional armor. In spite of the problems, sixty-four Sheridans were deployed in January 1969. Officially designated an armored reconnais-

M551 airborne assault/armored reconnaissance vehicle with 152mm gun-missile launcher was introduced in Vietnam on a trial basis. Despite mixed reactions, it saw extensive service. National Archives.

sance/airborne assault vehicle, rather than a tank, this sixteen-ton tracked vehicle was armed with a somewhat flawed 152mm main gun capable of firing either antitank guided missiles or conventional rounds with combustible cartridge cases, although the missile-firing capability was never used in Vietnam. In addition to a multipurpose HEAT round, canister and beehive rounds were provided for the main gun. Two machine guns, a 7.62mm coax and a pedestal-mounted .50-caliber on the turret, completed the armament. Powered by a 225 horsepower diesel engine, the amphibious-capable vehicle had a cruising range of 373 miles and a maximum speed of forty-three miles per hour.[23]

The first Sheridans to arrive in Vietnam were issued to the 1st Squadron, 11th Armored Cavalry, where they were substituted for ACAVs in the scout sections, and to the 3d Squadron, 4th Cavalry, the 25th Infantry Division's cavalry squadron, where they replaced M48A3 tanks. Initial evaluations noted both advantages and disadvantages of the new Sheridan, depending in part on what they were compared to. They obviously packed far more firepower than the ACAV, with the canister round fired by the main gun proving particularly devastating. They were also maneuverable and fast, although they proved to be more vulnerable to enemy mines and RPGs than a tank. In addition, there were persistent problems with incomplete combustion of the main gun shell casings and with malfunctions of the electrical firing system, especially in wet weather. A number of common difficulties with the system's durability, such as overheated engines, turret electrical power failures, and failure of the gun's recoil system, were also encountered. These problems were very disturbing and offered little consolation to the crew. Of even more concern to crews were the facts that the 152mm combustible-case ammunition could be detonated by mines and that RPGs easily penetrated the hull's aluminum armor. Fear of these catastrophic events, coupled with the fact that the tankers had to fight in a hot, cramped crew compartment, caused significant fatigue. It is ironic to note that all the problems experienced in Vietnam in 1969 were identified years before the M551s were deployed. On balance, the Sheridan was judged to be a significant enough success that additional systems were fielded in Vietnam—a total of some two hundred by late 1970.[24]

While the Sheridan proved deployable, another new system introduced in Vietnam was an undisputed failure. The M114 command and reconnaissance vehicle was tracked and lightly armored, weighing just seven tons. Powered by a 120 horsepower engine, it had a cruising range of 375 miles and a maximum speed of thirty-seven miles per hour. Armament consisted of two pedestal-mounted machine guns—a .50-caliber and a 7.62mm—with no shielding for the gunners.[25] When the M114 was introduced into South Vietnamese units, it soon became clear that it "could not move cross-country and had difficulty entering and leaving waterways," absolutely devastating failings in that environment. By November 1964 the M114 had been replaced by M113s and withdrawn from Vietnam.[26]

Other armored vehicles employed in Vietnam included the M56 Scorpion self-propelled 90mm antitank weapon, sometimes called the SPAT, which was employed by Company D, 16th Armor, 173d Airborne Brigade. Armored vehicle launched bridges (AVLBs) with scissors bridges provided another important capability. The M60 AVLB (built on an M60 tank chassis) could span a sixty-foot gap, while an improvised lighter system mounted on an M113 could lay a thirty-foot bridge. The M578 light recovery vehicle and the superb M88 tracked recovery vehicle (VTR) proved indispensable. The M728 combat engineer vehicle (CEV), which mounted a 165mm demolition gun, a heavy-duty boom and winch, and a dozer blade, was both useful and versatile. Finally, there was the M42 Duster, a dual 40mm self-propelled tracked antiaircraft weapon recycled for ground support use, an aging system that nevertheless put out a high volume of very effective fire. All of these vehicles contributed significantly to the capabilities of the forces in the field.

No discussion of armor in Vietnam would be complete without at least some mention of the advisory effort. U.S. advisers were on the ground with South Vietnamese forces long before American units deployed and, as has been noted, they were instrumental in passing on to U.S. units important lessons learned by the ARVN in terms of mobility factors, tactics, and equipment modification.

Gen. Creighton Abrams devoted most of the year he spent as deputy commander in Vietnam to helping South Vietnamese forces improve their capabilities. When he succeeded Westmoreland as MACV commander in July 1968, he took seriously his responsibilities as senior adviser to the Vietnamese. General Starry, then a colonel, remembers having dinner with General Abrams on the night before he took command of the 11th Armored Cavalry. "He was very concerned by the fact that many U.S. commanders were still in the frame of mind that 'you little guys [meaning the South Vietnamese] get out of our way,'" Starry recalled. "He said to me after the change of command: 'Don't push yourself on the Vietnamese. They're going to have to learn to pick up the combat load, and you're going to have to help them learn that.'"[27]

As the war continued for year after year, Vietnamese armored forces expanded and upgraded their capabilities. M24 tanks, left over from the days of French influence, were replaced by M41A3s. Eventually the number of armored cavalry squadrons expanded from four to eighteen, one with each ARVN division and seven separate. In 1971 three tank regiments equipped with M48A3s were also formed. The entire MACV Staff, said Maj. Gen. Stan L. McClellan, had been opposed to giving the South Vietnamese the M48A3. When General Abrams asked McClellan for his view, the latter recalls advising the MACV commander that he was "strongly in favor and gave a short analysis of RVNAF capability to use the M48 and the obvious need to up-gun their armored force." He says that Abrams, after briefly pondering his comment, said, "Okay, go ahead." That proved to be a crucial decision. "In this case," concluded McClellan, "the heroic stand of the RVN 1st Tank Regiment (M48) during the 1972 Quang Tri NVA invasion was the single factor which caused the attack to fail. The enemy would have taken Hue on the first day except for the determined and effective defense by Vietnamese-manned M48 tanks."[28] The decision also reflected the confidence Abrams had in the Vietnamese tankers, most of them people he knew personally. General Abrams, observed his aide, Maj. Tom Noel, "had good rapport with many of the Vietnamese armored commanders who did a helluva job all over Vietnam."[29] The American advisers who worked with those forces made an important contribution to the conduct of the war.

By the autumn of 1967, when a substantial armored force had been built up in Vietnam, the Army Chief of Staff noted that "the doctrinal missions of armored cavalry are reconnaissance, security, and economy of force." In Vietnam, though, the 11th Armored Cavalry and the divisional armored cavalry squadrons were "performing the following additional missions: convoy escort, search and destroy (mounted and dismounted), cordon and search, search and clear, route clearing, and base defense reaction force." Two tank battalions were by that time also in country and, observed the Chief of Staff, "tank units are often tasked to link up with airmobile infantry. Tank units with attached infantry are also performing search and destroy, convoy escort, and security missions similar to those assigned to armored cavalry units."[30] Tank and armored cavalry units also provided protection for land clearing teams and supported the pacification program. On some occasions tanks were even employed in an indirect fire role, supplementing available artillery.

The tanklike role carved out by the ACAV has already been noted. Another important departure from established doctrine was that, with the advent of airmobile infantry, armored units—traditionally used as the exploitation force—were often used to fix the enemy while airmobile infantry deployed as the maneuver element.[31] Later, as pressure increased to hold down American casualties during disengagement, more and more combat elements of whatever type sought to fix the enemy when contact was made while firepower of every description was used for exploitation. Doctrine was also developed where none had previously existed, notably for cooperation between air cavalry and ground units. Given the often extremely dense foliage, for example, units on the ground frequently relied on air cavalry to guide them to their objectives. When armored forces were available, insertions of airmobile infantry often included plans for an armor link-up. And air cover flown by air cavalry became an important security provision for armored convoys on the move.

While armor doctrine held that units should be employed intact, not broken up and parceled out piecemeal, that principle was widely and persistently violated in Vietnam. This was due in part to the early restrictions on the deployment of armored units imposed by Generals Westmoreland and Johnson, restrictions that resulted in only two divisions deploying with their tank battalions. Once it became clear that armor could operate effectively in Vietnam, everybody wanted some, but there was just not enough to go around. Divvying up available armor forces thus became the norm. The 4th Infantry Division's 2d Battalion, 34th Armor, for example, arrived in Vietnam in September 1966 and was promptly split in three, with its individual tank companies being sent to widely separated areas. "The battalion," observed J.C. Pimlott, "never fought together during its period of service in Vietnam."[32] Similarly, the 1st Battalion, 69th Armor, located in the Central Highlands, habitually deployed one tank company in the vicinity of Bong Son, all the way across the country on the coastal plain.

Such wide dispersion of limited armor assets, often far from their next higher headquarters, put enormous strain on a logistical support system that was at best only marginally capable of supporting them. While hard-pressed staffs worked to develop makeshift support systems, even those arrangements were continually being undermined by the frequent redeployment of the supported units. In one extreme case, a tank battalion near Saigon retained responsibility for supporting one of its tank companies deployed in the far reaches of Military Region 1, some 750 kilometers to the north. One key part of the problem became just finding the supported unit, which had often been further divided and dispersed to multiple locations.

The 3d Squadron, 5th Cavalry, while based at Bear Cat in Military Region 3, was ordered north to Da Nang in Military Region 1 to work with the 1st Marine Division. The move was made by LST and, upon arrival, one troop was sent still farther north to work with an Army brigade operating near Hue. By agreement, the Marines were to supply the squadron with only food and fuel, while repair parts had to come from the Army. Unfortunately, however, the supported Army brigade was an airmobile unit and thus possessed no capability to support an armored force. "The problems of supply and repair of vehicles that resulted," said the squadron's executive officer, "were a nightmare for us."[33] These difficulties were further compounded by the fact that the centralized inventory system for repair parts simply broke down.[34] Many battalion executive officers and motor officers were reduced to wandering through depots searching for—and often finding—tank parts that did not show up in the records.

Armored units were frequently assigned the mission of route security, a much more difficult and repetitive task in Vietnam than in a more conventional combat environment. When General Johnson was Army Chief of Staff a civilian defense official once proposed sending a partly trained unit to Vietnam anyway, arguing that it could complete its training in a secure rear area. "Mr. Secretary," Johnson patiently explained, "there are no secure rear areas in Vietnam."[35]

Deployed units quickly adapted tactics and techniques, whether sanctioned by existing doctrine or not, for dealing with a determined and elusive enemy. The greatest threats to armor were enemy ambush—"a tactic in which he excels," admitted a U.S. Army report[36]—and mines. Ambushers habitually employed recoilless rifles and RPGs, both very hazardous to light armored vehicles and even tanks.

Counterambush techniques were stressed, although time after time units paid a heavy price in initial casualties amongst men and vehicles, then sought vainly for the rapidly withdrawing enemy forces that had inflicted them. Units coming under attack were taught to "herringbone," meaning that alternating vehicles angled right and left so as to bring effective fire on the ambushers while providing some armored protection for those thin-skinned vehicles that could find shelter between the tanks and ACAVs. Instantaneous return of a high volume of fire was stressed.

Many units also sought to discourage the setting of ambushes and planting of mines along routes they were securing by conducting at irregular intervals, both day and night, what were dubbed "thunder runs." These involved sending out columns of armored vehicles that would, without warning, unleash all the firepower they possessed at some suspected or likely ambush or mining site, all the while continuing to march. The theory was that such unpredictable onslaughts would discourage would-be attackers from getting into position.

These and other battle tactics often resulted in substantial casualties being inflicted on the enemy, but it is also true that throughout the conflict enemy mines and RPGs continued to be effective. Some mine rollers were sent out for use with armored vehicles, but they had little success. The problem for the mine roller, Lt. Gen. John H. Hay Jr. observed dryly, "was to survive the mine it detonated."[37] So unsatisfactory were the systems tried in Vietnam that General Starry judged them "not as effective as some 1945 equipment."[38]

A survey covering one six-month period at the height of the war found that throughout Vietnam some three-quarters of all tank and APC losses were caused by mines. These findings were duplicated by a similar survey conducted a year and a half later.[39] The fact is, throughout the war mines continued to be discovered and cleared by running over them with armored vehicles—a solution viewed as less than ideal by their crews.

Likewise, the enemy's RPG-2 antitank grenades, and later the more lethal RPG-7, continued to take a heavy toll, particularly of armored personnel carriers. Said General Abrams in August 1969, "the B-41 RPG-7 is the best hand-held antitank gun in the world."[40]

"Although many measures to defeat these weapons were tried," acknowledged General Starry, referring to both mines and the RPGs, "no adequate means was ever found."[41]

Despite many innovations and the occasional repudiation of an existing doctrinal precept, a 1966 U.S. Army "Vietnam Training Report" concluded "armor doctrine has been reaffirmed in Vietnam," citing successful application of the classic attributes of firepower, mobility and shock effect.[42]

The 1st Battalion, 69th Armor, was the first U.S. Army tank battalion deployed to Vietnam, arriving there in March 1966 as part of the 25th Infantry Division. Soon posted to the Central Highlands with the division's 3d Brigade Task Force, the unit there saw the 1st Platoon of Company B involved in a small but significant battle.[43] The action, later referred to as the Battle of Landing Zone (LZ) 27 Victor, took place in the western reaches of Pleiku Province, southwest of Duc Co and adjacent to the Cambodian border. There the tank platoon was assigned to assist in providing perimeter defense for a position occupied by a Republic of Korea (ROK) Army outfit, the 9th Company, 3d Battalion, 1st Cavalry Regiment, commanded by Captain Lee. There, on the night of 9-10 August 1966, an NVA battalion frontally attacked the dug-in infantry unit and its supporting tanks and, although repulsed repeatedly, kept attacking throughout a long, dark, and bloody night, suffering some 197 casualties in the process.

The platoon's five M48A3 tanks were stationed at strategic points around the perimeter of the position. At sunset the tankers—led by 2d Lt. Charles E. Markham—set a 50 percent alert, looked out over the single strand of concertina wire encircling the position at the heavy stands of elephant grass beyond, and gave a final check to the lay of their guns. Shortly before midnight, alerted by a member of the ROK 9th Company to the sound of digging nearby, S.Sgt. Wallace T. Ferneyhough's tank crew used its searchlight to illuminate the area and conducted a reconnaissance by fire with its coaxial machine gun. Within seconds, the entire tree line to the southeast erupted with heavy enemy automatic weapons fire. Although three tank crewmen were lightly wounded in this exchange, they managed to mount their tanks and return fire.

The volume of incoming fire continued to build, including heavy concentrations of small arms, mortars, and recoilless rifles in addition to the automatic weapons. Then came numerous assaults by small groups attempting to penetrate the defensive lines. It appeared that the defenders' initial recon by fire had served to disrupt a planned coordinated attack. Despite an overwhelming advantage in numbers, the sustained enemy assault managed to get only a single soldier through the defensive wire, and he was killed with a bayonet by a ROK defender.

Nearly continuous illumination was maintained over the battle area throughout the night, first by the two tanks mounting searchlights, later by mortars fired from within the 9th Company's position, and finally by U.S. and ROK artillery and a U.S. Air Force flare ship. The tanks fired every type of main gun ammunition they had (with the exception of HEAT), but canister rounds and the coaxial machine guns were used most extensively. Coordination of fires between tanks was excellent, with the commander of one tank alerting the commanders of adjacent tanks when targets moved toward their sectors of fire. One tank flicked on its searchlight periodically to draw fire, while another tank engaged the enemy thus revealed. It was impressive teamwork, especially for soldiers in their first combat action. Meanwhile, U.S. and ROK batteries provided extensive artillery support. Some 105mm fire was called in to within thirty meters of the perimeter. Heavier artillery was also used to interdict probable enemy routes of withdrawal.

The engagement finally ended at 4:30 A.M. when the surviving enemy withdrew, leaving the ground outside the perimeter literally covered with dead. As the defenders swept the area they recovered some 350 RPG-2 antitank rockets, five 60mm mortars, a heavy machine gun, forty-five AK-47 rifles and nineteen SKS carbines, twelve antitank rocket launchers, several satchel charges, and a large quantity of ammunition, packs, and other individual gear. Enemy documents retrieved and prisoner of war interrogation reports indicated the operation had been a planned coordinated attack against the position by a battalion from the 88th NVA Regiment.

The enemy's system of passing on "lessons learned" must have been effective, for after LZ 27 Victor they generally avoided infantry assaults on positions occupied by U.S. armor.[44] Instead, most of the subsequent combat engagements involving American tanks, armored cavalry, and mechanized infantry resulted from meeting engagements, counterambushes, or the reinforcement of embattled friendly units in contact. Enemy appreciation for how effective the combined U.S.-ROK force had proven itself was mirrored by the U.S. leadership with the subsequent award of the Presidential Unit Citation to both the U.S. tank platoon and the ROK infantry company. "We'd not had our annual training test" when the 1st Battalion, 69th Armor, deployed to Vietnam, wrote Sergeant Zumbro, "but we were to find out that the Viet Cong and the NVA were more than willing to provide one."[45] At LZ 27 Victor, B Company's 1st Platoon passed the test.

While the bulk of 1st Battalion, 69th Armor, operated in the Central Highlands, concentrating on keeping open Route 19E, the main supply route from the coast, one company was dispatched to the Bong Son plain in the coastal region north of Qui Nhon, there to support elements of the 1st Cavalry Division. This was typical of the wide dispersion of units that was the lot of armor in this war, and it did not stop at company level. The A Company tankers were parceled out by platoon, and platoons often sent sections of two or three tanks to work with the airmobile infantry. In one operation toward the end of 1966 or early in 1967 this fragmentation process reached a new level.

The 3d Platoon, like the rest of the company, was shorthanded, working with three-man crews while supporting squad-size elements of the 1st Cavalry Division in somewhat helter-skelter operations against dispersed elements of Vietcong in the vicinity of LZ English.[46] Company A tanks all had nicknames—"Ape," "Assassin," and the like—and Tank Three-Three was called "A-Go-Go." Its driver, Sp5 Kellen Wilson, was somewhat of a celebrity in the company, and for good reason. His family was in the restaurant business, and growing up in the midst of that Wilson had acquired the ability to, as his fellow soldiers put it, "make C rations fit for human consumption," working with special cooking oils, herbs, and spices his family sent from home. On a given day Wilson was going to distinguish himself in another way.

Called in to help out a half platoon of pinned-down infantry fighting in an abandoned village complex, the "A-Go-Go" took fire from a concealed enemy machine gunner. Its tank commander was wounded, reducing the crew to two men. After the injured man was evacuated by helicopter, Wilson pulled into position to shield another tank while it repaired a mine-damaged track. There a sniper zeroed in on his loader, who in turn was "dusted off," leaving Wilson as the only crewman left. Just then an infantry squad leader called for a tank to take out a gun emplacement that was holding him up. To the surprise of others on the net, Wilson answered the call, got the infantry's location, and roared off in "A-Go-Go."

Wilson soon reached the squad's position, where he crawled from his driver's seat into the turret, loaded the main gun, and recharged the coax, then jumped into the gunner's position. Then he commenced methodically to fire the 90mm, get up to reload, return to the gunner's seat, fire again, and so on until he had neutralized that particular enemy position. Next, using the coax, he took out an enemy machine gun that had another squad pinned down. Then someone else called for help, so Wilson dropped back down into the driver's seat and drove to a new location, repeating his performance as a man for all positions. This went on for four hours. In the process Wilson not only earned the Silver Star, but demonstrated the fragmentation of armor units carried to its ultimate extreme: a one-tank, one-man operation.

In the autumn of 1966 the Army, seeking definitive data on the performance of armored units in Vietnam, convened a study group headed by Maj. Gen. Arthur L. West Jr. to look into the matter. This was to be no academic exercise, but rather a firsthand evaluation on the ground. Underscoring how thoroughly that was the case, during the course of the study five members of the team were wounded, including West himself.[47]

The study—called "Mechanized and Armor Combat Operations in Vietnam" (MACOV)—was to cover in depth all aspects of doctrine, tactics, techniques, materiel, organization, and force mix of all U.S. Army mechanized infantry and armor units in Vietnam. A huge team, including forty-eight field data collectors and an evaluation staff of fifty-one, was assembled for the study, which was carried out during the period January-March 1967.[48]

The study team developed the trafficability data cited earlier, then turned its attention to operations and doctrine. "With the emergence of the M113 as a fighting vehicle," stated MACOV, "armor, scout, and mechanized units are engaging the enemy in mounted combat, while current doctrine prescribes this form of combat only for tank units." Also: "The employment of air cavalry has developed far beyond the limits of current doctrine." These observations were made approvingly, and in effect MACOV codified the modifications of doctrine that armored forces had developed in the field. Those changes had, wrote Lt. Gen. John H. Hay, quoting the MACOV study in his monograph on tactical and materiel innovations during the war, "evolved due to the nature of the enemy in Vietnam, the concept of area war and the balanced combined arms structure of the armored cavalry squadron."[49]

During the course of the study, General West, an experienced and highly decorated soldier who had, until being very seriously wounded, commanded an armored infantry battalion in World War II, also formed some negative views of what he had seen in Vietnam—observations that were not confined to the armored force. "Currently," he wrote while the study was in progress, "most battalion commanders and all brigade and division commanders command and control from the relative safety of a helicopter. We are teaching many bad habits that could cost us dearly in a war of the future where we do not have absolute control of the skies. Also Vietnam looks just a bit different from the air than it does from the ground." General West was also critical of the failure to continue combat operations at night. "The bulk of the US operations and all of the ARVN regular unit operations are conducted during the hours of daylight," he noted. "At night, our units go into a tight perimeter defense. At first light they move out and, if it is a mechanized or armor unit, about the first thing that happens is that the lead tank and/or APC hits mines."[50]

General West and several other members of the MACOV team had a chance to brief General Abrams, then the Army Vice Chief of Staff, when he visited Vietnam during the study. West and Abrams had commanded battalions side-by-side in the 4th Armored Division during World War II and maintained a close friendship afterward, so it was no surprise that the briefings took place in West's room at the Rex Hotel in Saigon and lasted until the small hours of the morning. Those briefings, West said later, "served as sort of a primer course for Abrams on Vietnam."[51] Whatever the validity of that assertion, when Abrams later took command in Vietnam the tactics changed at once, including an emphasis on the conduct of multiple small patrols and ambushes, both day and night.

The completed MACOV study, forwarded to the Secretary of the Army and the Chief of Staff, had some beneficial effects. For one thing, General Starry later observed, it was the first means by which "the potential of armored forces was fully described to the Army's top leadership." Among the interesting things they were told by MACOV was that "armored cavalry was probably the most cost-effective force on the Vietnam battlefield."[52] The capability of armored forces to operate throughout South Vietnam, weather and terrain notwithstanding, was also documented. Furthermore, and perhaps one of the most significant outcomes, "General Westmoreland . . . later commented that the study had prompted him to ask for more armored and mechanized units in troop requests."[53]

In late March 1967, just as the MACOV study group was completing its work, a classic case of "cavalry to the rescue" was acted out at a place called Fire Support Base (FSB) Gold.[54] There, near Suoi Tre, south of the Fishhook in Tay Ninh Province, the 2d Battalion, 12th Infantry, was hit with a rare daylight attack mounted by five battalions of the 272d Vietcong Regiment. Assault after assault struck the isolated position, penetrating the lines in three places and forcing the infantry and the artillery manning the outpost to pull into a tighter and tighter perimeter. Ammunition was running low, and the artillerymen had leveled their tubes and were firing beehive rounds at point-blank range. The enemy had closed to within hand grenade range, and things looked pretty grim.

Just then tanks from 2d Battalion, 34th Armor, and ACAVs from the 2d Battalion, 22d Mechanized Infantry, burst from the wood line to the enemy's rear and began raking the attackers with the devastating fire of more than two hundred tank cannon and machine guns. Within minutes the siege was lifted and the enemy driven off, leaving behind the bodies of more than six hundred Vietcong. No account of frontier days in the Old West could offer a more thrilling example of mounted forces galloping to the rescue.

Armored forces played an important, even pivotal, role in defeating enemy attacks in the 1968 Tet Offensive. "Rapid movement was imperative in the early stages of the enemy attack," emphasized General Starry, "and the armored units were the first ground forces to reach the battlefield in almost every major engagement, although the winning of the battles eventually involved all forces."[55]

Saigon was the enemy's primary objective. In the fierce fighting that raged in and around the capital, and especially "in the critical approaches, . . . cavalry and mechanized infantry decided the fate of the city."[56]

A dramatic example was the fighting at Tan Son Nhut Airport. In the early morning hours of 31 January, three enemy battalions attacked a mixed force of defenders that included National Police, Vice President Ky's security guard, and paratroopers from a nearby base camp who were soon joined by two airborne companies fortuitously standing by at the airport terminal waiting to be airlifted to the north.[57] An emergency request for reinforcement was passed to Lt. Col. Glenn K. Otis's 3d Squadron, 4th Cavalry. Launching from their base at Cu Chi, some fifteen miles distant, elements of the squadron drove hard in another thrilling gallop to contact, this time in the dark of night. "On its way to Tan Son Nhut," wrote South Vietnamese J2 Col. Hoang Ngoc Lung appreciatively, "the U.S. armor column was guided by air-dropped flares and took cross-country short cuts, bypassing the embattled area of Hoc Mon and probably ambush sites. At daybreak, the column entered Tan Son Nhut and inflicted serious losses to the enemy force, which was compelled to fall back."[58]

The next morning Lt. Col. Hugh J. Bartley, commander of the 3d Squadron, 5th Cavalry, was aloft in his command and control helicopter. From there he could see "that Saigon, Bien Hoa, and Long Binh were literally ringed in steel. . . . Five cavalry squadrons had moved through the previous day and night, converging on the Saigon area. When dawn broke, they formed an almost-continuous chain of more than five hundred fighting vehicles around the outskirts of the metropolitan area."[59]

Armored units contributed significantly to turning back the widespread enemy attacks in every part of the country. The 3d Squadron, 5th Cavalry, galloped to the rescue at Bien Hoa Air Base, devastating a waiting ambush by taking it in the rear while en route to the objective.[60] The 11th Armored Cavalry executed a twelve-hour forced march to reach Long Binh as that vital base was under assault.[61] At Kontum, in the Central Highlands, A Troop, 2d Squadron, 1st Cavalry, was assaulted repeatedly by an estimated three enemy battalions. The hugely outnumbered troop held its ground, fighting so courageously that it earned a Valorous Unit Award for the action.[62] At Pleiku, also in the Central Highlands, tanks of the 1st Battalion, 69th Armor, helped drive enemy forces out of that city. On board some of the tanks were air force people who had been catching a ride back to their station at Pleiku Air Base. So urgent was the need to get into action that the tankers didn't have time to off-load their Air Force passengers, and in the ensuing fight one of them was even pressed into service as a loader. That airman came away from the experience with a new appreciation for the tank. "You can really express yourself with one of these things!" he exulted.[63]

Meanwhile at Lang Vei Special Forces Camp, near Khe Sanh in the I Corps area, the enemy made his first use of tanks in this war when, on the night of 6-7 February 1968, an attacking force from the 304th NVA Division employing eleven PT-76 amphibious tanks overran the camp during a night of furious fighting. All but one of twenty-four U.S. personnel in the camp became casualties, including ten killed—one of whom was posthumously awarded the Medal of Honor. The indigenous forces also suffered heavily: 219 killed or missing and seventy-seven wounded. The Marines at Khe Sanh Combat Base, who had been tasked to prepare a contingency plan to rescue the small garrison at Lang Vei if it got into trouble, refused to help.[64]

Within days of the onset of the Tet offensive General Westmoreland, quite obviously converted from his earlier skepticism about the ability of mounted forces to fight effectively in Vietnam, submitted an urgent request that an armored brigade be sent to augment his forces. "The Army is behind the power curve with respect to meeting demands for trained manpower," cabled back the Chief of Staff, General Johnson. Multiple, increasing, and short-reaction demands, he informed Westmoreland, constitute a "steady leak in the reservoir that sustains your forces." But, Johnson added, "my job is to replenish the reservoir. I am making all possible representations to do this."[65]

Five months later, the 1st Brigade, 5th Infantry Division (Mechanized), arrived in Vietnam, bringing into battle the 1st Battalion, 77th Armor; 1st Battalion, 11th Infantry; 1st Battalion, 61st Infantry (Mecha-

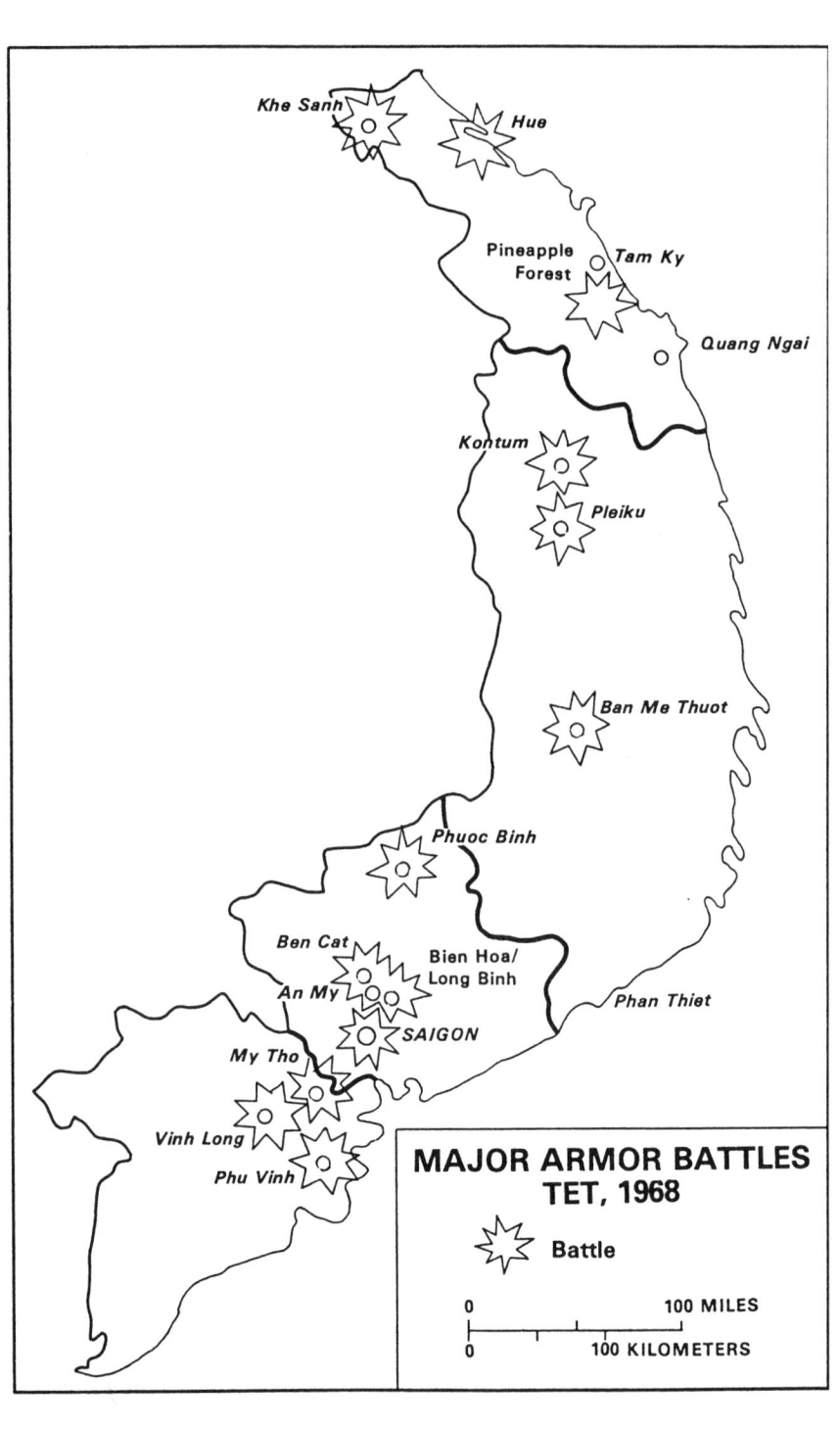

Khe Sanh

Hue

Pineapple
Forest

Tam Ky

Quang Ngai

Kontum

Pleiku

Ban Me Thuot

Phuoc Binh

Ben Cat

Bien Hoa/
Long Binh

An My

Phan Thiet

SAIGON

My Tho

Vinh Long

Phu Vinh

MAJOR ARMOR BATTLES
TET, 1968

Battle

0   100 MILES

0   100 KILOMETERS

nized); 5th Battalion, 4th Artillery (155mm self-propelled); and Troop A, 4th Squadron, 12th Cavalry.[66] Thus the last major U.S. unit to be deployed to Vietnam was a classic armored combined arms team.

Maj. Gen. George S. Patton, who as a colonel commanded the 11th Armored Cavalry in Vietnam, once observed that "nearly all actions in the war commenced with a classic movement to contact."[67] The 3d Squadron, 5th Cavalry, was involved in just such a meeting engagement on the coastal plain in Quang Tri Province in late June 1968.[68] In the pattern of wide dispersal that had become typical for armored units in this war, the squadron was operating far from its parent division, then based at Bear Cat, several hundred miles to the south. The squadron was itself further fragmented, with its command post and two troops operating out of Wunder Beach, on the South China Sea east of Quang Tri, and a third troop some eleven miles north providing security for Marine logistical operations at the mouth of the Cua Viet River.

To ease the logistical and maintenance problems inherent in such dispersal, the squadron commander, Lieutenant Colonel Bartley, rotated his troops through the assignment in the north, changing over about once a week. The troop being relieved was tasked to conduct area reconnaissance of the coastal strip traversed on the way back to the squadron base in the south. One day in late June 1968, A Troop was thus moving south, reconnoitering as it went. Approaching within about 150 yards of a village known as Binh An, the unit began taking small-arms fire, then the lead tank was struck by an RPG.

The platoon in contact deployed and returned fire while the troop Commander, Capt. Stewart McLaughlin, ordered his other two platoons to seal off the village by moving to blocking positions to the north and south of it. In the process, the three NVA soldiers who had fired the first shots were cut off and captured. During interrogation, one of the men revealed that a battalion some three hundred strong was positioned in the village.

Squadron headquarters put its light observation helicopter up over the scene of the contact and soon received reports of villagers streaming south carrying their possessions. Clearly they anticipated a destructive battle and were trying to get out of the way, expectations that were to prove very well founded. Lieutenant Colonel Bartley saw it the same way and ordered B Troop, conducting routine operations in the vicinity of Wunder Beach, to cease work there and move to Binh An at once. B Troop covered the half dozen miles or so quickly, moving along the hard sand at the water's edge until it encountered A Troop formed up in a semicircle facing to the southeast. B Troop joined the cordon and,

Col. George S. Patton (right), commander of the 11th Armored Cavalry Regiment, the largest mechanized unit in Vietnam, receives a commendation from Gen. Creighton W. Abrams Jr. (left) *ARMOR* magazine.

Donn A. Starry as a colonel in command of the 11th Armored Cavalry Regiment in Vietnam. He later served as the chief of armor and commanded V Corps in Germany before taking command of TRADOC. *ARMOR* magazine.

with the sea to the east, the village of Binh An was effectively surrounded. Soon artillery and naval gunfire from a cruiser and two destroyers on the gun line began pounding the village. The bombardment was interrupted only long enough for a loudspeaker team to broadcast a surrender appeal, to which there was no response. Meanwhile, two companies of infantry from the 1st Cavalry Division had arrived to augment the encircling forces and were interspersed with the cavalry.

In the late afternoon forces in the northern sector began sweeping south. Meanwhile, the armored vehicles at the south end buttoned up for protection against any machine-gun fire that might reach their positions. Once through the village, the assaulting elements reversed course and swept back to the north, resuming their positions in the cordon as the artillery resumed firing. As darkness came on, tank searchlights and night vision devices were used to watch for any enemy attempting to break out of the encirclement. Periodically during the night small groups or individuals were stopped or turned back by gunfire, while others were captured. As would later become apparent, the cordon was leak proof.

By morning the end was near. After a period of intensified artillery bombardment, buttoned-up armored forces from the north and south advanced toward one another, collapsing the cordon on the encircled enemy. They met, returned to the perimeter, and then repeated the process, this time with supporting infantry. Those enemy remaining alive came out with their hands up and were taken prisoner.

The operation was characterized by timely response to enemy contact, establishing and maintaining an effective cordon, the integration of armored cavalry, infantry, artillery, and naval gunfire, and a deliberate approach that saved the lives of friendly forces. In the process 233 NVA soldiers were killed in action and forty-four captured, along with a substantial number of individual and crew-served weapons. Among the enemy dead were the battalion commander and his staff and all of his company commanders. Friendly losses were one killed and nine lightly wounded.

Those were the immediate results. The larger effect of this and other operations in Quang Tri Province during the spring and summer of 1968, as later described by Col. Michael D. Mahler, "were quiet nights, infrequent contact with the enemy, and clear roads in an area that had been hostile territory for years." Highway 1 had been paved, SeaBees and Army engineers had built new and better bridges, "and local farmers and fishermen were out traveling those roads and bridges to market as they had not been able to do a year before." This was, testified Mahler, who had been there as executive officer of the 3d

Squadron, 5th Cavalry, "one accomplishment that could actually be seen and touched in a war where progress was not always easy to quantify."[69]

In March 1969 Ben Het Special Forces Camp, situated in a remote part of the Central Highlands near the Cambodian border, was the scene of a small but significant action—significant because for the first and only time in the war North Vietnamese and U.S. armored forces fought one another.[70] Company B, 1st Battalion, 69th Armor, was operating in the area in an effort to take some of the pressure off the small camp, which had been undergoing heavy and sustained shelling. Captain John P. Stovall, commanding the tank company, occupied strong points and staked out critical bridges along the ten kilometers of road that linked Ben Het with Dak To, and put one platoon of M48A3 tanks at Ben Het itself, where he eventually established his own command post. There they joined three companies of Civilian Irregular Defense Group (CIDG) troops, three 175mm guns, and two M42 Dusters, along with a twelve-man U.S. Special Forces A Team.

PT 76 tank destroyed 3 March 1969 at Ben Het Special Forces Camp, South Vietnam, by the 1st Platoon, B Company, 1st Battalion, 69th Armor, 4th Infantry Division. Enemy element believed to be from 202d NVA Armor Regiment. R.W. Wall Collection, Patton Museum.

Shortly after dark on the night of 2 March 1969, the tank platoon Sergeant reported hearing engine and track noises west of the camp. The next night, at almost the same hour, Ben Het began receiving recoilless rifle fire, followed by a heavy mortar bombardment. Soon engine sounds were heard again, and when something abruptly caught fire near a cluster of antipersonnel mines, "three tanks and some kind of open tracked vehicle were illuminated by the fire." Surprisingly, no infantry accompanied the enemy armor.

The U.S. tanks took these targets under fire using HEAT ammunition, while the camp's 81mm mortars put up illumination rounds. In the light from the magnesium flares at least two enemy tanks could be seen to take main gun hits. Both were on fire, with their ammunition and fuel going up, and the open tracked carrier was also burning. At that point Captain Stovall was standing on the back deck of one of his tanks, shielded by the turret, when a main gun round slammed into the vehicle's front slope. He and the tank Commander were blasted off the tank, and the driver and loader were both killed instantly. Other crewmen took their places and soon got the tank back into action. Meanwhile, the enemy withdrew without ever mounting a final assault. Ground patrols the next morning confirmed the destruction of two enemy tanks and one tracked carrier from what was later determined to be the 16th Company, 202d NVA Armor Regiment.

In late April 1970 U.S. and South Vietnamese commanders were given the green light to move forces across the border into Cambodia to interdict enemy lines of communications and clean out base areas adjacent to the boundary. Prohibitions on such operations had for years given the enemy a free ride, permitting him to operate from a secure base area and providing him a haven in which his forces could recuperate, refit, and retrain. Now, even though it was to be an incursion of limited duration and depth, the U.S. and ARVN forces had their governments' blessings to seek out and engage the enemy in his sanctuary. U.S. armor was to have an important role, and would operate in larger aggregations than had usually been the case during this widely dispersed war. On 29 April ARVN forces attacked into the Parrot's Beak, followed a day later by U.S. formations that entered the Fishhook. Both regions were riddled with enemy base areas.[71]

Brig. Gen. Robert M. Shoemaker, assistant commander of the 1st Cavalry Division, headed up an impressive task force rich in armor elements, including the 11th Armored Cavalry; 2d Battalion, 47th Infantry (Mechanized); and elements of the 2d Battalion, 34th Armor; as well as his own division's 3d Brigade (Airmobile) and the ARVN 3d Airborne

Brigade. This force was to drive north and west into the objective area. Subsequent attacks involved the 1st Battalion, 5th Infantry (Mechanized); 2d Battalion, 22d Infantry (Mechanized); and 3d Squadron, 4th Cavalry.[72] The 1st Squadron, 9th Cavalry, 1st Cavalry Division, also played an important role, discovering what turned out to be the largest cache of the war. This find was so extensive that it was dubbed "The City," and "yielded more than 1,500 weapons, millions of rounds of ammunition, and tons of supplies" and "took several weeks to search and evacuate."[73]

Lt. Gen. Michael Davison, commander of II Field Force, Vietnam, was in overall charge of the forces entering Cambodia. He said General Abrams had issued him a mission-type order—"I want you to go into Cambodia and clean out those supply points"—and then given him free rein to accomplish that mission. Abrams only asked him one question, Davison later recalled: "Are you really capturing all that crap they report in *Stars & Stripes?*"[74] Indeed they were. The final tally for the sixty-day operation, according to Lt. Gen. Phillip B. Davidson, a former MACV J2, included "23,000 individual weapons, enough to equip 74 full-strength NVA battalions; 2,500 crew-served weapons, 25 battalions' worth; 16,700,000 rounds of small-arms ammunition, the amount the Communists expended in one year; 14 million pounds of rice; 143,000 rounds of mortar, rocket, and recoilless rifle ammunition, and about 200,000 rounds of antiaircraft ammunition."[75]

The settled view of the operations in Cambodia is that the enemy did not fight to defend his stockpiles, but rather withdrew deeper into Cambodia, taking with him what he could and abandoning the rest. However accurate that may be as a generalization, it is also true that the forces entering Cambodia continued to experience the ambushes and mine warfare so familiar to them in Vietnam. Time after time, advancing columns took casualties from unseen enemy who then melted away into the jungle. The stockpiles turned up in Cambodia were acquired the hard way.[76]

The Cambodian incursion had been intended to deprive the enemy of his stockpiles of food and war materiel, disrupt his lines of communication, and buy time for Vietnamization to progress as American withdrawals continued. It did all of those things, and President Richard M. Nixon called it "the most successful operation of the Vietnam war."[77] Armored forces were a big part of that success.

By early 1969, with the change of administrations in Washington and the growing capability of South Vietnamese forces, it became clear that the time was approaching when U.S. forces could begin to withdraw. Planning for that eventuality was very closely controlled in MACV, with a small task force headed by Col. Donn Starry reporting directly to General Abrams. The instructions issued by Abrams were clear and succinct: "Do it right, do it in an orderly way . . . [and] save the armor units out until last, [because] they can buy us more time." It was a totally asymmetrical situation. "The armor units," said Starry, "specifically excluded from the buildup until late 1966, would anchor the withdrawal of American combat units from Vietnam."[78] Indeed, Colonel Starry, a key player in the MACOV study, would later command the 11th Armored Cavalry Regiment and lead it into Cambodia in May 1970.

The first twenty-five thousand men were withdrawn—"redeployed" as the euphemism had it—in July and August 1969, and succeeding increments departed on a more or less inexorable schedule. Following General Abrams's guidance, from the beginning armored units—tanks, mechanized infantry, air and armored cavalry—were stripped out of departing units and held in country. When the 4th Infantry Division left the Central Highlands in December 1970, for example, the 1st Squadron, 10th Cavalry, remained behind, reassigned to control of I Field Force, Vietnam. Later it was placed under the operational control of the 173d Airborne Brigade and then, when that outfit departed in August 1971, went to work for the headquarters of Military Region 2, finally heading for home itself in November 1971. The same thing happened with other divisional units. Not until April 1970 did the first battalion-sized armor units pull out, so that by the end of 1971 armored units constituted more than half of the U.S. maneuver battalions still in country. The last tank battalion—the 1st Battalion, 77th Armor—left in August 1971. Almost all the air cavalry units remained until early in 1972. In April 1972 the last U.S. armored unit—the 1st Squadron, 1st Cavalry, an armored cavalry outfit that had served with Task Force Oregon, the 23d Infantry (American) Division, the 11th Infantry Brigade, the 101st Airborne Division, the 196th Infantry Brigade, and the 23d ARVN Division—left for home. For the armored force, the war had ended.[79]

Eventually General Abrams, the quintessential armor commander, was about all that was left of the American armored force in Vietnam. He had in effect sent his Army home before him. Soon a new generation of armor leaders would emerge, many of them having practiced their trade under Abrams's tutelage. Among those who rose to four-star rank are Wallace H. Nutting, Glenn K. Otis, Crosbie Saint, and Donn Starry, all from ground cavalry outfits, and Robert M. Shoemaker from the air cavalry. Others who became prominent general officers include Julius

W. Becton, Frederic J. Brown, David K. Doyle, Richard G. Graves, Richard D. Lawrence, Paul S. Williams Jr. and John W. Woodmansee.[80]

Individual soldiers in armored units had also distinguished themselves during the war, earning nineteen of the 155 Medals of Honor awarded. No one had a more difficult war than the infantry foot soldier, or risked more for the sake of others than the medevac crews, but much of what armored troopers were assigned to do was frustrating, dangerous, or just plain hard work, relieved by moments of triumph. They are remembered for their service and sacrifice by this inscription on the Armored Forces monument: "In the U.S. Army Vietnam, the air-mobility of helicopter-borne infantry was augmented by the ground-mobility and firepower of the mechanized infantry battalions, armored cavalry regiment and squadrons, tank battalions, and armored artillery. In hundreds of combat actions these armored units of the infantry divisions demonstrated again the importance of mobile armor-protected firepower."

NOTES

1. The Army units listed are: the 11th Armored Cavalry Regiment; Armored Cavalry Squadrons: 1/1, 2/1, 1/4, 3/4, 3/5, and 1/10; Mechanized Infantry Battalions: 2/2, 1/5, 2/8, 1/16, 2/22, 4/23, 2/47, 1/50, 5/60, and 1/61; Armored Artillery Battalions: 5/4, 8/4, 3/6, 8/6, 7/8, 3/13, 6/14, 7/15, 5/16, 3/18, 5/22, 1/27, 6/27, 2/32, 6/32, 2/35, 1/39, 1/40, 1/82, 1/84, 2/94, and 2/138; Armored Cavalry Troops: A/12, E/1, E/17, and F/17. Marine Corps units are: Tank Battalions: 1 and 3; Amphibian Tractor Battalions: 1 and 3; Armored Amphibian Company: 1.

2. Ernest B. Furgurson, *Westmoreland: The Inevitable General* (Boston: Little, Brown, 1968), p. 84. Presumably Westmoreland at some point also attended parachute school to earn the jump wings he always wore.

3. As quoted in Gen. Donn A. Starry, *Armored Combat in Vietnam* (New York: Arno Press, 1980), p. 56.

4. Ibid.

5. Ibid., p. 57.

6. Ibid., p. 58.

7. Gen. Michael S. Davison, Oral History Interview, USAMHI. Even after the decision was made to deploy the 11th Armored Cavalry, Westmoreland first tried to get its M48A3 medium tanks replaced with M41 light tanks and then, unsuccessful in that initiative, to get a mechanized infantry brigade substituted for the armored cavalry regiment. "He stated flatly he had no need for two more tank battalions, which the 132 tanks of the regiment in fact represented." In a compromise, the regiment's tank strength was reduced to fifty-one, with APCs substituted for the rest. See

Shelby L. Stanton, *The Rise and Fall of an American Army: U.S. Ground Forces in Vietnam, 1965-1973* (New York: Dell, 1985), p. 70.

8. U.S. Army, videotape: "Vietnam Training Report Number 5: Armor Operations," 1966.

9. Findings of the MACOV Report as per U.S. Army, videotape, "Vietnam Training Report Number 14: MACOV," 1967.

10. Ibid.

11. Ralph Zumbro, *Tank Sergeant* (Novato, Calif. Presidio Press, 1988), p. 180.

12. Army of the Republic of Vietnam, "FM 3-1 (Draft): Armor Operations in Vietnam," Mar. 1966.

13. Starry, *Armored Combat*, p. 64.

14. The Marine Corps also deployed two tank battalions, along with other armored elements.

15. *History and Role of Armor* (Fort Knox: U.S. Army Armor School, 1968), pp. 30-31.

16. Michael D. Mahler, *Ringed in Steel: Armored Cavalry in Vietnam, 1967-68* (Novato, Calif: Presidio Press, 1986), p. 103.

17. *History and Role of Armor*, p. 32.

18. John A. Cash, John Albright, and Allan W. Sandstrum, *Seven Firefights in Vietnam* (Washington: OCMH, 1970), p. 41.

19. J.C. Pimlott, "Armour in Vietnam" in J.P. Harris and F.H. Toase, eds., *Armoured Warfare* (New York: St. Martin's Press, 1990), p. 153. Also see Lt. Gen. John H. Hay Jr., *Tactical and Materiel Innovations* (Washington: Department of the Army, 1974), p. 133.

20. Simon Dunstan, *Vietnam Tracks: Armor in Battle, 1945-1975* (London: Osprey, 1982), p. 105.

21. Maj. Gen. Robert J. Sunell, as quoted in Maj. Steve D. Dietrich and Major Bruce R. Pirnie, *Developing the Armored Force* (Washington: CMH, 1989), p. 2.

22. James R. Arnold, *Armor* (New York: Bantam Books, 1987), pp. 9, 19.

23. *History and Role of Armor*, pp. 39-40.

24. Starry, *Armored Combat*, pp. 142-5. Also see "Why the Sheridans in Vietnam?" and "The Vietnam Tests" in House Committee on Armed Services, *Review of Army Tank Program*, Report: Armed Services Investigating Subcommittee, 91st Cong., 1st sess., 9 July 1969, (Washington: GPO, 1969), pp. 28-34.

25. *History and Role of Armor*, p. 39.

26. Starry, *Armored Combat*, p. 38. Unfortunately for the service," wrote General Starry, "the M114's lackluster performance was ignored by U.S. Army decision makers and the vehicle with all its inadequacies became standard issue for the Army everywhere but in Vietnam. It was not until 1973 that General Abrams, then the U.S. Army Chief of Staff, [acting on the strong recommendation of General Starry, who at the time was commander

of the Armor Center at Fort Knox] branded the vehicle a failure and ordered it retired from the Army."

27. Gen. Donn A. Starry, interview with author, 1 Sept. 1989.

28. "Recollections of Major General Stan L. McClellan on 4 September 1974," Enclosure to McClellan Letter to Mrs. Abrams, 8 Oct. 1974.

29. Oral History Interview, The Abrams Story Collection, USAMHI.

30. Chief of Staff's *Weekly Summary* (5 Sept. 1967), p. 31.

31. Starry, *Armored Combat*, p. 71.

32. In Harris and Toase, eds., *Armoured Warfare*, p. 152.

33. Mahler, *Ringed in Steel*, p. 125.

34. Starry, *Armored Combat*, pp. 183-84.

35. As quoted in David M. Barrett, *Uncertain Warriors: Lyndon Johnson and his Vietnam Advisers* (Lawrence: University Press of Kansas, 1993), p. 115.

36. Videotape, "Vietnam Training Report Number 5."

37. *Tactical and Materiel Innovations*, p. 134.

38. Starry, *Armored Combat* p. 223.

39. Ibid., p. 79. The specific figures were 73 percent of APC losses and 77 percent of tank losses during the period Nov. 1968-May 1969, and 75 percent of all combat vehicles lost in Dec. 1970.

40. Comment in a 5 Aug. 1969 MACV briefing for visiting Ambassador G. McMurtrie Godley.

41. Starry, *Armored Combat*, p. 47. One innovation that provided some protection for tanks and APCs in stationary positions was the erection of a piece of chain-link fence in front of the vehicle to serve as an RPG screen, causing the shaped-charge warhead to detonate before it hit the vehicle. When the unit moved on, the sections of fence could be rolled up and hoisted aboard, ready for use when the next position was occupied. See Hay, *Tactical and Materiel Innovations*, p. 109.

42. Videotape, "Vietnam Training Report Number 5."

43. This account is based on the Presidential Unit Citation awarded the tank platoon and on after-action reports submitted by its parent battalion.

44. Two exceptions occurred during Operation Junction City in Mar. 1967. At Prek Klok on the evening of 10 Mar. the enemy launched a ground assault against a defending mechanized infantry battalion from the 1st Infantry Division, a rash action that resulted in nearly two hundred enemy deaths at a cost of three friendly killed. And, on the night of 19 Mar. at Ap Bau Bang, tanks and APCs of the 9th Infantry Division's 3d Squadron, 5th Cavalry, fought a hard battle against repeated enemy assaults and, with the help of a counterattack mounted by a nearby cavalry troop, drove off the attackers and accounted for 227 enemy killed at the expense of three friendly deaths and a number of APCs destroyed. See Jim Mesko, *Armor in Vietnam* (Carrollton: Squadron/Signal Publications, 1982), p. 31. Also see Arnold, *Armor*, pp. 60-61.

45. Ralph Zumbro, *Tank Aces* (New York: Pocket Books, 1997), p. 268.

46. This account is based on Zumbro, *Tank Sergeant,* pp. 60-63, and interviews with Sgt. Maj. Charles R. Hazelip, M.Sgt. Robert M. Kinzel, Sp4 Robert M. Holt, and Sergeant Zumbro.

47. Maj. Gen. Arthur L. West Jr., Oral History Interview, The Abrams Story Collection, USAMHI.

48. This discussion is, except where otherwise cited, based on the final report, "Evaluation of U.S. Army Mechanized and Armor Combat Operations in Vietnam," Headquarters MACV: MACOV Evaluation Team, 29 Mar. 1967. The MACOV report notes that four team members were wounded as a result of hostile action, differing slightly from the five mentioned by General West in his oral history.

49. Hay, *Tactical and Materiel Innovations,* p 115.

50. Maj. Gen. Arthur L. West Jr., Response to Generalship Study Questionnaire, The Franklin Institute Research Laboratories, Feb. 1967, submitted in the form of a personal letter to Gen. Bruce C. Clarke, U.S. Army Command & General Staff College Library, Fort Leavenworth, Kans.

51. West, Oral History Interview.

52. Starry, *Armored Combat,* p. v.

53. Ibid., p. 85.

54. This account is based on Mesko, *Armor in Vietnam,* p. 31; Starry, *Armored Combat,* p. 102; and Harry G. Summers Jr., *Historical Atlas of the Vietnam War* (Boston: Houghton Mifflin, 1995), p. 118. Summers's account differs in identification of both the friendly and enemy units involved.

55. Starry, *Armored Combat,* p. 115.

56. Ibid., p. 118.

57. Col. Hoang Ngoc Lung, *ARVN: The General Offensives of 1968-69* (Washington: CMH, 1981), p. 60.

58. Ibid., p. 61.

59. Mahler, *Ringed in Steel,* p. 116.

60. Ibid., pp. 112-13.

61. Arnold, *Armor,* p. 76.

62. David B. Stockwell, *Tanks in the Wire* (Canton, Ohio: Daring Books, 1989), p. 85.

63. Zumbro, *Tank Sergeant,* p. 143.

64. Stockwell, *Tanks in the Wire,* pp. 143, 155.

65. Johnson to Westmoreland, message WDC 2526, 18 Feb. 1968, Box 142, Harold K. Johnson Papers, USAMHI.

66. Shelby L. Stanton, *Vietnam Order of Battle* (Washington: U.S. News Books, 1981), p. 77.

67. "Foreword" to Dunstan, *Vietnam Tracks,* p. 7.

68. This account is based on Mahler, *Ringed in Steel,* pp. 211-6; Ralph B. Garretson Jr., "The Battle of Binh An," *ARMOR* (July-Aug. 1969): pp. 25-28; and interviews with Brig. Gen. Hugh J. Bartley and Colonel Mahler. The memories of various participants differ on some particulars.

69. Mahler, *Ringed in Steel,* p. 216.

70. This account is based on Starry, *Armored Combat,* p. 153; Stanton, *Vietnam Order of Battle,* p. 93; Capt. Gerald R. Cossey, "Tank vs. Tank," *ARMOR* (Sept.-Oct. 1970): pp. 16-19; and Zumbro, *Tank Aces,* pp. 304-10, quotation from p. 307.

71. Lt. Gen. Phillip B. Davidson, *Vietnam at War: The History, 1945-1975* (Novato. Calif.: Presidio Press, 1988), p. 625.

72. Starry, *Armored Combat,* p. 174.

73. Ibid., pp. 170-72.

74. Gen. Michael S. Davison, interview with author, 25 Feb. 1988.

75. Davidson, *Vietnam at War,* p. 627.

76. The realities of the Cambodian incursion as experienced at the troop level are graphically and brilliantly portrayed by Keith William Nolan in his book *Into Cambodia: Spring Campaign, Summer Offensive, 1970* (Novato, Calif.: Presidio Press, 1990).

77. Richard Nixon, *RN: The Memoirs of Richard Nixon* (New York: Grosset & Dunlap, 1978), p. 467.

78. Starry, *Armored Combat,* p. 164.

79. Stanton, *Vietnam Order of Battle,* pp. 124, 129; Starry, *Armored Combat,* pp. 164-65, 180; and Dunstan, *Vietnam Tracks,* p. 185.

80. Starry, *Armored Combat,* pp. 227-37.

SUGGESTIONS FOR FURTHER READING

The most authoritative single work is Gen. Donn A. Starry's volume in the United States Army's "Vietnam Studies" series of monographs entitled *Mounted Combat in Vietnam* (Washington: Department of the Army, 1978), later reprinted commercially as *Armored Combat in Vietnam* (New York: Arno, 1980). General Starry's is one of the few monographs in this series that was completed after the war was over and thus constitutes a comprehensive rather than truncated look at the topic it addresses. Interestingly, General Westmoreland, then serving as Chief of Staff, rejected recommendations that such a book be written and that Donn Starry should write it. When General Abrams succeeded Westmoreland as Chief of Staff, he directed Maj. Gen. William R. Desobry, the Armor Center commander, to write a history of mounted combat in Vietnam. Starry, who replaced Desobry in 1973, ultimately took on the task and finally produced the cited study five years later. At one point it was the best selling of all the Army's Vietnam Studies monographs and has been reprinted several times commercially. It has also been widely translated for study by foreign armies. Simon Dunstan's *Vietnam Tracks* (London: Osprey, 1982), while largely derivative, contains much useful material. Interesting and informative first-person accounts include Ralph Zumbro's *Tank Sergeant* (Novato, Calif.: Presidio Press, 1986) and Michael D. Mahler's *Ringed in Steel* (Novato, Calif.: Presidio Press, 1986).

Zumbro has also incorporated some Vietnam material in his more recent book *Tank Aces* (New York: Pocket Books, 1997). John A. Cash provides an excellent description of the Battle of Lang Vei, in which the enemy used tanks for the first time, in *Seven Firefights in Vietnam* (Washington: OCMH, 1970). Detailed information on the deployment, equipment, and service of individual units may be found in Shelby L. Stanton's comprehensive *Vietnam Order of Battle* (Washington: U.S. News Books, 1981).

11

AirLand Battle

Richard M. Swain

The long ninth decade of the twentieth century proved to be the heyday of the tank in American notions of land warfare. Driven largely by the presence of the Soviet armored threat to NATO, the years from the end of U.S. military involvement in Vietnam to the Gulf War were marked by the creation of the most powerful armored force in U.S. history. It was better equipped, better trained, and generally more soundly schooled than any armored force that preceded it. Ironically, by the time this army was tested in battle, it was already undergoing its dismantling. The battle that validated the efforts of the long ninth decade was fought in Iraq by an armored force on its way to the scrap yard, its long-time Soviet rival overcome by that state's own internal political and economic contradictions. In the wake of the Cold War, the U.S. Army was downsizing and simultaneously turning its search for qualitative superiority to a narrow focus on technical rather than conceptual superiority. The polymorphic nature of post–Cold War problems and the concomitant shift to continental U.S. basing brought into question the continued centrality of heavy armored striking forces as the basis of U.S. military power.

Creation of the Desert Storm army was the result of an extraordinary effort to create a land force qualitatively unmatched anywhere in the world. Such qualitative superiority was required because it was accepted that any U.S. force in the world almost certainly would be outnumbered. Quality was to redress quantitative inferiority. The effort to achieve qualitative superiority involved the enlistment and retention of highly skilled and disciplined volunteer soldiers; the development of the world's best fighting equipment; the evolution of a near science of realistic, performance-oriented training to standard; and the development of a sequential and progressive professional education system designed to hone leader skills. VII Corps's victory in the desert in 1991 rested to a marked extent upon the skills and leadership of a noncommissioned officer corps unique in the world for its professionalism and level of responsibility.

Tying all of these human and material investments into a focused and harmonious whole was a doctrine of war on land called AirLand Battle. This chapter traces the evolution of that body of beliefs and practices from 1973 to 1986—the year in which the last edition of the army's capstone doctrinal manuals to employ the trademark was published. The 1986 *Field Manual (FM) 100-5, Operations,* codified the doctrine the army took to the Persian Gulf in August 1990.

Doctrine, as an authoritative and formal declaration of how a military organization intends to fight, is a creation of the nineteenth century general staff culture. The purpose of doctrine is to unify or harmonize the individual efforts of members of an organization in the performance of their collective tasks. It guides training, organization, and acquisition.[1] Because it is directive in nature, it can also be used to change an organization's focus. However, the use of doctrine to perform this relatively short-term function abruptly may produce or exacerbate existing instabilities in the organization. And, when a proposed modification of published doctrine is not accepted by the institution's membership, its failure can call for its rectification in a sort of dialectic process of thesis, antithesis, and synthesis.

Doctrine is a matter of choice.[2] Someone ultimately has the authority to choose between alternatives. Others have the authority to vote or veto, and still others to advise. To that extent, the development of doctrine is hierarchical and doctrinal choice can be influenced heavily by the personal idiosyncrasy of those with authority to choose or advise. No matter how the choice is made, however, doctrine development may be seen conceptually in two ways. It may be viewed as a bottom-up, building block process—take care of the fundamentals and the larger issues will solve themselves, for example. It may also be viewed as a process of relating inferior means to superior ends—a top-down process in which the ends are first well defined and qualified, then appropriate ways and means are developed. The choice of doctrine in either case is affected by institutional and extrainstitutional contexts—social, historical, cultural, and political. Choices are influenced heavily by the range of available or anticipated technologies or tools. The evolution of doctrine in the 1970s and 1980s reflected the presence of all of these influences.

Finally, formal doctrine is of little value if no one reads it. Schools are important to the infusion of doctrine into the body of the Army if doctrine is written in the professional schools, as it generally is. The viability of a doctrine depends largely on how well it is explained and accepted by the officers who must train and command the wider army in its use. The process of infusion takes place largely in schools and, in the late twentieth century U.S. Army, in combat training centers where, through criticism of unit performance, doctrine is either validated or sometimes skewed in interpretation, away from the original drafters' intentions.

United States participation in the lost war in Vietnam ended in 1973. In August that year the U.S. Army in the United States underwent a significant change. The Continental Army Command (CONARC) was reorganized into a Forces Command (FORSCOM) and Training and Doctrine Command (TRADOC), both commanded by four-star generals. Training and Doctrine Command would combine under a single headquarters the Army training base for individual training and education, development of organizational models (force design), articulation of materiel requirements (as opposed to production of end items), and the development of Army doctrine. The reorganization of CONARC was the result of a study titled STEADFAST, conducted by the assistant vice chief of staff, Lt. Gen. William E. DePuy.[3] DePuy was subsequently selected for promotion to General and appointed to head the new Training and Doctrine Command, headquartered at Fort Monroe, Virginia.

William E. DePuy was an extraordinary officer. Incisive, tough-minded, ruthless in pursuit of his goals, and constitutionally unable to tolerate fools gladly, he was often prone to consider opponents—or simply those less astute than he—as fools. DePuy enlisted in the South Dakota National Guard as an engineer in the late 1930s.[4] He later enrolled in the Reserve Officers's Training Program (ROTC) at South Dakota State University, where he earned a degree in economics and was commissioned in the infantry in 1941. DePuy fought World War II in Europe as an officer in the 90th Infantry Division. He commanded a battalion, served on regimental staffs, and finished the war as the division G3, retraining the division for movement to the Pacific to participate in the invasion of Japan. DePuy emerged from the war with a Distinguished Service Cross and three Silver Stars for gallantry, and with certain beliefs about the nature of training embedded firmly in his mind. Subsequent experience in Vietnam as General Westmoreland's operations officer and then as Commander of the 1st Infantry Division only confirmed what he already believed.

First, DePuy was contemptuous of the training he and other company grade officers received before their introduction to combat in World War II. He called his division "a killing machine," adding, "of our own troops."[5] He learned from bitter experience the value of indirect approaches in battle, and developed a lifelong concern for the importance of suppression as the sine qua non of battlefield movement.[6]

Whereas army officers as a class tend to concentrate on the study of generalship, DePuy, a quintessential "leg" infantryman, applied his impressive and keen intellect to the study of the army's fundamental organization, the infantry squad. As a battalion Commander a decade after the war he wrote simplified squad training pamphlets that emphasized tactical overwatch as the basis of battlefield movement. DePuy based his ideas on his observations of armor unit training in Maj. Gen. Hamilton Howze's division in Europe.[7] Later, as a Colonel, DePuy composed a conceptual essay on the nature of the infantry squad that remains a classic of military writing.[8] One of his first concerns when he took command of TRADOC was to replace the common foxhole with a more sophisticated arrangement that provided frontal cover and arranged interlocking fires between individual positions, a notion he had worked on for years.[9]

DePuy was the ultimate Cartesian. He sought to reduce all combat functions to their lowest denominator, define them with precision, and then regurgitate them. One briefing slide he often used reduced the tactical problem facing the army in Germany to a single tank company team (a company-sized unit consisting of two tank and one infantry platoons) being required to service (destroy) sixty Soviet tanks within 7.3 minutes.[10] The expression of larger issues in such a focused manner coincided with his belief that the overwhelming requirement for success in battle was excellence in individual and tactical unit training. One knew where he intended to begin when he arrived at TRADOC: training companies and battalions for combat by emphasizing disciplined performance at squad, platoon and company level. He had, according to Gen. Paul Gorman, DePuy's Deputy Chief of Staff for Training, "what he termed 'a fire in the belly' over the issue of training commanders and soldiers for close combat, a deep, abiding concern that the Army's mission was in that respect unaccomplished."[11] Finally, DePuy was convinced that the capital ship of the battlefield was the tank.[12] In European and Middle Eastern armies, it was.

There were three other circumstances that influenced General DePuy's situation at TRADOC after he took command. The first was the lost war, its results, and their implications. The second was the health of the chief of staff, Gen. Creighton Abrams. The third was the impact of the 1973 Arab-Israeli war on the U.S. Army.

The lost war had ruined the army as a combat organization. It had also produced an officer corps that was aware of the futility of its recent efforts and determined not to repeat the experience. Most company and field grade officers were open to new ideas and driven to study the past, the present, and even the putative enemy, for whatever lessons each might offer. At the same time, new ideas from those in authority would be questioned and debated, rather than adopted quietly on the basis of superior authority alone. The Army's senior leadership had lost much of its professional legitimacy in Vietnam.

The American diplomatic drawing-in that followed the war in Vietnam and the Watergate affair simplified the military's task in one sense. The unlikelihood of immediate overseas adventurism meant that the Army could focus its energies on preparing for the most likely—and most hoped-to-be-avoided—threat: an all-out confrontation with Warsaw Pact forces in Europe. The notion that what would work in Europe would very likely work elsewhere was taken largely for granted.[13] The fact that the most widely recognized alternative locales for conflict were the Middle East or Korea seemed to underwrite this view.

At the same time, the combination of any imaginable scenario in Europe and the abolition of the draft—a further result of domestic disgust with the conduct of the Vietnam War—meant that the pre-Vietnam structure of a conscript Army, maintained as a foundation for a larger emergency force, was also abandoned. The Army would be all volunteer and would have to be prepared for come-as-you-are wars. It might be dependent on reserve components to fill out active force shortages, but it would have to be prepared to fight with what it had rather than depend on its ability to stave off defeat long enough to mobilize and equip a large citizen force as the nation had in World War II. This change too was fraught with consequences recognized by General DePuy. His army would have to be prepared to win both its first *and* last battles.

In 1974, General Abrams's illness and death in office created an institutional vacuum in the Army hierarchy into which the strong-minded DePuy would insert himself.[14] Gen. Frederick C. Weyand, Abrams's Vice Chief and successor, made clear at the outset his was only an interim appointment dedicated to carrying through Abrams's program.[15] The lack of a strong hand at the top gave the "Barons" (four-star generals outside Washington) a degree of latitude with which to work that might otherwise not have been available. DePuy would be free to work doctrine and training unhindered by the Army staff. However, he would also have to sell what he did to the Army's field operating commands, particularly FORSCOM and U.S. Army Europe (USAREUR). He addressed the former through joint TRADOC and FORSCOM doctrine conferences, called "October Fests," and the latter, at least in part, by outflanking the Americans in Europe by paying particular attention to the requirements and views of the German allies.

This structural ambiguity, as much as DePuy's temperament, resulted in the production of a doctrine identified personally with the TRADOC commander and adopted largely without reference to the Army staff.[16]

The third factor, was the event of a war in the Middle East that seemed analogous to that for which the army was preparing in Europe. The metaphor provided by the war along the Suez Canal and on the Golan Heights, and later in Lebanon in the early 1980s, may have been too limited a model from which to derive a doctrinal paradigm. Middle Eastern events, narrowly read, cast long shadows as the U.S. Army worked its way back from Vietnam.

There was one final circumstance that is also important. The fundamental acquisition decisions governing what to buy during the next ten years had already been made. The Army, prior to the creation of TRADOC, decided to focus its materiel acquisition efforts on five weapons systems: a state-of-the-art tank, an infantry fighting vehicle to accompany and support it on the battlefield, an advanced attack helicopter, a new assault (troop carrying) helicopter, and a new air defense system. General DePuy, not coincidentally, was the driving force behind that decision as head of a departmental committee while serving as Assistant Vice Chief of Staff.[17] Ironically, it became his job as TRADOC commander to decide how the army would organize and train to fight the selected systems, and to make the case for the "Big Five" to the Congress and the Army itself.

One way to get the Army back on its feet while waiting for the new weapons systems to arrive was to undertake tough training to standard, a task for which Bill DePuy had prepared all his life. He had at TRADOC a former subordinate from the division he had commanded in Vietnam, Brig. Gen. Paul Gorman. Gorman was the creator of the Army Training and Evaluation Program (ARTEP) and Skill Qualification Tests (SQT), which established clear standards for performance of all collective and individual combat tasks.[18] Gorman also led a revolution in training technology, introducing the army to laser simulations that permitted realistic training engagements in which "combat" losses could be accurately scored.

The Gorman training revolution actually preceded the development of a new family of doctrinal publications, a shortcoming he made up for by issuing interim and narrowly focused pamphlets that indicated the direction doctrine seemed to be going in their guidance for the performance of particular actions or tasks. These were called training circulars and appeared in a variety of formats depending on the intended audience: small comic books for enlisted men and junior officers, and prototype manuals for higher staffs. Gorman's training programs produced by far the most important and lasting change to the place of doctrine in the U.S. Army. Once the ARTEP was established as a testable compendium of tactical tasks, the application of doctrine could be enforced.[19] Plans were made to create what grew into the Combat Training Centers to evaluate (the cover story was they only *trained*) FORSCOM units routinely, according to TRADOC standards, on instrumented ranges capable of following every combat vehicle or Soldier throughout an engagement. Doctrine thus became part of the Army's discipline.

The October War in 1973 came as a shock to the Army, which was undergoing the long, painful process of recovery in a period of social unrest, restrictive budgets, and massive, rapid force reductions. The conflict pitted three mechanized armies looking much like those facing each other in Europe in a series of battles that suggested a revolution in military affairs had occurred while the U.S. Army was preoccupied with Vietnam.[20]

Training and Doctrine Command was quick to learn what had happened and to respond accordingly as it redrew the army's training and doctrine regime. The high-intensity tank battles on the Golan Heights became the metaphor for modern combat. Soon General DePuy himself was on the road sharing the wisdom TRADOC distilled from the recent conflict. Not surprisingly, for a man who believed war was fought from the bottom up, General DePuy and TRADOC began with individual weapons lethality and aggregated the war as a whole from that fundamental fact. Also not surprising, the lessons confirmed much of what DePuy already believed.

In June 1973, four months before the war, General DePuy told an audience at Fort Polk, Louisiana, that future wars would be "short, violent, and important." They would be short because the politicians would step in and stop any conflict quickly, lest it grow out of control. (One might add: Or it would confront NATO with the choice between escalation to nuclear war or surrender within days at best.) It would be important, because otherwise the U.S. would not be involved. It would be violent, because U.S. forces would likely be outnumbered and confronted by an enemy equally well equipped. Indeed DePuy said, pointing to possible intervention in the Middle East in support of Israel: "We have got to have one American tank battalion at least equal to five Arab tank battalions. One American infantry battalion has to be worth five of theirs and I really mean that. We have to be that much better."[21] In short, the army would have to be prepared to fight outnumbered and win. The difference would be in training, for near parity in equipment was assumed.

In a letter to General Abrams written almost immediately after the 1973 war, DePuy concluded that "during the next ten years battlefield outcome will depend upon the quality of the troops rather than the quality of the tanks."[22] He offered as proof that standard simulations of the Golan Heights battles had indicated the Israelis would lose, when in fact the outnumbered side had prevailed. Initially DePuy seems to have been most concerned by the threat to the synergy of air and ground forces, previously a major Israeli advantage that the Arabs enjoyed some success disrupting at the outset of the war. He warned that the army could not take for granted the immediate availability of close air support in any future war.

By spring 1975, a more detailed briefing on the lessons of the war also supported views DePuy had developed over a number of years.[23] The briefing was illustrated by numerous graphical displays pointing to the new lethality of air and ground combat systems. It was characteristic of General DePuy to rely heavily on techniques of operations research to get at the underlying calculus of battle. He drew three major lessons from the war. First, the battlefield environment was far more lethal than ever before. Second, fighting demanded a highly trained and integrated combined arms team. Third, tactical training could make the difference between success and failure.[24]

Movement was essential for success in battle, he argued, and suppression was required for movement. (Suppression of enemy air defenses, he observed, was vital to ground movement as well.) Clever use of terrain for cover and concealment was essential. "What can be seen, can be hit. What can be hit can be killed," was the new mantra. "The tank today is the single most important weapon on the mechanized battlefield . . . However, the tank can't do it alone."[25] Indeed, the proliferation of man-portable antiarmor weapons, most notably the "suitcase" Sagger wire-guided antitank missile, in engagements along the Suez Canal came as a shock not just to Israeli tank commanders, but also to a U.S. Army that had tied its future to the heavy tank.

General DePuy then began to describe a defensive concept for thinly deployed Europe-based troops in which units, deployed initially on line, would close in on the flanks of an enemy penetration, thickening the defense at the point of attack, until it was stopped and destroyed or thrown back. "What this means," DePuy said, "is that the defending force must possess the ability to move. It must engage in an *active defense* of the sector." [26] [Emphasis added] These tactics, which the tank battles on the Golan Heights appeared to foreshadow, had grown out of efforts made in 1973 to rewrite army tactical doctrine at the Armor Center. The basic defensive concept seems to have come from the second of DePuy's inner circle, and ultimately his successor at TRADOC, Maj. Gen. Donn Albert Starry.

Starry, like DePuy, was a strong-minded man. He enlisted in the army in World War II but went to West Point and missed participation in that war. If DePuy was a quintessential "leg" infantryman, Starry was "Mr. Armor." He had worked in unconventional warfare in Korea and, like DePuy, shared connections with the military wing of the Central Intelligence Agency (CIA). Starry commanded the 11th Armored Cavalry Regiment in some of the Vietnam War's most brutal fighting, including the U.S. and South Vietnamese invasion of Cambodia in 1970. He was a protégé of General Abrams. When Abrams sent him out to Fort Knox in the summer of 1973 (the same time DePuy was creating TRADOC, within whose jurisdiction Fort Knox fell), Starry took Abrams's admonishment to "Go out and get the Army off its ass" as a personal assignment.[27]

Starry began exploring the requirements of the tactical problem in Europe at once. He began with the notion that what mattered most in battle, given some reasonable balance of forces, was seizure and retention of the initiative—a view inspired in the 1950s by Robert Helmbold's analytic work for the BDM Corporation.[28] Helmbold's research challenged the notion that superiority of numbers alone insured success. Starry's conclusion was that what seemed to be most important in that regard was not numbers but control of the initiative in battle.

Starry went to the Golan Heights in January 1974 and found in the example of the Israeli defense there the same paradigm facing the NATO allies in Europe: conducting a successful tactical defense within the strategic and operational strictures of a forward defense posture.[29] The Israelis were forced to defend forward on the Golan because there was no maneuver depth behind them. In Europe, the Germans required forward defense. Too many population centers were located on the border to allow German commanders to contemplate trading space for time.[30] That NATO lacked the forward-deployed forces necessary to make such a cordon defense viable simply did not matter. NATO depended too much on German participation in the alliance to challenge what was, for the nation divided in its center, the fundamental premise of national survival.

The chief purpose of NATO strategy was deterrence, not war fighting. The alliance's chief deterrent was an increasingly dubious linkage between tactical defense and a strategic nuclear exchange between the United States and the Soviet Union under conditions, as Raymond Aron implied, in which the two superpowers were united in a conspiracy to ensure there was no war in Europe.[31] NATO, relying on the threat of

Gen. William E. DePuy. Fort Leavenworth Command and General Staff College.

Gen. Donn A. Starry. Fort Leavenworth Command and General Staff College.

Gen. William R. Richardson. General Officer Management Office, Office of the Chief of Staff, Army.

Gen. Paul Gorman. Fort Leavenworth Command and General Staff College.

Lt. Gen. L.D. Holder. Fort Leavenworth Command and General Staff College.

Brig. Gen. Huba Wass de Czege. Fort Leavenworth Command and Gereral Staff College.

nuclear war as the primary deterrent, was unwilling to match the Soviet numbers across the border. Moreover, replacing nuclear deterrence with a viable conventional defense was no more acceptable to many of those who might be called upon to execute it.[32] The result was the requirement placed on DePuy and Starry to develop a suitable tactic for ground defense alongside the German corps in circumstances where operations and strategy were essentially bankrupt a priori.

In August 1974, Starry assigned a group of officers at Fort Knox to come up with an effective means of defending a sector of the inter-German border through use of a war game known as Hunfeld 1.[33] The result of this exercise was a defensive concept for conducting an active defense that called for organizing the battlefield into three zones: the covering force area, main battle area, and rear. The compatibility of Starry's and DePuy's views on tactics is evident in the short essays Starry wrote for *ARMOR* magazine in 1974 and 1975, titled "The Commander's Hatch." Starry was the right man to do what DePuy wanted done.

General DePuy did not get around to focusing on doctrine until July 1974, after the lessons of the 1973 war had begun to solidify and the work of Starry and others was well underway. DePuy dispatched a letter on 23 July to the commandants of the TRADOC schools. The letter became known as the "pot of soup letter" because of a homey little story with which he opened it.[34] DePuy included a concept document with this letter to indicate the lines upon which he was thinking.[35] He invited criticism, though it is not at all clear he intended to invite any fundamental reorientation. The first paragraph made it explicit that the focus of his interest was on tactics. A careful reading indicates the tactics he addressed were minor tactics, the business of those concerned with fighting weapons systems as opposed to employing units—or, in his schema, squad- to division-sized units, with the emphasis on companies and battalions. DePuy followed up this letter with a conference for school commandants at Camp A.P. Hill, Virginia, in December.

Initially the task of writing combined arms doctrine belonged to the commander of the Combined Arms Center (CAC) and commandant of the Command and General Staff College (CGSC) at Fort Leavenworth, Lt. Gen. John Cushman. Cushman, another highly intelligent officer, had come to Fort Leavenworth from command of the 101st Airborne Division, an organization in which he had commanded a brigade in Vietnam. The CGSC was undergoing a sea change in 1974, as the new Commandant sought to shake up the institution. He essentially turned the school's course content on its head, changing the focus along the way to mid- and high-intensity conflict and from division and corps operations downward to division, brigade, and battalion. All of this was part of General DePuy's program to refocus the various levels of officer education on the level of command student officers were most likely to encounter immediately upon graduation.[36]

General Cushman took a Leavenworth draft of *FM 100-5*, later known as the "A.P. Hill Draft," with him for review in December. Cushman clearly had no idea where he was headed. In a television tape made for the CGSC on the eve of the conference, he described doctrine as he understood it. Cushman was and is something of a romantic and idealist, no less capable of arbitrary dictatorial action than General DePuy. He called doctrine "a mysterious kind of thing . . . a truth, a fact, or a theory that can be defended by reason."[37] Cushman was no less dedicated to revolutionizing doctrine than DePuy, but his approach was humanist rather than scientific or arithmetical; qualitative rather than quantitative. His script railed against scholasticism, pedantry, dogma, and preoccupation with rules. He founded his beliefs on a statement in George C. Marshall's *Infantry in Battle* that: "Tactics has certain principles which can be learned, but it has no traffic with rules."[38] "Tactics," Cushman concluded, "is a thinking man's art."

DePuy wanted to wrench the Army out of its malaise by hitting it with a jolt of reality and tough standards at the lowest levels of command. Cushman sought to establish some rather general philosophic truth. It can well be argued that in other times, the balance of justice would be on Cushman's side, but DePuy was not acting in other times. The Army was broken, and strong medicine was needed. Worse, Cushman's doctrine writers did not know how to write doctrine. The manual, a copy of which still resides in the Combined Arms Research Library at Fort Leavenworth, can best be described as drivel; eighty-eight double-spaced, typewritten pages of vacuous, platitudinous rambling. In December, at Camp A.P. Hill, General Cushman had his orals in front of his boss and those junior school commandants over whom he was supposedly the integrating authority. He failed.[39]

DePuy's response was characteristic. He simply wrote Cushman and Leavenworth out of the doctrine business so far as *FM 100-5* was concerned. He, General Gorman, and the tactical school commandants began drafting the manual in Cushman's presence but without his direct participation. Starry's concepts ultimately prevailed. DePuy took the manual back to his headquarters at Fort Monroe and put its fleshing out in the hands of a group of relatively junior officers known as the "Boat House Gang" because they operated out of a boat house at Fort Monroe.

That General Cushman did not anticipate such abject failure is indicated by the fact that the purpose of his TV tape was to announce the issue of the draft manual to the CGSC for use in instruction, something he would hardly have contemplated had he anticipated the reaction he received. That General DePuy considered the writing of doctrine the business of general officers is a matter of no less significance. In an army and society in which, just then, generals lacked the traditional professional respect and legitimacy they might ordinarily expect, this philosophy proved detrimental in the long run.

Starry's role—and DePuy's decision in the fall of 1974 to lock the CAC and CGSC staffs out of the writing process—were both the result of a fundamental difference over the purpose of doctrine and the source of a good many of the problems that resulted when the manual and doctrine were finally unveiled in July 1976.[40] DePuy's goal was precisely to write a prescriptive tactical doctrine, whereas the authors and Commandant at Fort Leavenworth believed combined arms doctrine, or at least its highest formulation, should be broader in scope and content. In 1974, abstract, timeless principles were the last thing Bill DePuy was worried about.[41] That Leavenworth officers persisted in their belief and behavior suggests they either did not understand their new commander or that they simply refused to accommodate themselves to his wishes. While one may argue that their disagreement, if that is the case, was honest, there is no evidence that they were prepared to present a credible counterargument—a fact that certainly undermined their position.

In either case, the Leavenworth staff and faculty were left out of the doctrine drafting process and, as a consequence, instructors there never quite understood the concepts that it was their responsibility to infuse into the Army.[42] For that reason, as much as for its unorthodox content, the July 1976 edition of *FM 100-5* became both the most read and most attacked doctrinal statement in the history of written doctrine in the U.S. Army.[43] Because of the DePuy training initiatives, particularly initiation of the ARTEP, doctrine had become too important to ignore. If it was tried and not accepted, it had to be changed. Learning that in an institution as hierarchical as the Army generated much heat over the next six years as critics inside and outside the service did an unusual thing: They argued about official Army doctrine.

When the new manual was published in July 1976, the supreme allied Commander in Europe (SACEUR), Gen. Alexander Haig, who had been DePuy's G3 in the 1st Infantry Division in Vietnam, created something of flap when he reviewed it. Haig opined that the manual's focus on the special circumstances in Europe could deflect attention from sound tactical principles. "I would not want for that reason," Haig wrote, "to see decisive offensive action discounted altogether, and thereby to dampen the aggressiveness, initiative, and willingness to dare upon which any successful military action ultimately depends."[44]

DePuy responded that Haig had obviously gotten the wrong message. He denied, somewhat implausibly, that the doctrine was oriented on Europe, and replied with some asperity that, "One does not breed offensive spirit through patently unworkable doctrine or over ambitious planning."[45] DePuy further pointed out the doctrine had deliberately been designed to be consistent with German practices. Notably, he referred to the defensive form championed in the manual as "the active defense."

Haig, recognizing his old boss was annoyed, replied that his comments were intended to be constructive and not to cause difficulties. "Actually, my concerns attach less to the particular tactical methods," Haig wrote, "than to the possibility that the unwary might infer from them a doctrinal or strategic perspective which—intended or not—might adversely affect both deterrence in Europe and our military posture generally."[46]

DePuy and Starry had made a significant effort to incorporate the lessons of the 1973 Arab-Israeli war and to harmonize what they did with the doctrine of the German army, particularly the German tactical manual *HDv 100/100*.[47]

The new U.S. manual was published in a loose-leaf binder, indicating it was subject to change, and came with a camouflage cover. It was a "how to fight" manual that began by emphasizing the truisms about contemporary war DePuy had emphasized in his 1973 talks: winning the first battle; fighting outnumbered and winning; the importance of weapons effectiveness; the importance to train to fight and win now; and the need to convince the U.S. Army that it *could* win—a note more characteristic of General Starry's concerns than DePuy's. The manual's second chapter consisted of a compendium of weapons effects tables. It closed with a discussion of tactical nuclear weapons effects and depth objectives for Soviet advances. The final sentence summed it all up: "Battle will be fought on a scale and at a tempo rarely seen in all history."[48]

Chapter 3 spoke of how to fight. General DePuy had removed the traditional principles of war from the manual and replaced them with battlefield dynamics. Although he and General Starry believed later they were misunderstood in the charge that they had produced an attrition-centered doctrine. The "Battlefield Dynamics" emphasized the concentration of forces and weapons at critical times and places; battle "controlled and directed so that the maximum effect of fire and maneuver is concentrated at decisive locations"; use of concealment, suppression, and combined arms (General DePuy's influence); and crews and teams "trained to use the maximum capabilities of their weapons."[49] "The defense," the manual said, "is a race for time to detect the enemy's main thrust and to concentrate combat power." [50] Concentration was to be achieved by lateral movement of uncommitted or less committed forces.

The manual was designed to address battle successively at division, brigade, and battalion levels:

> Generals commanding corps and divisions *concentrate forces.*
> Colonels and lieutenant colonels commanding brigades and
>   battalions *control and direct the battle.*
> Captains and their companies, troops, and batteries *fight the*
>   *battle.*[51]

If he was successful in implementing this program, DePuy had surely solved the problems encountered by his 90th Infantry Division in World War II.

The chapter on offensive operations indicated that a commander should attack only "if he expects the eventual outcome to result in decisively greater enemy losses than his own, or result in the capture of objectives crucial to the outcome of the larger battle."[52] The chapter broke down the role of each of the elements in a coordinated, combined arms attack. An important assumption, taken from the Germans and Israelis, was the notion that on the modern battlefield infantry would normally fight mounted, as accompanying forces in a tank-centric armored force. From this presumption came the development of the Bradley-based infantry squad, and the possibility of moving proponency for "armored infantry" from Fort Benning to Fort Knox served to liven up TRADOC branch politics for much of the period.

The following chapter on defense was scarcely longer than the eleven pages on offense, but it got most of the attention. The doctrine divided the defense into three zones: the covering force area, main battle area, and rear area. The covering force was now to be a fighting element. It would initiate the battle and fight to force the enemy to commit to a course of action before arriving at the main battle area. As a result, it had to be reinforced with sufficient combat power to be effective in the initial battle. The battle itself would eventually be passed to the main battle area and would be fought in depth throughout the defensive zone. The attempt would be made to defeat the enemy as far forward as possible, but a premium was placed on the *elasticity* of the defense, implying that the measure of success was destruction of the enemy forward of the rear boundary. The chapter pointed to the importance of attack as the only means to seize the initiative, but it warned that "sweeping counterattacks which expose our forces to heavy losses . . . must be the exception."[53] What was vital was maintenance of the coherence of the defenses through utilization of successive battle positions in depth. "Counterattacks should be conducted," the manual concluded, "only when the gains to be achieved are worth the risks involved in surrendering the innate advantages of the defender."[54] It was admonitions such as these, true though they might be, that gave the manual its reputation for defensive mindlessness and caution.

The book went on to address special topics, retrograde operations, intelligence, the air-land battle (close air support), electronic warfare, tactical nuclear operations, chemical operations, combat service support, NATO operations (contributed by USAREUR) and what were called, special environments. The manual did contain admonitions calling for mission orders.[55] The treatment of armored forces spoke of the ability to "maneuver rapidly, to break through defenses, to strike deep into the enemy's rear, to encircle his flank, and *to win decisive battles.*"[56]

On the other hand, the book dispensed with more traditional general patterns of the defense—the area defense and mobile defense—and it was all but silent on the use of reserves, which were expected to consist of uncommitted forces rather than forces withheld from combat to achieve specific purposes.[57] Indeed, the chapter on the defense warned explicitly that a division withholding a brigade in reserve—the traditional solution—would be defeated in detail before it could bring its forces into play.[58] References to desired force ratios—3:1 to defend, 6:1 to attack—and emphasis on the elasticity of the defense were far more common. Confronted by a Soviet breakthrough attack, a division commander in Europe was expected to concentrate six-eight of his ten-eleven maneuver battalions on a twelve-kilometer front, or one-fifth of an average European division sector. This, of course, entailed significant risk on the flanks.

General DePuy submitted the manual to General Weyand for approval in February 1976.[59] He claimed that it reflected the views of a variety of constituencies: the major commands, selected corps and divisions, and the German and Israeli armies. He indicated the start of an initiative to bring army and air force concepts into harmony and a program in TRADOC to develop derivative "How to Fight Manuals" at the various branch centers and schools.

With the manual's publication in July, the army set out to understand the doctrine. The officers called upon to execute it, no less external critics, voted on it as well. The temperature was about to go up in classrooms and army messes.

By then, Donn Starry was in Europe as V Corps Commander. He began trying to convince his corps that Active Defense worked and, more important, that they could win by executing it. It was not an easy task, for the corps he took command of in February 1976 was dispirited and defeatist in its thinking.[60]

Starry found two problems in V Corps regarding its combat edge. First, it was not thinking about fighting. He set about fixing that by a series of terrain walks on which commanders participated in a dialog with their corps Commander on the ground where they would fight and about how they could defend against the first Soviet front. Starry writes that he was shocked to find how few Commanders had walked the ground on which they might have to do battle. He was also surprised how little many of them knew of any doctrine at all. Finally, he was surprised to find many officers who had no appreciation at all for terrain.

Terrain walks with the chain of command and instruction in war-gaming problems, to include computation of "battle calculus" with a model provided by the BDM Corporation, were vehicles General Starry used to school his subordinates on all these matters.[61] By September he could see confidence growing as his command began to understand the doctrine he and General DePuy had provided the army. There was a growing belief that they could stop the first Soviet echelons. Whether the credit for this growing confidence in the possibility of success owed more to the concept or the new corps Commander's insistence that unit commanders focus on fighting is a subject of legitimate debate.

The second problem the new V Corps commander had is more difficult to assess. General Starry identifies it as the "Leavenworth malaise"—an active resistance to the new doctrine that had its origin at Leavenworth.[62] In support of this thesis he offers the case of an irreconcilable brigade Commander related to a Leavenworth instructor. This miscreant refused to try the new doctrine because his brother-in-law at Leavenworth had told him it was not doctrine and would not be approved.[63] The story has the ring of truth, but this explanation overlooks a broader and more likely set of general problems. The middle ranks of the officer corps in the mid- to late-1970s were skeptical of their senior leaders—albeit not without cause after the loss in Vietnam and the various failures of leadership and integrity that war spawned. Second, whatever resistance there might have been as a result of institutional obstructionism, rather begs the more likely explanation that the Leavenworth faculty, kept isolated from the development of the new techniques (unlike the faculty at Fort Knox) simply did not understand the doctrine it was supposed to be teaching. Starry points out that Fort Knox had taught Active Defense since 1974. Leavenworth had as well, but not with much confidence or skill.[64]

Nor did the 1976 *FM 100-5* lack ambiguity. It was not until the publication of a new division manual, *FM 71-100*, produced by Fort Knox in September 1978, that a fully authoritative statement of how the new concepts were to be executed was articulated. In the meantime, an interim Training Circular that served even the light infantry of the 82d Airborne Division acted as a doctrinal stopgap.[65]

Moreover, there were serious problems with the doctrine regarding its execution. What was to come after the defeat of the enemy's first echelon front? No less important, how would the doctrine fare if transferred to another set of circumstances? Badly or mechanistically taught, Active Defense could look like little more than a delay, and that doctrine, like dogma, could not adequately be explained. [66]

Starry recognized the first problem and made it the foundation of the revision of the 1976 manual upon his return to the United States to assume command of TRADOC in 1977. The fact was, though, there was a vast inchoate dissatisfaction abroad in the army with the doctrine as written and as taught, recognizing that these two manifestations often were not the same. This persistent criticism had to be addressed. Critics did fix on some of the platitudes and mantras unfairly, but that was simply an argument against their inclusion. Midgrade officers, congressional staffers, and various others voice arguments and criticism that no doubt seemed obstructionist at the time, particularly as they often presented them with a decidedly prosecutorial tone. Nevertheless, the discourse was fruitful to the extent that it represented an officer corps working to understand the dynamics of combat as much as had the participants in General Starry's V Corps staff rides.

These years were some of the richest for professional dialogue in the U.S. Army's history. The voices of critics also found a place in the next version of *FM 100-5*, a revision initiated in 1979 by General Starry as TRADOC commander. It was, in fact, the synthesis of these two streams of thought that ultimately produced the notions that together became AirLand Battle doctrine.

Reaction to the newly published 1976 doctrine arose almost at once and drew attention, perhaps disproportionate, to its demerits due to the ham-handed way with which initial and, indeed, subsequent criticism was addressed. The first critic to breach the TRADOC defenses was William Lind, a minor congressional staffer in the office of Sen. Robert Taft Jr. of Ohio. Taft was a one-term senator and later ambassador to NATO. Bill Lind was bright and aggressive and fresh from Dartmouth and Princeton. He wrote a critique of the new *FM 100-5* for *Military Review* that the powers at TRADOC apparently tried to stifle, only to be exposed in *Armed Forces Journal*. The *Journal*'s editor not only questioned the way Lind's essay was handled, but referred to the manual (favorably reviewed elsewhere in the same edition), as untested theory, "laid down by TRADOC," noting, "There are suggestions that the manual is not an Army manual: that TRADOC put it together, got a lower-level DA staff chop, then presented the Chief of Staff with a *fait accompli*."[67] The Army relented and Lind's essay was published in *Military Review* the following spring.[68]

Publication of the Lind essay opened the floodgates to an outpouring of criticism, much of which overlooked many of the restraints within which DePuy and Starry had developed their doctrine, most notably a European defense strategy of questionable war-fighting utility. From the TRADOC point of view, area defense would not work on the extended frontages resulting from the paucity of deployed forces. Similarly, a traditional mobile defense was not possible in the absence of depth. Finally, there would be insufficient forces deployed forward to network a major Warsaw Pact armored attack in the event large mobile reserves were retained by the defender. Thus, Active Defense.

Lind and other critics tended to ignore these restrictions. On the one hand, they raised the very valid point—one that DePuy and Starry both had recognized from the outset—of whether it was appropriate to generalize the European situation into the doctrine of a global army.[69] However, Lind went further, suggesting that the military should simply announce that the conventional defense of Europe was impossible. Along with scoring rhetorical points against such mantras as "win the first battle," and "fight out numbered and win"—which he took as admonitions rather than requirements—Lind raised the more substantive point that the doctrine was silent about the second battle. He wanted to know how the defenders were expected to deal with the second, third, and fourth fronts, if the Soviets launched a full-scale attack. The answer was that NATO had only two choices at the time: surrender or nuclear response.

Lind also raised the dichotomy with which he and other reformers simplified their criticisms, which they called *maneuver warfare*, in contrast to the official variety, which they characterized as *firepower-attrition*. The essential difference, as laid down in the *Military Review* article, was that the former depended on psychic dislocation as the defeat mechanism in battle, whereas the latter relied on killing men and things. Finally, Lind questioned the Army's capability of executing the doctrine it had adopted under the pressures of combat, particularly considering its requirement for continuous command and control. Lind's article lit the fuse to a run of criticism of the 1976 doctrine that ultimately led to its revision, and informed, if it did not guide, the modifications made in its successor.

Lind was not the only critic to raise substantive questions about the viability of the Army's Active Defense doctrine.[70] In 1978, Donn Starry, as TRADOC commander, sought to respond to the various critics in an essay drawn from a speech made to an Inter University Seminar Symposium conducted at Fort Leavenworth.[71] Starry welcomed criticism. He laid down the constraints or context within which the 1976 manual had been composed and which still existed to a great extent. He explained the nature of Soviet offensive doctrine that Active Defense was designed to counter—a doctrine of mass, momentum, and continuous combat. Finally, he articulated a concise expression of what Active Defense was supposed to involve. He said it was an imperative to see deep to find the following echelon, move fast to concentrate forces, strike quickly before the enemy could break the defense, and finish the fight before the second echelon closed with the defenders. All this was to be accomplished while using the defender's natural advantage—terrain—to multiply the strength of the defense.[72]

It is noteworthy that the goal expressed does not necessarily require the techniques adopted in the Active Defense, and more so, that nowhere did the 1976 manual display with prominence a single core idea with such clarity.

Starry then went on to explain why the Active Defense had looked appropriate under the circumstances. He admitted that the most difficult point to get across was the European army's inability to support retention of a subtracted reserve at every echelon. (He had pointed out, in a letter written to a U.S. division commander in March of that year that, when he arrived in V Corps, a total of 56 percent of the tanks in the corps were held in reserve at all echelons of the command. He admitted, however, that under his command both V Corps divisions had deployed a brigade in depth on major avenues of approach, with their commitment subject to the corps commander's approval.[73]) Starry acknowledged that the shortage of reserves made the Active Defense a technique that permitted no mistakes in execution, which was one of Lind's main criticisms. He also acknowledged that the manual was criticized for underemphasizing the offensive. He responded: "We need to change that, but printing the offense chapter in blood isn't the way to do it. It has to be instilled in soldiers' minds."[74] The way to do that was with training, employing individual skill qualification (the lowest level of the Gorman system) and the ARTEP. The critics were not satisfied.

A year later, in June 1979, Starry was told by the newly designated Army Chief of Staff, Gen. Edward C. "Shy" Meyer, that it was time to revise *FM 100-5*.[75] The correspondence between Meyer and Starry is instructive. Meyer criticized the 1976 manual for its European focus and indicated that the "capstone doctrinal manual" should be more comprehensive. At the same time, he acknowledged the accomplishment of the authors, principally DePuy and Starry, who he said had initiated a sustained doctrine development process and taken on the problem of turning the Army's attention from the last battlefield to the next. "However," he added, "FM 100-5 has performed an even more valuable function. In my experience, I can't think of any other doctrinal publication that has caused such profound and widespread dialogue across the entire spectrum of basic tactical doctrine than has FM 100-5. It has caused people to think aloud for a change."[76]

Finally, in addition to broadening the applicability of the doctrine, Meyer wanted it to expand its horizon from company through brigade to encompass the corps and theater battlefields. It was a direction that ultimately led to adoption of an operational level of war, the notion that doctrine limited to consideration of tactics alone was no longer sufficiently robust.

Starry replied later that month.[77] He wrote that he agreed in principle and repeated the explanation that the original attempt was only a partial solution. It had been driven by the desire to focus the Army on the most challenging problem it faced, and in recognition of the need to agree with German allies in Europe on an approach to battle that was compatible with German doctrine. Starry acknowledged the problem of the defensive orientation, and admitted that he and DePuy had not known in 1975 how to deal with the problem of the second echelon. The problem was that one might successfully defeat the first echelon, only to be hammered successively by a second or third echelon. Eventually, the attacker was likely to prevail if he could continue to run fresh forces into the fight.

"We know what needs to be done about that now," Starry wrote. Then he went on, responding to the maneuver warfare theorists: "However, our critics who trumpet that we should steer away from a firepower based strategy aren't all correct—we're going to have to kill a whole lot of them—just to get their attention. And we should make no mistake about that! So to say it's all maneuver and that maneuver will solve all, is to ignore the very real problems with space and depth, especially in Europe, and with logistical support of highly mobile operations."[78]

Meyer's letter had been drafted in the office of the DCSOPS, which Meyer left to become chief of staff, under direction of one of his principal assistants, Maj. Gen. William R. Richardson. Richardson was soon on his way to Fort Leavenworth to take command of the CAC and become Commandant of the CGSC. Richardson carried forward, and, in some cases, reshaped initiatives Starry began as TRADOC commander. Richardson provided unusual long-term continuity of vision from a succession of positions, first as CAC Commander, then as DCSOPS, and finally as the replacement for Starry's successor (Gen. Glenn Otis) at TRADOC.

Richardson's appointment to command at Leavenworth was important because one of the lessons Starry had learned from his experience with the 1976 *FM 100-5*—and subsequently in V Corps—was that how TRADOC went about writing its doctrine mattered as much as its content.[79] As a result, he returned responsibility for *FM 100-5* to Leavenworth, where General Richardson personally guided a small group of doctrine writers who carried out the revision of *FM 100-5*. The result was nearly a counterrevolution. The authors not only addressed Starry's concerns about dealing with the second echelon of any Soviet or Soviet-like mechanized attack, but they reinvigorated the basic doctrine, making it more offensive, stressing the importance of initiative, and restoring such traditional concepts as the principles of war and basic forms of attack to discussion of tactical forms. Last of all, after General Starry left TRADOC to assume command of the U.S. Readiness Command, they would be directed to add the concept of levels of war to the basic doctrine, thus coming to terms with the difficult problem of translating political objectives into military actions and conducting sustained operations on land.

No lesson was more important than the one that indicated method was as important as result. Recognizing that too much public involvement by the TRADOC Commander led to complaints about individual doctrinal preferences being foisted on the army, Starry distanced himself from the process. He charged General Richardson, as CAC commander, with responsibility for writing the manual. Finally, he brought to TRADOC an energetic and personable brigadier general, Donald R. Morelli, to serve as Deputy Chief of Staff for Doctrine. Morelli's job was to respond to TRADOC's critics and deal with consultants on doctrinal matters.[80]

The process of revising *FM 100-5* did not begin as a counterrevolution. Indeed, it seems to have been intended to be a process of updating what already existed. The original drafting team, like its predecessors, was instructed to stay close to German tactical doctrine. They set out doing that but apparently made little headway. To energize the team, General Richardson brought in two very gifted officers, both veterans of the faculty at West Point, Lt. Col. Huba Wass de Czege and Maj. Leonard "Don" Holder.

Wass de Czege was a light infantryman, the son of a Hungarian novelist who fled his homeland in 1956. Wass de Czege graduated from West Point and later returned to teach in the Social Science Department. He was energetic, charismatic, and highly imaginative. In outlook and vision, he was an idealist or romantic. Holder, by contrast, was a laconic Texan, an Army brat whose father was killed in Vietnam while commanding the 11th Armored Cavalry Regiment. Don Holder graduated from Texas A&M, became a cavalryman himself, and taught military history at West Point. In 1991 he commanded the 2d Armored Cavalry Regiment in Desert Storm. Where Wass de Czege was a romantic and an enthusiast, Holder was a realist. He quite likely was one of the Army's most competent tacticians, and he was a man of awesome integrity. The third member of the writing team, Lt. Col. Richmond B. Henriques, was present from the start. He had been one of the drafters of the NATO tactical manual, *Allied Tactical Publication (ATP) 35.* Henriques had tried to do a revision of the 1976 manual according to German and NATO doctrine and found it impossible to accomplish while taking cognizance of the views of the various critics. A new doctrine was called for, and he said so.[81]

With Wass de Czege in the lead, the three officers wrote a new *FM 100-5* under the immediate supervision of General Richardson and, at Fort Monroe, General Starry. Holder moved in and out of the project as it progressed. At one point he was assigned to General Richardson's Extended Battle Concept Team, a group responsible for briefing the field on how deep attack systems would be built and deep battle conducted.[82] Holder drafted early conceptual parts of the manual and, in the end, undertook the final revision—which incorporated comments from the field in a much edited and reduced final draft—while General Richardson and Lieutenant Colonel Wass de Czege were on a trip to China.

The manual that resulted, *FM 100-5 (1982),*[83] abandoned Active Defense as a prescription for defense against breakthrough attacks. What is more important, it abandoned the narrow focus on the direct-fire battle and emphasis on target servicing that had dominated its predecessor. The deep battle was addressed and the young authors achieved the tone that their generation looked for, a style of war marked by the offensive, maneuver, and surprise—with due attention paid to human as well as material factors of battle. The final vital addition, the inclusion of an operational level of war intermediate between strategy and tactics, came from General Otis, Starry's immediate successor, at Morelli's urging. The operational level of war was the domain of operations and campaigns, as distinct from battles and engagements; the employment of military actions to achieve strategic goals.[84]

General Starry had become deeply aware of the second echelon problem while he walked the terrain assigned to his V Corps. He had inspired confidence in his subordinates, convincing them that they could handle the first echelon of a Soviet attack. However, he recognized that the echelonment problem was his.[85] In his 1978 speech and essay, quoted above, he said he had learned that each Commander from brigade to division had to be aware of the corresponding second echelon force headed toward his close battle. The difficulty was that, to do this, commanders had to depend on resources they did not control. Commanders had to rely on another service (the Air Force) to locate the enemy advance and to interdict it. Starry continued to work on this problem in TRADOC, particularly with the Field Artillery School, while *FM 100-5* was being rewritten.[86] In 1981, shortly before departing for Readiness Command, Starry addressed what he had learned in a *Military Review* article titled "Extending the Battlefield." [87]

Starry laid down the principles of what became known as the deep battle, part of a new conceptual division of what is now the AirLand battlefield, which generally is divided into deep, close, and rear battles. Unable to find tactical depth to the rear, Starry's concept called for engaging Soviet follow-on echelons far in advance of the covering force while simultaneously fighting the main battle. The responsibilities of the various headquarters were measured in time rather than distance. The brigade was responsible for all forces within a distance of twelve hours of the forward line of troops, the division out to twenty-four hours, and the corps to seventy-two hours.[88] The principal difference in their respective goals was that the corps' objective was to reduce the closure time of the follow-on forces, while the division's deep battle inevitably merged with the close (direct-fire) fight. Starry put it bluntly: interdiction was the key to battlefield success.[89]

The difficulty again was that, except for limited missile systems such as the Lance, the corps commanders were and would remain largely dependent on Air Force assets for the target acquisition and interdiction they required at extended ranges. The AH-64 Apache attack helicopter promised some organic relief, but its survivability in the European environment, with its dense mix of visual and electronic air defense

systems, remained problematic throughout the 1980s. Starry emphasized the need to find procedures for working with the Air Force. Tactical Air Command (TAC) seemed to be responsive to TRADOC
initiatives. Indeed, after the Gulf War, some Air Force leaders seemed
to believe they had been too responsive, a consequence of the division
of Air Force thinking into separate tactical and strategic communities
during that period.[90] TRADOC and TAC had undertaken cooperative
engagement under General DePuy and Air Force Gen. Robert J. Dixon,
a process Starry continued with Gen. Bill Creech, Dixon's successor.[91]

   In 1981, Lt. Gen. Glenn Otis, the DCSOPS, and his Air Force counterpart, Lt. Gen. Jerome F. O'Malley, signed an agreement that at least
provided for some commonality of terms and concepts for apportionment and allocation of air support, although without seeking agreement on availability—a decision reserved for the joint commander in
accordance with circumstances.[92] The Army, however, understanding
its requirement, began to assume it would be met. Corps Commanders
thus believed they would be free to fight their own deep battle.

   The Leavenworth authors were largely responsible for the manual's
new tone. The new *FM 100-5* made its concept explicit in a second chapter
that set the tone for the whole manual, as surely as the second chapter
of the 1976 manual had done with its charts and graphs of weapons
effects. The key difference was that the new manual was descriptive
and conceptual. The core concept was initiative.

   The 1982 manual was divided into three sections of unequal length:
"The Army and How It Fights," "Offensive Operations," and "Defensive Operations." The preface was much less perfunctory than its
predecessor's. It asserted the manual's status as the "Army's keystone
How to Fight manual," claimed consistency with NATO doctrine and
strategy, and gave as its purpose the explanation of "how the Army
must conduct campaigns and battles in order to win. It describes U.S.
Army operational doctrine involving maneuver, firepower, and movement, combined arms warfare, and cooperative actions with sister services and allies. It emphasizes tactical flexibility and speed as well as
mission orders, initiative among subordinates, and the spirit of the
offensive."[93] The spirit of the new manual clearly was more positive.

   The manual was focused at a higher level than the 1976 manual,
which had concentrated on brigade level and below. By 1982 there was
a foundation of new subordinate tactical manuals upon which *FM 100-
5* could build, as well as for which it might set a new direction. Chapter
1 of the 1982 manual, "Challenges for the Army," opened with the observation that tactical success was not always sufficient, that the manual
focused on "winning battles and campaigns," rather than the first battles

of the old text. AirLand Battle doctrine, so named because Donn Starry had concluded that the Army could only fight with Air Force support, particularly for deep interdiction, was intended to be worldwide in application and to be broad enough in its scope to encompass nuclear, chemical, and conventional battlefields.[94] This attempt to normalize consideration of nuclear weapons use (about which the enemy also had a choice) led to more problems with the NATO audience.

Retention of the initiative, deep attack, and decisive maneuver, came to be the doctrine's hallmarks. To this end, a second chapter, "Combat Fundamentals," laid down a set of concepts that were to serve as the framework for battle. "AirLand Battle doctrine," it declared, "is based on securing or retaining the initiative and exercising it aggressively to defeat the enemy. Destruction of the opposing force is achieved by throwing the enemy off balance with powerful initial blows from unexpected directions and then following up rapidly to prevent his recovery."[95]

"The best results are obtained," the concept continued, "when initial blows are struck against critical units and areas whose loss will degrade the coherence of enemy operations, rather than merely against the enemy's leading formations."[96]

Thus focused, four tenets were postulated to characterize AirLand Battle doctrine: *Initiative, Depth, Agility,* and *Synchronization.*[97] The first was intended to embody an offensive spirit. Depth was to be measured in both time and space. Agility inferred a relative ability to act faster than one's enemy, and Synchronization, which proved by far the most questioned after the Gulf War, was described as "more than coordinated action. It results from an all-pervading unity of effort throughout the force. There can be no waste."[98] The idea was to capitalize on complementary as well as supplementary effects of combined arms. The object was to fix the enemy on the horns of a dilemma where his response to one threat would in turn expose him to the effects of another. It was not an unimportant consideration when doctrine began to emphasize initiative and agility that the new family of combat vehicles—the M1 Abrams tank and M2 and M3 Bradley fighting vehicles—began to appear in tactical organizations. Greater capabilities were required to fight AirLand Battle than Active Defense and they were embodied in the vehicles that began appearing in unit motor parks and field exercises.

According to Don Holder, the four tenets were selected by the principal authors and overseers: Wass de Czege, Holder and Henriques, and Richardson and Starry.[99] An initial set— Initiative, Violence (Holder's favorite), Integration, and Depth—evolved into the final four.[100] Initiative and Depth were obvious choices given General Starry's history and intentions, although they were unlikely to have been contrary to the views of any of the inner circle. Violence fell out as a lower level generalization (to be replaced, writes Holder, ever the pragmatist, "by the 'pixie-like' Agility"[101]). According to Holder, General Richardson made the change from Integration to Synchronization.[102] The notion of synchronization led to a certain formalism of planning that could carry over into execution, certainly an unintended effect in a doctrine predicated upon initiative.

Gen. Bill DePuy was remarkably open-minded about the need to revise the doctrine he had foisted on the Army in 1976. DePuy addressed himself to the notion of synchronization and its attendant benefits during the next several years and achieved no small success in selling its importance to the Army. He explained the concept as he understood it in a handwritten note in 1982 with a clever formulation:

> The value of *Synch:*
> Finding an enemy howitzer (by radar) is *interesting*
> Finding several batteries is *valuable*
> Silencing several batteries by counter-fire is *important*
> Silencing several batteries which otherwise would suppress our
> defense just as the enemy launched his attack is *crucial*
> Stopping the attack so that our counterattack into the flank
> destroys
> the enemy force is *decisive.*[103]

DePuy, who spoke with authority in army councils until his death in 1992, continued to explore the notion of synchronization and how it was achieved. Attracted by the Israeli experience in the Bekaa Valley in the early 1980s, he turned increasingly to a systems management matrix as a solution for the synchronization model: events aligned along a time line. The Army adopted the method in its officer instruction and combat training centers, and, by the time of the Gulf War, it pervaded the planning shops of higher staffs.[104] Unfortunately, there was no block on the matrix for opportunism, and the enemy, of course, had his own event time line. Insofar as these matrices then dominated execution, they tended to keep things synchronized and slow.

The manual replaced diagrams of weapons effects with historical examples. Grant's Vicksburg campaign (somewhat truncated) and the World War I Battle of Tannenberg became the classic studies of campaigns and major operations for a generation of army officers. The manual also contained the traditional principles of war, albeit relegated to an appendix without much discussion, and, more prominently, the notion of levels of war: strategic, operational, and tactical. By incorporating the levels of war, the army acknowledged that the conduct of campaigns and major operations comprehended more than just fighting battles. Col. Harry Summers drove this lesson home on the first page of his semiofficial study of the Vietnam War.[105] This need to connect tactical events to strategic end states was a fairly immediate lesson learned by reflecting on the Army's Vietnam experience. In the 1970s, officers at Leavenworth and the Army War College began reading translations of Soviet publications about operational art and studying German World War II mechanized campaigns as models. However, more than one critic pointed out that the spatial conditions on the eastern front in 1941-45 hardly compared to those along the inter-German border in the 1980s.[106]

The notion of incorporating the operational level of war in U.S. doctrine seems to have originated at the Army War College and was seized upon by General Morelli.[107] Both General Starry and the Leavenworth authors rejected it as being too much for the army to take on in 1982. However, when General Otis succeeded General Starry at TRADOC in 1981, Morelli raised the issue again. This time he succeeded in having it included in the manual—largely, it seems, because differentiating between the tactical and operational levels of war helped Otis explain why the relationship between fire and maneuver changed according to which set of activities one was addressing.[108]

Coincidentally, with adoption of the operational level of war, a new school, the School of Advanced Military Studies (SAMS) was founded at Leavenworth in 1983 and made speculation about operational art its main business for the first five to eight years of its existence. Huba Wass de Czege was SAMS's first director. Don Holder was its third. Its second director, Col. Rick Sinnreich was the final author of the 1986 edition of *FM 100-5*. That manual was written at SAMS by Holder, Wass de Czege, and Sinnreich, largely as an effort to clarify the notion of operational art in army doctrine.[109]

To return to *FM 100-5 (1982)*, two more lists finished the chapter on "Combat Fundamentals": a discussion of the "Dynamics of Battle" and "Combat Imperatives." The "Dynamics of Battle" section included "Combat Power"—a construct reminiscent of J.F.C. Fuller that was derived from maneuver, firepower, protection, and leadership.[110] Also featured was a list of "Combat Imperatives": Unity of Effort, Use of Strength Against Weakness, Identification and Sustainment of a Main Effort, Sustainment, "Move Fast, strike hard and finish rapidly," Use of Terrain and Weather, and Protection of the Force. Subsequent chapters addressed such subjects as terrain and weather, battlefield environments (nuclear, chemical, electronic and smoke), combat service support, and tactical intelligence. The book was something of a book of lists.

The first section of the manual closed with a chapter on the "Conduct of Operations." AirLand Battle was again described as "a nonlinear [meaning defense in and throughout the zone] view of battle," conducted over an expanded area, "stressing unified air and ground operations throughout the theater."[111] It recognized, it declared, the nonquantifiable elements of combat power, especially maneuver and the human element: *"courageous, well-trained soldiers, and skillful effective leaders."*[112]

The new manual was decidedly traditional in its descriptions of the forms of maneuver. Offensive maneuver was categorized as frontal attack, penetration, envelopment, or turning. Defense was described for structure rather than form, either traditional (area or mobile) or active defense. It split the difference regarding the employment of reserves. Brigades and higher echelons were expected to maintain reserves equal to about a third of their strength. For battalions and companies, it depended on the circumstances. To finesse the argument on defensive forms, the manual discussed "Defensive Techniques" that applied at the brigade and battalion level (the territory of the 1976 manual). *"Army doctrine,"* the new manual declared, *"does not prescribe a single technique for defense."*[113] The style chosen could vary from static to dynamic, depending on circumstances, with large-unit operations combining both of these general characteristics. Reference to the covering force battle was decidedly permissive about the degree of resistance to be offered, although the covering force was still expected to fight. The chapters on the offense concluded with an admonition on pursuit operations. Retrograde operations and breakout from encirclement both had their own chapters in the defensive section of the manual. A fourth section addressed joint, contingency, and combined operations.

The manual stated baldly that, "The Air Force is an equal partner in air-land battle,"[114] reflecting a view that ground operations were simply not possible absent air superiority and supporting air to ground operations, both tactical and operational. The manual promised that corps commanders would be supported with close air support, battlefield air interdiction, and tactical air reconnaissance missions.[115] Where General DePuy had warned after the 1973 Arab-Israeli war that ground commanders could not predicate their actions on the availability of integrated air support, the era of good feelings between TAC and TRADOC led to precisely the other assumption. The solution that General Starry had seen to the second echelon problem was to disrupt the continuity of the succeeding attacking echelons' advance (or delaying the arrival of reserves in the offensive) and thus break up the tempo of the enemy's actions long enough to provide time to receive the follow-on attack in a balanced stance. To this end, corps Commanders were encouraged to engage forces beyond the forward line of troops out to seventy-two hours or two hundred kilometers.[116] To execute this they would have to depend on Air Force assets being responsive to their requirements.

Setting aside the question of the viability of attacking aircraft against the Soviet air defense regime, and competition for assets committed against strategic targets, European corps Commanders were likely to find themselves in competition with their theater commander for offensive air assets. Gen. Bernard Rogers, the SACEUR since 1979, had his own plan for deep interdiction called "Follow-on Forces Attack."[117] NATO army groups, to which the U.S. and German corps were subordinated, were likely to have their own priorities for limited offensive air assets as well. There was an inherent contradiction in the habit of writing doctrine for national service forces that would fight as part of a coalition, no less a concept of land combat dependent on the support of a separate service that had its view of war.

The Army and Air Force tried to solve this problem in a variety of ways. The Army's contribution was the development of long-range tactical missiles capable of breaking up enemy concentrations and cooperating with the Air Force in the development of joint intelligence systems.[118] These became the Army Tactical Missile System (ATACMS) and Joint Surveillance, Target Acquisition, Reconnaissance System (JSTARS) of the Gulf War. When it arrived, the AH-64 Apache attack helicopter provided some capability for deep air attack under direct corps control, and the ATACMS could reach out about a hundred kilometers. Still, in the Gulf War, the theater Commander, an Army officer, retained control of the theater deep battle, to the discomfort of the corps Commanders, who believed their ability to shape their battlefield environment according to AirLand Battle doctrine, as they understood it, was thus limited.[119]

By 1982 the Big Five weapons systems were coming into the Army inventory. The Reagan Administration was investing heavily in defense. Many of the difficulties of the 1970s were history. Recruiting standards had been raised, as had soldier pay. Easy administrative discharges for troublemakers and random drug testing were cleaning up the Army. The problematic side was that President Ronald Reagan's aggressive talk about "The Evil Empire," which coincided with the introduction of intermediate-range missile systems into Europe, was frightening America's NATO allies. AirLand Battle doctrine, with its emphasis on deep battle and consideration of the integrated battlefield, as well as more speculative future warfare concepts, played very badly before European alliance political audiences.[120]

In a 1985 letter to General Richardson at TRADOC, General Rogers reflected that the 1982 *FM 100-5* and a concept for future war developed by General Morelli at TRADOC, called AirLand Battle 21, had been nothing but trouble for NATO. The two publications, the manual and the concept, tended to get confused, contrary to General Starry's intentions. Rogers was critical of the notion Starry had imposed in *FM 100-5* under which nuclear weapons appeared to be considered just another battlefield *environment*.[121] This was somewhat ironic. In the 1970s, misperceptions about the efficacy of Active Defense had led to concern that the U.S. was going to be too defensive to deter. This led in part to European insistence on the deployment of intermediate-range nuclear weapons. Now, with a new set of characters, the more offensive tone of American doctrine proved even more upsetting, as did the actual fielding of those systems.[122]

In 1984 Don Holder was brought back to Leavenworth to spend his senior service college year revising the 1982 *FM 100-5* with the SAMS director, Huba Wass de Czege. Wass de Czege and Holder were joined by another bright, forward-looking officer, Lt. Col. Richard Hart Sinnreich. Sinnreich, a field artilleryman, had been involved with the debates about Army doctrine for some time. He had worked for Alexander Haig at Supreme Headquarters, Allied Powers Europe (SHAPE) at the time of the SACEUR's exchange with DePuy, then had been instrumental in developing the extended battlefield concept for General Starry while assigned to the Field Artillery School. Sinnreich advised General Richardson, when the latter was DCSOPS, on the Army's acceptance of the 1982 *FM 100-5,* and later Lt. Gen. Carl Vuono, the CAC commander, on how to solve the doctrinal problems with NATO. After Don Holder went on to duty at Fort Hood and Wass de Czege took command of a light infantry brigade in California, Sinnreich, by then the SAMS director, wrote the final draft of the 1986 *FM 100-5.* The manual that resulted owes a good bit to the final author's abilities as a disciplined and concise writer. It is by far the clearest and best written of the three manuals discussed here.

In 1987 Don Holder reflected that the primary changes in the 1986 manual were the restoration, after twenty years, of the forms of area and mobile defense and clarification of the concept of operational art.[123] (He might have mentioned the addition of infiltration as an offensive form, too.) This is a fair, if modest assessment. The 1986 manual was the final statement of AirLand Battle doctrine. Discussion of the operational art, the creative act of campaign planning, replaced talk of an operational level of war.[124] Notions of *branches* and *sequels* entered the official lexicon. Branches were anticipated alternative course of action intended to provide flexibility and agility, and sequels were follow-on actions intended to exploit success or respond to failure.[125] From these notions grew the proliferation of contingency plans and "playbooks" that marked Gulf War operations. Concepts taken from Carl von Clausewitz's *On War*—especially center of gravity, continuity, and point of culmination—were threaded throughout, as was the notion of lines of operation from Antoine Henri de Jomini's *The Art of War*. These concepts no doubt influenced the SAMS writers. Synchronization was redefined as "the arrangement of battlefield activities in time, space and purpose to produce maximum relative combat power at the decisive point."[126] The "Combat Imperatives" of the 1982 manual were retitled "AirLand Battle Imperatives" and increased from seven to eleven. The new manual was deliberately intended to serve as an instructional tool, retaining the characteristic of reducing principles to lists.

The section of the manual that dealt with air-ground operations, the *Air*, in *Air*Land Battle, was subtitled, "Tactical Air Operations."[127] Notably, the discussion acknowledged the importance of strategic bombing but warned that strategic effects were delayed and thus attacks against enemy operational forces remained a necessity. The manual favored the use of airpower for deep operations, interdiction of follow-on forces, and reflected existing joint doctrine and practice in incorporating what the army called "Battlefield Air Interdiction (BAI)"—deep targets flown in support of ground operations—into the larger air interdiction effort conducted under the direction of the theater air Commander. The doctrine provided only for coordination at the service-component level to achieve unity of effort. No conclusions appear to have been drawn about the implications for tactical ground operations if anticipated air support was not forthcoming. It is notable that, given the arrangement of the command structure in the Persian Gulf War, the Air Force could claim to have observed precisely the provisions of Army doctrine in the conduct of the AirLand Battle, at least insofar as targeting for the deep battle.

The manual went a long way to make peace with the SACEUR, General Rogers, stating explicitly that, "in Europe U.S. troops operating in the framework of FM 100-5 will execute NATO's forward defense plans in compliance with ATP 35A." The preface also made explicit the function of *FM 100-5* to serve as "the Army's principal tool of professional self-education in the science and art of war." Tactics, techniques, and procedures were now the stuff of subordinate manuals. The manual was founded on the central notions of the 1982 manual, identified as the operational level of war, the importance of the initiative, multiservice and coalition operations. The tenets—Initiative, Agility, Depth, and Synchronization—were also reemphasized. Alhough it did not say so, the manual refined and clarified discussion of the notions of deep, close, and rear battles, now named deep, close, and rear operations. It made clear the priority to be placed on the close operations, at the expense, if need be, of the deep.[128] The manual was, indeed, the Army primer on combat operations.

By 1986, however, the period covered by this essay was ending. The Cold War was clearly ending by 1989, only three years after the publication of the final manual. By 1990, when Saddam Hussein attacked Kuwait, the Army for which AirLand Battle doctrine was written, no less the problems for which it was designed, was disappearing. It was Saddam Hussein's misfortune to present the U.S. Army with the target General DePuy had forecast at Fort Polk so many years earlier. The Army showed the following February that it had achieved his goal of being five times better, man for man or tank for tank, than the Arab army it finally confronted.

Bill DePuy and Donn Starry had succeeded in wrenching the Army's attention from defeat in Vietnam to fighting armored warfare outnumbered. They did so by viewing doctrine narrowly, focusing the Army's central doctrine on the limited field of battalion- and brigade-level tactics in Europe, and combining doctrine with disciplined, evaluated training to standard. The Army officer corps responded, as humans often do to high-handed treatment, by opposing the new doctrine. It failed to notice it was becoming more disciplined, more knowledgeable, and better skilled. All the while, it was becoming better equipped. It tried out the 1976 manual's principles with Donn Starry on the ground for which the doctrine was designed and learned that defeating the first Soviet echelon was only the beginning of a solution. It took on the deep battle problem, read Clausewitz, studied German and Soviet operations, argued back and forth, and trained more. In 1979 the DePuy-Starry thesis met the Leavenworth antithesis. The result was AirLand Battle, a new conceptual synthesis that, when executed by the Big Five weapons systems operated by disciplined, well-trained soldiers, carried the Army to victory in the Gulf War.

Obviously, individuals and circumstances influenced the nature of the development this chapter has described. Bill DePuy was the way he was because of his experience in World War II. To an extraordinary extent this quintessential infantry officer remained focused on small-unit tactics as the key to success on any battlefield. His focus was understandable because it fit the immediate requirements of the army he had to rebuild and the tools immediately available to that army to meet the immediate military problems it had ignored for so long.

Donn Starry was greatly influenced by his experience visiting the battlefields of the Arab-Israeli war. He had the advantage of working out the problems with the 1976 doctrine in the field in Europe. Starry came to the task as a cavalryman who had not been marked by the experience of attrition-based combat in Europe in World War II or Korea, where he personally had fought a different kind of war. Even in Vietnam, where infantry combat predominated, Starry had commanded the premier armored formation, the 11th Armored Cavalry Regiment. Apart from that, as the school Commandant responsible for development of doctrine for the armored force from 1973-76, it is not surprising that it was Starry who was working on what DePuy wanted for the Army's principal doctrinal concept in the post-Vietnam era. It is notable that once Starry became TRADOC commander he expanded the horizon of doctrine and refocused the Army school system beyond his predecessor's principles of immediate utility. Starry combined respect for the calculus of the direct-fire battle with an appreciation for timeless principles of war—not the least of which is maneuver and its associated shock action—and the professional value of historical precedent. He also understood the importance in large organizations of respecting bureaucratic boundaries when you can. Like DePuy, Starry was fortunate in his successors, particularly Bill Richardson, who was able to extend Starry's vision into the future and, as Starry had with Bill DePuy's, build on it and make it better.

By the mid-1980s the officers who wrote AirLand Battle doctrine were teaching it at Fort Leavenwoth. Its origins rested securely in the work done by General Starry and officers at the Armor Center at Fort Knox. Although it remained focused on the armored force in Europe, it had inevitably grown into a broad combined arms doctrine. This fact is attested to by the prominence not just of officers like DePuy, Starry, Otis, and Richardson, in its development, but even more so by the principal authors of the 1982 and 1986 manuals: Henriques (infantry), Wass de Czege (infantry), Holder (cavalry), and Sinnreich (field artillery).

To the extent doctrine is what the Army does, more than how it talks about it, the application of AirLand Battle doctrine by Lt. Gen. Crosbie E. "Butch" Saint at III Corps in the late 1980s likely served as the most important external effort following publication of the 1986 manual.

Saint, who had commanded the 11th Armored Cavalry and 1st Armored Division in Europe, and Don Holder, his G3, wrote an auxiliary manual for the force Saint identified as "The Mobile Strike Corps." Titled the *III Corps Maneuver Booklet,* the manual supported Saint's belief that reserves, at least as understood by most Commanders, were wasted assets.[129] Saint trained his corps for mobile offensive warfare in Europe as the Allied Forces Central Europe (AFCENT) reserve corps, and undoubtedly influenced the veterans of that corps sent to the Persian Gulf three years later. Saint was the USAREUR commander during Desert Storm.

Other heavy units went to the National Training Center and trained on the weapons DePuy had forecast. They trained using concepts based on manuals derived from the principles laid down in successive volumes of *FM 100-5,* and units were evaluated against doctrinal standards of performance according to the vision of Paul Gorman. At Fort Leavenworth, SAMS graduated majors educated in the principles of operational art, and doctrine was written in TRADOC schoolhouses according to a system established by Donn Starry and reinforced by Bill Richardson. Some of the anticipated problems of armored warfare, most notably the threat of man-portable antiarmor weapons, were overcome technologically, with better training and, in the Gulf War, by an enemy less adept than the one the Army had prepared to fight during the Cold War. A measure of the doctrine's success in the Gulf is the extent to which its principles were retained in the 1993 edition of *FM 100-5,* which was produced under the supervision of Gen. Frederick M. Franks Jr. Franks commanded the VII Corps in the desert and assumed command of TRADOC soon afterward.

AirLand Battle doctrine was designed to counter a continental enemy who fought in successive echelons of combat units. The post-Vietnam apotheosis of the U.S. Army's armored force occurred largely in response to the fact of this continental armored threat. The collapse of the Soviet Union without the attempt of military adventurism in Europe, and the outcome of the Gulf War, provide two measures of the success of the restoration of American armored power.

Today the Army faces a new set of challenges, as well as a much less well-defined set of tactical and operational threats. It remains to be seen if its leaders will develop a doctrinal response to these circumstances that will be as effective as AirLand Battle doctrine was in its day.

Notes

1. A formulation given me by Gen. Donn A. Starry on many occasions.

2. A central proposition of Maj. Paul H. Herbert, *Deciding What Has to Be Done: General William E. DePuy and the 1976 Edition of FM 100-5, Operations*, Leavenworth Papers, No. 16 (Fort Leavenworth, Kans.: Combat Studies Institute [CSI], 1988). Herbert's remains the best account of the DePuy doctrinal revolution.

3. See John Romjue, *From Active Defense to AirLand Battle: The Development of Army Doctrine, 1973-1982* (Fort Monroe, Va.: TRADOC Historical Office, 1984).

4. Lt. Col. Romie L. Brownlee and Lt. Col. William J. Mullen, eds., *Changing an Army. An Oral History of General William E. DePuy, USA Retired* (Carlisle Barracks, Pa.: USAMHI, n.d.), 2, et seq. Biographical information that follows is taken from Brownlee and Mullen.

5. Brownlee and Mullen, *Changing an Army*, p. 202.

6. See Chapters II and III, ibid., especially pp. 84, 90-91.

7. Ibid., p. 108. Pamphlets are copied in Gen. William E. DePuy, *Selected Papers of General William E. DePuy*, Col. Richard M. Swain, comp. (Fort Leavenworth, Kans.: CSI, 1994), pp. 1-16.

8. Col. William E. DePuy, "11 Men, 1 Mind," *ARMY* (Mar. 1958): pp. 22-24, 54-60.

9. Brownlee and Mullen, *Changing an Army*, p. 109. See also, "Briefing by LTG DePuy [At Fort Polk, La.]," DePuy, *Selected Papers*, pp. 64-65.

10. Gen. Paul Gorman, *The Secret of Future Victories* (Fort Leavenworth, Kans.: CGSC Press, 1993; originally published by the Institute for Defense Analyses [IDA] as IDA Paper P2653, Feb. 1992), pp. 111-34.

11. Ibid., pp. 111-12.

12. Paul Herbert quotes DePuy as saying, "The infantry['s] . . . role is to support the tanks. That has been quite hard for the infantry school to stomach . . . the decisive offensive weapon is the tank . . . The Israelis, Egyptians, Syrians, Russians, British, Germans and I accept it." See Herbert, *Deciding*, p. 115 n27.

13. Gen. William E. DePuy, "FM 100—5 Revisited," *ARMY* (Nov. 1980): p. 12; and Gen. Donn A. Starry, "A Tactical Evolution—FM 100-5," *Military Review* (Aug. 1978): p. 4.

14. Observation by Dr. Roger Spiller, George C. Marshall Professor of History at the Command and General Staff College (CGSC), Fort Leavenworth, Kans., to author.

15. He said as much when he addressed the CGSC class soon after assuming the chief's chair. The author was present as a CGSC student.

16. A fact acknowledged later by Gen. Edward Meyer, who in 1976 was Deputy Chief of Staff for Operations (DCSOPS). It was Meyer who directed

revision of the 1976 manual in 1979. Gen. Edward Meyer, DCSOPS, to Gen. Donn Starry, commander TRADOC, 21 Mar. 1981, in CGSC Archives, Combined Arms Research Library (CARL), CGSC, Fort Leavenworth, Kans. (hereafter CGSC Archives), file: CGSC 82, DTAC—026/022, General Shy Meyer's Comments on Feb. 1981 DRAFT of FM 100-5.

17. Orr Kelly, *King of the Killing Zone* (New York: W.W. Norton, 1989), p. 88. *King of the Killing Zone* tells the story of the M-1 Abrams tank. Also see, Lt. Col. Jack Goldstein, "The Army's Big Five: Irons in the Fire," *ARMY* (May 1973): pp. 18-23.

18. Gorman, *Secret of Future Victories*, pp. 111-39, 42.

19. See comment by General Starry in Herbert, *Deciding*, p. 114 n4.

20. Gen. William E. DePuy, commander, TRADOC, to Gen. Walter T. Kerwin Jr., vice chief of staff, 24 Mar. 1977, reprinted in DePuy, *Selected Papers*, p. 213. DePuy wrote in part, "There has been a revolution in weaponry measured in terms of range and lethality, coupled with amazing advances in target acquisition and night vision."

21. Gen. William E. DePuy, "Briefing by General DePuy, 7 June 1973 [At Fort Polk, La.]," in DePuy, *Selected Papers*, p. 59.

22. Gen. William E. DePuy, commander, TRADOC, to Gen. Creighton W. Abrams, 14 Jan. 1974, reprinted in DePuy, *Selected Papers*, p. 71.

23. One version of the briefing is reproduced ibid., pp. 75-111.

24. Ibid., p. 76.

25. lbid., p. 86.

26. Ibid., pp. 93-94.

27. Gen. Donn A. Starry to Dr. Richard M. Swain, 7 June 1995, p. 6. (Hereafter Starry Letters). Letter in possession of author. A copy has been given to the CARL at Fort Leavenworth, Kans.

28. Ibid., pp. 3-4. Helmbold had analyzed the outcomes of a thousand battles to challenge the adequacy of Lanchester's Laws to explain ground combat.

29. Gen. Donn A. Starry, commander, TRADOC, to Maj. Gen. C.P. Benedict, commander lst Infantry Division, 13 Mar. 1978, CGSC Archives, file: CGSC 82, DTAC-041, Starry Letters. Starry makes clear to General Benedict the influence of the battle on the Golan to his thinking about war in Europe.

30. Lt. Gen. Hans-Henning von Sandrart, chief of staff of the German army, "Forward Defense—Mobility and the Use of Barriers," *NATO'S Sixteen Nations* (Feb.-Mar. 1985): p. 38. Von Sandrart indicated that 30 percent of the West German population and 25 percent of its industry were within a hundred kilometers (sixty miles) of the inter-German border.

31. Raymond Aron, *On War*, trans. Terence Kilmartin (New York: The Norton Library, W.W. Norton, 1968), pp. vii, 1-5.

32. For example see, John Keegan, "The Specter of Conventional War," *Harpers* (May 1982): pp. 46-53.

33. Col. Edwin G. Scribner, Memorandum: Doctrine Development by TRADOC, May 1973-Dec. 1979, CGSC Archives, file: CGSC 82, DTAC-038, Scribner Paper on Doctrinal Process, 2. (Hereafter Scribner Memorandum.)

34. Memo is at Appendix C, Document 1, to Romjue, *Active Defense*, pp. 80-81; or DePuy, *Selected Papers*, pp. 121-35.

35. DePuy, *Selected Papers*, pp. 122-35.

36. Gen. William E. DePuy, commander, TRADOC, to Gen. (Ret.) Bruce C. Clarke, 18 Aug. 1976; and Gen. William E. DePuy, commander, TRADOC, to Gen. Bernard Rogers, chief of staff, 27 Apr. 1977, reprinted in DePuy, *Selected Papers*, pp. 197-98, 219-20.

37. Lt. Gen. John Cushman, text of TV Presentation on FM 100-5, CGSC Archives, file: CGSC 82, DTAC—030, Lt. Gen. John Cushman Various Views. Page two of the five-page script is missing.

38. Ibid. The statement is modified somewhat (as Cushman acknowledged). Marshall's book (actually Maj. Edwin F. Harding's) spoke of the *Art of War*. See *Infantry in Battle* (Washington: The Infantry Journal, 1939), p. 1.

39. Two firsthand accounts of the A.P. Hill process are the Scribner Memorandum, and Starry Letters. Herbert, *Deciding What Has to Be Done*, is the best account of the process in the whole and is favorably disposed to General Cushman's position.

40. This difference of concept is the subject of Herbert, *Deciding What Has to Be Done*.

41. General DePuy told Paul Herbert that he rejected the Leavenworth draft because "we didn't think that [it] conveyed to the Army the sense of urgency, that it was too scholastic, wasn't enough how to retrain, reorient and refocus an Army." As quoted in Herbert, *Deciding What Has to Be Done*, p. 116 n13.

42. A criticism made by Col. Don Holder in a 1987 interview with Dr. Michael Pearlman at Fort Leavenworth, Kans. Holder attended CGSC in the class of 1977. He was a principal author of three of the succeeding four FM 100-5s, 1982, 1986 and 1997, and served as CGSC commandant 1995-97. CGSC Archives, file: CGSC 1987, SAMS-012, Interview: Col. Don Holder, 15 May 1987. (Hereafter Holder Interview, 15 May 1987.)

43. Something General Starry had recognized within less than two years from the doctrine's promulgation. Starry, "A Tactical Evolution," p. 3.

44. Gen. Alexander Haig, supreme allied commander, Europe (SACEUR), to Gen. William E. DePuy, commander TRADOC, 10 Sept. 1976, CGSC Archives, file: CGSC 82, DTAC-010, Al Haig's Comments on FM 100-5.

45. Gen. William E. DePuy to Gen. Alexander Haig, SACEUR, 13 Oct. 1976, CGSC Archives, file: CGSC 82, DTAC-010, Al Haig's Comments on FM 100-5.

46. Gen. Alexander Haig, SACEUR, to Gen. William E. DePuy, commander, TRADOC, 20 Oct. 1976, CGSC Archives, file: CGSC 82, DTAC-010, Al Haig's Comments on FM 100-5.

47. See Starry's comment as quoted in Herbert, *Deciding What Has to Be Done*, p. 118 n13. Also see Gen. William E. DePuy, commander, TRADOC, to Gen. Fred C. Weyand, chief of staff, 18 Feb. 1976, which outlines the steps taken by General DePuy to gain acceptance for the manual. Reprinted in DePuy, *Selected Papers*, pp. 179-83.

48. Headquarters, Department of the Army, *Field Manual 100-5, Operations* (1 July 1976), p. 2-32. (Page numbers in the manual are expressed as chapter and page numbers [e.g., chapter 2, page 32 here]).

49. Ibid., p. 3-3.

50. Ibid.

51. Ibid., p. 3-4.

52. Ibid., p. 4-3.

53. Ibid., p. 5-7.

54. Ibid, p 5-14.

55. Ibid., p. 3-2.

56. Ibid., p. 2-6.

57. Ibid., p. 5-3 speaks of reserves.

58. Ibid.

59. DePuy to Weyand, 18 Feb. 1976, in DePuy, *Selected Papers,* pp. 179-83.

60. Starry Letters, p 13.

61. Ibid., p. 14.

62. Ibid.

63. Ibid., p. 15.

64. See especially, Maj. Richard Hart Sinnreich, "Tactical Doctrine Or Dogma?," *ARMY* (Sept. 1979): p. 34. It is notable—given the close association with the official army—that this was the cover essay of the Association of the United States Army's (AUSA) journal.

65. Department of the Army, TRADOC, Training Circular 7-24, *Antiarmor Tactics and Techniques for Mechanized Infantry* (30 Sept. 1975). See Maj. Theodore T. Sendak, "The Airborne-Antiarmor Defense," *Military Review* (Sept. 1979): pp. 43-51.

66. Sinnreich, "Tactical Doctrine," pp. 14-19.

67. F. Clifton Berry Jr., "Doctrine Developed in a Vacuum?," *Armed Forces Journal International* (Oct. 1976): p. 4. Philip Karber, a consultant specializing in Soviet military affairs, responded to Lind's various criticisms, which were presented in an article by John Patrick, a 1969 Dartmouth graduate then seeking work in national security or foreign policy fields. Berry himself wrote an uncomplimentary review of the manual.

68. William S. Lind, "Some Doctrinal Questions for the United States Army," *Military Review* (Mar. 1977): pp. 54-65. The essay was written originally for the *Marine Corps Gazette,* for whom Lind regularly wrote, and was requested by *Military Review.*

69. An observation made early on by a CGSC student during academic

year 1974-75. See Maj. Wesley K. Clark, "Winning the First Battle: Another Look at New Tactical Doctrine," CGSC Archives, file: CGSC 82, DTAC-029, Wes Clark on Tactical Doctrine.

70. See particularly, Sinnreich, "Tactical Doctrine," pp. 15-19; Col. (later Maj. Gen.) Robert E. Wagner, "Active Defense and All That," *Military Review* (Aug. 1980): pp. 4-11; and Col. (later Gen.) Wayne A. Downing, "U.S. Operations Doctrine: A Challenge for the 1980s and Beyond," *Military Review* (Jan. 1981): pp. 64-73.

71. Starry, "A Tactical Evolution," pp. 2-9.

72. Ibid., p. 17.

73. Gen. Donn A. Starry to Maj. Gen. C.P. Benedict, 13 Mar. 1978, p. 2.

74. Starry, "A Tactical Evolution," p. 10.

75. Lt. Gen. Edward C. Meyer, DCSOPS, to Gen. Donn Starry, commander TRADOC, 13 June 1979, CGSC Archives, file: CGSC 82, DTAC-007, Re. General Meyer on 1976 *FM 100-5*.

76. Ibid.

77. Gen. Donn A. Starry, Commander, TRADOC, to Gen. E.C. Meyer, DCSOPS, 26 June 1979, CGSC Archives, file: CGSC 82, DTAC-024, Re. Individual Comments on 1981 DRAFT.

78. Ibid.

79. Starry Letters, p. 22.

80. Ibid. For an account of how Morelli worked with outside critics and advisers, see Alvin and Heidi Toffler, *War and Antiwar. Survival at the Dawn of the 21st Century* (Boston: Little Brown, 1993).

81. John L. Romjue, Re: Memorandum for Record, interview with Lt. Col. (Ret.) Richmond B. Henriques, 14 May 1984, in CGSC Archives, file: CGSC 1984, DTAC-049, MRF: Lt. Col. Richmond B. Henriques, TRADOC, 14 May 1984.

82. See Holder Interview, 15 May 1987, p. 26.

83. Headquarters, Department of the Army, *Field Manual 100-5, Operations* (20 Aug. 1982).

84. Ibid., pp. 2-3.

85. Starry Letters, p. 16, et seq.

86. See "Implementing the AirLand Battle," *Field Artillery Journal* (Sept.-Oct. 1981): pp. 20-27.

87. General Donn A. Starry, "Extending the Battlefield," *Military Review* (Mar. 1981): pp. 31-50.

88. Ibid., p. 41.

89. Ibid., p. 42.

90. See Gen. Ronald R. Fogleman, "Aerospace Doctrine; More than Just a Theory," *Airpower Journal* (Summer 1996): p. 41.

91. Starry Letters, p. 26.

92. Memorandum is in Appendix C 6, Romjue, *Active Defense*, pp. 100-108.

93. *FM 100-5, Operations* (1982), p. i.

94. See Starry Sends Messages, 091538Z July 80 and 291350Z Jan 81, Documents 4 and 5 in Romjue, *Active Defense,* pp. 96-99.

95. *FM 100-5, Operations* (1982), p. 2-1.

96. Ibid.

97. Ibid., pp. 2-2-3.

98. Ibid., p. 2-3.

99. Communication with the author.

100. In a 1984 interview Wass de Czege attributed the initial formulation to Maj. Gen. (Ret.) John Barrity of BDM corporation. See Interviews: John L. Romjue, Re. Memorandum for Record, with Lt. Col. (Ret.) Richmond B. Henriques, and Re. Memorandum for Record, with Col. Wass de Czege on the Development of AirLand Battle Doctrine, 16-17 Apr. 1984, CGSC Archives, file: CGSC 82, DTAC-041, Starry Letters, p. 2.

101. Communication with author.

102. Wass de Czege attributes the synchronization decision to General Starry, who did not like the term integration. The term synchronization came from an article by General DePuy, and was supported by General Morelli. According to Wass de Czege, it was General Richardson who added Agility. See Romjue, Re. Memorandum for Record, interview with Col. Wass de Czege, p. 3.

103. Gen. William E. DePuy, copy of handwritten document in CGSC Archives, file: CGSC 82, DTAC 035-003, DePuy Comments on DRAFT 100-5, April 1981.

104. Gen. William E. DePuy, "Toward A Balanced Doctrine," *ARMY* (Nov. 1984): pp. 18-23; and "Concepts of Operation: The Heart of Command, the Tool of Doctrine," *ARMY* (Aug. 1988): pp. 26-40. For the influence of matters in Lebanon, see his review of Chris Bellamy's "The Future of Land Warfare" in *Parameters* (Dec. 1987): pp. 106-108.

105. Col. Harry G. Summers, *On Strategy, A Critical Analysis of the Vietnam War* (Novato, Calif.: Presidio Press, 1982), p. 1. This is the commercial edition of an earlier government publication.

106. See Holder Interview, 15 May 1987, and the references in Downing, "U.S. Operations Doctrine," p. 73. Downing's notes are interesting because he was innocent of Army schools. He attended the Armed Forces Staff College and the Air War College. See William E. DePuy, "Technology and Tactics in Defense of Europe," *ARMY* (Apr. 1979): pp. 14-33, which contains a diagram of an eastern front operation laid graphically on West Germany.

107. By 1980 the notion of operational art was abroad in the Army. Edward N. Luttwak, a military critic well known in Army circles, published an essay, "The Operational Level of War," in *International Security* (Winter 1980-81): pp. 61-79, that followed up on a study he had presented to the Army earlier. By then, officers who were interested in the topic were famil-

iar with one or more Soviet works that addressed Soviet notions of operational art, particularly V. Ye Savkin's *The Basic Principles of Operational Art and Tactics* (A Soviet View) (Washington: GPO, 1972). The many published essays of the War College's Cols. Harry Summers and Wallace Franz had given currency to German views as well. Don Holder wrote a memo in Aug. 1981 recommending the term not be included in the 1982 edition. CGSC Archives, file: CGSC—82, DTAC-041, Starry Letters. On Morelli, see Holder Interview, 15 May 1987, pp. 51-53. Also see Starry Letter, 7 June 1995, p. 29, and Romjue interview with Wass de Czege, Apr. 1984, p.s 5.

108. See Memorandum by Lt. Col. Don Holder dated 13 Oct. 1981, TAB-B to a Staff Action, Subject: Cdr. TRADOC's Directed Changes to FM 100-5, dated 29 Oct. 1981 in CGSC Archives, file: CGSC 82, DTA—026/026, General Otis's Copy of FM 100-5 DRAFT, July 1981.

109. Holder Interview, 15 May 1987, p. 51-52.

110. The formulation is obviously very close to that articulated by Fuller in various works. Col. Wass de Czege had written an essay in the seventies on a Combat Power Model that was very similar to Fuller's. General Richardson remembered the essay. See Romjue interview with Wass de Czege, Apr. 1984, p. 1.

111. *FM 100-5, Operations* (1982), p. 7-1.

112. Ibid.

113. Ibid., p. 11-9. It is, of course, because the new manual focused at division level and above that the deletion of Active Defense as the prescription for defense in battalion and brigade zones could be accomplished so easily. General DePuy pointed out with some justification, however, that the 1976 manual had, in fact, mentioned the words active defense only once, with *active* as an adjective. All of which only shows how instruction and follow-on documentation matters. DePuy, "FM 100-5 Revisited," p. 13.

114. *FM 100-5 Operations* (1982), p. 7-6.

115. lbid., p 7-11.

116. Ibid., p. 7-15.

117. Boyd D. Sutton, John R. Landry, Malcolm B. Armstrong, Howell M. Estes, and Wesley K. Clark, "Deep Attack Concepts and the Defense of Central Europe," *Survival* (Mar.-Apr. 1984): p. 54.

118. Leonard Famiglietti, "AF, Army Chiefs Sign 31-Point Agreement," *Air Force Times*, 4 June 1984.

119. Tom Clancy with Gen. Fred Franks Jr. (Ret.), *Into the Storm: A Study in Command* (New York: G.P. Putnam's Sons, 1997), pp. 156, 258-59, 540-41.

120. Lt. Col. Richard Hart Sinnreich, Memorandum, Subject: "Marketing" AirLand Battle Doctrine in NATO, 28 Dec. 1984. Copy in author's possession.

121. General Rogers, SACEUR, letter to General Richardson, commander, TRADOC, 26 Sept. 1985, CGSC Archives, file: CGSC 86, SAMS-007, EUCOM Comments on DRAFT 100-5.

122. See the detailed analysis and comparison of Follow-on Forces Attack (NATO) and AirLand Battle Doctrine in Sutton, Landry, Armstrong, Estes , and Clark, "Deep Attack Concepts," pp. 50-69.

123. Holder Interview, 15 May 1987, p. 14.

124. Department of the Army, *Field Manual 100-5, Operations* (5 May 1986), p. 9. (Pagination scheme for the 1986 manual abandoned the chapter-page convention.)

125. Ibid., pp. 30-31.

126. Ibid., p. 17

127. Ibid., pp. 47-50.

128. Ibid., p. 19.

129. Headquarters, III Corps, *III Corps Maneuver Booklet* (Fort Hood, Tex.: May 1987). On the surface, Saint's view appears to conflict directly with that of J.F.C. Fuller, who believed the skillful use of reserves was the acme of generalship. Saint's point appears to rest on what he saw as a weakness in officer training (he often said the reserve was no more than "what was left over") and in the need in armored maneuver warfare to employ all forces available nearly continuously.

12

"Lethal beyond all expectations": The Bradley Fighting Vehicle

Diane L. Urbina

This chapter describes the doctrinal and developmental history of the Bradley Fighting Vehicle (BFV). The Bradley was one of the "Big Five" post-Vietnam systems developed in the 1970s, fielded in the 1980s, and deemed "lethal beyond all expectations" during the Gulf War in the early 1990s.[1] Akin to Watty Piper's book *The Little Engine That Could*, the Bradley program has remained intact through perseverance and temerity on the part of program participants. The Bradley overcame strong congressional prejudice, three general officer reviews, a major redesign effort, program cancellation, and bashing in the mainstream media in the early 1980s. It was developed during a period of upheaval and change within the Army that included conversion to an all-volunteer force, extensive reorganization, and significant changes to doctrine and tactics. However, the program survived and the Bradley became a world-class weapon system in large part because Gen. William E. DePuy—then serving as TRADOC commander—believed the BFV was instrumental to a combined arms doctrine and Maj. Gen. Stan R. Sheridan, the Bradley program manager, orchestrated its development.

When the first production vehicle rolled off the FMC Corporation production line in May 1981, after nearly two decades of effort, the Bradley represented a truly "hybrid" vehicle. The post-Vietnam development of *panzergrenadier* tactics, of which the BFV was an integral part, was based on the integration of Armor School doctrine and Infantry School tactics. This combination and the development of the Bradley became the impetus for two single-purpose schools to initiate a common dialogue to "combine arms" and train to fight mounted or dismounted.

The program to develop a fast, lethal infantry fighting vehicle began in 1964 in the woods of the Pacific Northwest with a large wooden mock-up of a prototype vehicle known as the Mechanized Infantry Combat Vehicle (MICV). The MICV program was initiated in response to a series of studies conducted by the U.S. Army Combat Developments Command (USACDC) to identify the battlefield characteristics of an infantry fighting vehicle. At the time, the Army had no vehicles in its inventory that could satisfy the requirement for infantry fighting in a mounted role.[2]

The M113 APC, introduced in 1960, was little more than a battlefield taxi designed to shuttle eleven soldiers across hostile battlefields with a modicum of armor protection,[3] and the USACDC eventually decided that it lacked sufficient armor and firepower to survive and fight with the new main battle tank then being developed.[4] Until the early 1970s, it was envisioned that the MICV would replace M113 APCs in mechanized infantry units and be a companion to the main battle tank. It would provide the infantry with a vehicle from which they could fight on the move. It was later stated that the MICV was intended to replace only some of the M113 family in service.[5]

In June 1964, Pacific Car and Foundry (PACCAR) received an R&D contract to build one prototype and five pilot vehicles "which would provide cross-country mobility and a mounted fighting capability for mechanized infantry and armored units, weapons, and selected supporting elements of mechanized units."[6] The design of the MICV, which was designated the XM701, included a Hispano-Suiza 20mm cannon mounted in a two-man turret, firing ports for an eleven-man infantry squad, grenade dispensers, and an overpressure system for protection against nuclear, biological, and chemical (NBC) threats. It also had a stove and toilet to allow twenty-four-hour occupancy, and the same 425 horsepower diesel engine, transmission, and suspension used in the M109 and M110 self-propelled howitzers, also produced by PACCAR.[7]

Founded in 1905 in Renton, Washington, PACCAR first built railroad cars to haul logs out of the Pacific Northwest. During World War II, PACCAR began building military vehicles, including nearly a thousand Sherman tanks and tracked recovery vehicles. Its military products division quickly expanded to include the production of self-propelled artillery, tracked cargo carriers, portable bridging systems, and tracked landing vehicles. Test facilities included a 140-acre driving area, a firing range, and a fording tank.[8] The XM701 MICV pilot vehicles had a swimming capability similar to that of the M113 APC. Following initial testing in the fording tank, the XM701 was taken to the southern end of Lake Washington to swim in thirty to forty feet of water. Photographs taken in June 1965 show an XM701 MICV pilot vehicle swimming at the Lake Washington beach park property, with sailboats from the Renton Sailing Club in the background.[9]

Pilot vehicles were delivered to the Army in 1965, tested, and then rebuilt and shipped to Germany for comparison with the German-designed *schutzenpanzer* vehicle. The Army eventually rejected the XM701 because, at twenty-six tons, it was considered too heavy to be airlifted in a C-141 transport aircraft and too slow to keep up with tanks. However, the pilot vehicles were used as a baseline to develop requirements during concept formulation for a proposed vehicle designated the MICV-70. In 1966, the Army evaluated three gun-control stabilization systems and two water propulsion systems for use on the MICV-70. Then, in February 1967, a TRW turret-mounted cannon was evaluated for use on the MICV-70.[10]

In August 1965, the Army Vice Chief of Staff directed that a comprehensive analysis be made of the MICV to determine if it would be feasible for adoption within the 1966-1972 time frame. Three MICV studies were conducted during the period 1966-68 by Cornell Aeronautical Laboratory. The first was a parametric design/cost effectiveness study, which evaluated fifty-three concepts from industry and government and concluded that the MICV should be a fully-tracked aluminum vehicle with a stabilized cannon and a twelve-man crew. The second study integrated vehicle design characteristics with tactical doctrine and operational concepts. This resulted in the determination of optimum vehicle characteristics. A third study evaluated six parameters that influenced the vehicle's size, weight, and cost.[11]

In 1968 the U.S. Army Materiel Command (AMC) established a MICV project management office in Warren, Michigan. Lt. Col. Richard H. Sawyer was named project manager (PM), becoming one of the first PMs within AMC. In August, the Vice Chief of Staff appointed an ad hoc committee to review the MICV-70 qualitative materiel requirements (QMR) and make recommendations for change. The committee was led by Brig. Gen. George W. Casey, then deputy director for doctrine at USACDC.[12] Casey, a West Point graduate, was a veteran of service in Korea, Europe, and Vietnam. He was assigned to the 1st Cavalry Division during his first Vietnam tour in 1966, where he served as 2d Brigade Commander and Division Chief of Staff. According to Maj. Bruce Pirnie, Casey became aware during these assignments "that U.S. tactics were founded on the '2½-mile-per-hour infantryman.' In other words, infantry officers thought too exclusively of the dismounted role. Armored personnel carriers were used as 'battlefield taxies' to move troops to the line of departure. By contrast, Casey thought that mechanized infantry should exploit its light armored vehicles to operate more closely with tanks. To accomplish this change, he envisioned a successor to the M113 APC which would have better armor protection, firing ports, and a more lethal main weapon, perhaps a rapid-firing cannon."[13]

In September 1968 General Casey submitted a final report recommending the development of a vehicle for the infantry and a cavalry variant that would be smaller, lighter weight, and less costly. The report also included recommendations calling for a three-man crew plus nine infantrymen, and a two-man turret with a stabilized, rapid-fire cannon and 7.62 coaxial machine gun.[14]

Casey's report, which resoundingly confirmed the need for the vehicle, inspired Lt. Gen. George Sammet Jr., who later became AMC's deputy commander, to become a proponent of the infantry fighting vehicle. Sammet, now retired, recalls: "Early on I predicted George would become the Chief of Staff of the Army. He was that talented. So when he told me, 'build it,' I threw my weight behind it. Unfortunately, while piloting his own helicopter in a torrential tropical rainstorm in Vietnam, he flew it into a mountain. The Army lost a potential Chief of Staff, the Bradley lost a great proponent, and I lost a great friend."[15]

The MICV-70 QMR was approved in October 1968 by Gen. William Westmoreland, who had recently turned over command of U.S. forces in Vietnam to become Chief of Staff. The following month, the Assistant Chief of Staff for Force Development, Lt. Gen. Arthur Collins, directed that a comprehensive review of all alternative vehicle systems to the MICV-70 be conducted before action was taken on the MICV PM's request to enter contract definition. Several weapons were included in the study to provide a backup for the primary weapon for the MICV, and an in-house design was selected as being the most cost effective. Cornell Aeronautical Laboratory conducted an alternative cost effectiveness study that supported General Casey's report but recommended a heavier vehicle. Congress rejected the idea because of limited funding. Lieutenant General Collins then proposed an austere MICV design featuring a one-man turret, with the commander located in the hull, in order to reduce its cost. The QMR for the MICV was then revised to permit development of the one-man turret concept. This change turned out to be a disastrous decision that took several years to reverse. However, the immediate result was approval to enter engineering development and the award of a $29.3 million contract to FMC Corporation on 22 November 1972 to build seventeen pilot vehicles. The contract award was based on a competition between full-scale wooden mock-ups built by PACCAR, Chrysler Corporation, and FMC.[16]

In November 1967 the Soviet-made BMP-1 was unveiled during a parade in Red Square. The vehicle was developed as a replacement for the BTR-50P, and was fitted with a 73mm cannon, an AT-3 Sagger antitank guided missile. It had a crew of three and carried an eight-man infantry squad. The BMP-1 and new Soviet tanks were part of an overall improvement of Soviet forces facing the North Atlantic Treaty Organization (NATO). Two years later, the Germans displayed their Marder infantry combat vehicle, which was equipped with a 20mm Rheinmetall automatic cannon. The marder had a crew of three and carried a six-man infantry squad.[17]

Despite discussions within the U.S. Army regarding specific MICV requirements, the appearance of the BMP-1 confirmed that the Army needed an infantry fighting vehicle of its own to counter the Soviet threat. The BMP-1 thus provided the necessary impetus to sustain the MICV program through post-Vietnam funding cuts.

Following Vietnam, the program was swept along in a ten-year vortex of change driven by major reorganization in the U.S. Army, the impact and analysis of the 1973 Arab-Israeli War, intense debate regarding doctrine, and finally, the acceptance of the infantry fighting vehicle by the infantry community.

According to Maj. Paul Herbert:

> In 1973, the U.S. Army began to emerge from one of the most traumatic periods in its history. From the 1968 Tet offensive in Vietnam, to the twin shocks of the Cambodian invasion and the killing of antiwar demonstrators at Kent State University in 1970, to the 1971 investigation into the My Lai incident, to the withdrawal from Vietnam and the shift to an all-volunteer armed force in 1973, the U.S. Army increasingly found itself the focal point of public criticism. Racial tensions and drug abuse among soldiers compounded the sense of defeat that, however gilded, attended the withdrawal from Southeast Asia. The Army's theoretically highest-priority unit, the Seventh Army in Europe, was probably at the lowest state of readiness in its history, the victim of a personnel replacement system in Vietnam that used other major commands as

replacement pools, resulting in drastic shortages of officers and noncommissioned officers. Public disillusionment with the war in Vietnam became a general sentiment against all war and all military institutions, especially the army. The U.S. Army in 1973 was in danger of losing its institutional identity and pride of purpose.[18]

In July 1973 the USACDC was abolished and Continental Army Command (CONARC) was divided into two commands: Training and Doctrine Command (TRADOC) and Forces Command (FORSCOM). The former assumed responsibility for training and the development of doctrine and materiel requirements, while the latter assumed command of CONUS-based combat units. This reorganization and the subsequent appointment of Gen. William DePuy as TRADOC Commander resulted in the profound changes to the Army's tactics, doctrine, organization, training, and weapons described in detail in the previous chapter.[19]

There was a certain irony in General DePuy's appointment as TRADOC Commander, given his blunt and opinionated personality. In order to attract an all-volunteer army, great emphasis had been placed on changing the public's perception of an insensitive army plagued with drug abuse and racial tension. The new all-volunteer army's motto—"Today's Army wants to join you"—developed by a New York advertising agency, was met with mixed reactions. Many in the army felt that the pendulum had swung too far in favor of human relations, with not enough emphasis placed on the war-fighting mission. DePuy, who addressed a TRADOC leadership conference at Fort Benning the same year the new slogan was announced, left no doubt as to his approach to leadership by saying, "We are not in this business to be good guys. Nice warm human relationships are satisfying and fun, but they are not the purpose of an Army." DePuy, who relieved fifty-six officers during a one-year period in Vietnam, did not suffer fools gladly. His leadership style left no doubt as to his expectations, an approach that was instrumental in transforming the post-Vietnam army.[20]

General DePuy began his first year at TRADOC by integrating combat development functions. While that effort was underway, the October 1973 Arab-Israeli War occurred. The Arab-Israeli War was important to the U.S. Army because the combat actions were similar to those U.S. troops in Europe were most likely to engage in.[21]

According to General DePuy:

The three major lessons in the war are: First, that modern weapons are vastly more lethal than any weapons we have encountered on the battlefield before. Second, in order to cope with these weapons it is essential we have a highly trained and highly skilled combined arms team of armor, infantry, artillery and air defense backed by the support required to sustain combat operations. Third, the training of the individual as well as the team will make the difference between success and failure on the battlefield. . . . The Arab-Israeli War dramatized the lethality of modern antitank weapons, including most particularly the high velocity tank cannon and the long range antitank guided missile. With one exception (the Battle of Kursk in 1943), there has never been a comparable loss of tanks in such a short period of time. If the rate of loss which occurred in the Arab-Israeli War during the short period of 18 to 20 days were extrapolated to the battlefields of Europe over a period of 60 to 90 days, the resulting losses would reach levels for which the United States Army is not prepared in any way.[22]

Following the Arab-Israeli War, General DePuy began several major efforts: development of the Army Training and Evaluation Program (ARTEP); matching materiel development with equipment requirements to include the continued development of the "Big Five"—the BFV, the M1 Abrams tank, the AH-64 Apache attack helicopter, the Patriot air defense missile system, and the UH-60 Black Hawk troop-carrying helicopter—and finally, developing the doctrine of "active defense."[23]

General DePuy, in planning the reorganization of CONARC and USACDC into TRADOC and FORSCOM, recognized that while AMC, the army's procurement agency or materiel developer, was reducing costs by improving contracting procedures, not enough was being done by the combat developer to combine the development of doctrine with equipment specifications.[24]

In 1975, DePuy advised the Chief of Staff: "As you know, better than I do, the Army has always had a difficult time explaining just why it needs a particular weapon system and even more difficulty in explaining how that particular weapon system fits in with all of the other army systems and organizations and finally difficulty in answering the inevitable question as to whether some other combination or alternatives might not be better or more effective." DePuy believed that TRADOC had to articulate a specific scenario in which the equipment would be used. This approach would relate specific improvements in capabilities and demonstrate compatibility with existing Army equipment. DePuy then began developing a doctrinal approach that integrated materiel requirements.[25]

DePuy believed that the Army's tactical doctrine should conform to NATO's need to conduct a forward defense. The goal of this doctrine was to provide the Army with the tools needed to defeat a Warsaw Pact attack along the inter-German border. One of the key elements of this concept was the combination of tanks and mechanized infantry or *panzergrenadiers*, which was the responsibility of the German Armor School—a fact not lost on the U.S. Army's Armor School Commandant, Maj. Gen. Donn Starry. The linchpin in the German *panzergrenadier* concept was the ability to fight while mounted. "Of course," wrote General DePuy, "one of the problems in adopting *Panzergrenadier* tactics for our mechanized infantry lies in the inadequacy of the M113 as an armored fighting vehicle. The M113 cannot keep up with tanks cross country without scrambling the rifle squad inside. The 50-caliber machine gun is not adequate for the suppression of enemy antitank rocket weapons, such as the RPG7. The 50-caliber cannot be fired usefully on the move. Lastly, the 12.5 machine gun will penetrate the M113 and 152mm fragments will penetrate the top armor."[26]

Confusion or misunderstanding in some circles regarding the role of an armored personnel carrier versus a mechanized infantry fighting vehicle was a long-standing problem in the Army. As late as 1986, Army Chief of Staff Gen. John Wickham felt compelled to issue a White Paper comparing the maneuver capability, firepower, and protection of the Bradley to that of the M113.[27] The paper's bottom-line conclusion was that even the beefed-up version of the M113 employed in Vietnam—the armored cavalry assault vehicle (ACAV)—was woefully inadequate as an infantry *fighting* vehicle.[28]

In the early 1970s, Maj. Gen. Thomas M. Tarpley was Commandant of the Infantry School, the proponent school for the MICV. With the infantry having been immersed in the Vietnam War for the previous ten years, Tarpley was not as quick to dismiss the Vietnam experience as a deviation from traditional warfare, as did his Armor School counterpart, General Starry. Starry, in lockstep with General DePuy's doctrine, was given responsibility for revising the Army's doctrine into a combined arms concept. The Infantry School seemed unable to articulate how the MICV would be used in a mechanized war due to its unfamiliarity with the tactics and doctrine of armored vehicles. Starry was at no such loss, and with DePuy's support quickly began setting about doctrinal reform. Aware that doctrine for the employment of mechanized infantry was the purview of the German Armor School, Starry claimed proponency for developing similar doctrine for the U.S. Army.

At one point, Tarpley feared the Infantry School would lose the mechanized infantry mission entirely to the Armor School. It is interesting to note that during the mid- to late-1970s, during the BFV's critical development phase, the PMs were armor rather than infantry officers. The Infantry School ultimately retained the mechanized infantry mission, and the field manuals governing its employment were numbered differently to clearly indicate they were neither armor nor infantry in origin.[29]

Following the award of the 1972 MICV engineering development contract, FMC's Ordnance Division began development and fabrication of four pilot vehicles. After the Vietnam War, FMC began developing two candidate combat vehicle systems with private company capital. One of the systems was a scout vehicle and the other was the MICV-70 or XM723. The scout vehicle program produced two candidates: a full-tracked version and a wheeled version. The scout program was canceled in 1974 after General Starry operated both vehicles during testing and found both severely lacking, including the 20mm cannon, which according to Starry, "wouldn't punch a hole in anything." When DePuy saw the vehicles the following day he said, "I don't know what I want, but I don't want those vehicles," and canceled the program.

However, the requirement for a scout vehicle remained valid. In the spring of 1975, after a general officer combat vehicle review at TRADOC, it was determined that the MICV could be modified to meet

MICV XM723 developed by FMC's Ordnance Division. U.S. Army.

the scout requirement and that designing a single vehicle for both the infantry and cavalry would be cheaper in the long run.[30] In addition, it was clear to all concerned that the MICV in either role needed a better weapon. After further study planners agreed to mount the TOW anti-tank guided missile system on the MICV despite the fact the decision would result in more time lost in the redesign process.

By 1974 the XM723 MICV-70 program had begun to experience delays due to serious technical problems with the suspension and trans-mission. In May, the MICV PM and FMC decided to realign the pro-gram, which included a "get-well" plan resulting in increased contractor testing prior to government testing, and suspension of all other indirect efforts until the technical problems were corrected.[31] The plan to inte-grate the Vehicle Rapid Fire Weapon System Successor (VRFWS-S) or Bushmaster, which was being developed concurrently with the MICV, was also indefinitely deferred. The MICV PM was directed by AMC to plan for the M139 product-improved weapon system, a 20mm cannon with dual feeder, for the MICV through production.[32]

Testing of the prototype MICV began in the summer of 1975. The transmission was still experiencing failures, but the suspension prob-lems had been corrected. A back-up transmission program was initi-ated with Allison, as was a transmission reliability improvement program with General Electric, the transmission subcontractor. Gov-ernment testing of a MICV mounting a 20mm cannon began in October 1975.[33]

In April 1975, General DePuy wrote Maj. Gen. Gordon Sumner Jr., director of the Near East and South Asia Region in the Office of the Assistant Secretary of Defense, that even though the secretary of de-fense still had doubts regarding the MICV and Bushmaster, DePuy was convinced that there was no other alternative but to press on. "The Secretary of Defense has been told that the MICV is too expensive. Yet it is less expensive than the MARDER, about the same as the BMP and no more than the much inferior Food Machinery Armored Infantry Fighting Vehicle. Somewhere we have gotten off the track with the Secretary, or somebody has gotten him off the track. There is too much at stake for the Army on the next battlefield to break off our efforts on behalf of the MICV." DePuy later advised Gen. Fred Weyand, then serv-ing as Army Chief of Staff: "The Army must have a MICV. We need it to fight the way we must fight to survive and win on the mechanized battle-field in conjunction with tanks. We need it because the M113 is not even in the same league with BMP and MARDER—it will be driven off the battlefield. This is one of those issues which goes to the heart of the Army's capability. Therefore, we must win this one—the earlier the better."[34]

An Air Force C-5 Galaxy disgorges a preproduction M2 Bradley. Patton Museum.

Brig. Gen. Stan Sheridan was appointed MICV PM in the summer of 1975. Lt/ Gen. George Sammet selected Sheridan for the position because he stood head and shoulders above his peers. Sammet recalls: "I wanted someone that could sell the program. Someone with the political skills to sell the program. And I had no doubt at all that Stan was the person that could do the job. He was a skilled politician and proved himself outstanding in every way during the M60 program."[35]

Shortly after his arrival, Sammet assigned Sheridan responsibility for three previously project-managed systems: the MICV, the scout; and the Bushmaster. He was also in charge of developing all vehicular-mounted armament (the product-improved M139 20mm automatic cannon, the 25mm automatic cannon, 25mm ammunition, the XM714 fuse family for 20-30mm cannon, and the firing port weapons), as well as a family of ancillary vehicles that included the carrier for the Multiple Launch Rocket System (MLRS). The most significant aspect of this

First firing of the TOW missile from the IFV-CFV (Bradley) in Fort Irwin California in July 1978. Col. Carl Zillian (left), USA (Ret.), TOW Program Director, Hughes Aircraft Company, and Brig. Gen. Stan Sheridan, gunner. Stan Sheridan.

revised management structure was that it facilitated the development and integration of the 25mm armament system into the MICV. The critical path was through the development and selection of the 25mm armament system and its integration into the MICV's primary weapon station.

Sheridan was also faced with coordinating the efforts of major contractors. This included competition between the candidate externally powered and self-powered 25mm cannons and the ammunition, all being developed on the West Coast; the Allison transmission, Cummins engine, fuse family, and stabilization system being developed in the Midwest; and the General Electric transmission being developed on the East Coast.[36]

Stan Sheridan was born in 1928 in Hollywood, California. Educated at West Point, Sheridan spent two tours in Vietnam. His first was spent as a technical intelligence and explosive ordnance demolition adviser to the Vietnamese Joint General Staff. During the second he commanded the 1st Battalion, 69th Armor, and served as deputy commander of the 1st Brigade, 4th Infantry Division. Sheridan had also been a tank tester at Aberdeen Proving Ground, Maryland (1955-56), a project officer in the early days of the space program (1959-62) and a tank action officer on the Army Staff (1965-68). In 1971, as a highly decorated combat veteran, he became the PM of the M60 tank family, followed by an assignment as commander of the 2d Armored Division's Support Command at Fort Hood, Texas.[37]

Personnel working on the MICV program staff found Sheridan to be an athletic, purposeful officer who was capable of sudden ferocity as the need arose. His single-mindedness to the task at hand, and swift, sure decisions built a loyal following. Two signs hung behind Sheridan's desk that reflect his temperament and management style. The first read: "The number of people having any connection with a project must be restricted in an almost vicious manner. Use a small number of good people, lock the others out." The second was a picture of a vulture sitting on a tree limb, with the admonition: "Patience my ass, I'm going to kill something!"

General Sheridan found a powerful ally in Charles B. Salter, a career Tank Automotive Command (TACOM) civilian and experienced manager and combat-vehicle engineer who had served as Sheridan's deputy PM for the M60 tank program. Sheridan again selected Salter to serve as his deputy in the fall of 1975. The two were close personal friends for many years, and as alter egos, became a single voice within the MICV program. Although Sheridan was an armor officer and Salter a civil service employee, infantrymen were well represented by both. Salter, who served as a nineteen-year-old combat infantryman during the Battle of the Bulge in World War II, truly understood the difficulties and requirements of combat and kept the MICV team focused on the infantry soldier. Together, Sheridan and Salter made a powerful and winning team. Although Sheridan (and later Brig. Gens. Philip L. Bolté and Donald P. Whalen) managed the program and made the tough decisions, Salter provided the technical expertise.

Based on previous experience with developmental improvements on the M60 tank family, Sheridan and Salter believed that integrating overall system reliability during vehicle development was imperative. Their vehement advocacy of reliability as part of the design process was based on the premise that failures identified later in the program could not be fixed by "designing in" reliability after the fact. Salter emphasized reliability to such a degree that it became a program watchword, driving both government engineers and civilian contractors to set and achieve reliability goals never before obtained for combat vehicles. Many within the Army believe that their emphasis on reliability was the key factor leading to the Army's acceptance and fielding of the vehicle on schedule. The BFV's stellar performance in Operation Desert Storm five years after Salter's early death was proof positive of his sound advice and hard work.

Shortly after Sheridan's arrival it became apparent that the PM's office (PMO) was not sufficiently staffed or technically prepared to manage such a diverse range of programs and technical approaches. Undeterred, Sheridan methodically began building a strong team that, in addition to government personnel, included people from the major contractors, including FMC, General Electric, Hughes Aircraft Company, Hughes Helicopter Company, Cummins Engines, Honeywell, and Ford Aeroneutronic. Sheridan believed that the PMO and contractor needed to work hand in hand, sharing success and failure equally—a unique perspective for a materiel developer in 1975. This approach has recently been recognized in the acquisition community through the use of Integrated Product Teams (IPT). The purpose of the IPT is to develop a program jointly, with the government and contractors working together to reduce the review and revision time required by both parties.[38]

In order to begin the process of integrating the vehicle and armament programs, ancillary vehicles, and multiple contractors into a cohesive program, Sheridan developed a management information system, a tool he picked up in the late 1950s during his work with the space program. The purpose of the system was to identify task responsibility on a short- and long-term basis. A key feature of the system was the Program Evaluation and Review Technique (PERT) chart, which integrated the schedules of the prime and submanufacturers and the government and identified the critical path to successful program completion. Commitment to a schedule, particularly from system engineers, became the basis for a baseline schedule. Eventually a master schedule, which graphically portrayed all of the vehicle and armament schedules on a single chart, was developed. The chart was updated monthly, and more than 125 copies were disseminated to contractor sites. The master schedules were displayed in "control rooms." Sheridan developed identical control rooms at other government installations and at selected contractor and subcontractor facilities. The identical rooms contained the master schedule as well as other program information. In addition to consolidating program information, it provided continuity for program members moving back and forth between the East and West Coasts.[39]

Later, Sheridan instituted Milestone Control System (MICOS) meetings, a management tool that is still being used by the BFV Systems PM today. Monthly meetings were held for the government personnel involved in the program. The date of the meeting was the same from month to month, and attendance was mandatory. Participants included everyone in the PMO, the Department of the Army System Coordinator, and representatives from the Infantry School. On odd months, contractors were invited. During the meeting, milestone accomplishments were reviewed for every aspect of the program, including delays, changes, and future milestones. At least 150 slides were reviewed during a typical eight-hour meeting. Results of the meeting were summarized in a MICOS newsletter and disseminated to everyone involved in the program.[40]

In November 1975, Sheridan wrote the first of his *MICV System Messages,* which he developed to acquaint user representatives with program progress and used throughout his tenure as PM. Two mobility test rigs, a prototype, and ten engineering-development MICVs were built. Prior to the initiation of developmental testing, the vehicles were run for more than fifty thousand miles and more than forty-five thousand main gun rounds were fired. Although the MICV met the majority of its mobility, firepower, and protection requirements, there were still problems with transmission reliability, main armament considerations, air transportability, stowage, and weight growth.[41]

During the fall and winter of 1975, the Government Accounting Office (GAO) conducted a study of the MICV program. Issues identified included the expansion of the MICV program in view of performance and schedule problems, the incompatibility of the Bushmaster and MICV development schedules, the total cost of the MICV, and the

MICV's contribution toward protecting the main battle tank in view of its apparent vulnerability to enemy weapons. Included in the GAO study was a discussion challenging the need for the MICV. In the August 1974 issue of *Military Review,* Richard Ogorkiewicz, a British authority on military vehicles, challenged the current doctrine requiring mounted attack by mechanized infantry accompanying tanks into battle. According to Ogorkiewicz, "if the MICVs were really to operate alongside battle tanks, they would have to be as well armored as battle tanks." He added "that it is doubtful whether it is necessary or desirable for MICVs to operate as far forward as battle tanks, except in special circumstances. One reason cited is the relatively little firepower it could add to the tanks."[42]

During early February 1976, General Sheridan stopped the MICV's government engineering testing at Aberdeen Proving Ground, Maryland, when it became apparent that the demonstrated reliability in both the automotive and primary weapon subsystems were significantly below requirements. Sheridan later said, "I think that we all realized from the start that the MICV one-man turret design was unfightable, with the commander in the hull behind the driver, the gunner in a one-man turret, and the very crowded crew compartment, more an arms room than a fighting compartment."[43] The "arms room" concept was developed during the late 1960s, when USACDC deleted two members of a ten-man squad and substituted machine guns, light antitank weapons, grenade launchers, flamethrowers and mortars.[44]

Five months later, Secretary of the Army Martin R. Hoffmann challenged the concept of the MICV and the requirement for it. Hoffmann believed that the MICV would prove to be a weapon system "incapable of engaging the perceived threat which would include massed tanks on the battlefield."[45] In 1976, the MICV program objective was to provide accurate suppressive fire while moving, overwatch protective fire for dismounted infantry, and the capability to defeat the BMP at long ranges, rather than a tank-killing capability. Although TRADOC initially envisioned the MICV's mission as providing suppression in support of infantry, it became apparent that increases in antitank capabilities were required with the advent of the highly accurate TOW and Dragon missile systems. There was an immediate dilemma in having to develop an interim and final antiarmor vehicle, as the MICV was still years from being fielded. The M113A1 was proposed as a near-term solution as a mobility base for antitank weapons, with the MICV as a mid- to long-term approach. General DePuy's first priority was the MICV armed with the Bushmaster, and his second priority was developing a MICV equipped with a combination TOW-

Bushmaster armed turret. A TOW missile on the MICV would provide a tank-killing capability without increasing the existing force structure.[46] This increased capability would also compensate for a 16 percent increase in MICV program acquisition costs due to inflation.[47]

A task force was created in the summer of 1976 to determine if the vehicle being developed would meet the Army's future needs. Brig. Gen. Richard X. Larkin, a West Point graduate, combat-experienced infantry officer, and assistant Commander of the 4th Infantry Division (Mechanized) at Fort Carson, Colorado, headed the task force.[48]

General Sheridan welcomed it as an opportunity to voice his design and fightability concerns, present alternate designs based on operational testing, and revalidate the infantry/cavalry requirement. The team assembled at Fort Benning, where a prototype MICV was available. To resolve the conflict between infantry support versus tank killing capability, Larkin recommended the MICV be armed with the 25mm Bushmaster cannon and the M113A1 with an elevating TOW launcher.[49]

Larkin was also concerned about the MICV's vulnerability. However, any effort to add additional armor would further degrade the vehicle's already questionable swim capability. During his final briefing to the Army Chief of Staff, Larkin depicted the MICV's ability to swim water obstacles by showing it slung under a helicopter.[50] Although Sheridan supported dropping the swim requirement altogether, the user community still felt it was needed.

The recommendations Larkin presented to Chief of Staff Gen. Bernard Rogers in October 1976 included the development of two vehicles on a common chassis: one for the infantry and one for the cavalry. Larkin also recommended adopting the two-man TOW-Bushmaster Armored Turret (TBAT-2), which would increase the space for 25mm ready ammunition, accommodate an external TOW missile launcher, and put the commander back in the turret where he belonged.[51]

Sheridan later wrote: "The Larkin Task Force gave us the opportunity we were looking for to correct the design deficiencies and make the vehicle fightable . . . the real step toward success was the approval to go with a two-man turret for the commander and gunner, leaving only the driver in the forward hull; at the same time we redesigned the crew compartment so that the six soldiers in it could fight from the vehicle as comfortably as possible. All of this combined to give us a soldier friendly and soldier fightable design that would ultimately spell success for the vehicle and the program."[52]

Almost a month later, Secretary Hoffmann approved the task force's recommendations. There would be a nine-man squad, a two-man turret, 14mm armor protection, a swim capability with full barrier, a 25mm

Bushmaster cannon, an armored, nonelevated TOW missile launcher with two tubes, and one basic vehicle for infantry and cavalry use. The infantry/cavalry vehicles were designed to be identical externally; internal differences included crew size (nine troops in an infantry fighting vehicle, and five in the cavalry version), a requirement for a motorcycle in the cavalry vehicle (which was later eliminated), and the amount of armament, food, and ammunition. The basis of issue for the infantry vehicle was established as four per platoon, thirteen per company, and forty-one per battalion.[53]

A contract was awarded to FMC the following week for full-scale development of an infantry/cavalry fighting vehicle (IFV/CFV) meeting the specifications described above. Within three months, the 20mm MICV program was terminated. The developmental effort awarded to FMC consisted of two phases. The first phase involved development of the two-man turret mounting the TOW and 25mm cannon, and redesign of the hull. The second phase, begun the following year, would complete the developmental effort.[54]

In January 1977, following two general officer reviews, the Senate Armed Services Committee expressed doubts about the survivability of a combined infantry/cavalry vehicle partnered with the XM1 tank. Sheridan became increasingly beleaguered in the summer of 1977, briefing a steady stream of staff visitors following this pronouncement. One visitor was Justus "Judd" White, a principal staff member from the House Armed Services Committee. Sheridan began his briefing, which showed the BMP-1 as a primary threat, and found White amused. White announced that the primary threat to the MICV was not the BMP-1, but the U.S. Congress. White, a graduate of the University of Mississippi, had served five years in the infantry and had no doubts at all regarding the importance of the IFV/CFV concept. Sheridan found White to be an invaluable, if not timely, ally. In December, President Jimmy Carter canceled the program when procurement funds were deleted from the FY 1979 budget, and a third general officer review was initiated.[55]

A study group led by Brig. Gen. Fred K. Mahaffey was also created. General Mahaffey, then serving as assistant commandant of the Infantry School, supported the Crizer Task Force, concluding in August 1978 that an eleven-man squad was required to perform mechanized infantry tasks. He also concluded that there was no better alternative to the IFV/CFV program based on operational effectiveness and cost.[56]

The Crizer Task Force, which convened in October 1977, was headed by Maj. Gen. Pat W. Crizer and reported to the Vice Chief of Staff. Its task was to reevaluate the specific requirements for the infantry/cavalry vehicle and determine if there were more survivable vehicle alternatives. This seven-month review included U.S. and foreign vehicles. Crizer considered and rejected the German Marder because it had an unstabilized cannon in a one-man turret with the Commander in the hull, lacked a swim capability, could not be transported in a C-141, and cost more than $500,000. The Soviet-made BMP-1 was rejected due to its small size, one-man turret, and unstabilized cannon. An IFV/CFV based on the XM1 chassis was also considered and rejected because it would cost in excess of $900,000. The Crizer Task Force concluded in its recommendations to Congress in April 1978 that a more survivable vehicle would not increase effectiveness and would be offset by the vehicle's high cost. It also recommended that two additional infantrymen be added because the primary purpose of the vehicle was to "facilitate the dismounted employment of infantrymen."[57]

At the time President Carter canceled the program, Sheridan and the Army as a whole were forbidden by the Secretary of the Army from lobbying Congress for its reinstatement. During this period, White recruited the recently retired General DePuy to assist with lobbying efforts. Ray Tower, the president of FMC, also began lobbying efforts, meeting with congressional staffers, House Armed Service Committee members, and other prominent members of Congress. Their efforts were successful, and on 21 February 1978 the Secretary of Defense announced that the program would maintain options for FY 1980 procurement while the Mahaffey and Crizer studies were being conducted. The program was eventually restored, but there was an unusual twist: a statutory provision, MICV Section 206b, which provided that the "Secretary of the Army shall structure the development program for the MICV to provide for initiation of production of such vehicle not later than May 31, 1981." The program now faced a congressionally imposed deadline for initial production deliveries.[58] It became Sheridan's "pact with the devil" when he agreed to the May 1981 production deadline to save the program, and White inserted the appropriate language for congressional approval.

In January 1979, Sheridan was succeeded by Brig. Gen. Philip L. Bolté. Bolté, who graduated a year ahead of Sheridan at West Point, served on the Armor Engineer Board in the early 1960s; as assistant project manager for the XM1, charged with developing main armament for the tank in the late 1970s; and, just prior to his transfer to the IFV/CFV program, as Deputy Commander of the Army's Test and Evaluation Command at Aberdeen Proving Ground, Maryland. The fact that he had been awarded a Silver Star and Combat Infantryman's Badge in both Korea and Vietnam helped Bolté, an armor officer, when dealing with the Infantry School. He was known for his easygoing and personable nature, as well as for a keen intellect.

Brig. Gen. Philip L. Bolté, Bradley Fighting Vehicle Program Manager, with his father, Gen. Charles L. Bolté (Ret.), former Army Vice Chief of Staff. U.S. Army. Philip L. Bolté.

Shortly after Bolté's arrival, FMC Corporation announced a significant cost increase from $350,000 per vehicle to more than $500,000 due in part to the two-man turret design. The other cost driver was a requirement to use unrealistic inflation figures, which were half the actual inflation rate. This expectedly came to the attention of Congress, and Bolté appeared several times before both the House and Senate Armed Service Committees.[59]

The selection in January 1979 of the Hughes Helicopter Company's XM241 externally powered 25mm cannon over the Ford Aeroneutronic Corporation's XM242 self-powered 25mm cannon as the main armament for the IFV/CFV marked the end of an interesting and heated competition by the two Southern California companies. Originally, the army had selected Ford to develop an Americanized version of the Swiss Oerlikon KBA-B02 25mm self-powered cannon along with a family of Americanized 25mm ammunition. TRW Corporation designed the self-powered cannon in the late 1960s and eventually sold the rights to Oerlikon to develop the cannon to European standards. Ford bought the patent rights to manufacturer the cannon in the United States and established a test firing facility in the canyons east of San Juan Capistrano in Southern California. The requirement to Americanize the ammunition was necessary because of U.S. safety standards, required performance improvements, and differences in European and U.S. manufacturing techniques.

Early in this development (1975), the Hughes Helicopter Company (HHC) approached General Sammet at AMC with an offer he could not refuse. HHC had, in its own "Skunk Works," developed a 25mm externally powered cannon that it called the "chain gun." Hughes's management proposed to Sammet that it be allowed to compete with Ford's self-powered 25mm in a "winner take all" competition. The army advised HHC that it had no money to fund development of a second cannon, but HHC was so confident in its chain gun that it offered to participate in a cannon competition at no cost to the government. The army accepted HHC's offer and in February 1976 awarded the company an army contract for its participation in the competition in the amount of one dollar, thus making the process completely legal. The contract called for delivery of fifteen externally powered XM241 chain guns for government competitive testing. Eighteen months later HHC was ready for competitive testing. After a year of testing both cannons, the army conducted a two-month "shoot off" between the two companies using Ford's Americanized ammunition. While HHC won the cannon competition, Ford did not walk away empty-handed, having retained the ammunition portion of the program.[60]

FMC continued on course with the development of the prototype vehicles. Eight were delivered between December 1978 and February 1979. Formal contractor testing began on the first three vehicles, followed by formal government testing of three vehicles in June 1979. Due to budgetary restrictions, the prototype vehicles were limited to eight, which resulted in competing requirements for both developmental and operational testing, as well as other demands. This required manipulation of test schedules to meet the January 1980 milestone review and type classification actions. Although normally somewhat flexible in nature, these dates were now inviolable as a result of the congressionally mandated May 1981 initial production date. To accommodate this time frame, Bolté convinced the commanding general of the Operational Test and Evaluation Agency to compress operational testing so that the contractor could fix problems identified in contractor testing. Bolté demonstrated his personal flexibility when, again faced with a shortage of test vehicles, he borrowed the MICV monument located in front of Infantry Hall at Fort Benning, shipped it to Aberdeen Proving Ground, and blew up a mine underneath the vehicle. The monument was subsequently repaired and returned to Fort Benning.[61]

The program was very nearly stopped again during testing as a result of toxic fumes testing in conjunction with the Human Engineering Laboratory at Aberdeen Proving Ground. The test officer, contrary to common sense, insisted on measuring the possibility of toxic fumes when the entire basic load of ammunition was fired with the vehicle stationary and without any wind. Predictably, fumes from the weapon firing, extracted from the vehicle by the blower, were drawn back inside. Bolté in turn recruited a senior medical officer from nearby Fort Detrick, Maryland, who got the program back on track by measuring toxic fumes during operational testing. The doctor concluded that the fumes the soldier would be exposed to within the vehicle during normal operation were less than what he would get from smoking a cigarette.[62]

Prototype qualification testing was conducted from June 1979 to June 1980. The purpose of the testing was to verify that the IFV/CFV designs were ready for the milestone review and type classification. During this phase, which was conducted at Fort Carson, the prototype vehicles were paired with M60A1 tanks. It quickly became apparent that the tank's mobility was no match for the more mobile IFV and CFV.[63]

The 25mm cartridges, fuse, and link, were type classified Standard A on 22 November 1979, clearing them for introduction into the Army's inventory.[64] Then, in January 1980, the Defense Systems Acquisition Review Council (DSARC) met for barely more than an hour and approved type classification and production of the IFV/CFV, the externally powered 25mm cannon, and the firing-port weapon.[65] Bolté paved the way for this quick action by visiting all of the DSARC's working-level staff members prior to the collective meeting to iron out differences beforehand.

The significance of the DSARC's decision was twofold. First, it marked the first time in the Army's developmental history that a combat vehicle of any type had been accepted and approved for production by the Office of the Secretary of Defense (OSD) the first time around. This was due in large part to the fact the program greatly exceeded its reliability goals during testing—which was in no small way attributable to Salter's dedicated efforts. Secondly, it marked the end of the research, development, and engineering phase of the program and the beginning of vehicle production.[66]

A low-rate initial production contract for one hundred vehicles to be followed by four hundred more was awarded to FMC Corporation in February. Bolté retired in June and was replaced by Brig. Gen. Donald Whalen. Whalen was born at Fort Bragg, North Carolina, in 1935 and graduated from West Point in 1957 as a field artillery officer. He saw action as a battery commander with the 1st Brigade, 101st Airborne Division, and was awarded the Silver Star and a Purple Heart. He also served as the S3 of a heavy artillery battalion in the Central Highlands. He graduated from the CGSC in 1968, finishing first in the class—just ahead of Colin Powell, who later became chairman of the JCS. Prior to arriving at TACOM he had been a staff officer in OCRD, commanded a field artillery battalion in the 1st Infantry Division, served as PM of the Lance Missile System for three years, and commanded the army's largest laboratory, the Large Caliber Lab at Picatinny Arsenal.

During his tenure as PM from July 1980 through September 1983, Whalen focused on the system's transition from initial to full-rate pro-

Brig. Gen. Donald Whalen, General Officer Management Office, Office of the Chief of Staff, Army.

duction, providing logistical support, activating the training base, fielding the Bradley in FORSCOM's 2d Armored Division at Fort Hood and in USAREUR's 3d Infantry Division in Germany, cost containment, and initiating product improvement programs to support future vehicle growth.

Whalen recalls that

> By 1980, the anticipated production cost of a Bradley had grown to many times more than earlier estimates, and senior army leadership was seriously concerned at the potential impact of this growth on the program's continuation. Their strongest charge to the incoming Program Manager was to get these costs contained. Special ASARC and DSARC level reviews occurred during the summer to address this issue. This concern led to extremely tough and sustained negotiations with FMC over the price of the initial production contract for 100 vehicles. Believing that this contract would set the baseline for all future production, the PMO and TACOM were determined to reach a settlement that protected the government's interests and insured that the program would continue. While these negotiations eventually reached a successful conclusion, vehicle costs nevertheless remained higher than had earlier been hoped. Only gradually over succeeding years has there been widespread realization, as costs of the Abrams, Bradley and other modern systems remained both consistent among themselves and higher than originally expected, that the superb capabilities of this new generation of combat vehicles could not be achieved at the price of earlier and far less capable systems.[67]

In October 1980 the OSD approved the start of a TOW II production program and on 8 May 1981, three weeks prior to the congressional mandate, the first production vehicle was delivered to the army.

In October 1981 the M2 infantry and M3 cavalry fighting vehicles were named the Bradley Fighting Vehicle in honor of Gen. Omar N. Bradley, who died six months earlier. In 1977, General Sheridan wrote a letter to General Bradley requesting permission to name the vehicle in his honor. He did not receive a response, and was subsequently asked by the Chief of Staff, General Weyand, to "cease and desist until the program was a success." However, there was duplicity in Sheridan's motive: "I didn't figure any one in the World would cancel the program while Bradley was still alive."[68] During this same period, Sheridan also designed the MICV logo: a MICV superimposed over a crossed cavalry saber and infantry rifle. Later, when the vehicle became the IFV/CFV, the logo was updated to reflect the new vehicle. When the IFV/CFV was designated the Bradley Fighting Vehicle, the vehicle's image was removed and five stars were added to the logo, representing General Bradley. That logo is still in use.[69]

The formal naming ceremony, which took place at Fort Myer, Virginia, on 20 October, was conducted with traditional military flair. A Bradley Fighting Vehicle, complete with a nine-man rifle squad from Fort Benning formed up next to it, was positioned in front of the reviewing stand. Mrs. Bradley turned over the christening of the vehicle to Army Chief of Staff Gen. Edward C. Meyer, after failing to break the champagne bottle with several swings. The vehicle moved into full-scale production shortly after the ceremony.[70]

Sheridan, by then a major general and in his early fifties, attended the ceremony. As he watched the festivities he finally felt that his obligation to Congress had been fulfilled. Although he had left the program two years before, he remained in close touch with the PMO, both officially and unofficially, through Salter, Bolté, and Whalen. For Sheridan, then two years away from retirement, the development of the BFV and its acceptance by the Army culminated a career of combat vehicle development that began as a first lieutenant at Aberdeen Proving Ground in 1955.

In 1983 the M2 Bradley became the workhorse vehicle of U.S. mechanized infantry forces and the M3 picked up the reconnaissance and security mission in armored units, replacing the M113 APC as fast as production allowed.

Later in the decade, as a result of a series of live-fire vulnerability tests, the army began adding a number of survivability enhancements. These included steel appliqué armor, exterior attachment points for adding reactive or passive armor, interior spall liners for the crew compartment, and relocating ammunition to less vulnerable areas. This high-survivability vehicle was classified M2A2/M3A2, and included an increase in engine power to accommodate the weight increase.

On 26 February 1991 a total of 2,200 Bradleys were in the Persian Gulf area. A total of 1,730 were deployed with mechanized infantry and cavalry units, and 470 were held in reserve. During the brief war, Bradleys performed well according to after-action reports, users, and vehicle mechanics. The reports and user comments noted that the vehicles exhibited good reliability (more than 90 percent readiness rates), lethality, mobility, and range. According to the crews, the Bradley, especially the A2 model with its enhanced steel plates and spall liners for the crew compartment, provided excellent protection against small arms and artillery fires. Unlike the M113 APC and the M109 self-propelled 155mm howitzer, Bradleys had no trouble keeping pace with the Abrams tank.[71] However, a somber gap in the application of ground mobility doctrine was still evident because in some instances troops dismounted before it was necessary, leaving their Bradleys behind to provide covering fire.[72]

After proving itself in Operation Desert Storm, the Bradley system continues to be upgraded and derivatives are still being added to the Army's force structure. The upgrades include digital and core electronic architecture to meet Army requirements for Force XXI. By the end of the last production run in 1995 6,724 Bradleys had been produced and it became the armored vehicle of choice for new missions, such as peacekeeping operations in Somalia, Haiti, and Bosnia.

Why has the BFV been such a lightning rod over the years? Part of the answer lies in the fact that it is a hybrid vehicle whose mission has evolved as doctrine and requirements changed. There has been long-standing confusion or misunderstanding what the Bradley is and is not designed to do. It is *not* an M113 APC, designed as a battlefield taxi to transport troops or serve as an arms room, although it does transport infantrymen and their weapons. Nor is it a tank. It lacks the tank's heavy armor, although it has a tank-killing capability and is a partner to the tank in combat. Finally, it is *not* a boat, although for many years it was required to swim.[73] Those characteristics are secondary to the Bradley's primary infantry mission, that of mounted or dismounted mobile warfare. It is designed to provide speed of movement and reaction on the battlefield, as well as suppressive fire and antitank capabilities. Considering *those* requirements, the BFV is the finest vehicle of its kind in the world today.

NOTES

I am greatly indebted to Maj. Gen. (Ret.) Stan Sheridan, Brig. Gen. (Ret.) Philip Bolté, and Brig. Gen. (Ret.) Donald Whalen for being extraordinarily supportive of my efforts and for providing their invaluable perspectives on the development of the BFV. These major players provided personal interviews, critiqued the manuscript, provided counsel and encouragement, and opened their papers for documentation of this project. I have also received invaluable guidance and encouragement from Gen. (Ret.) Donn Starry, Mr. Judd White, Lt. Gen. (Ret.) George Sammet, Mr. Brent Sherman, Ms. Shirley Saunders, Mr. Ben DeMarko, Mr. Fred Perry; Mr. John Romjue,

Mr. Edward Rupnick, and Mr. Arthur Durante of the U.S. Army Infantry School. Others who have contributed greatly include former Bradley PMs and their staffs, the FMC Corporation, Col. (Ret.) Roy Dunnaway, Col. (Ret.) Jerry Houston, Col. (Ret.) Robert Noce, Col. (Ret.) Richard Sawyer, Mr. Ernie Leonard, Mr. Mike Cleary, and countless others associated with the development of the Bradley.

For an excellent academic study, see William Blair Haworth Jr., "The Bradley and How It Got that Way: Mechanized Infantry Organization and Equipment in the United States Army," (Ph.D. diss., Duke University, 1995). Also see Richard P. Hunnicutt's forthcoming technological history of the Bradley systems, to be published by Presidio Press.

NOTES

1. General Accounting Office (GAO), "Operation Desert Storm: Early Performance Assessment of Bradley and Abrams," GAO/NSIAD-92-94 (Washington, January 1992), pp. 2-19.

2. Bruce R. Pirnie, "From Half-Track to Bradley: Evolution of the Infantry Fighting Vehicle," Analysis Branch, CMH, Washington, D.C., p. 8.

3. TACOM, "M113 World Class Systems Data Book," AMCPM M113/M60, p. 2.

4. Pirnie, "From Half-Track to Bradley," p. 7.

5. TACOM, "Annual Historical Summary for 1970," PM MICV, p. 4.

6. Ibid, 1964, p. 234.

7. "Carco's New Personnel Carrier Seen In Test," *Record-Chronicle*, 23 June 1965, p. 3. Also Edward J. Rupnick, former Pacific Car and Foundry engineer, interview with author, 1996.

8. Brochure: "Transport Mobility," Pacific Car and Foundry Company, n.d., pp. 1-10.

9. "Carco's New Personnel Carrier," p. 3.

10. U.S. Army Weapons Command, "Annual Historical Summary for 1967," pp. 211-12.

11. TACOM, "Annual Historical Summary for 1 July 1973–30 June 1974," PM MICV, pp. 1-2.

12. Ibid., p. 4.

13. Pirnie, "From Half-Track to Bradley," p. 11.

14. Ibid., p. 11.

15. Lt. Gen. (Ret.) George Sammet Jr. to Diane L. Urbina, 21 May 1996. Letter in author's possession.

16. Pirnie, "From Half-Track to Bradley," pp. 11-12.

17. Ibid., p. 10.

18. Paul H. Herbert, *Deciding What Has to Be Done: General William E. DePuy and the 1976 Edition of FM 100-5, Operations,* Leavenworth Papers, No. 16, (Ft. Leavenworth, Kans.: CSI, 1988), p. 5.

19. Ibid., pp. 21-22.

20. Herbert, *Deciding What Has to Be Done*, p. 12; Gen. William E. DePuy, Keynote Address, TRADOC Leadership Conference, 22 May 1974, reprinted in DePuy, *Selected Papers*, p. 120; and Gen. (Ret.) Donn Starry, interview with author, 1996.

21. Herbert, *Deciding What Has to Be Done*, pp. 25, 30.

22. DePuy, "Implications of the Middle East War on U.S. Army Tactics, Doctrine and Systems," reprinted in DePuy, *Selected Papers*, pp. 75-111.

23. Herbert, *Deciding What Has to Be Done*, p. 37.

24. Ibid., p. 27.

25. Ibid, pp. 28-29.

26. Gen. William E. DePuy, TRADOC commander, to R.W. Komer, 24 Apr. 1975, reprinted in DePuy, *Selected Papers*, pp. 157-58.

27. "Soldiers and Civilians of the U.S. Army, Army Chief of Staff Office, 1986," reprint provided by the Infantry School, Fort Benning, Ga.

28. Campbell and Real, *FMC Growing Orbit: The Story of FMC Corporation* (Chicago: FMC Corporation, 1992), pp. 108-10.

29. Herbert, *Deciding What Has to Be Done*, p. 37; and Maj. Gen. (Ret.) Stan R. Sheridan, interview with author, 1996.

30. Starry interview.

31. TACOM, "Historical Summary for 1 July 1973–30 June 1974," p. 17.

32. Ibid., p. 13.

33. *MICV Systems Message,* AMCPM MCV, 5 Nov. 1975.

34. Gen. William E. DePuy, TRADOC commander, to Maj. Gen. Gordon Sumner Jr., 29 Apr. 1975, reprinted in DePuy, *Selected Papers*, pp. 161-62.

35. Sammet to author, 21 May 1996.

36. TACOM, "Historical Summary for 1976," pp. 4-5.

37. Sheridan interview.

38. Ibid.

39. TACOM, "Historical Summary for 1976," p. 29; and Col. (Ret.) Jerry B. Houston, former PM MICV staff member, interview with author, 1996.

40. Ibid.

41. *MICV Systems Message.*

42. TACOM, "Historical Summary for 1976," p. 48.

43. Sheridan interview.

44. Pirnie, "From Half-Track to Bradley," p. 10.

45. TACOM, "Annual Historical Summary for 1977," PM MICV, p. 7.

46. TRADOC, "Annual Historical Summary for 1975," p. 172.

47. TACOM, "Historical Summary for 1976, Appendix A," p. 2.

48. TACOM, "Historical Summary for 1977," p. 7.

49. Pirnie, "From Half-Track to Bradley," p. 16.

50. Ibid., p. 12.

51. TACOM, "Historical Summary for 1977," pp. 7-8.

52. Sheridan to author, 13 May 1996.

53. TACOM, "Historical Summary for 1977," p. 8.

54. Ibid.

55. Justus "Judd" P. White, former principal staff member of the House Armed Services Committee, interview with author, 1996.

56. Pirnie, "From Half-Track to Bradley," pp. 18-19.

57. Ibid., p. 19.

58. White interview.

59. Philip L. Bolté, former PM BFVS, to author, 24 May 1996. Letter in author's possession.

60. TACOM, "Historical Summary for 1976," p. 13.

61. Bolté to author.

62. Ibid.

63. TACOM, "Annual Historical Summary for 1979," PM BFV, pp. 7, 16-17.

64. TACOM, "Annual Historical Summary for 1980," PM BFV, p. 12.

65. Ibid., p. 23.

66. Bolté to author.

67. Donald Whalen to author, 29 Dec. 1997. Letter in author's possession.

68. Sheridan to author.

69. Ibid.

70. Houston interview.

71. Norman Friedman, *Desert Victory: The War for Kuwait* (Annapolis, Md.: Naval Institute Press, 1991), p. 393.

72. GAO, "Operation Desert Storm: Early Performance Assessment of Bradley and Abrams," p. 2-3. However, some problems were identified, such as leaking radiators, unreliable heaters, and misdirected exhaust. In addition, crews recommended a laser range finder, improved sight magnification and resolution, and an increase in reverse speed to match the Abrams's twenty mph.

73. Mr. Arthur Durante, U.S. Army Infantry School, Fort Benning, Ga., interview with author.

13

The Abrams Tank System

Robert J. Sunell

At the conclusion of the ground war in the Persian Gulf on 26 February 1991, the 3,113 Abrams tanks in the region maintained a readiness rate of 90 percent or higher. Through the course of the hundred-hour ground war, it was quite evident that the Abrams was exhibiting outstanding reliability, lethality, mobility, and survivability. Several Abrams M1A1s reported minimal frontal damage despite hits by 125mm smoothbore rounds fired from Iraqi T-72s. Not a single Abrams was destroyed or penetrated by the enemy during the war. Army observers and tank crews alike were impressed with the power and accuracy of the Abrams's 120mm smoothbore gun. In addition, U.S. tankers raved about the power and performance of the turbine engine.[1]

How did the U.S. Army come to develop such an imposing combat fighting vehicle, considered today one of the best tanks in the world?

It is impossible to cover all the details of the Abrams tank story in a single chapter. To cover it from concept formulation through production would require, at a minimum, an entire book. I therefore have elected to cover only those events that were major turning points during the history of the Abrams tank program.

The Abrams story is more than just a story about a tank; it is also the story of the people and organizations that made that tank a reality. Numerous hands and minds helped shape the design of this great fighting machine, so I must limit myself to a discussion of only those key personnel who were critical to the Abrams project.

The M1 tank project grew out of the failure of two other attempts to build a replacement for the U.S. Patton tank series. The first was the

The MBT70. Patton Museum.

joint United States-West German MBT70 project launched by Secretary of Defense Robert S. McNamara. No sooner had Congress provided funding for this joint venture than a series of developmental problems began to delay it. By the end of 1969, the MBT70 was mired in international management dilemmas, conflicting national priorities, persistent cost overruns, and technical problems with the turret-mounted 152mm gun-missile system, novel automatic loader, variable compression ratio diesel engine, and enhanced adjustable hydro-pneumatic suspension system.[2]

In January 1970, when the U.S.-German alliance disintegrated, a Congress decided to pursue a unilateral austere version, the XM803, for the United States. The following year, however, both the Senate and House Armed Services Committees found the XM803 too costly and sophisticated to justify further development. Many committee members believed that tanks were nearing the end of their combat usefulness, their extinction threatened by a new generation of antitank guided missiles (ATGM).[3] Senator Thomas F. Eagleton best represented the congressional mood when he said that tanks were too vulnerable and had grown less valuable in the United States's overall defense effort.[4] Shortly after Congress terminated the Cheyenne helicopter program due to cost overruns, Army Chief of Staff William Westmoreland unilaterally eliminated the XM803 program. The West Germans, meanwhile, pursued development of what became the Leopard 2 tank.

The Army defended its request for the tank as its principal land-based weapon system by arguing that tanks were to be used not only against enemy tanks but in various mutually supporting roles in combination with mechanized infantry, artillery, and tactical airpower. Despite the Army's arguments, many congressional leaders were adamant that tanks would no longer be able to survive on a European battlefield. In the wake of the MBT70/XM803 imbroglio, Congress encouraged the Army to continue to develop the M60A2, with its dual gun-missile launcher system, stating the that XM803 offered no advantage over the M60A1E2 except a slight edge in cruising range and speed.[5]

In December 1971 a joint House-Senate conference committee issued a report on the Department of Defense (DOD) appropriations bill, noting: "The committee continues to feel that the MBT70/XM803 is unnecessarily complex, excessively sophisticated and too expensive, and that the Army has failed to satisfy the recommendation of the committee report on the fiscal 1970 bill. For these reasons the Committee has recommended that all funds for the MBT70/XM803 be deleted from the budget and the program be terminated."[6]

A dissenting view came from Gen. James H. Polk, former USAREUR Commander in Chief, who argued that the defense legislators were misguided in authorizing huge amounts for missiles and aircraft to support the armor-infantry team at the expense of developing a better tank for the Army's close-combat forces. He strongly believed U.S. tankers deserved the best tank, a tank capable of dominating any future battlefield. "It is clear that the lessons of history are going unheeded as we drift into a runner-up spot in the quality of our armor," he noted. Polk was very critical of the M60 modernization program, citing the M60's high target silhouette compared to Warsaw Pact tanks. He was also critical of the M60A2's gun-missile launcher. One serious problem with the tank's Shillelagh guided missile was that it was a line-of-sight weapon that required the gunner to keep the target in the crosshairs of his gun sight as the missile flew. The gunner's view could be degraded by trees, smoke, heavy rain, blowing sand, hills, and darkness—all factors that could cause him to miss his intended victim. Polk had supported the XM803 program, believing it to be innovative because of the tank's potential to fire and hit targets while on the move and because increased long-range accuracy would allow enemy vehicles to be destroyed at extended ranges. The XM803's spaced armor, self-sealing fuel tanks, and blowout vents for ammunition storage areas were other innovations Polk supported, ideas that were later adopted in the XM1 system.[7]

Nevertheless, the Cold War M60 modernization program continued. The Soviet Union's massive tank production capacity and the resultant Warsaw Pact threat to western Europe had been the reason Congress earlier supported the joint United States-West German effort to build a modern battle tank. This threat had continued to be a major concern as the Warsaw Pact's numerical superiority continued to grow. The M60, tank for tank, was almost on a par with the existing generation of Soviet-fielded tanks, but with the next Soviet tank improvement, the M60 was in danger of falling behind. This prospect opened the door for the new U.S. tank program, which began as the XM1 and eventually became the Abrams tank system.

The first step was the Armed Services Committee's addition of $20 million to the army's Research Development Test and Evaluation (RDT&E) budget for "the purpose of initiating a prototype program to build a limited number of tanks of two designs for test and evaluation as rapidly as possible."[8]

Vietnam War costs had virtually stalled heavy force modernization in the United States, allowing the Soviet Union to increase its lead at what some observers considered to be at an alarming rate. The Army was forced to react. On 20 December 1971 the USACDC commander was directed to establish a special task force to determine requirements for a new tank. On 15 January 1972, he established the Main Battle Tank Task Force (MBTTF). The task force was to be based at Fort Knox, and the Armor Center commander, Maj. Gen. William R. Desobry, was designated its chairman. The commanding officer of the USACDC Armor Agency at Fort Knox was made the deputy chairman. Also assigned to Fort Knox was a team of twenty-two officers and four civilians from USACDC, CONARC, AMC, and the army comptroller's office.[9] The task force had a clear mandate: "Lay out a new MBT program for the Army, write the critical performance specification for the MBT, write the documents essential to the process, and be prepared to sell the Chief of Staff, the Secretary of the Army, the Defense Department bureaucracy, all of this in six months."[10] This would have been a very ambitious task even under the best of circumstances.

In mid-March 1972, two months into the study, the Army Chief of Staff directed that the new MBT should be fielded in six years, a significant reduction from the normal ten-year development cycle. This reduction was based on the assumption that previous work on the canceled MBT70/XM803 designs would be relevant to the new program, thus saving considerable development time. However, some critics—including some members of the task force—questioned this assumption.

By early August 1972 the MBTTF had completed its work, outlining the new tank on paper. It was expected to weigh about fifty tons and employ a crew of four, be powered by either a conventional diesel or an unconventional turbine engine, and mount a 105mm main gun, a 50-caliber commander's machine gun, and a coaxial Bushmaster cannon of 20mm or 30mm. Top speed was to be forty-five miles per hour. This was the task force's priority ranking:

1. Crew survivability
2. Surveillance and target acquisition performance
3. First- and subsequent-round hit probability
4. Time to acquire and hit a target
5. Cross-country mobility
6. Complementary armament integration
7. Equipment survivability
8. Environmental impact
9. Silhouette
10. Acceleration and deceleration
11. Ammunition storage
12. Human factors
13. Ease of production
14. Range
15. Speed
16. Diagnostic aids
17. Growth potential
18. Support equipment
19. Transportability[11]

When the task force turned its work over to the Department of the Army (DA) it became the responsibility of the Army staff to convince the Washington bureaucracy of the MBT's viability. One of the leaders in this effort was Brig. Gen. Robert J. Baer, who was assigned to the Army staff's Office of Research and Development. Baer was subsequently appointed to serve as the new tank project's program manager (PM), and would carry the program through the decision-making process.

A disagreement over the vehicle's weight then surfaced as an emotional issue. General Desobry's goal was agility, and to achieve this he was—despite the task force's designation of survivability as the number one priority—adamant that the tank's weight should not exceed fifty-two tons. (How he arrived at this number is not now clear.) However, Lt. Gen. William DePuy, the Army's Assistant Vice Chief of Staff, argued for a higher weight limit. He briefed the decision makers, arguing that the application of a new British armor, called Chobham, could increase survivability but at the expense of increased vehicle weight.

Chobham armor's layers of ceramic, steel, and titanium plates, laminated between layers of ballistic nylon would resist penetration by both kinetic- and chemical-energy ammunition.

Although U.S. tank designers were aware of Chobham armor development, it had not been incorporated in either the MBT70 or XM803 due to restrictions on the transfer of technology. Chobham armor had been a closely held secret; few knew what it was or what level of protection it provided. While the Army's Ballistic Research Laboratory (BRL) was experimenting with Chobham armor, Gen. Creighton Abrams, the Chief of Staff, made the decision to increase the tank's weight to fifty-eight tons to accommodate what he called "Special Armor." Abrams reportedly said that without this special armor, "The new tank would not be much better than the old M60—warmed over potatoes from last year and the year before."[12]

This was only partly correct. Many key components employed in the M1 family were a result of the efforts of Maj. Gen. Stan Sheridan, who served as the M60 tank PM between 1970 and 1974. One of these components was the thermal-imaging system that increased the tank's night fighting capability.

General Abrams's decision to adopt Chobham armor was right on target. Proponents of the new tank used the M60 and its product improvements as the baseline when evaluating the need for a new tank. Special armor proved to be the major difference between the two. And the added weight penalty was overcome when the turbine engine and suspension components were developed to provide the cross-country agility General Desobry had demanded.

General Baer left the Pentagon in September 1972 and established his XM1 program management office (PMO) near TACOM headquarters in Warren, Michigan. He was joined there by most of his initial design team, including a number of veterans of the MBT70/XM803 program.

The nature of XM1 development would be very unusual. Normally the Ordnance Department produced combat equipment and provided it to the user for test and evaluation. Under the new process, the PMO provided user requirements to the contractors, who then built prototypes. The government assisted, but essentially it was a hands-off process until a contractor was selected for full-scale engineering development (FSED).

The request for proposals was prepared by Baer's office and released on 23 January 1973. The responding contractors were General Motors and Chrysler. They were given requirements with broad performance bands, then left to develop an XM1 tank design. An exception was the special armor. The contractors were given the available government information on its composition and capabilities. It was the contractor's task to determine its best application, production plans, and, eventually, performance parameters.

General Abrams gave Baer and Maj. Gen. Donn Starry, Desobry's successor as commander of the Armor Center, their marching orders. He held them both responsible for producing the tank within the cost thresholds established for the program: $507,000 in FY 1973 dollars. He demanded a simple, reliable, fightable tank. General Baer later said: "The best guidance I received at the very start was from General Abrams. He said we were not out to build the best of everything here—what we are looking for was the best system. We could not afford the best of everything, as he put it. And the thing he emphasized was that we build a tank which the soldier would be confident in, comfortable with, and one which he felt could accomplish the mission."[13]

Armed with guidance from the chief—and knowing full well a watchful Congress and a host of yet unidentified critics waited in the wings—General Baer as the developer and General Starry as the user representative began crafting the XM1 tank. From the outset they promised unity—that no one would separate them. Indeed, Starry assigned noncommissioned officers from the Armor School to Baer's PMO so the experienced user voice would be heard during development.

Beginning in June 1973 the government launched the first phase of the XM1 program's validation competition between the two contractors. This phase required each contractor to deliver a full-up integrated tank, an automotive test rig, and a ballistic hull and turret to Aberdeen Proving Ground for testing. After completion of testing, the next phase—FSED by the single-source contractor—could begin. This would be followed by the production phase. However, the journey was to be long and difficult.

An oft-overlooked chapter in the history of the Abrams tank was the revalidation of the MBT requirement in the wake of the October 1973 Arab-Israeli War. General Abrams dispatched Baer and Starry to Israel to learn firsthand the tactical and equipment lessons learned by the Israeli Defense Force's (IDF) Armour Corps.

The MBT requirements documentation specified a 105mm gun. This reflected the user's enormous dissatisfaction with the gun-missile launcher on the M551 Sheridan and M60A2, and the availability of a new British-designed rifled 105mm gun. In the meantime, the OSD was deep in negotiations to buy a German-designed 120mm smoothbore cannon featuring caseless ammunition. General Abrams's first question to Baer and Starry was, "Do we need a larger gun based on tank-on-tank experience in Israel, and if we do, will it delay the XM1 program?" The two generals reported back to their chief that eventually a larger gun was required, but that improved ammunition for the 105mm gun would suffice for some time. Starry then recommended to General Abrams that he be permitted to reopen the requirements at Fort Knox and make certain that these and other lessons from the Yom Kippur War were clearly reflected. General Abrams agreed. Starry then assigned the task to Brig. Gen. Glenn Otis, commander of the Armor Training Center, a part of Starry's command.

Meanwhile, media accounts of the Yom Kippur War were touting the lethality of ATGMs and questioning the tank's ability to survive in the future. Both Generals Baer and Starry returned from Israel with data that showed IDF tank losses to ATGMs at no more than 8-10 percent. In America, ATGM zealots had grossly exaggerated the problem. It was with these considerations in mind that the Tank Special Study Group (TSSG) headed by General Otis began its work in the summer of 1974.

General Otis's study group was composed of experienced armor officers, operational research analysts, cost experts, and engineers. After a summer of extensive work, the group revalidated the majority of MBT requirements. The Bushmaster cannon was replaced by a coaxial machine gun, more emphasis was placed on improving the penetration capability of tank gun rounds, and the turret envelope was enhanced to accommodate a future 120mm main gun. Survivability remained the highest priority.

At this point it is worth discussing the validation contracts let in June 1973. The army had agreed to a cost-plus incentive fee contract. It had also agreed to give contractors full freedom to develop prototypes without interference. However, if they wanted to remain in the competition, cost had to be an overriding consideration. As some critics noted, this had the makings of a disaster, with cost overruns and "gold plating" leading to an unacceptable product. These concerns were reasons to delay or perhaps terminate the program. The validation phase was thus not without risk. This was well known in the PMO, where thirty-five of its members were survivors of the canceled MBT70/XM803 programs. According to General Baer, "These great engineers both military and civilian, who had worked so hard for so many years on the MBT70/XM803 with years of criticism about their program, had reason to be cynical."[14]

Fortunately this concern was unfounded, at least according to the TACOM historian, because "the complete validation phase prototypes were delivered to Aberdeen Proving Ground for the Combined Development/Operational Test (DT/OT-I), which ran from 31 January to 7 May 1976. During this same period, the competing automotive test rigs were driven more than 3000 miles over a variety of terrain." As a result, he noted, "Both prototypes were displayed in public for the first time in February 1976." The Aberdeen tests indicated that both validation contractors' prototypes met the specified requirements. The Chrysler tank, with its gas turbine power plant, demonstrated a slight advantage in acceleration. Nevertheless, the performance of both prototypes was deemed satisfactory. The operational portions of the test included two weeks of simulated combat operations with the tank operated by crews from Fort Knox and Fort Hood. After the validation phase, the process of selecting a single-source contractor for the FSED phase began. It was scheduled for July 1976.[15]

The army finally had a full-up competition between two great companies. Baer's office offered no suggestion of a clear-cut leader in the competition. General Baer stated at a dinner attended by both validation representatives that the American people and their army were the real winners. He added that he would be satisfied with whichever contractor the selection board recommended for the full-scale engineering phase of development. As a former member of Baer's staff, I can still recall the words. He believed that during all the deliberations to determine which contractor would be the single source, the program management team kept hands off during the selection process. He offered the last caveat to ensure that everyone would know that the PMO provided information when requested but did not comment on or express opinions in the final source selection process. The analysis and conclusions were the responsibility of the Source Selection and Evaluation Board, headed by General Otis.

The source selection process was an emotional period for both Chrysler and General Motors. Only one contractor would continue with the program. Those in the XM1 PMO waited for this decision so the winner and loser could be properly notified. Those in the selection and evaluation board had to present facts and conclusions to the source selection authority, which then made the decision and defended it before the DOD, the Congress, and the public.

It has been well documented in oral histories, books, articles, and recent interviews with the prominent participants that General Motors was the first choice for the FSED contract. Yet General Motors was not selected. How this came about is an interesting and important part of the XM1 story.

General Otis, who headed the 1974 revalidation study group at Fort Knox, now moved to direct the selection and evaluation board.

Rear view of (from left) the General Motors XM1, Chrysler XM1, and M60A1 tanks. R.P. Hunnicutt.

Otis was a highly qualified, experienced armor officer. Schooled in analytical techniques of evaluation and decision, he was highly respected for his ability to stay focused on the important issues. Otis's fifty-person board began the deliberation phase of the process near TACOM headquarters. Its mission was to evaluate the performance of the competing tanks as they were being tested at Aberdeen Proving Ground. They received enormous amounts of technical data as well as observations from crews and testers. Eventually the board's evaluations and conclusions would be presented to Secretary of the Army Martin R. Hoffmann, who was the source selection authority.

Both tanks performed well at Aberdeen. The turbine engine's performance, which enhanced agility, favored Chrysler. But armor protection and survivability—still the number one priority—favored General Motors. The Otis Board pondered these and the many other variables that came from the long hours and miles of testing conducted at Aberdeen. "There were no doubts by those who sat in on the many briefings presented by the board that if a vote had been taken GM would have been the winner," wrote Orr Kelly, author of *King of the Killing Zone*.[16]

On 20 July 1976 Otis presented his board's findings to Hoffmann, who accepted them. Otis then went on to give the same presentation to William Clements, the deputy defense secretary, who rejected Hoffmann's decision. With the concurrence of Secretary of Defense Donald Rumsfeld, Clements directed that the program be extended for about four months for further analysis of the power plant and main gun options. This action was the result of a tug of war between the Army

Front view of (from left) the Chrysler XM1, M60A1, and General Motors
XM1 tanks. R.P. Hunnicutt.

and DOD that began in 1973, and reflected concerns General Abrams
voiced to Generals Starry and Baer regarding gun caliber.

In June 1973 Secretary of Defense–designate James Schlesinger and
the West German defense minister agreed to investigate the possibility
of standardizing their new tanks' main guns. The following December
defense officials from the United States, West Germany, and the United
Kingdom signed an agreement to develop a common tank cannon. In
1974 DOD directed the army to negotiate a Memorandum of Understanding with West Germany to evaluate the Leopard tank as a possible
candidate for the next generation of U.S. tanks. In September, three
Leopard test units arrived at Aberdeen.[17]

Meanwhile, both validation contractors were directed to address
the concerns over the power plant and main gun option. General Motors
had to rebid the turbine engine in its prototype, and Chrysler had to
rebid the diesel-powered piston engine in its prototype. Both were also
required to redesign their turrets to accommodate the larger German-
designed 120mm gun.

There was considerable concern at this turn of events in the XM1
PMO and at the contractors' facilities. Was their project about to become another MBT70/XM803? Was there a plan to use the XM1 as an
international bargaining chip? These real-world concerns and many
others were allayed when the Army secretary flew to Warren and personally assured Baer's staff that, regardless of the decision to rebid the
engines, the XM1 program was still on schedule.

Since the canceling of MBT70 and XM803 programs, the mood in
Congress had changed, and members had become very supportive of

a new tank program. In many respects this was due to Baer's intensive effort to ensure that members of Congress who were interested in the tank were kept informed of both the good and bad in the selection process. There had been no smoking guns and no cover-ups. However, when DOD decided to rebid for commonality and add a German-built 120mm gun to the program, a number of congressional observers complained that they seemed to be the last to be informed.

A House Armed Services Committee subcommittee chaired by New York Democrat Samuel Stratton convened new hearings. Stratton had more than a passing interest in this program: all U.S. tank guns were produced at Watervliet Arsenal, which was located in his district. The hearings were far from agreeable and somewhat hostile. It took considerable lobbying by Rumsfeld and Clements to persuade Representative Stratton not to overturn their decision. Hoffmann, Ed Miller (his research and development deputy), and Baer were key witnesses.

The decision to rebid for commonality and to provide for the German gun also disappointed the contractors. According to General Baer in an interview with the TACOM, both contractors had been hoping for a clean decision. Both of them believed that their prototype probably would be the winner. Furthermore, Baer noted: "Chrysler's officials, given the conditions surrounding the delay, certainly recognized it was to their advantage and so they were more receptive to it than was General Motors. The General Motors people almost immediately began to realize that the problems that they faced were going to be difficult ones, and could be detrimental to them."[18]

General Motors had reason for concern since Clements had demonstrated an obvious preference for the new technology reflected in the turbine engine. It was difficult for General Motors to competitively bid the cost of installing a turbine engine in their prototype, which had been built with a diesel-fueled piston engine. However, there was no time for the contractors to debate the decision. Both teams worked around the clock to redesign their prototypes.

During the rebuilding phase, it appeared that Chrysler had a greater sense of urgency because information had leaked indicating that there had been a preference for the General Motors prototype. Considering subsequent discussions with participants, some believe today that General Motors did not give its best effort in the relook phase of the program.

At a news conference on 12 November 1976, DOD announced that Chrysler had won the FSED contract. According to published reports, cost reduction efforts at Chrysler had cut its bid to $196 million, compared to $232 million from General Motors. The increase in the latter was, no doubt, partially due to the higher initial cost of the Avco-Lycoming AGT-1500 gas turbine compared to the Teledyne Continental AVCR-1360 diesel engine. However, Chrysler's diesel-powered proposal for the FSED program was priced at $186 million. The five-month delay in awarding the contract had permitted further refinement of the design and provided a better method of determining manufacturing costs. For example, the design-to-unit cost for Chrysler's tank decreased from $574,000 in the original FSED bid to $422,000 for the final proposal. (The unit cost for Chrysler's diesel tank design was $376,000. All of these prices were in 1972 dollars for comparison with earlier estimates. The actual unit price of the winning tank was expected to be about $754,000 in 1976 dollars.)[19]

Although the FSED award had been made, the XM1 was still not a sure thing. First, the program had to survive another challenge by a group in DOD that was eager to sell the Airborne Warning and Control System (AWACS) to Germany. In return, it wanted the United States to consider standardizing on the German Leopard 2. The consequence was a head-to-head test between the Chrysler prototype and the Leopard 2 AV (Austere Version or American Version) conducted at Aberdeen Proving Ground.[20] The result of this might have meant an unprecedented decision to buy a foreign tank for the U.S. Army. As it turned out, both countries' prototypes were too far advanced. In addition, too many local vested interests threatened. Thus national pride, service parochialism, and technical problems prevented the tank standardization program from becoming a reality.[21]

The DOD's consideration of the German tank was not a popular decision in the XM1 PMO or with Chrysler, which had just won the FSED contract. First, there was the matter of national pride and tradition. After World War I the United States had gradually set in motion its own native tank program. Second, the German tank design did not emphasize the army's number one design priority: survivability of the tank and crew. A tank's overall combat effectiveness is a function of firepower, mobility, and survivability. Military experts in the United States and West Germany disagreed on the relative importance of each function. The Germans placed more emphasis on firepower and mobility, whereas American experts gave more weight to survivability, thus emphasizing armor protection. In addition, the Germans were not privy to Chobham armor's capabilities. Last, there was an honest belief by members of Baer's staff, who were very knowledgeable in all MBT programs, that the XM1 was a better tank than the Leopard 2. They were convinced theirs was a superior concept that, when it matured, would become the best tank in the world.

The Germans sent a hull and turret for ballistics testing and a completed vehicle to test its durability and maneuverability. The Americans were confident of winning because their special armor was superior to the armor used in the German tank, but they were surprised when the Leopard 2 AV—with its diesel engine—was equaling and in some cases surpassing the XM1 in maneuverability. Only by chance was it discovered that the German vehicle did not have its full complement of special armor protection, decreasing its weight considerably. This explained its superior performance. The Germans withdrew the Leopard 2 AV from the tests in May 1977, and another threat to the XM1 program became history.[22]

The program manager at the Tank Main Armament Systems (TMAS) at Picatinny Arsenal in Dover, New Jersey, the site of AMC's Large Caliber Gun Systems Laboratory, was Brig. Gen. Philip Bolté. He was responsible to General Baer for both ammunition and armament for the new tank. Bolté had been selected to head the U.S. group on an international team that was evaluating 120mm tank guns. The following account of that process is derived largely from his recollections.

Recall that late in 1973, when the defense ministers of Germany, the United Kingdom, and the United States had signed an agreement to seek a common main armament for the next generation of tanks, the development of the XM1 was already underway. Meanwhile, Germany and England had instituted a planning program for a common main battle tank; however, it was in its earliest stages. Thus, there was already a timing problem.

Generals Baer and Starry had told General Abrams that with ongoing and potential ammunition improvements, the 105mm gun would be adequate for the near future. In the language of the report, the 105mm was "the least gun adequate" for the XM1. The significance of "the least gun adequate" was that the smaller the gun that met the requirements, the more other advantages accrued. More ammunition could be stowed in the tank. Retaining the 105mm gun—already standard on M60 tanks—meant that a family of ammunition and ammunition stocks already existing could be used, thus avoiding logistics problems. Finally, with the XM1 design well underway, a change in main armament could delay the program significantly, increasing the cost. There was no enthusiasm within the army for the new agreement, but it was nevertheless committed to test several foreign designs.

The British were developing a rifled 110mm gun that they were eager to have adopted by other countries. Over the years, multinational adoption of the British 105mm gun had reaped great financial rewards from both direct sales and licensing agreements. The British believed the new 110mm gun had the potential of repeating that success.

Germany, meanwhile, had been developing a 120mm smoothbore gun. The Germans believed that a significantly more powerful gun would be required for future tanks, and sought to gain economic benefit by having the gun adopted by other countries, especially the United States.

The effort to seek a common main armament was formalized as the Tank Main Armament Evaluation Program. Working groups from the three nations would evaluate such factors as gun performance, production, cost, and vehicle interfaces. These groups all set about to ensure that all factors were considered and to reach agreement to the extent possible.

When the results were in, there was no doubt that the 120mm German gun was the most powerful. The 110mm British gun, which had the disadvantage of being new, performed less well. Even in the fairly early stages of the program, it became obvious that the British candidate was the loser. As a consequence, the British switched to a developmental rifled 120mm gun and accelerated their development program.

Meanwhile, the Baer-Starry visit to Israel in 1974 had resulted in improved American ammunition for the 105mm baseline gun, specifically long-rod penetrators for kinetic energy ammunition, which put the United States ahead of Germany and the United Kingdom in that critical area. As a result, the U.S. guns proved a good match for German ammunition in the key area of kinetic energy armor penetration, lessening the performance gap between the two candidates. In addition, the 105mm ammunition had the traditional full-metal cartridge case, whereas the German gun used combustible fiber that was far less durable. In the view of the U.S. armor community, General Starry in particular, the German caseless rounds had all the disadvantages so dramatically demonstrated by the Sheridan's ammunition in the Vietnam War. In Starry's view, caseless ammunition was simply not crew-safe.

As the program progressed, national positions became more emotional and firmly fixed. The joint German-United Kingdom tank program was concluded. It became quite clear the British would have their own large rifled gun on their Challenger II tank, while the Germans would mount their 120mm smoothbore gun on the Leopard 2. Thus, the evaluation program evolved from an attempt at international standardization into weapon selection for the XM1.

Ultimately, DOD authorized initial production of the XM1 with the 105mm gun so as not to delay the program, with the German 120mm smoothbore gun to be installed after initial production. This was consistent with the Baer-Starry recommendations to General Abrams. To at least give lip service to gaining further commonality between the XM1 and Leopard 2, the Germans agreed quid pro quo to "consider" using the American gas turbine engine in their Leopard 2s, but there is some doubt that such a move was ever seriously considered. German officers later confirmed this suspicion.

Meanwhile, the British pursued their rifled 120mm gun program independently, hoping to inspire greater interest among other nations. The rifled 120mm gun was incorporated in the Challenger II tank.

As a result of the American selection of the German gun, it soon became obvious that the 120mm would become the new free world standard tank main armament. Under license or independent development, 120mm smoothbore guns have since been incorporated in Israeli, French, and Japanese tanks, and in those M1s and Leopards used by other armies.

The foregoing discussion helps to illustrate some of the issues faced by program managers during the FSED phase of the MBT program. But regardless of these events, the XM1 eventually passed both the user operational tests and systems performance tests.

In the meantime, Maj. Gen. Donald M. Babers succeeded General Baer as the XM1 PM in July 1977. General Babers was an experienced ordnance officer with extensive field and acquisition experience with combat vehicles. According to Baer, he was exactly the right person for the job. Babers had served in TACOM for ten years. He had been a contracting officer, maintenance engineer, director of procurement and production, and a project manager on two occasions—for the M561/XM705 trucks and for M60 tank production. Before his assignment as the XM1 PM, Babers was TACOM's Deputy Commander.

Just as General Baer had many obstacles to overcome during his tenure as program manager, there was no shortage of hurdles for Babers. First, he had to establish the tank industrial base before production could begin. The FSED portion of the contract had to be completed, followed by both development and operational testing. Additionally, he was in charge during the tank armament selection process described earlier. New equipment fielding teams had to be trained. An adequate supply of repair parts, maintenance procedures, and training publications had to be produced. All of these tasks would be accomplished under the watchful eye of Congress and an occasionally hostile press.

When discussing the tank industrial base it is important to understand that it includes vendors in forty-one states and Canada. It included assembly facilities for engines, transmissions and final drives, gun mounts, slides and breeches, fire control components, and special armor. These vendor parts and assemblies were shipped to tank assembly plants in Warren, Michigan, and Lima, Ohio. Hulls and turrets were fabricated at the Lima plant and transported to Warren for final assembly. Besides welding hulls and turrets, the Lima plant also produced the tank's special armor.

When Babers took over as PM, the Lima facility was being used as a storage facility for defense industrial plant equipment, and considerable work was required to turn it into a modern manufacturing and assembly facility. Since no one in the PMO or TACOM had any experience establishing such a facility, a task force headed by Lt. Col. Jim Evans was set up within the PMO. Evans's task force consisted of industrial engineers, costs analysts, and management personnel from the PMO and TACOM. The task force also contracted with Peter Ponta, a retired Ford Motor Company executive with experience setting up industrial space for automobile production. Ponta stayed with the PMO after the task force completed its study, providing valuable assistance as the M1 industrial base was established.

Chrysler was also involved in this effort. As prime contractor for both the FSED phase and the initial production phase, Chrysler was responsible for the vendor base. The PMO closely monitored the cost of major assemblies, such as engine and transmission, final drives, fire control, and computer. Although not accomplished until later, it was Babers's intent from the beginning to have the government buy these major assemblies and provide them to the contractor—but only after Chrysler and its subcontractors resolved production problems.

A thorough review by the task force and Babers's staff estimated that it would cost $1.3 billion to provide the facilities to produce the M1 tank. Babers submitted these cost estimates to the army staff and to OSD, where this hefty amount generated fears that Congress would again question the program. After many discussions, DA and OSD costs analysts came up with a revised figure of $800 million, which was submitted to Congress. Ironically, after many reprogramming actions, the final cost turned out to be slightly more than $1.3 billion. (An additional $400 million was added to the program due to the requirement to build a facility for the production of depleted uranium armor.)

When asked about the Lima facility, General Babers stated: "It was a massive effort, for sure. We had to build 250,000 additional square feet of manufacturing space, modernize existing facilities, and acquire and install industrial plant equipment and special tools/test equipment."

Furthermore, he recalled: "There was no spare time, and as the first tanks rolled down the line, it is no secret that many of the production processes were not mature. Many processes had to evolve and develop during the first year or two of production . . . again its one of the risks of going into a highly compressed program."[23]

Cost was not Babers's only problem. There were also a number of political considerations, such as the controversy over the selection of the machine tools required for fabricating the turret. General Babers was looking for a single machine that could perform the same functions that forty-one separate machines had performed on the M60 tank turret. "Since the Buy American Act had to be followed, we went through all the necessary calculations and a Mitsubishi-made machine beat out all American competitors," Babers said. The issue became more intense, he noted, when the decision was announced to buy the Japanese machine tools, as opposed to those made by a Cincinnati machine tool company. The congressman from that city, Rep. Thomas Luken, took great exception to the Mitsubishi choice. He argued that at stake were American jobs and industry, and the security of the American people. As a result, Babers said, "We had some very interesting meetings with Mr. Luken and several members of the DA staff. There was no doubt in my mind that the right decision was made in adopting that [the Mitsubishi] machine."[24]

Fabricating the M1's special armor also presented a unique and difficult challenge. Chrysler had to take the BRL design and duplicate it in a production process. Because the armor's composition was classified, it had to be made in a secure building separate from the Lima tank plant. Through long trial and error special processes and machines were tested. Production samples were shipped to the BRL for penetration tests. At first, repeated failures occurred, but by the time the first tanks rolled off the assembly line the facility was producing qualified armor. Babers remembers many stressful days watching this process mature.

The thirty-six month FSED contract required Chrysler to produce eleven pilot tanks, which were manufactured at the Warren plant. The tanks—still called the XM1—were delivered to the government between February and July 1978. The tanks were shipped to Aberdeen Proving Ground for the second phase of the development test (systems performance against design specifications) and to Fort Bliss, Texas, for further operational testing. These tests, which continued until September 1979, where intended to prove the tank's suitability for the army, complete the design process, and incorporate any changes into the low-rate initial production tanks.

Another problem Babers faced was the long lead time required to produce some items. Normally this would be a sequential process. The M1 program, however, required advanced production to meet a shortened acquisition cycle early in the program. In some cases, this required the ordering of long lead time items before the tank's design was completed. Babers recognized that to maintain the schedule, which was an important benchmark for Congress, he and the Army had to accept some risk. So funds for the procurement of long lead time hardware—such as fire control components, special castings, power train components, and other items—were released to Chrysler beginning in November 1977, three months before the government's receipt of a single FSED prototype. As it turned out, the pilot tank was delivered on schedule and in February 1978 the user developmental and operational tests began. Although some developmental problems surfaced during the tests at Forts Knox and Bliss, there were no weaknesses that would merit terminating the program.

Maj. Gen. Thomas Lynch, the Fort Knox Commander during operational testing at Fort Bliss and throughout the decision processes, kept the program on track. An experienced armor officer with combat experience in both Korea and Vietnam, Lynch understood tanks and training, and, like General Babers, was committed to fielding a survivable and Soldier-friendly tank. Before the tests began, Lynch sought to have them conducted at Fort Knox rather than at Fort Bliss because the tank was being developed for the European theater, and the Kentucky terrain more closely approximated that found in Europe than did the West Texas desert. In addition, as the chief of armor, he was responsible for providing the operational training packages and overseeing the tests.

Another experienced and respected armor officer headed the Army's Operational Test and Evaluation Agency (OTEA) and was responsible for Operational Testing Phase (OT) II. He was Maj. Gen. Julius Becton Jr., who made the decision to conduct the test at Fort Bliss. This proved to be extremely fortuitous: the problems discovered during the Fort Bliss tests served the army well during Operation Desert Storm, the ultimate test of the M1 in combat to date.

The army and the PM also accepted risk of another kind when it was decided to conduct operational testing before the completion of development testing at Aberdeen Proving Ground. Again, sequential development would have slowed down the accelerated acquisition schedule. But in this case, the acceptance of risk caused considerable problems.

The turbine engine was ingesting dust and dirt at an alarming rate because of faulty air filter seals, resulting in engine failures during the operational tests. Tracks were being thrown and were breaking while maneuvering in the sandy desert areas at Fort Bliss. There were also problems with the turret hydraulics and the fire control systems. In addition, integrated logistical support was not yet in place, nor was there an adequate training package available to prepare the troops who would do the testing. Lynch and Babers halted the test for several weeks to solve these problems, and to make sure there was adequate training for the supporting troops. Engineers from the XM1 PMO and Chrysler worked around the clock to resolve these issues.

Rumors of problems were rampant during the XM1's operational testing at Fort Bliss. There is a saying in the army that if something is going well very little is said, but if things sour, the problems are known throughout the Pentagon within minutes. The XM1's failures were no exception.

General Babers said that he and his staff "knew there was risk (in conducting an operational test before development tests were completed) and that as these problems surfaced it would be grist for our critics' mill and it would be necessary to go before Congress, lay out the problems and solution and convince them that the program was under control, and it would continue."[25] That is exactly what happened.

It was not just Congress that had to be convinced that the XM1 program was on schedule. Many senior members of the active Army were beginning to question its relative worth based on the early results of operational testing at Fort Bliss. But Generals Lynch and Babers were convinced that these problems were being solved, which was the case. On 30 March 1979, upon conclusion of the Fort Bliss tests, General Lynch sent the following important and critical message to the Army's elite regarding the status of the XM1 tank:

1. this report is intended to update you on xm1 tank status, as i see it from the armor center.
2. after a lengthy professional debate, the issue of light vs heavy tanks was tabled in 1972 when the armor community deliberately closed ranks and stood shoulder to shoulder to get the xm1 program started and on track.
3. the xm1 tank is in full-scale engineering development (fsed) and has completed operational test (ot II). development test II will continue through august 1979. the present xm1 represents six years of intensive tank development backed up the sunk costs and technological

dividends of both the mbt-70 and xm803 tank development efforts.

    a. the fsed xm1 has proved to be significantly better than the m60a1 tank in terms of hitting performance, survivability, agility and maneuverability, ot ii demonstrated that our soldiers can be readily trained on the xm1. i am, however, disappointed in the reliability demonstrated to date and very concerned with the lack of ram growth.
    b. notwithstanding these problems, the fsed xm1 is a more combat effective tank than either the m60a1 or the m60a3. it is also more than a match for the soviet t-62, t-64 and t-72. we need this tank now for its proven performance and its potential for capability. our cost to move the xm1 into low rate initial production, at least through initial fielding, will be reduced availability and increased maintenance effort.
    c. as a practical matter, the xm1's growth potential can only be realized through tank production and subsequent product improvements. we can afford neither the operational risk nor the cost inflating delays that would result from an attempt to seek perfection with prototypes.
4. we should recognize that the xm1 tank is not the best that money can buy, but that it is the best that the army can now afford. stated another way, "imperfection" prior to fielding could kill the xm1 program through inflation costs and possible perceptions of program difficulties, and risk continued exposure to an increasingly potent threat.
5. as armor leaders, we all must remain actively involved in the tank development process and the armor center intends to set the pace. we seek to optimize the combat effectiveness of the xm1 tank, and we need your help to do it right. we invite debate, active exchange, discussion and challenge within professional arenas: but to get this tank we must have your unqualified support up front. "in short, you can draw saber and join the charge, dismount and defend the wagons or quietly hold the horses. whatever you choose to do, do not cheer for the hostiles."
6. your professional thoughts are always sought and appreciated. the 1979 armor conference is scheduled for 15-18 may. you are most cordially invited to attend.

As General Lynch stated in his message, he was not satisfied with the reliability demonstrated to date and was very concerned with the lack of reliability-availability-maintainability (RAM) growth. "As a result three of the tanks that were tested at Fort Bliss were refurbished by the contractor and shipped to Fort Knox in June of 1979 for additional durability and reliability testing. During these tests, (6,000 miles) the reliability-availability-maintainability-durability (RAM-D) goal was 272 mean miles between failures. The XM1 reached 326 mean miles between failures."[26] Although this did not close the RAM-D issue, it did demonstrate growth potential that would continue as the tank matured.

Another interesting sidebar to the history of the Abrams tank was the concern over the turbine's early problems during the operational tests at Fort Bliss, which created an opportunity for diesel proponents to resurrect their program by bringing congressional pressure to bear. Congress required the army to consider the air-cooled AVCR-1360 diesel, the same engine bid by General Motors in the XM1 program, as a back-up to the AGT-1500 turbine engine in the M1. Was this a pork barrel political decision due to contractors pressuring selected congressmen? When asked that question, General Babers replied:

> I'll give Congress credit for being honorable and trying to make the decision that was in the best interest of the Army. There is no doubt that there was a small but vocal constituency for the diesel engine which we were directed to mature. The extent to which industry pressure was exerted on members of Congress, I can only speculate. I can say that those of us with responsibilities in the tank program were firmly convinced we had a turbine engine that was going to get the job done. We believed that the low risk approach was to stick with the turbine engine, and I think that time has proved our position was correct.[27]

In my personal opinion, having participated in a portion of the XM1 program, there is no question that the diesel "excursion" was pure political pork. Congressional interference cost the program and taxpayers millions of dollars initially and even more annually as this back-up program continued into the mid-1980s before it was concluded. Those dollars could have been spent on an auxiliary power unit, improved track, or other product improvements, rather than on research for an engine that the Army never intended to install in a tank. This was an example of the kinds of distractions that can plague managers of high-dollar programs such as the MBT.

On 7 May 1979 the Secretary of Defense approved the low-rate initial production of 110 XM1 tanks, with the first two being delivered at a special acceptance ceremony on 28 February 1980. At that time the new tank was named the Abrams, in honor of the late Gen. Creighton Abrams. Mrs. Abrams attended the ceremony, accompanied by her three sons, and proudly christened the first production tank.[28]

As stated earlier, General Babers had many obstacles to overcome during the FSED phase of the program and then during low-rate initial production at Lima. Still more obstacles arose during the user's operational and developmental tests. A series of GAO reports raised questions about the durability of the power train, but after an extensive review of the XM1 program conducted from February through November 1979, the GAO reported to Congress that the tank was making steady progress. However, it noted, the turbine engine's superior performance had not yet been demonstrated. The GAO recommended that a full-scale diesel engine development program be initiated if the turbine problems persisted.[29] A year later, the GAO was still critical of the power train's durability, and this time suggested that production be limited until the turbine engine's problems were resolved.[30] This occurred even though a blue ribbon panel and extensive testing found that the turbine engine was showing demonstrated performance improvement that eventually exceeded the reliability performance of the diesel engine.

In spite of the criticism, the M1 tank, although evolutionary, included many revolutionary innovations that would make it one of the most survivable tanks ever produced. First was the special armor package that provided unprecedented crew protection from both chemical and kinetic energy weapons. Second, the ammunition compartment was designed to blow out and away from the crew should a detonation occur. This was during live-fire tests. Third was the fire control system, which included a ballistic computer that stabilized the tank cannon, allowing the crew to shoot on the move. This ballistic computer system—along with the Abrams's exceptional thermal sight—played a key role later in the Gulf War. The fourth innovation was the turbine engine and the mobility it provided. The enhanced transmission and suspension system provided high cross-country speed with acceptable crew comfort. The M1 achieved the priorities described by the Desobry task force in 1972, and was accomplished within the $570,000 cost threshold.

Meanwhile, the turbine engine's endurance problem was not immediately resolved, and the next two program managers would contend with it until the final power train durability tests were successfully completed in 1984.

In June 1980 Maj. Gen. Duard D. Ball took over the Abrams program. Like his predecessor, General Babers, Ball was an experienced ordnance officer with a wealth of knowledge regarding new equipment training and fielding and integrated logistical support. He had the background required to take the M1 program into full production. Unfortunately, General Ball died from cancer before he was able to participate in TACOM's oral history program. Consequently, the information provided here regarding his tenure came from discussions with his former deputy, Col. William Sowers, Generals Babers and Peter M. McVey, and my own recollections of those events.

General Ball's mission was to complete the fielding of the low-rate initial production tanks and, after considering the results of the user's operational and developmental testing, oversee whatever engineering changes were required before full-scale production. One of his main concerns was to improve the turbine engine's reliability. In addition, he had to refine the new equipment training program and continue developmental testing at Aberdeen; Yuma, Arizona; the Alaska Cold Region Test Center; and at White Sands Missile Range, New Mexico, where electromagnetic radiation and nuclear vulnerability evaluations were conducted. He also had to complete the next operational testing phase (OT III), which began in September 1980 at Fort Hood and Fort Knox, and continued until May 1981.

General Babers had started the policy of direct contracting of major assemblies, which continued throughout the program's life as a way of reducing costs. The PM and TACOM's procurement division would contract directly with major vendors for assembly line items such as engines, transmissions, final drives, fire control, and track. These items would then be delivered to the prime contractor.

Engine problems persisted during low-rate initial production. A considerable amount of time was spent on this problem by both Chrysler and government personnel at the government-owned plant operated by Avco-Lycoming in Stratford, Connecticut. The company was facing a major task in making the transition from producing handmade engines in support of the FSED vehicle to building production line engines in support of low-rate initial production. Avco-Lycoming's problem was compounded by the government-owned facility in Stratford. "When the workers arrived to begin setting up shop, they found paint hanging in strips from the ceiling and so many puddles on the floor that visitors had to borrow galoshes to walk through the high-ceilinged, hangar-like plant. Up above, birds flew freely in and out the myriad broken windows, and their droppings were a constant annoyance."[31] Recall that creating the facilities to produce the tank industrial base was a major effort of the tank PMO, and the Stratford plant was part of that $1.3 billion price tag.

It took many years and several changes of management before engine production and quality stabilized. Workable engines were so scarce when the first production line tanks were nearing completion that the few available engines were used again and again in initial running tests to avoid closing down the entire operation. In the meantime, engineless tanks were placed in storage. At one point in early 1983 there were thirty-five tanks awaiting engines.[32] Given the scope of the problem, it is easy to understand the attention it attracted from the army, DOD, and Congress.

Another major effort was the establishment of a New Equipment Training Team (NETT) to train Soldiers and their leaders to fight and maintain the new tank. Colonel Peter McVey was selected to lead this effort, and in September 1980 his team set up at Fort Hood to train the first battalion to receive initial production tanks. The team consisted of four officers, twelve enlisted subject-matter experts, and the 2d Battalion, 5th Cavalry, of the 1st Cavalry Division. The NETT program provided instruction on operating the tank and included live-fire exercises. Maintenance training from operator through battalion level was the responsibility of the training team, while higher levels of maintenance were taught at the Ordnance School.

While NETT and operational tests were underway, the documentation for full-scale production was working its way through the acquisition channels in the Army, DOD, and Congress. By February 1981 production was approved for 7,058 tanks at a rate of thirty per month and a unit cost of $1.5 million. The M1 was type classified as standard on 17 February 1981 as a 105mm gun, full-tracked combat tank.[33] The number of tanks per month would double and nearly triple before the collapse of the Soviet Union brought an end to the urgent requirement for additional modern tanks.

The NETT experience at Fort Hood proved the principle that set the standards for all future training on the Abrams series. In July 1981, McVey's team—now expanded to forty-four—moved to Vilseck, Germany, and began fielding the M1 tank to U.S. forces in Germany. The first unit to receive the M1 there was the 3d Battalion, 64th Armor, of the 3d Infantry Division (Mechanized).

Introducing the M1 in Germany was a courageous decision by General Ball. The first battalion to be equipped was issued the remaining low-rate initial production tanks. The repair parts list, although established during operational and developmental testing at Forts Bliss and Knox, was not complete, and not all the vendors were yet running at top speed. Consequently, General Ball made the decision to support the initial fielding with parts taken from the production line. This was not a secret; before the decision was made to field the M1 in Germany, Ball and McVey briefed key leaders of the risks involved.

McVey left the NETT in October 1982 and was followed by two more experienced tankers: Cols. Jim McWain and Tim Donovan.

The U.S. Army then moved in good faith to transition German cannon technology into the U.S. industrial base. Although the licensing agreement with the German company Rheinmetall GmbH was signed in February 1979, difficulties occurred in the transfer of data, causing a one-year delay in fielding the M1A1 tank. When asked about this problem by Joseph Mahalik, the TACOM command historian, General Babers replied: "There's a lot of 'black magic' associated with technical data packages, whether they be of German or U.S. design. It's not unusual for the U.S. Army or the other services to buy a technical data package, award a contract to somebody other than the firm that developed the package, and be faced with subsequent difficulty in translating that technology into new hardware that functions as it did for the original developer."

The uncertainty, Babers told Mahalik, was "in transitioning from technical data packages to actual production. . . . When you then throw in another language and several thousand miles between the developer of the technical data packages and the implementor of an American production base, you've built in another set of problems."[34]

In spite of accusations and problems, there was no foot dragging by the U.S. Army or the Germans. However, history has demonstrated that technical data packages are difficult to duplicate and put into production. This problem was first experienced during World War I, when the United States attempted to build the French Renault light tank.

Aggravating the gun problem was a controversy over the breech mechanism. Army weapon developers intended to use an American design to simplify the German design, which included over 125 moving parts. Although not confirmed, it is suspected that General Babers was concerned about American parochialism. The ground rule for the breech redesign, which he laid out for the Benet Laboratory at Watervliet Arsenal, was: "If you can design a breech that is markedly at a risk that is no higher than the German breech, we will consider it for production."[35] As it turned out, the German breech was retained.

During General Ball's tenure as PM the prime contractor management of the tank program changed when General Dynamics bought the Chrysler Defense Division, which produced the Abrams tank. The new organization became the General Dynamics Land System Division (GDLS). Tank program personnel remained, however, with Dr. Philip Lett Jr. becoming the vice president for R&D.[36] The General Dynamics takeover brought an end to a historical era; an era that began when Chrysler developed and produced its first tanks before the United States entered World War II.

Although there was some initial organizational turbulence with General Dynamics's management reorganization, holdover Chrysler employees, including Lett, Lou Felder, Joe Yeats, and Carmelo Milia, had all had been with the tank program since its inception. They were a stabilizing influence. As a result, the flow of tanks and the development of the M1A1 were not affected.

In May 1983 I became the PM after serving earlier as deputy PM for Generals Baer and Babers during the XM1 competition and the beginning of the FSED phase. My task was to continue the production and fielding of the M1 tank. We were also completing the power train durability test that had held the program hostage since the end of the FSED phase, and completing development of the 120mm gun and ammunition. In addition, the new charter called for developing the improved 105mm and 120mm ammunition necessary to dominate upgraded Soviet tanks. But most important was another important task: to carry through its formative stages a highly classified program to create an even more advanced special armor that would increase the tank's survivability.

On the Army side, the tank program was also about to undergo a major reorganization, establishing a PM for all U.S. Army tank systems. The new organization was to be headed by a Major General, with Colonels assigned as program or product managers for individual tank systems. The reorganization was designed to manage the resources and requirements of all tank programs more efficiently.

Col. Joseph Raffiani Jr. took over as the M1A1 PM in June 1983 and served until June 1987, Col. William R. Rittenhouse became the M1 PM, Col. William M. Kearney was the M60 series PM, and Col. Donald R. Kenny remained as the TMAS PM at Picatinny Arsenal. Under the reorganization, these managers became part of the U.S. Army Tank Systems PM's organization.

Although the M1 was being produced and issued to the field, problems with the power train still plagued the program. Congress had mandated that before a production rate of sixty a month could be exceeded, the power train durability requirement had to be met. Durability tests conducted at Aberdeen consisted of seven tanks each accruing

4,125 miles. In December 1983 the last test was satisfactorily completed, demonstrating that the durability requirement had been achieved.[37]

Track life still failed to meet M1 durability requirements of 2,000 miles, with the actual mileage achieved being only about 800 to 1,200 miles, depending on the terrain. The M1's lightweight track, which was designed to reduce weight, simply could not withstand the heat generated by the tank's high-speed travel. German-made Diehl track (its rubber pads removed for employment in combat) could have satisfied this need, but it was rejected because the road wheels and sprockets had to be replaced before mounting the track. The track problem was solved in February 1988 when the army awarded FMC Corporation a $140 million contract to produce M1 tank track with a guaranteed 2,100-mile life span. Several years passed before the entire tank fleet was fitted with the new track as replacement was accomplished through attrition.

Meanwhile, planning for production of the Improved M1 (IPM1) was underway. The improvements included an upgraded transmission and final drive, an improved suspension system, and enhanced survivability. These improvements were also designed to accommodate the heavier 120mm gun. However, as it turned out, both the M1 and the IPM1 mounted the 105mm gun. The production run for the M1 ended after 2,374 were produced.

The next product improvement to the Abrams tank was the M1A1, with work progressing toward a rollout in the summer of 1985. Originally, the improved tank was to include the 120mm gun, a Commander's independent thermal viewer, a driver's thermal viewer, a hybrid individual NBC protection system, a microclimate cooling system, a fast fueling capability, an enhanced smoke dispenser, and a data bus to accommodate future digital improvements.

A serious problem was revealed during live-fire testing of the M1A1. When the test tank was hit in the ammunition compartment, which was loaded with combustible-case 120mm ammunition, there was a catastrophic detonation that would have killed the entire crew. This confirmed General Starry's serious concerns about combustible-case ammunition voiced a decade earlier. The problem stemmed from two failures: the absence of antifratricide brackets to hold the 120mm rounds in the ammunition compartment, and failure of the ammunition door between the crew compartment in the turret and battle storage. The door bulged, permitting excessive heat and poisonous gases to enter the crew compartment. This problem had to be quickly solved.

General Otis—then serving as USAREUR commander and the senior armor officer on active duty at the time—was advised of plans to correct the problem. The plan was to install German designed ammunition racks developed for the Leopard 2, which included antifratricide bars. Seals were redesigned to strengthen the ammunition doors, and the overhead blowout panels on the ammunition compartment were also redesigned. These panels direct the blast from exploding ammunition away from the crew. A major effort by General Dynamics and army engineers solved the problem, and this was confirmed by additional tests. As a result, development of the M1A1 remained on schedule. It was type classified during the summer of 1985 after successful completion of all operational and developmental testing. On 29 August the first production vehicle was delivered to the army at the Warren tank plant.

The M1A1, as delivered, added the 120mm gun, a Microclimate Cooling System (MCS) that cooled the crewmembers but not the entire crew compartment, and an NBC overpressure system, as well as other crew survivability items. But it did not have the Commander's independent thermal viewer, a feature that tests at Fort Knox had demonstrated resulted in an improvement in first- and second-round hits and a significant improvement in third-round engagements. The armor community wanted this feature included in the M1A1 but it was delayed until the M1A2 block improvement program.

The development and subsequent use of depleted uranium (DU) armor is yet another interesting part of M1A1 tank history and a real Army success story. This account comes from my own personal experience and the recollections of General Raffiani, who was the responsible PM.

From the beginning, crew survivability continued to be the number one priority in the Abrams tank program. Research and development on better armor protection was being conducted under highly classified conditions by the army's Ballistics Research Laboratory at Aberdeen Proving Ground. One of the so-called black programs explored the use of depleted DU armor technology. Test results had demonstrated significant improvements in protection compared to older technologies, so it was decided to complete the research and development of a DU alloy that could be added to the M1A1's composite special armor as soon as it could be worked into production. Very few in the Army chain of command were aware of the DU armor project. Indeed, a decision by James Ambrose, then serving as undersecretary of the army for R&D, is what set the program in motion.

Completion of the R&D phase was challenging but relatively straightforward. The two main challenges were moving the armor into full production with consistent protection results while operating in a very high security environment and at the same time dealing with potential radiation hazards.

Where could DU armor be produced? Undersecretary Ambrose and Troy Wade from the Department of Energy decided that the Idaho National Engineering Laboratories, located in Idaho Falls, Idaho, an Energy Department facility, offered physical security, low-level radiation safety, and a highly skilled workforce with the necessary security clearances.

When I first visited this facility it was necessary to fly to Boise, Idaho, with the announced intention of visiting the state adjutant general, Maj. Gen. Jim Brooks. The cover story was that we were meeting to discuss M60A3 fielding plans and construction of a test track to evaluate the Israeli steel track on the desert rocks of the Gowen Field National Guard training facility. The rocks and terrain there were similar to that found in Israel's Golan Heights.

Because U.S. intelligence agencies did not want anyone to know that members of the Tank Systems PMO were doing business in an Energy Department facility, all participants were required to adjust their identities. After our arrival in Boise we flew to Idaho Falls to join John James, the Energy Department PM, for briefings and a tour of the intended facility.

The structure to be used for the production facility was an old eight-story hangar-type building. It had been designed to house another classified Energy Department project that had been abandoned, and the only occupants were great horned owls and much trash. This facility would effectively hide the work from Soviet satellites. Although there were serious doubts about both the cost and risks involved with this program, it was accomplished thanks to the outstanding work of John James and his Energy Department associates as well as Colonel Raffiani and his team. One of the biggest problems they faced was concealing the facility's purpose from the multiple contractors involved with remodeling it so that the DU armor secret was not compromised.

The transition from R&D to the production of DU armor was far more difficult than anticipated. While it was relatively easy to make a few armor packages by hand, it was a much bigger challenge to make quality packages at a high production rate as DU armor had never been fabricated in that fashion before. In addition, the production equipment and manufacturing processes had to be developed at the same time. And all of this had to be accomplished in total secrecy.

Another concern was the fear that depleted uranium posed a health hazard. This perception persisted in the minds of some even though a form of depleted uranium called "staballoy" was already in use in tank ammunition because of its high density and pyrophoric (fire-causing) effect on impact. Staballoy penetrators for armor-piercing ammunition were stored in tanks both in and out of the United States. Physicists, the Army Surgeon General, and members of the safety community had conducted extensive tests and concluded that staballoy did not pose a health problem. Secretary of Defense Caspar W. Weinberger, after being advised that all problems and issues had been resolved, finally approved the production of DU armor.

In January 1986 Peter McVey replaced me as U.S. Army Tank Systems PM. It marked the fourth time he was involved with the Abrams system, and he remained with the program until 4 October 1993—a total of eight years.

During his tenure McVey saw many changes in his charter and in the organization of the PMO. First, he was the tank systems PM. Later, in 1986, "he became a capstone program manager for close combat vehicles which included tanks and light combat vehicles. In 1987, McVey became one of the Army's first Program Executive Officers initially for close combat vehicles and then heavy force modernization. From that assignment he became the Program Executive Officer for armored system modernization that included the Abrams tank systems, the Bradley fighting vehicle systems, and future combat vehicle systems."[38]

General McVey confronted his first major problem when he arrived in Detroit. The United Auto Workers Union (UAW) had been on strike against General Dynamics seeking restoration of the money concession its members had made when Chrysler was in financial trouble. This bitter struggle was eventually settled, but tank production fell six months behind schedule. As it turned out, the delay was fortuitous as the time was used to solve the ammunition compartment problem. Although it had been solved during development testing, the fix was not repeatable in the production vehicle. It was another case of having to make the transition from handmade articles to a production line—something that is not an easy task even within the same company.

The ammunition door problem identified during development testing was caused by the gap tolerance in the ammunition doors. When the production engineers began working on the vehicle they said the tolerance of the gap would have to be increased by 50 percent to make it easier to manufacture. This became a major issue since the original ammunition door gap was designed to keep gasses that accumulated after a detonation outside the crew compartment. This was obviously a "no-go" situation. McVey halted production and directed that General Dynamics meet the survivability requirement. The problem was solved—as it had been in the development phase—by the hard work of contractor and government engineers. There was no real impact from this delay in producing the M1A1 because the line was still turning out the initial production M1s that had been delayed by the UAW strike.

With the ammunition compartment problem solved, the M1A1 was being produced and would eventually replace the other variants in the active army. Early M1A1s, whose production began in August of 1985 and was disrupted by the UAW strike, were produced without DU armor. That protection was introduced into production in May 1988. This variant was called the M1A1 Heavy. The production run ended after a total of 2,329 M1A1 and 2,140 M1A1 Heavy tanks had been produced for the army.

As production of the M1A1 variants continued, the next Abrams variant, the M1A2, was being designed. Colonel Raffiani, the PM, and Lt. Col. Jim Barbara led the management team for the revolutionary product improvement of this already technically advanced tank.

McVey and the chief of armor, Maj. Gen. Frederic Brown, had explored the virtue of digital versus analog electrical systems. Specifically, digital control is more precise. With digital electronics, gun calculations and position become more precise and crisp. A gun laid on target with digital electronics will be quick and accurate. In other words, digits eliminated the "slop" from the system.

Maj. Gen. Thomas Tait, who replaced General Brown at Fort Knox and became the principal user representative, had significantly different views from those of his predecessor. Tait emphasized target acquisition, speed of delivery, very accurate fires, and improving tank ammunition rather than developing new digital capabilities. Tait's message to the developer was clear and unwavering: "Do not take your eyes off the objective. The Abrams tank must be able to provide accurate protective fire first on the battlefield, it must survive on the battlefield, and we must be able to execute violently."[39]

The M1A2 program was approved in 1988, shortly before the demise of the Soviet Union. It appeared that the Cold War was over, and it was time for a peace dividend, according to the Congress. Major questions now arose: Why do we need an Army? Why do we need these expensive machines? Who is the enemy now?

Personnel at TRADOC, McVey, and Col. John E. Longhouser, the new M1 PM, had to answer the numerous questions being asked and to keep the tank industrial base open. Lt. Gen. Don Pihl, the assistant to the acquisition executive for the Army, led the effort to defend the M1A2 in the DOD and Congress. After being informed that the M1A2 would be canceled and that there would be no new production of the Abrams tank after completion of the current production run of M1A1 Heavy tanks, Pihl secured DOD approval to complete pilot vehicles and procure sixty-two new production M1A2 vehicles. His effort probably saved the industrial base and the tank program.

Nevertheless, it was time to "bite the bullet." The armor system modernization plan called for a new tank in the year 2002, but the Pentagon's leadership and the president's Budget Office preferred converting M1A1s to M1A2s. McVey noted that this "was not greeted with any enthusiasm by anybody, especially Longhouser, who thought it would be a nightmare to manage, much harder than new production. Regardless, two plans were produced, first to convert M1A1s, followed by converting M1s into the M1A2 configuration."[40] Only a few saw this as a viable way to get M1A2s and maintain the base until the future system arrived.

Faced with the end of production, the Army—supported by the OSD and contractors involved in tank production—began an aggressive plan to sell M1A2 tanks to overseas customers to keep the production line open and to provide revenue for additional tanks.

The M1A2 was now being prepared for a low-rate initial production decision. In July 1990 Col. John Caldwell replaced Longhouser as the Abrams tank PM. Caldwell had just completed a tour with the JCS, where he was familiar with the planning that later determined which units would be deployed to meet the Iraqi invasion of Kuwait. This information proved helpful, as McVey and the Program Executive Office (PEO) were to play an important role in the Gulf War effort.

Because of the Cold War and the need to have first-line equipment facing the Soviet threat the M1A1 tank had been deployed first in Europe, followed by issue to CONUS-based units. There was one exception: the 3d Armored Cavalry Regiment at Fort Bliss. This unit was issued the first production M1A1 tanks, which, on the eve of the Gulf War, were three years old.

As Middle East tensions increased, NATO and USAREUR were in the midst of complying with the Conventional Forces Agreement (CFA), which reduced the number of forward-deployed tanks and other weapons systems. However, the United States was not yet able to plan on using Europe-based forces for Operation Desert Shield in early July and August of 1990, shortly after the Iraqi invasion. This meant that forces deploying to Saudi Arabia would have to come from the United States with their 105mm M1s. Although this was still a very good tank, it was not as survivable as the M1A1 with its 120mm gun.

Why, everyone wanted to know, couldn't the M1s be replaced by

M1A1s in Saudi Arabia? McVey and Caldwell, along with Col. Dave Bird, the European Abrams fielding team chief, set out to find an answer. They subsequently devised a plan that called for pulling tanks and ammunition from reserve stocks in Europe. McVey's team briefed the plan to key leaders in Saudi Arabia and at the Pentagon. General Carl Vuono, then serving as army chief of staff, made the decision to delete the M1 tanks scheduled for deployment and replace them with 120mm M1A1s drawn from outside the United States.

The Commander in Chief of U.S. forces in the Gulf, Gen. H. Norman Schwarzkopf, feared that the change in tanks might add training delays to his timetable. He feared it would require too much time to make the crewmen proficient with the new tanks and recommended that the crews stick with their older M1s. General Vuono and Gen. Colin Powell, chairman of the JCS, ruled that the more capable tank was the way to go. They were essentially brand-new tanks, and the M829A1 and M830 enhanced 120mm rounds extended the lethality and range of the gun system. The tank's heavy armor also increased survivability.[41]

By the time Operation Desert Storm started, all armor units were equipped with the M1A1 tank except one brigade of the 1st Infantry Division. It was a remarkable feat accomplished by the entire Army, but

M1A2 Abrams main battle tank with 120mm gun. General Dynamics Land System.

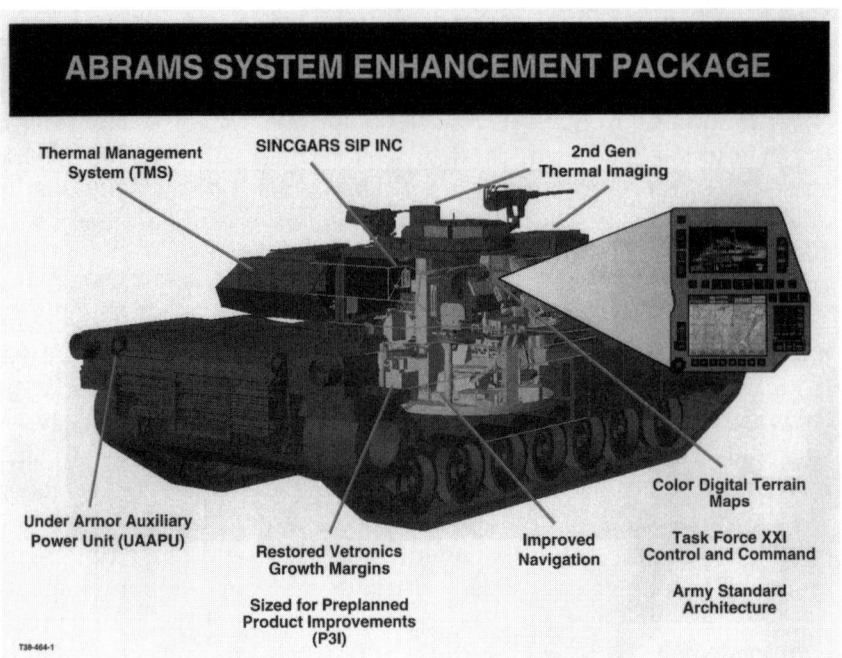

ABRAMS SYSTEM ENHANCEMENT PACKAGE

Thermal Management System (TMS)

SINCGARS SIP INC

2nd Gen Thermal Imaging

Under Armor Auxiliary Power Unit (UAAPU)

Restored Vetronics Growth Margins

Sized for Preplanned Product Improvements (P3I)

Improved Navigation

Color Digital Terrain Maps

Task Force XXI Control and Command

Army Standard Architecture

a great deal of credit must be given to the armor team that put the plan together.

Meanwhile, the M1A2 development program continued along its rocky path. Following Desert Storm, the M1A2 completed its early user tests at Fort Hunter Liggett, California. The results were mixed. Integration of the commander's independent thermal viewer reduced target acquisition time dramatically: "4 seconds faster in the 1st engagement, 4 seconds faster in the 2nd engagement and 16 seconds faster in the 3rd engagement—this was truly going to be a fighter's tank."[42]

However, several systems did not work as well. The command and control and Inter-Vehicle Information System (IVIS) did not perform as expected. The new radio system experienced severe range restrictions when using digital overlay. The digital systems also caused some safety and software problems, including a problem with gun jump—uncontrolled movement of the main gun tube from the gunner's aiming point. The result was a pause in testing while changes were made and crews retrained. The tests then resumed and communicators, contractors, and government engineers continued to correct problems. When they were all satisfactorily resolved, Maj. Gen. William Forrester, head of the OTEA, issued a letter to the M1A2 PM confirming that the tank was now acceptable to the user.

McVey and Caldwell, working the halls of the Pentagon, sought approval for limited production of the M1A2. David Latson, a deputy PM for the Abrams system recalled, "On 9 January 1990 a memorandum signed by George E. Damson, acting Assistant Secretary of the Army (Research, Development and Acquisition) approved acquisition of long lead items and material for 62 tanks."[43] The first two M1A2s rolled out of the Lima plant on 1 December 1992.

The rollout ceremony attended by Chief of Staff Gen. Gordon Sullivan and Mrs. Abrams was not the end of the story. Another test was required before full production could begin. This was the operational test and evaluation conducted by the Operation Evaluation Command from September to December 1993. This test showed the commander's independent thermal viewer and the position location system to be winners, but there were safety problems with uncontrolled gun jump and in the communication supporting the IVIS. Further training in the use of the IVIS was required, which caused concern about its effectiveness.

The last sidebar to the Abrams tank story involves the gas turbine engine, which was plagued by a stormy start but eventually became a winner. By the time of the Gulf War, engine readiness rates in the high 90 percent range were common most of the time. While the Abrams's range was limited because of its high fuel consumption and the crew needed to frequently clean air filters in the sandy desert environment, the army found these drawbacks manageable. However, it was found that the army's standard recovery vehicle, the M88A1, could not recover the seventy-ton Abrams.[44] In mid-1996 production of the gas turbine engine at the Stratford engine plant was terminated. Once the sixty-two M1A2 tanks were built, there would be no more new tank engines for the army. All future engines would be rebuilt at the Anniston Army Depot and would go into earlier M1s that had been converted to M1A2 configuration.

General McVey retired on 30 September 1993. He had been the longest serving PM in the history of Abrams development, and his continuity was a major factor in the program's success. McVey was replaced by General Longhouser, who was responsible during his tenure for completing the operational test and evaluation, implementing the conversion program, supporting the foreign military sales program, and continuing with what was left of the Abrams system management program. Of course he also had the difficult task of fighting for army programs as the "peace dividend" movement continued to take its toll on the tank industrial base that had been developed so diligently during the Cold War. The Detroit tank plant had been closed, AGT-1500 engine production was ending, and the last M1A2 was delivered to the government at the Lima plant on 11 March 1994.

General Longhouser and Colonel Caldwell continued to correct problems discovered in operational test and evaluation, and shepherded the acceptance process for the conversion program through the Pentagon. David Latson, the deputy PM, recorded that "the Abrams upgrade program was given approval by the Army Systems Acquisition Review Council to proceed into full rate production on 8 April 1994. George Dausman, the Army Acquisition Executive, signed the decision memorandum on 20 April 1994, approving the program for production and deployment with delivery of 998 converted vehicles. The memorandum also stipulated that production [upgrading] was anticipated beyond that plan."[45]

On 23 July 1994 Col. Christopher V. Cardine replaced Caldwell, who was subsequently promoted to major general and continued to support the Abrams program from the Pentagon. Cardine had not previously been involved with the Abrams, but had been part of the Bradley fighting vehicle program. Cardine had also been a member of the PM team charged with correcting the OT II test problems identified between 1979 and 1981. He was an armor officer, well schooled in mounted warfare and in the acquisition process.

Preplanned product improvements, which began in the late 1970s, will continue for the life of the program. The new and converted M1A2s differed dramatically from the M1A1. Effectively putting two sets of eyes on potential targets, instead of one, the Commander's independent thermal viewer made shorter engagement times a reality while giving the Commander a 360-degree, all-weather, day-night target surveillance capability under most battlefield conditions. In my personal opinion, this feature is, in itself, the most important feature of the M1A2.

In addition, the M1A2 has the Commander's Integrated Display Unit (CID), a single unit that can display either the Commander's independent thermal viewer or the command, control, and communications functions. From the integrated display unit the tank Commander can send orders to other tanks, send orders requesting supporting artillery, or fire the main gun by using the independent thermal viewer and the Commander's control handle.

Although not yet fully perfected, the CID will allow the tank Commander to enjoy perfect "situation awareness" of the battlefield. In other words, the platoon leader will be able to see on his CID both his location and the location of other tanks in his platoon. Likewise, his company Commander can see both himself and his platoon leaders, and the battalion Commander can see the locations of his company Commanders. Commanders will be able to tailor what appears on the screen to their needs. Their situation awareness will depend on the individual preferences of Commanders, the tactics being employed, and procedures that are being developed as the system matures. The IVIS concept of digital command and control—what the Army calls "digitization of the battlefield"—offers the potential for dramatically improved command and control of the tank in combat. Cardine says that as a result, "the M1A2 has become a heavily armored, lethal computer, rather than a conventional tank. This architecture provides the tank with functional growth potential."[46]

As this is written, Colonel Cardine, supported by the Fort Knox users, is planning the next stage of product improvements. This version, Latson noted, "is now titled the M1A2 System Enhancement Program (SEP) and includes: a color display for the tank commander integrated into the commander's display unit, an under-armor auxiliary power unit and, a thermal management system to cool the vetronics and crew."[47]

In addition, the Army is currently developing processor and memory upgrades to the digital system to meet a common digital operating environment.[48]

The engineering programs started in 1997 will test prototype tanks with enhanced electronic changes in 1998 and 1999. A tentative M1A2 SEP will enter production in 1999 at the rate of 120 per year. All 1,079 M1A2s will be converted to this configuration and are to be fielded in the third quarter of FY 2000.

The active Army fleet will contain about 2,700 tanks through the year 2015. It is expected that 998 M1s will be upgraded to M1A2 standards, joining the original 62 M1A2s. In addition, in an attempt to keep the industrial base open, both McVey and Longhouser and their contractors successfully consummated foreign military sales to Egypt, Kuwait, and Saudi Arabia.[49]

The Abrams tank system is an amazing story of the development of a machine and of the people and organizations who created what is likely the most effective tank in the world today. Although I have concentrated on the continuity among the several military PMs, the M1 tank system would never have been successful without AMC's dedicated civilian employees, especially those in the Ballistics Research Laboratory who worked so hard on both the armor and the ammunition for both the 105mm and 120mm gun systems.

The Abrams tank is expected to remain in service until at least 2025. In addition, eight hundred original M1s are being converted into a heavy assault bridge launcher, called the Wolverine, to increase maneuver force mobility by allowing units to cross antiarmor obstacles. Also, as a result of the lessons of Operation Desert Storm, an M1 chassis-based system called the Grizzly Breacher, equipped with a full-width mine-clearing blade and a power-driven evacuating arm, is currently in the design maturation phase of engineering and manufacturing development.[50]

NOTES

1. GAO, "Operation Desert Storm: Early Performance Assessment of Bradley and Abrams" GAO/NSIAD-92-94, Jan., 1992, pp. 21-24.

2. Richard M. Ogorkiewicz, "A Battle Tank for the 1970s: A U. S.-German Design with Innovations," *The Engineer,* 2 Feb. 1968, pp. 198-200; and Fred Schreier, "The XM1 Program—A Synopsis," *International Defense Review* (Mar. 1977): p. 460. For the Army's defense of the XM803 project, see Meritte W. Ireland, "MBT70/XM803," *ARMOR* (July-Aug. 1970): pp. 33-34. Colonel Ireland was the chief, Technical Coordination Division, Office of the MBT Project Manager.

3. Brooke Nihart, "Main Battle Tank Still in Trouble with Congress, OSD," *Armed Forces Journal,* 21 June 1971, pp. 36-37; House of Representatives, *Authorizing Appropriations, Fiscal year 1972, For Military Procurement, Research and Development, and Reserve Strength, and for Other Purposes,* Report No. 92-232, 92d Cong., 1st sess., 26 May 1971, pp. 24-25; and Senate Report No. 92-359, 92d Cong., 1st sess., 7 Sept. 1971, pp. 36-37.

4. *Congressional Record-House,* 15 June 1971, H 5196; and *Congressional Record-Senate,* 30 Sept. 1971, S 15548-50. Also see Thomas F. Eagleton, "Fact Sheet on Amendment # 445 regarding the Main Battle Tank," *Congressional Record-Senate,* 30 Sept. 1971, pp. 1-4.

5. *Congressional Record-Senate,* 30 Sept. 1971, S 15548.

6. Lt. Col. (P) Larry E. Willner, "The Birth of the Main Battle Tank," Student Report No. 75-028, 1975, The Industrial College of the Armed Forces, p. 2.

7. Gen. James H. Polk, "We Need a New Tank," *ARMY* (June 1972): pp. 8-14.

8. Willner, "Birth of the Main Battle Tank," p. 5.

9. Ibid., p. 8.

10. Ibid., p. 9.

11. Eric C. Ludvigsen, "XM1 Face-Off to Have Strong Impact on Future Ground Forces," *ARMY* (Feb. 1976): pp. 32-35; and Orr Kelly, *King of the Killing Zone* (New York: W. W. Norton, 1989), p. 7.

12. Kelly, *Killing Zone,* p. 128.

13. Lt. Gen. Robert J. Baer (Ret.), interview with Joe Mahalik, TACOM command historian, 25 Sept. 1985, p. 5.

14. Ibid., p. 32.

15. Richard P. Hunnicutt, *Abrams: A History of the American Main Battle Tank,* vol. 2 (Novato, Calif.: Presidio Press, 1990), p. 181. Hunnicutt's excellent study concentrates primarily on the technical aspects of U.S. tank development from the early 1950s to the end of the 1980s.

16. Kelly, *Killing Zone,* p. 149.

17. Comptroller General of the United States Report to the Congress, *Department of Defense Consideration of West Germany's Leopard as the Army's New Main Battle Tank,* PSAD-78-1, 28 Nov. 1977, pp. 3-7.

18. Baer interview, p. 43.

19. Hunnicutt, *Abrams,* p. 195.

20. For a debate on this issue see Comptroller General of the United States Report to the Congress, *Critical Considerations in the Acquisition of a New Battle Tank Department of Defense,* PSAD-76-113A, 22 July 1976; and Comptroller General, *Department of Defense Consideration of West Germany's Leopard.*

21. "Tanks: How not to standardize," *The Economist,* 2 Oct. 1976, p. 59; and "A starter for Carter," *The Economist,* 6 Nov. 1976, p. 17.

22. Comptroller General, *Department of Defense Consideration of West Germany's Leopard,* pp. 8, 15.

23. Lt. Gen. Donald M. Babers (Ret.) interview with Joe Mahalik, TACOM command historian, 24 Sept. 1987, p. 22.

24. Ibid., p. 22; and James P. Herzog, "Two lawmakers fight separately for tank contract," *Cincinnati Post,* 13 Feb. 1978, p. 11.

25. Babers interview, pp. 15-19.

26. Hunnicutt, *Abrams,* p. 197.

27. Babers interview, p. 9.

28. Hunnicutt, *Abrams,* p. 202.

29. Comptroller General of the United States Report to the Congress, *XM1 Tank's Reliability Is Still Uncertain,* PSAD-80-20, 29 Jan. 1980.

30. Comptroller General of the United States Report to the Congress, *Large-Scale Production on the M1 Tank Should-be Delayed Until Its Power Train Is Made More Durable,* MASAD-82-7, 15 Dec. 1981.

31. Kelly, *Killing Zone,* p. 160.

32. Ibid.

33. Hunnicutt, *Abrams,* p. 204.

34. Babers interview, p. 18.

35. Ibid., p. 19.

36. Hunnicutt, *Abrams,* p. 214.

37. TACOM Annual Historical Review, 1984, author's files.

38. Maj. Gen. Peter M. McVey, interview with Thomas F. Kornacki, TACOM historian, 25 Sept. 1993, p. v.

39. Babers, telephone interview with author.

40. Unpublished notes provided to the author by General McVey.

41. Ibid.

42. Ibid.

43. David Latson, deputy program manager for Abrams tank systems. A summary dated 24 Dec. 1996 titled, "The Abrams Program," printed from his computer files for author, p. 4.

44. GAO, "Operation Desert Storm: Early Performance Assessment of Bradley and Abrams," pp. 21-30, 38-39.

45. Latson, "The Abrams Program," p. 6.

46. Christopher V. Cardine, comments to author, Oct. 1997.

47. Ibid.

48. "Decisive Operations: M1 Abrams Tank," *ARMY* (Oct. 1997): pp. 259-60.

49. Latson, "The Abrams Program," p. 12. The following is the planned distribution of the Abrams tank system worldwide as of Jan. 1997: M1, 2,374; IPM1, 894; M1A1, 4,492; M1A1 (National Guard), 58; M1A1 (USMC), 221; M1A2, 62; M1A1 (Egypt), 555; M1A2 (Saudi Arabia), 315; M1A2 (Kuwait), 218. Total: 9,189.

50. "Engineer Equipment," *ARMY* (Oct. 1998): pp. 275, 277.

Bibliographic Note

The Abrams tank system progressed through tortuous research and development processes because of dedicated individuals both in and out of the army. All their contributions to the program were noted in the above Abrams history. There were many others whose contributions were critical and deserve credit for their contribution to the Abrams history. A special thanks is extended to TACOM Security for the information concerning depleted uranium armor.

This chapter would be incomplete if it did not list another group of individuals who played significant roles in the Abrams tank development: the dedicated personnel of Chrysler and General Motors, the two firms that contended for the XM1 contract, and General Dynamics, which bought the program from Chrysler. Dr. Phil Lett, lead engineer for Chrysler, stayed with the program from its inception, and remains a senior consultant for General Dynamics. Retired army Cols. Joe Yeats, Lou Felder, and Carmelo Milia, who started with Chrysler and continued with General Dynamics, made significant contributions to the program. Mr. Fred Best was the General Motors leader through the most significant days and did a superb job. The General Motors production planning work was of particular note.

Others who helped this project were appointed U.S. government officials: Donald Rumsfeld, William Clements, Norman Augustine,

Howard Callaway, Martin Hoffmann, Ed Miller, Clifford Alexander, and James Ambrose—all provided support that helped keep the program on course. (This may be questioned by some, as political and international winds were not always favorable.) The contribution of Justus (Judd) White, in helping all players walk the tank program through Congress, was enormous. Orr Kelly, author of *King of the Killing Zone,* Richard Hunnicutt, who wrote *A History of the Main Battle Tank,* and Joe Mahalik, who assembled the tank history in the TACOM Oral History Program, did an impressive job in telling the Abrams story.

The following individuals read and commented on this manuscript: Donn Starry, Glenn Otis, Robert Baer, Donald Babers, Robert Harrington, Thomas Lynch, Peter McVey, Joseph Raffiani, John E. Longhouser, Philip Bolté, Joseph Ameel, Michael D. Jackson, Richard L. Knox, Christopher Cardine, and Judd White. Their comments were deeply appreciated.

A sincere thanks to Jon T. Clemens, managing editor of *ARMOR,* and George F. Hofmann, who made editorial changes and additions to the original text.

14

The Approach of Mounted Warfare in the Marine Corps (1970-95)

Kenneth W. Estes

The fluid and often confused nature of Vietnam War engagements left little legacy for the Corps. Interservice rivalries, especially over the control of airpower and lesser-scale spats over the command of large ground formations, left many senior Marine Commanders wary of the future American way of war. Marines felt reassured that their emphasis on small-unit tactics and leadership had been rewarded. However, there was less certainty over the impact of heavy weapons systems in the long and agonizing campaign that resulted in such a seemingly indifferent outcome.

On the other hand, the traditional Cold War enemy remained in place and the Corps clearly returned to post–World War II concepts of mobilization and reinforcement of the European theater. With that renewed emphasis, one could anticipate a reassessment of the firepower and mobility resident in the Fleet Marine Force (FMF), and the modernization of armored fighting vehicles not deemed crucial to operations in Southeast Asia.

A new amtrac was the first major hardware item the Corps acquired in the post-Vietnam era, and it proved to be a thoroughbred. The Commandant had hoped to field it in FY 1967 and contracted in June 1962 for six LVTPX11 prototypes to be delivered by the end of 1963. The

The LVTP7. Naval Institute.

headquarters staff took direct control of the development program, charging the Landing Force Development Activity (later the Developmental Center) as the program manager. An ambitious Borg-Warner design using a Lycoming gas turbine, electric drive, band track, and improved armor was specified. Unfortunately, the program was canceled in mid-1963 when the design proved unfeasible because of propeller propulsion limitations and the unfortunate beam-length ratio. Smarting from these setbacks, the Corps turned to the FMC Corporation in its search for a more conventional tracked vehicle.[1]

Between 1967 and 1968 the Corps took delivery of fourteen prototypes of the LVTPX12—a nineteen-ton carrier for twenty-five troops that featured a torsion bar suspension and 400 horsepower Detroit Diesel engine linked to water jets. The vehicle was capable of driving forty miles per hour on land and eight miles per hour in water. Testing through 1969 revealed it to be highly superior in all measures to the LVTP5 family except in internal volume. It particularly distinguished itself in surf, crossing through plunging surf in excess of ten feet while fully loaded. Although prototypes included a 20mm cannon and 7.62mm machine gun in a power turret, the first production version delivered for testing in October 1971 featured the .50-caliber M85 machine-gun. The Hispano-Suiza 20mm cannon had caused problems and project officers feared the heavier weapon would be employed as an assault gun.[2]

The new LVTP7 began to reach the FMF in 1972, but only additional recovery and command variants were produced. No engineer or armored amphibians were built owing to budget limitations. The same year, the Corps retired the M103 and M67 tanks and established the four-company, all-medium tank battalion as standard. Hopes ran high that the XM803 tank would be purchased, giving the Corps a state-of-the-art tank in place of the aging M48. The failure of this last vestige of the Army's MBT70 program threw the Corps into a quandary. The M48, even if up-gunned, would not fare well in combat with Warsaw Pact armor. Once more, the standard Army tank had to be acquired, and the M60A1 replaced the Corps's M48s in 1974, followed in 1977 by the newest production M60A1 (RISE)(PASSIVE), which featured passive night sights, full stabilization, and the "reliability improved" engine.

However, much work remained to be done, for Marine Corps units and leaders lacked experience in the conduct of mechanized operations. The tank battalion and assault amphibian battalions returned to the division organization from force troops in 1976-77, marking the end of the "light" period brought on by the Hogaboom Board. Even with an improved amtrac, redesignated as the assault amphibious vehicle (AAV) in 1985, and the current main battle tank, modern notions of combined arms grew slowly in the FMF. Marines did poorly in the "Strafer Zuegel" NATO exercise in which a Marine Amphibious Unit (MAU) participated in an exercise in northern Germany. Restricted from using helicopters in the accustomed way, the MAU, a reinforced battalion with a platoon of M48A3s and LVTP7s, was hopelessly outmatched by fast-moving *panzertruppen*.

But early trial exercises with the 2d Tank Battalion and 2d Marine Regiment had begun in 1972, using M48A3 tanks and the LVTP7s at Fort Stewart, Georgia. In 1976 the FMF transferred a reinforced platoon of AAVs to the desert training base at Twenty-nine Palms, California, where the 3d Tank Battalion had returned from its former base in Okinawa, and the "Palm Tree" series of mounted fire support exercises began in earnest. Thus when Marine Corps units returned to the fall NATO exercises in 1976, it was with a complete 4th MAB, which included a tank battalion (two medium companies) with M60A1 tanks and a half battalion of LVTP7s mounted by troops now more familiar with basic tactics. Success in that year's Exercise NATO Bonded Item 76 carried over into an even better performance in Exercise NATO Bold Guard 78. A beefed up 4th MAB that included fifty-three tanks and more than a hundred LVTP7s, with fire support provided by a self-propelled artillery battalion backed by the full panoply of Marine aviation: helicopters, fighter-bombers, and missiles, participated in the latter exercise.

The Corps introduced the TOW and Dragon ATGMs into the FMF

in 1975. However, the TOW platoons were concentrated in a large AT company added to the tank battalion organization for improved maintenance support. These were part of the growing pains that Marine Corps tacticians experienced as combined arms skills grew haltingly in the FMF.

Much of the Corps's leadership viewed the change to mechanized operations on the European continent with considerable trepidation. The Corps retained its historic paranoia, fearing budget conscious observers in Congress and the Pentagon would view it as a superfluous second land Army. Many generals wanted to abandon what appeared to be a heavy armored concept of operations and embrace with new enthusiasm the amphibious capability the Corps had exemplified in World War II—an indispensable capability for the United States, which was viewed from within the Corps as a traditional sea power. But even the new assault amphibious vehicle had only improved its water speed by a few knots. An amphibious landing force for the 1980s had to do better, especially if it was to gain the notice of NATO strategists.

Accordingly, in August 1974 the Corps issued a requirement for a new Landing Vehicle Assault (LVA), to be produced by the mid-1980s. Compared to the automotive evolution represented in the LVTP7 series, the LVA called for revolutionary progress through the development of a planing hull with retracting suspension, powered through the water at twenty-five miles per hour by a high-powered but compact stratified-charge rotary engine. The Marine Corps Amphibian Vehicle Test Branch evaluated industry proposals in 1975, recommending against surface effect and hydrofoil designs, based upon their complexity, transition problems from high to low speed, and surf zone and river hazards. Estimating that technical feasibility could be achieved by 1988, they urged the project be continued with the fifteen hundred horsepower rotary engine as the primary power plant and a gas turbine developed as an alternative power source. Two developmental contracts were issued: one for an air cushion vehicle and the other for a planing hull design solution. Some 932 LVAs (seventy-seven with command and control modules) were planned to replace the 855 LVTP7 and seventy-seven LVTC7 models, with fifty-four retrievers to be replaced in kind. The concept of operation was for the LVA to land an infantry regiment from twenty-five miles offshore through calm seas with no more than one-foot swell, across eight-foot surf, and then continue to attack ashore at forty to fifty-five miles per hour as an APC using vehicular weapons. Tanks and other support would make the same landing in trace, using air-cushion landing craft already under development by the Navy to replace conventional landing craft.

In 1979, however, cost estimates had shot up so that the Corps could not proceed, given the unresolved technical risks. The LVA program was canceled and an immediate requirement emerged for an improved AAV, called the LVX, which would bring the LVTP7's land fighting characteristics in line with the Army's M2 Bradley while retaining the same level of performance in the water. The LVX would have to perform as a true infantry fighting vehicle, with provision for from thirteen to twenty-one infantrymen to fight from within the vehicle.[3]

Alas, the complexities of merging the two types of vehicles resulted in cost estimates no better than the LVA's. As a result, in 1985 the Corps again reversed itself and embarked on an AAV7A1 program to extend the life of the AAV7 through the end of the century. New engines, weapons stations, and suspensions—as well as a series of other components, including appliqué armor—kept the AAV7 in service and tactically viable.[4]

Yet doctrinal and organization questions remained at the center of the controversy. The tank, assault amphibian, and infantry battalions lacked any common means of operating under a modern combined arms concept. Infantry manuals dated from 1970 and the new tank and AAV manuals written through 1976 merely reflected the new vehicle characteristics. They offered no new approach to their employment. The Army had embarked on a new *FM 100-5* series, ultimately leading to the AirLand Battle concept. Yet, the dissimilarities of Army and Marine maneuver battalions prevented any simple copying. In the end, Marine units had to task organize the three units to mount infantry for extended periods in AAVs, and mix them into tank- or infantry-heavy combined arms task forces. Yet the AAV leaders insisted on maneuvering the vehicles, leaving the infantry mere riders, as if embarked in landing craft or helicopters. The tankers, on the other hand, wanted the infantry to come out of their mental foxholes and directly embrace the goals of combined arms, relegating the amtrackers to the role of simple custodians of their equipment. The 1975 Haynes Board suggested the creation of permanent heavy assault regiments, one on each coast, in which the infantry, tank, and AAV battalions lived and worked together to create the combined arms force deemed necessary for NATO operations. But none of the senior officers could bear to see such a decisive alteration of FMF organization, given the doubts and fears associated with Armylike organizations in the Corps.

Tentative doctrine emerged from the education center in the form of Education Center Publication (ECP) 9-3 (1978) and later Operational Handbook (OH) 9-3B (1980). These introduced the concept of Marine battalion-level combined arms task forces, directing the formation of mobile assault companies that effectively translated conventional mechanized doctrine applicable to Marine Corps line units. The way at last stood clear for effective tactical growth for Marine Corps combined arms and later a spirit of maneuver warfare using Marine Corps infantry units, mounted in AAVs, moving in concert with tank units, with a full array of artillery, naval gunfire, and air support. The combined arms exercise program at Twenty-nine Palms continued to refine these concepts, just as the Army worked to convert doctrine to practice at the National Training Center at Fort Irwin, California.

However, the doctrinal cohesion was never obtained in the Corps, despite lessons learned and repeated at Twenty-nine Palms, Camp Pendleton, Camp Lejeune, and excursions to Army bases for training. The principal factors preventing a true mechanized doctrine centered on the uneasiness of Corps leaders with mechanization, doubts about the strategic lift costs of so much heavy equipment, and apparent nostalgia for an almost mythical state of "lightness" that would leave the infantrymen supreme, devoid of technological burdens. After all, tanks and AAVs had not proven decisive in Vietnam or Korea, whence the current leaders had earned their professional reputations. Had not the Commandant, Gen. Robert Barrow, as a company Commander approaching Seoul, shrugged off the NKPA T34s with a few volleys of 3.5-inch bazooka rockets? Somehow, there had to be a way for Marines to remain different and not simply imitate the NATO armies.

Enter the light armor. Lighter armored fighting vehicles in the sixteen- to eighteen-ton range could be carried in the newest heavy helicopter, the triple-engine CH-53E, which began entering service in the late 1970s. If such vehicles could be made effective against Warsaw Pact armor, then the Corps could return to its long-cherished notions of vertical envelopment and discard the heavy mechanized regiments advocated by the Corps's handful of noisy armor advocates. In 1973 the Developmental Center proposed the design of a mobile protected weapons system (MPWS). Such a system would be a "light weight, highly mobile and agile, helicopter-transportable weapons system capable of supporting the infantry against armor, materiel and personnel targets, an assault gun in the force beachhead and for subsequent operations ashore." Several DOD agencies began to support such a project, and the Marine Corps received approval to lead a joint project, with the Army as junior partner, called the Armored Combat Vehicle Technology (ACVT) Program. The Army, ever reluctant to see programs threatening its main battle tank research and mindful of growing criticism in political and defense circles over past and future tank programs, nevertheless participated out of a need for a light combat vehicle for its airborne and airmobile forces.[5]

The ACVT Program centered upon two research vehicles and a field exercise. The forty-five-ton HIMAG chassis evaluated a fifteen hundred horsepower gas turbine, and the high-survivability test vehicle, light (HSTV-L) used a 650 horsepower gas turbine on a 20.5-ton chassis. Both vehicles offered minimum armor protection against small arms, relying on their "agility" to evade killing shots from antiarmor weapons on a smoke-filled and confused battlefield. The Marine Corps and the Army evaluated that tactical concept at Fort Hunter Liggett in an instrumented test dubbed the Advanced Antiarmored Vehicle Evaluation (ARMVAL). A new hypervelocity automatic 75mm cannon was envisioned as the primary weapon in these and other projects.

The Corps published its formal requirement for the MPWS in 1978, seeking a sixteen-ton vehicle that emphasized antitank capability. The requirement called for a fully developed and engineered vehicle ready to enter service in 1988. However, if a hybrid vehicle could be converted from existing light tank or armored car designs, the Corps hoped to field it in 1986.

Clearly, the MPWS could not be achieved in the near future, yet the Corps accepted more commitments with major NATO and theater commands facing mechanized opponents in Europe, the Far East, the Middle East, and Southwest Asia. The Corps aligned troops to the three brigade sets of equipment to be stored on board maritime prepositioning ships (MPS) and ready for operations by the mid-80s. The brigades had to meet DOD guidance calling for the equivalent of an Army mechanized brigade on the ground, together with the other components of a Marine air-ground task force. All the active tank and assault amphibian battalions would be required to man and equip these brigade sets. One element of Corps war planning continued to move toward the heavy assault regiment requirement, at the same time that the light force advocates searched for weapons to make their program effective under the same conditions of operation. Congressional testimony by key Marine Corps generals in 1980 confirmed that the Corps needed a near-term light armored vehicle (LAV) to keep the bulk of the infantry viable in the modern maneuver warfare envisioned in the war plans. Thus, the Congress came to the Corps's rescue with funds, beginning in 1981, for the purchase of LAVs available "off-the-shelf" in the world armaments industry. After a rushed series of tests, the Corps selected the General Motors–Canada LAV candidate, an eight-wheeler based on the Mowag design, as the basis for a family of vehicles.

The Marine LAV battalion started out as almost a clone of the

Marines around the world: A LVTP7 Amtrac crewman in Berbera, Somalia, Africa, spring 1982. USMC photo.

amtrac battalion, with a 25mm infantry carrier as its base vehicle, and variants for command, recovery, mortar, supply, and antiaircraft and assault gun missions. Infantrymen would be added only as an after-thought in the late 1980s, as the battalion moved in typology from a mobility/firepower unit to a motorized infantry and finally a mobile reconnaissance battalion, all in its first ten years of existence.

It seems clear that the Marine Corps generals who advocated the LAV battalion sought a miniature operational maneuver group. Such a

Marine Light Armored Vehicles. Naval Institute.

M60A1 Mine Clearing. USMC Museum and Historical Division.

force offered a theoretical rapid maneuver option to a Marine division still capable of mounting only one infantry regiment in AAVs, shuttling one regiment via helicopters, and retaining the third regiment as footmobile infantry. Ironically, the near-term solution to light armor became the final solution, for the technical promise of the MPWS faded in the face of main battle tank armor technology, the false hope that agility equaled protection, and the limited availability of heavy-lift helicopters.[6]

   After procuring the upgraded AAV7A1 and LAV vehicles in the 1980s, Corps planners focused their attention on the main battle tank. The armor threat represented by the Soviet T-72 and T-80 series tanks forced the NATO powers to field a new generation of tanks, including the Army's M1 Abrams. Four different Commandants had a hand in the long and painful deliberations that ultimately led to the Corps's decision to acquire the M1. Retaining the obsolete M60A1 or going to the industry's super-M60 option ran counter to the Army's logistics support priorities, which were transitioning to support of the M1 program. The M1 appeared and disappeared from Corps budget forecasts start-

USMC M1A1 Abrams. Patton Museum.

ing in 1981, when the Army first fielded it. Finally, in February 1985, the Commandant signed off on a decision memorandum that called for the purchase of 490 M1A1 120mm gun tanks to replace the 716 M60s then in the Corps's fleet of three active and two reserve battalions, three afloat MPS sets, and planned war reserve stocks. Changing to a four-tank platoon permitted some reductions, but mostly the depot and prepositioning requirements were shaved to meet the lesser requirement.[7]

Original planning for the M1A1 fielding called for training all battalions on the initial sixteen tanks at Twenty-nine Palms in 1991, and completing the active force issue in 1992. However, the Persian Gulf War altered that schedule.

In the wake of Iraq's seizure of Kuwait in August 1990, President George Bush ordered the U.S. Central Command (CENTCOM) to reinforce and defend Saudi Arabia and the other Persian Gulf states, in concert with a growing coalition determined to resist and ultimately expel the Iraqi forces. Among the first U.S. forces to arrive in Saudi Arabia for this purpose was the 7th MEB, which deployed to the key port and petroleum center of Al Jubayl, with its aviation based farther south on the Gulf of Bahrain. Off-loading heavy equipment, including fifty-three M60A1 tanks, twenty-eight LAVs, and about a hundred LVT7A1s from its dedicated Maritime Prepositioning Squadron 1, the brigade reported ready for operations on 25 August, a mere ten days after arrival in theater and weeks ahead of any U.S. Army armored force.

The brigade stood alone only for a few days, until follow-on elements from the I MEF began to arrive. Eventually growing to a force of more than ninety thousand Marines and sailors, the Marine Forces, Central Command (MARCENT), included two reinforced divisions, an enlarged aircraft wing containing the majority of the Corps's aircraft inventory, and the bulk of two force service support groups. More than twenty thousand Marines and sailors in the 4th and 5th MEBs remained afloat as the I MEF moved to the Kuwaiti-Saudi frontier.

Both the M60 and M1 series tanks saw their first American combat use with the Corps. The latter tank was hurriedly issued to companies from the 2d and 4th (Marine Corps Reserve) Tank Battalions as soon as they arrived in theater. The M60A1s also received "Blazer" armor appliqué kits, a Corpswide modification that had been delayed for the vehicles stored on board the MPS.

Turning to the mission and threat at hand, Marines prepared to fight what was widely perceived as a seasoned mechanized army, the seventh largest in the world. The ground units dug themselves in north of the Al Jubayl area and rehearsed antitank tactics, while Marine airmen scoured the skies, patrolling over land and covering the fleet units then in the Persian Gulf. The tank, LAV, and AAV units came into their own as the 1st and later the 2d Marine Divisions girded themselves for classic desert warfare. Unfortunately, no new progress in doctrine and training had occurred in the previous ten years, and the tank battalions had to "hold school" for the infantry regiments and their attachments on an ad hoc basis. In the end, each division and regiment developed its own procedures for joining AAV, infantry, engineer, tank, and antitank units into combined arms teams. All the active and most of the reserve armored vehicle battalions were sent to the force. These provided sufficient resources for each division to field two mounted regiments. In addition, the Army's 1st "Tiger" Brigade, 2d Armored Division, reported to the I MEF for operations, and eventually served with the 2d Marine Division. At sea, the growing amphibious forces maintained a vigilant stance, practicing when feasible under the difficult hydrography and limiting geography of the Persian Gulf.

After the allied coalition reached sufficient strength to assure the defense of the Persian Gulf states, the strategic posture changed. As the allies set about preparing for an offensive to liberate Kuwait, Marines began to grapple with the problems of overcoming an impressive system of field fortifications built by the Iraqi defenders facing the coalition. Marine commanders responded to these problems with an intensive period of training and the fielding of specialized engineer equipment designed to break the successive obstacles of mines, wire, ditches, and berms that they faced. For most Marines, this would be the most thorough training effort they experienced in their entire careers, as smaller and then larger unit battle drills were learned and rehearsed. The service support units struggled with long supply lines and harsh desert conditions to not only maintain the forces already present and arriving, but also to amass and store the sixty days worth of supplies commanders determined they would need for combat. In the end, an entire logistical support base for the ground combat element was excavated in the desert a short distance from the Kuwaiti border, almost under the enemy's nose. Aviation units worked hard to provide their sorties under the stiff maintenance and operational considerations, knowing that they would likely open the combat phase of the campaign long before any Marine placed his boot on Kuwaiti soil.

On 24 February the 1st and 2d Marine Divisions attacked into Kuwait, forcing their way through the Iraqi barriers and brushing aside frontline resistance. Mounted in a variety of tanks, AAVs, trucks, and LAV, the attacking regiments destroyed or captured whole battalions of Iraqi troops and swept through the burning oilfields toward the capital of Kuwait City. Artillery barrages and repeated strikes by fighter-bombers and attack helicopters supported the advancing regiments, halting only at night to prevent fratricide among the allied and Marine units.

A few counterattacks were beaten off with hardly any casualties, and the two Marine divisions, assisted by the Army's Tiger Brigade, pressed onward to the outskirts of the capital, cutting it off while Iraqi forces were attempting to withdraw. After just a hundred hours of combat, Marines dominated southern Kuwait and the capital was mopped up by neighboring Arab coalition forces.

In what Marines still refer to as "the drive-by shooting," the mounted regiments overran airfields and Iraqi battle positions without serious resistance, after first passing through the uncovered and poorly maintained minefields. The only armored vehicle casualties in the assault were in the mine-clearing teams, which lost a few dozer tanks while proofing the minefield lanes. The 1st and 3d Tank Battalions, with their M60A1s, claimed the destruction of 116 tanks and thirty-two APCs in their movement with the 1st Marine Division. The 2d Marine Division held the M1A1-equipped 2d Tank Battalion in reserve while the two M1A1 companies from the 4th Tank Battalion were attached to one regiment and the 8th Tank Battalion with its M60A1s was attached to the other regiment. In the only firefights in this zone, one of the 4th Tank Battalion's M1A1 companies killed or assisted in destroying fifty-

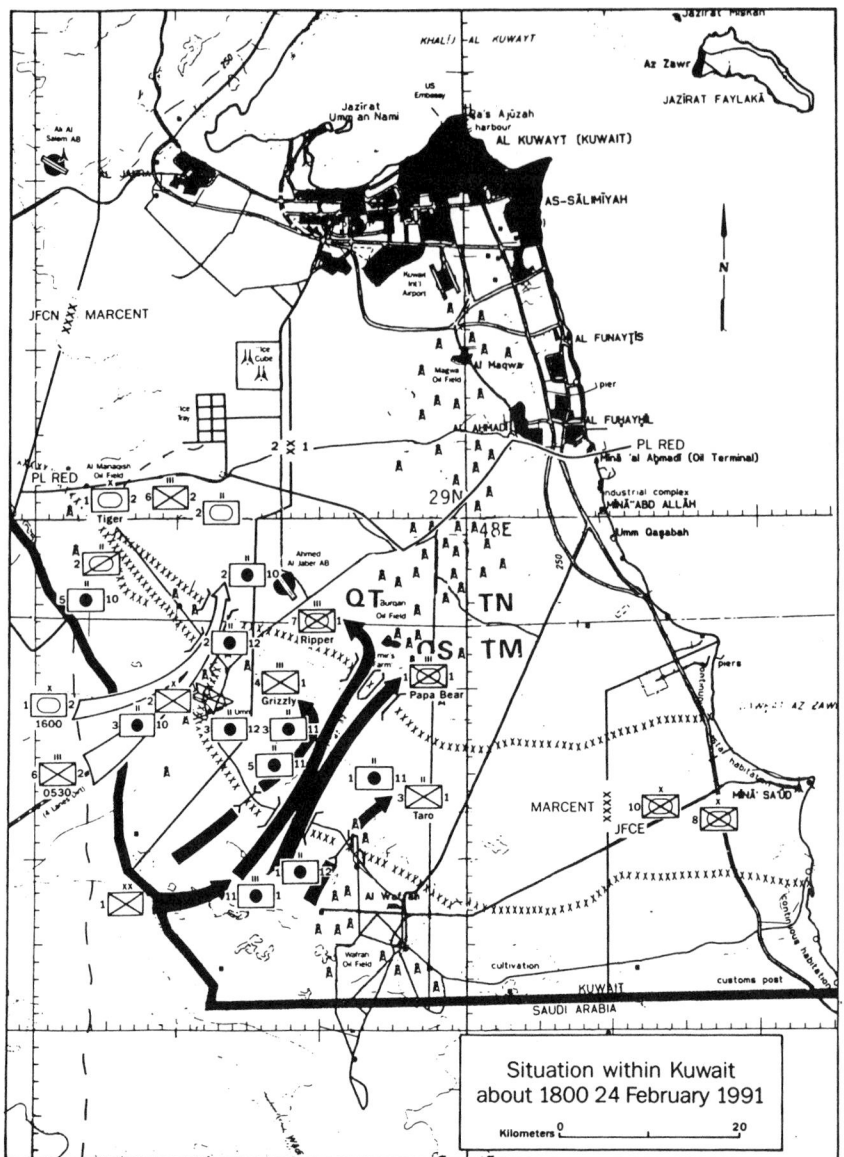

Situation within Kuwait
about 1800 24 February 1991

Kilometers 0 _____ 20

nine tanks and thirty-five APCs. The tanks fought at ranges of eleven hundred to twenty-six hundred meters, receiving not a single hit from an antitank weapon. The 580 AAV7A1 series vehicles performed equally well in mounting the bulk of four regiments and engineer teams and then waiting on board ships for the amphibious assaults that never took place.

Among the many emergency measures taken by the Corps to prepare for the campaign, the provision of adequate mine-clearing devices and systems responded to the last of the original 1949 Marine Armor Policy Board requirements. The mine roller and plow kits for both tanks performed well, and the line charge trainers and AAV7A1 launcher kits permitted the engineer teams to operate with great success against the large Iraqi barriers.

The performance of the LAV units in the Gulf War reflected the confused circumstances of their procurement and the doctrinal hybrid under which they operated. Finally organized with a few riflemen in the LAV-25 squads, the redesignated light armored infantry (LAI) battalions deployed to Saudi Arabia and generally operated in the style of U.S. Army armored cavalry squadrons in screening and minor reconnaissance missions. Deployed on a weak outpost line before the ground offensive began, two of the companies engaged Iraqi battalions on the frontier during the spoiling fight known as the Battle of Khafji. They scored impressive hits with their TOW missiles and held the old medium tanks and APCs in check with a large volume of 25mm fire. However, friendly TOW and aircraft fire claimed three of their own. Moreover, the I MEF and its divisions held no maneuver reserve in the vicinity of the outpost line to support the LAI battalions in the event they had to fall back. During the offensive, the LAI executed feints and maneuvered as a flank guard, and never served as the "eyes" for the mobile assault regiments—a role for which they were intended. The groping forward MEF units became so disoriented that the command posts of the I MEF, 1st Marine Division, and 1st Marine Regiment were each in turn menaced by intruding Iraqi tanks, most of which simply surrendered.[8]

Almost as fast as the Gulf War propelled the FMF into the heavy mechanized fighting that its leaders had sought to downplay, the Corps reduced its heavy component and began to theorize on the merits of a light force. Among other things, it sought technical improvements for the infantry and aviation components in order to reduce vehicle requirements.

At first, the Corps decided to resurrect the mobile assault regiments proposed by the Haynes Board, made more modern by mounting two infantry battalions in LAVs. Four such infantry battalions were subsequently created for duty in two combined arms regiments, which included the two surviving active tank battalions of the reduced post-war FMF. Another attempt was also made to develop a high-speed amtrac for amphibious assaults, designated the advanced assault amphibious vehicle (AAAV).

The Corps then abandoned many of the important parts of the ground force structure overhaul. The two combined arms regiments became too costly in face of the AAAV program's priority. The squeamishness of the Corps's leaders toward mechanizing ground units too much had reappeared all too soon in the aftermath of the Gulf War. Infantry units remained footmobile but capable of riding helicopters and AAVs to the battlefield, and thus were not capable of mounted combat. The three LAI battalions were converted to armored reconnaissance battalions, and two tank battalions and two assault amphibian battalions remained in place. Artillery lost the proposed MLRS, to be replaced by a support agreement with the Army. As partial compensation for the death of the combined arms regiment, a countermobility platoon augmented the assault amphibian battalion to carry engineers and line charges. In addition, the active and reserve assault amphibian battalions were expanded to four companies each. The infantry battalions finally were given the TOW missiles from the old tank battalion TOW company and regimental TOW platoons, and the Dragon ATGM gave way to the more effective Javelin system.

Antitank weapons developed to face the future generation of Warsaw Pact tanks finally began to reach the troops. The TOW IIB heavy ATGM with an improved top attack–type warhead went into production in 1992. The Javelin medium-range ATGM entered production in 1996. The Corps's Predator ATGM, designed for close-in antitank protection, had a highly successful demonstration and validity test and entered engineering and manufacturing development in 1994. A replacement for the AT-4 unguided rocket in the rifle companies, the Predator is a nineteen-pound, fire-and-forget, top attack missile effective to six hundred meters. That range may eventually be extended to a kilometer.

The Marine Corps tank program creaked along toward reaching the requirement originally funded in FY 1991-93, but then got lost in bureaucratic shuffling. With only 221 of the 490 M1A1 vehicles required, shortages existed in the Reserve, depot, and MPS stocks. Congress acted in the FY 1994 and FY 1995 DOD authorizations to transfer fifty and 132 tanks, respectively, from the Army to the Corps. The original money for the 490 M1A1s came to the Corps in the form of "fenced" funding in FY 1990-93 designed to keep the tank production plants running. The Com-

mandant, Gen. A.M. Gray, acted—against his staff's advice—to terminate the Corps buy at the halfway point and attempted to use the funding for other programs, with the result that both tanks and funds were
lost. Gray's successor, Gen. Carl Mundy, briefed Congress that he held
no concerns for the apparent shortfall in tanks, citing the experience of
the Army armored brigade assigned to the I MEF in Saudi Arabia as a
preferred alternative to spending more funds on Marine Corps tanks.
After much hesitation, he accepted the additional Army tanks, bringing
the Corps essentially to its planned objective in 1995.

The programmed replacement for the AAV7A1 generally revolves
around advanced technology hybrid skimmer vehicles that will carry
eighteen troops across seas with no more than three feet of swell at
twenty-plus knots and perform as state-of-the-art armored fighting
vehicles upon reaching the shore. The costs are high, especially after
the Corps kept the requirement for these vehicles on its MPS ship and
brought the overall requirement to over a thousand (instead of about
six hundred) such vehicles. That is a $7.6 billion program that remains
unfunded in an era of defense austerity. With its development extended
and quantity production delayed until 2007, an inevitable remanufactured
AAV7A2 incorporating Bradley Fighting Vehicle automotive upgrades
will be necessary as the Corps's personnel carrier for the next decade.

Almost half a century after the 1949 Armor Policy Board recommendations, the Corps enjoys a fully equipped tank battalion with
superior tanks, fully supported by organic scout, antitank, and tank-
launched bridge subunits and capable of coordinating all forms of fire
support. The assault amphibian battalion employs amtracs fully capable
of mounting infantry with mobility equivalent to the tank and under
armor protection. The light armored reconnaissance capability, dismissed
by the 1949 planners, exists in a lavishly equipped battalion.[9]

The Corps's infatuation with lightness continues to be an obstacle
to the proper operation of armored forces in the FMF (or, to use the
preferred 1990s term, Marine Corps Forces). Policy changes forecast
reductions in the numbers of battalions, including tank battalions, in
the FMF prior to the Persian Gulf conflict. However, lessons learned in
converting infantry regiments into ad hoc mechanized units, wrought
with such sweat and tears, found little hearing later in the Corps. Indeed, the trendy doctrine of "military operations other than war
(MOOTW)" seemed made to order for a slimmer, lighter Marine Corps
that is ever more anxious to gain a dominant niche in the postwar budget
stakes.

Incredibly, the enduring power of armored fighting vehicles to
accomplish offensive missions at great savings in casualties seems to be

forgotten. During the Corps's moderately successful involvement in Somalia (1992-93), the troop contingent originally destined for use (a reduced division) fell away to only a few battalions of infantry and light
armored vehicles. However, when the time came to disarm the major
rebel camps around Mogadishu, some containing M47 and T-55 tanks,
the rush was on to locate and stand up the few tanks that had been
unloaded in desultory fashion on the piers. A tanker cadre composed
of a Captain, a staff NCO, and four crewmen, fleshed out by some selected Marines, manned four M1A1 tanks, with a fifth vehicle kept ready
as a spare. The first operation saw these vehicles go into action armed
with only two hundred rounds of .50-caliber and four hundred rounds
of 7.62mm ammunition. The tank ammunition had been reloaded to
the MPS ships and some thirteen days elapsed before two pallets of
main gun ammunition arrived by air! On 10 January 1993, a "sort of"
tank-versus-tank combat occurred as the scratch platoon battled six M47s
using coax and .50-caliber fire. Incredibly, the rebels bailed out of their
obsolete but still lethal tanks, sensing that ranging machine guns had
their mettle and deadly 120mm rounds would soon follow. In the end,
the tanks operated for two and a half months, averaging twenty-five
hundred miles, and proving themselves highly useful in a light-force,
MOOTW deployment.[10]

   Thus, the continuing use of modern armored fighting vehicles and
organizations by the Marine Corps remains shrouded in doubt. Doctrinal gaps and institutional habits remain as obstacles for any effective
employment.[11]

The Corps became accustomed to Army materiel early in the history of
tanks, operating Sherman variants late into the 1950s. A limited purchase permitted the contingency fielding of technically superior tanks
in Korea with the M26, replaced a year later by the M46. The lack of a
serious threat on the battlefield and a new credo of lightness in combat
forces left most Marines content with this situation, with little appreciation of Corps weakness as the M48-M103 series served into the 1970s.
The M48 had served well in support of infantry in Vietnam, and tankers
gained all their combat experience in these vehicles. The tankers themselves, unlike the amtrackers, remained curiously wedded to their Vietnam vintage hardware. Many would complain later of the M60 and M1
purchases, stating, "we never should have given up the M48s."

   The old tankers thus presented some of the same obstacles as did
the infantry officers when it came time to modernize doctrine and accept modern concepts of combined arms. The infantry officers, generally dominating the Corps's leadership, could not face the implications

of large-scale mobile operations in which tanks and armored infantry would play key roles. The tankers, for their part, insisted that the tank battalion already was capable of combined arms operations as a maneuver battalion, even in an infantry regiment or Marine division devoid of doctrine. The tankers maneuvered, sometimes with their Ontos brethren, into the early 1970s in California and the East Coast without combined arms attachments. No armored vehicle battalion yet included functional cells for fire support coordination or command and control, nor did they practice the integration of infantry, engineers, artillery, or attack helicopters as required for combined arms operations.

The politics of armored fighting vehicle modernization and acquisition reflected these points and more. The infantry leadership after 1958 could not agree with the need for heavy combat vehicles, let alone what vintage. General agreement occurred over the need for a modern LVT, but its use beyond the high-water mark of the beach remained problematical. The problem of tank and amtrac modernization usually cited was fiscal. The purchase of a tank or amtrac often required 75 percent of the available funds for ground equipment acquisition for several years. Therefore, at any budget point, a tank or amtrac buy supplanted a dozen or so desired programs of interest to the Corps's leaders. On top of that, the evident personalization of such budget decisions, such as General Barrow's refusal to even consider a tank buy while he was Commandant, or the equivocations of Generals Gray and Mundy, could only exacerbate the established trends.

By the time of the Gulf War, all these attitudes could be detected. The Commandant and his staff made more than a hundred major decisions in the fall of 1990 as the FMF prepared to fight a mobile campaign in the desert. Ironically, the most difficult and tortuous decision centered on whether to accelerate the M1A1 acquisition process and equip Marine tank battalions then moving to or preparing to deploy to the Gulf. The Army posed no objections to lending the vehicles, as the M1A1's advantages were widely understood. But Marine generals feared that tankers were simply advancing their agendas with scare tactics, highlighting the evident weaknesses of the M60 series. Many thought that the current equipment was good enough, or that the TOW missile would defeat Iraqi armor. In the end, only 1½ of the Corps's 4½ tank battalions received the M1A1, although all could have been converted. In the late 1990s, many officers continued to believe that the antitank missile would kill all the tanks, hence removing any need for tanks in the Marine Corps. There is no concept in Marine Corps doctrine for the tank as the basis of offensive power.

Finally, it seems that the training and doctrinal process continues to lag in the Marine Corps when it comes to the use of armored fighting vehicles. Marine combat arms officers earn their spurs on deployments. In the 1960s, the problem of "BLT-itus" stemmed from the formation of Battalion Landing Teams (BLT) to perform nearly all deployments. In the BLT, the tank, amtrac, Ontos, artillery, and engineer officer attachments provided Commanders with technical and tactical expertise, leaving the infantry officer free to "take the hill." No growth in the tactical skills associated with various arms could occur under that system. The successor method bodes no better. Specialization is the chief characteristic of the Marine Expeditionary Unit (MEU) deployments. Typically, the tactical focus lies in a MEU battle drill playbook. Each rifle company is to specialize in either a raid, heliborne, or mechanized type of operation. Hence any exposure to armored or mounted operations for the majority of infantry officers is usually incidental.

Although Marines began to complain in the late 1960s about having to use Army hand-me-down equipment, that had not been the case historically for armored fighting vehicles in the Corps. The increasing wartime needs and changing Army production schedules had forced upgrades during World War II. After the war, memories of the need for effective armor support for the infantry, reinforced by the Korean War experience, was repeated twenty years later in Vietnam. The decision to equip the M48-M103 fleet with diesel engines in the early 1960s seems rational in view of the limited improvements offered by the M60 design, which continued the teething problems endemic to American tank production. The M48 performed well in its infantry support role in Vietnam, probably better than the M60 would have with its limited selection of main gun ammunition. Marine tankers gained their combat experience in the M48 and remained enamored of it and immune to its increasing obsolescence.

At the same time, the Corps's leaders may have lost their sense of need for modern armor. The World War II and Korean War generation of leaders knew how much tanks and amtracs had saved in casualties, and how they had facilitated the offensive use of infantry against all objectives. After the Vietnam War, however, the generals viewed tanks and amtracs required for amphibious assaults more as an expensive burden. Large-scale operations were the exception in Vietnam, and the need for armored vehicles to reinforce the offensive use of infantry and to reduce casualties generally was not appreciated. Instead, memories of disabled or mired tracked vehicles holding up operations remained current in leaders' minds.

Despite the increased role of Marine Corps forces in major war plans after the Vietnam War, little thought was given to the increased

Advanced Amphibious Assault Vehicle (AAAV)-Waterborne Test Rig (WTR). General Dynamics Land Systems.

Advanced Amphibious Assault Vehicle (AAAV)-Automative Test Rig (ATR). Computer enhanced. General Dynamics Land Systems.

value of armor in the large-scale operations envisioned. This factor may have stemmed from the limited tactical experience of the Corps's leaders. Unlike their Army counterparts, few Marine Corps Generals commanded major units for any period of time. Many of them, including some Commandants, had never commanded combat units larger than battalions. In addition, almost half the decision-makers were aviation generals who had neither experience with nor use for expensive land fighting systems. Thus, the continued maintenance of a modern tank fleet no longer found much favor in the post-1970 Marine Corps. Corps doctrine did not develop the means of fielding real combined arms units, nor were the tank and amtrac battalions especially fitted for the demands of modern armored warfare. Hope for dominant antitank munitions for infantry and aviation retarded the decision to field modern fighting vehicles. The Gulf War found the Corps at a point when most of these problems had been determined and corrective action undertaken, but early combat against a determined opponent in that campaign could have exposed serious weaknesses.

Today, the Marine Corps maintains a precarious balance in its armored fighting vehicle inventory, with modern units fielding capable weapons. The lessons of the Gulf War reside mainly in the archives and within the armored units themselves. Just as in 1945, one cannot speak of *armor* in the Marine Corps—just tank, amtrac, and now armored reconnaissance units, which may or may not be used in modern combined arms or limited military operations with either imagination or enthusiasm.

NOTES

1. CMC to BuShips, 29 Jan. 1962, NARS/Suitland/70A-5214/20; CMC to BuShips and Director, USMC Landing Force Development Activity, 25 Nov. 1963, Re: Director, Landing Force Development Activity, 22 July 1963, NARS/Suitland/69A-7645/19.

2. Archives, AVTB, Camp Pendleton; Steven Zaloga, *U.S. Amphibious Assault Vehicles*, (London: Osprey, 1987), pp. 39-41; Victor Croizat, "Fifty Years of Amphibian Tractors," *Marine Corps Gazette* (Mar. 1986), p. 75; and CG, Development and Education Center (MCDEC) to CMC, 25 Mar. 1969, Re: Service Test Final Report, demonstrated the following results for maintenance reliability of the LVTPX12, compared to the LVTP5 (P7/P5): mean time between failures 7.1 hrs/2.61; mean mileage between failures, 85.8/25.4; availability percentage 54/34; maintenance hrs per 110 miles 6.16/20; AVTB Archives.

3. LVA and LVX general data from Zaloga, *U.S. Amphibious Assault Vehicles*, pp. 41-43; LVA technical evaluation, AVTB, 20 Feb.1975; and LVA Operational Requirement, MCDEC, 21 Jan. 1975, AVTB Archives.

4. The penultimate rebuild of the now-venerable AAV-7 series—now weighing 28-31 tons—will be the "RAM" upgrade to enter service in 1999 with the Bradley engine and suspension, and a new transmission.

5. David R. Stephanson and Maurice A. Roesch III, "The Selection of a Light Armored Vehicle for the Marine Corps," Research Paper, Industrial College of the Armed Forces, 1982, p. 9.

6. Ibid.; Kenneth W. Estes, "LAV—Quo Vadis," *Marine Corps Gazette* (Dec. 1981): pp. 18-20; and Andrew Findlayson to author, 21 Apr. 1997.

7. Historical File, Tank Section, Ground Combat Weapons Branch, Marine Corps Systems Command, Quantico Va.; and CMC to Rep. Earl Hutto, 2 May 1994, author's files.

8. Charles J. Quilter II, "With the I Marine Expeditionary Force in Desert Storm and Desert Shield" (Washington: MCHC, 1993); Norman Friedman, *Desert Victory: The War for Kuwait* (Annapolis, Md.: Naval Institute Press, 1991); Historical File, Tank Section, Ground Combat Weapons Branch, Marine Corps Systems Command; and Stan Owen, "Marine Corps M1A1 Tanks in Operation Desert Storm" *Amphibious Warfare Review* (Summer/Fall 1992): pp. 50-54. Davis, "From the Sea," p. 648 notes that seventy-two

line charge kits were sent to Saudi Arabia prior to the start of the offensive. In a way reminiscent of World War I, many AAVs were equipped with two fascines, each made up of fifty-five six-inch plastic pipes, for use in crossing trenches and ditches.

9. Kenneth W. Estes, "A Well-Led Corps. . . ." *The Almanac of Seapower 1994* (Arlington, Va., 1994); "A New Day May Be Dawning for the Nation's '911' Force," *The Almanac of Seapower 1995* (Arlington, Va., 1995); "Force of Choice," *Armed Forces Journal International* (Sept. 1995); "Armor in the Amphibious Force," USNI *Proceedings* (Apr. 1986); and "What is Heavy, What is Light?" *Marine Corps Gazette* (July 1987): p. 30.

10. Maj. Michael F. Campbell, telephone interview with author, 27 May 1997. Campbell was formerly in charge of the 1st Tank Battalion Detachment, Task Force Mogadishu.

11. Lt. Gen. Martin R. Steele, HQMC, 1 May 1997. General Steele, the highest ranking Corps tanker in its history, reflected upon his twelve years' experience as a key staff officer in planning and briefing the fielding plans for the M1 series in the USMC.

15

The Hundred-Hour Thunderbolt: Armor in the Gulf War

Stephen A. Bourque

The 1991 Persian Gulf War represents the zenith of American armored warfare. Never in history had America's mounted forces arrived on the battlefield better prepared than they were in 1990. Seventeen years of intensive analysis and debate had revolutionized Army doctrine and given armor a focus that had never existed. Its equipment, doctrine, and training were now among the best in the world. Mechanized infantry, self-propelled field artillery, armored combat engineers, and attack helicopters complemented the protection and firepower of the M1A1 Abrams main battle tank. Multiple rotations at the National Training Center (NTC) and the Battle Command Training Program had created an almost veteran combat force. When Iraqi dictator Saddam Hussein attacked Kuwait in August 1990 and threatened the world's oil supply, American armor was ready to respond.

After a prolonged air campaign, the allied coalition's armored units attacked into Kuwait and raced across great expanses of desert to strike at the flanks and rear of the Iraqi defenders. In a little less than five days, the concentrated attacks by allied armored units forced Saddam Hussein to order a mass retreat from Kuwait to save his regime. This chapter examines the state of U.S. armored technology, doctrine, and training during the Gulf War. It will focus on several specific examples of armored operations that took place during this conflict: the deliberate attack, the movement to contact, the hasty attack, tank gunnery, and tactical maneuver.

As we have already seen, armor technology, doctrine, and training changed dramatically between the end of the Vietnam War and the beginning of the Gulf War. The Army placed most of its armored systems in "heavy" (i.e., armored and mechanized infantry) divisions.[1] The Persian Gulf War saw the first maneuver of U.S. armored divisions since World War II. In the wide-open expanses of the Arabian Desert, the moving division was an impressive sight. The cavalry squadron, equipped with M3 Bradleys and OH-58D scout helicopters, led the division. It acted as the division Commander's eyes and ears by scouting ahead to find the enemy force. Although the so-called Army of Excellence changes to the division TO&E had placed this squadron under control of the aviation brigade, few armor Commanders accepted the loss. Many division Commanders considered it, and controlled it, as a separate unit in its traditional cavalry role.

Following the cavalry squadrons in each division were the three maneuver brigades. In 1990, U.S. Army heavy brigades possessed more combat power than did an entire World War II division. Each armored brigade had two or three battalions of M1A1 Abrams tanks (fifty-eight each), and one battalion of M2A2 Bradleys (fifty-four each).[2] Each of these battalions also had a scout platoon, a mortar platoon, and a complement of fuel, ammunition, and maintenance vehicles. The division augmented each battalion with a package of other systems that included a fire support team, Stinger air defense missile teams, and engineer platoons. The tank and infantry battalions often exchanged companies so that most of them were combined arms task forces. Some tank battalions remained "pure" and were assigned the primary role of a reserve, counterattack, or exploitation force.[3]

A direct-support artillery battalion organized into three firing batteries moved along with each brigade. Along with the twenty-four M109A2 self-propelled artillery systems, this battalion contained over 250 other vehicles including fire support vehicles (FIST-V), M577 command tracks, ammunition trucks, fuel trucks, and tracked ammunition carriers. Its cross-country mobility allowed this battalion to keep up with the advancing armored and mechanized battalions and provide effective and responsive fire support with only a few moments notice. The remaining divisional artillery, controlled by the division artillery Commander, moved as a separate element. This collection of artillery batteries and battalions was a potent combat element. In Desert Storm, each of the divisions had (in addition to the direct support battalions with the brigades), one MLRS (Multiple-Launch Rocket System) battalion as the divisional general support artillery, and usually one field artillery brigade of three battalions of varying kinds.[4]

Dispersed throughout various parts of the division formation were the other combat support elements. The aviation brigade, with two battalions of AH-64 Apaches, a company of UH-60 Black Hawks, and a platoon of OH-58D scout helicopters, gave the division Commander a real reserve force. This brigade's mobility and lethality allowed him to influence the battle anywhere in the division sector. While the support vehicles would move with the division, most of the aircraft would leapfrog from one forward arming and refueling point (FARP) to the next. In addition, several engineer battalions with an incredible assortment of equipment from simple dump trucks and back hoes to combat engineer vehicles (CEVs) with dozer blades and 165mm guns to the M9 armored combat earth movers (ACE) moved with the division. Finally, signal units, intelligence units, air defense units, a chemical company, and several military police units, contributed to the division's combat power.[5]

The combat service support (CSS) organization provided the backbone for the division. Each maneuver battalion had an organic maintenance team equipped with M113 vehicles, M88 tank retrievers, and maintenance trucks. There were medical platoons with M113 ambulances, support platoons with dozens of heavy expanded mobility tactical trucks (HEMMT) and five-ton trucks, and mess teams with field mess trailers. Each unit had medics and aid stations to support the complex triage evacuation system that extended to the corps, with its mobile surgical hospitals (MASH). Within the maneuver brigade, the forward support battalion (FSB) commander controlled the support effort. The main support battalion (MSB) controlled logistics functions for the remainder of the division's units.[6]

On the move in the Arabian Desert, the armored division occupied frontages of 25 to 45 kilometers and a depth of from 80 to 150 kilometers. It had more than twenty-two thousand soldiers and 1,940 tracked vehicles. There were also more than seventy-two hundred wheeled vehicles in each division, joined by hundreds of other corps and miscellaneous vehicles that moved with the unit. In addition, more than a hundred divisional and hundreds of other aircraft flew in the unit's skies.[7]

Third Army employed two corps in the attack on the Republican Guard Forces Command (RGFC). Prior to the Gulf War, the U.S. Army had not published doctrine for field Army operations because the Army had not envisioned employing two corps together outside NATO, where Army groups operated according to NATO doctrine. Thus, when VII Corps was introduced into the theater, the theater commander U.S. Army

Central Command (ARCENT) had to rapidly establish an operational field army headquarters—in this case, Third Army. However, there was no published doctrine other than *FM 100-5, Operations.*

Third Army's two corps both had corps doctrine available to them, as well as the 1986 edition of *FM 100-5,* which said of a corps: "Corps are the Army's largest tactical units, the instruments with which higher echelons of command conduct maneuver at the operational level . . . corps plan and conduct major operations and battles." Gen. Donn Starry, the architect of AirLand Battle doctrine, had long envisioned the corps as the focal point to plan and execute the Army's tactical doctrine. The Army saw the corps as the bridge between the tactical and operational levels of war. Thus, in both VII and XVIII Corps, division conducted their battles and engagements in accordance with the corps plans, which in turn were major operations in accordance with Third Army and U.S. Central Command (CENTCOM) plans. In this way, plans were "nested," a term coined by Gen. William DePuy, author of the 1976 edition of *FM 100-5.*[8]

Any discussion of actions by smaller units must be seen in the context of the campaign plan, which was formulated in accordance with the army's capstone doctrinal manual, *FM 100-5.*

Other field manuals provided guidance on the various tactics and techniques needed to implement AirLand Battle doctrine at each level of command. However, in most cases, TRADOC schools had not finished these manuals and most units based their training on older manuals, draft editions, or service school student texts. Not until the eve of the Gulf War did the army publish *FM 100-15, Corps Operations* (September 1989), and *FM 71-100, Division Operations* (June 1990). Brigade, battalion, and most company field manuals remained in draft form until after the war. Soldiers in the field trained with the most current draft manuals they could obtain from the proponent service school. Fortunately, these drafts were within the context of fundamental military doctrine and AirLand Battle doctrine, with little apparent confusion at the unit level.

As a result of the Center for Army Lessons Learned (CALL) suggestions, small-unit tactics and techniques were constantly being revised. Starting in 1985, CALL published a continuous series of pamphlets that gave unit leaders practical tips on how to fight in a variety of circumstances and environments. Most of this information was gleaned from exercises at the NTC or other training events.[9] The CALL pamphlets became an essential part of most leaders' personal libraries.

In the 1980s the armor community redesigned its tank gunnery program. No longer content with the old set-piece "one tank–one target" scenarios of the post–World War II era, the new *FM 17-12 Tank Gunnery* series established a requirement for crews to engage multiple targets while firing buttoned-up and on the move. Tank crews learned to work together when engaging large numbers of enemy targets. Platoons, companies, and battalions learned how to engage large numbers of targets in both offensive and defensive situations.

All of this doctrine and training came together when units trained at the Combat Training Centers. Brigades stationed in the United States rotated through the NTC at Fort Irwin, California. Opened in 1982, the NTC provided the most sophisticated training environment in the world. Using a dedicated, highly motivated opposing force (OPFOR), a sophisticated instrumentation system that allowed umpires to accurately record vehicle "kills," and a cadre of proficient observer-controllers, the army ensured its units received the best training possible. In addition, Commanders had the opportunity to lead their units in a live-fire battalion attack and defense. By the end of a rotation, units knew they had received the best training possible.[10] Armored and mechanized battalions stationed in Europe trained at the Combat Maneuver Training Center (CMTC) at Hohenfels, Germany. While not as technologically sophisticated as the NTC, the CMTC gave USAREUR units the most intensive training they had ever received.

The army initiated the Battle Command Training Program (BCTP) in 1987 to provide rigorous training for division and corps Commanders and their staffs. VII Corps, for example, in 1990 alone had served as the higher headquarters for two BCTP exercises of its divisions—the 1st and 3d Infantry Divisions—as well as conducted a BCTP war-fighting seminar in Stuttgart in September 1990. Just prior to deployment in early December, the corps conducted a special one-day BCTP seminar in Stuttgart, followed later by a three-day plan rehearsal with BCTP cadre in King Khalid Military City in early January 1991.

Battle-focused training that the army called Mission Essential Task Lists (METL) was ingrained in its leaders and soldiers, published in *FM 25-100,* and allowed units to make rapid adjustments to the vastly different conditions they encountered in the Arabian Desert.[11]

In the early morning hours of 2 August 1990, Iraq's Medina Armored Division, Hammurabi Armored Division, and Tawakalna Mechanized Division crossed the Kuwaiti border and headed toward Kuwait City.[12] Almost simultaneously, commando units attacked the capital by helicopter and amphibious landings. Iraq controlled most of Kuwait by seven that evening. By 6 August, Iraq had more than eleven divisions and two thousand tanks inside Kuwait.[13] President George Bush and his advisers believed that Saudi Arabia and its large oil reserves were vulnerable to a continued Iraqi advance. He therefore launched a diplomatic offensive designed to convince the Saudi rulers they were in danger and to line up political support from other nations around the world.[14]

Meanwhile, Gen. H. Norman Schwarzkopf and CENTCOM began preparing for the defense of Saudi Arabia. After Schwarzkopf and Secretary of Defense Richard Cheney worked out the arrangements, the royal government invited American troops into the kingdom. On 7 August, Air Force fighter aircraft and the Army's 82d Airborne Division headed for the Persian Gulf, beginning Operation Desert Shield. The entire division was in Saudi Arabia by 24 August, including its Vietnam-era M551 Sheridan-equipped 3d Battalion, 73d Armor.[15]

Realizing that the 82d Airborne Division could not stop the massive Iraqi armor force assembling in Kuwait, President Bush ordered more units to the Persian Gulf. On 13 August the 24th Infantry Division (Mechanized) from Fort Stewart, reinforced by the 197th Infantry Brigade (Mechanized) from Fort Benning, began sailing out of Savannah, Georgia. When its first units landed on 27 August, Lt. Gen. Gary Luck, the XVIII Corps commander, had enough armor and tank killing systems to block an Iraqi attack toward the kingdom's eastern ports. The first U.S. Marine Corps units began landing at Dhahran on 14 August, and the 101st Airborne Division (Air Assault) began moving to Saudi Arabia on 17 August. By early October, the Fort Bliss–based 3d Armored Cavalry Regiment and Fort Hood–based 1st Cavalry Division were in Saudi Arabia, providing General Luck the armored combat power he needed to stop any Iraqi attempt to cross the border. His force included more than 763 tanks, 227 attack helicopters, and almost fifteen hundred armored fighting vehicles blocking potential Iraqi avenues of approach.[16] In addition, more U.S. Marine forces and the British 7th Armoured Brigade, with its Challenger main battle tanks and Warrior infantry fighting vehicles, arrived to bolster the kingdom's defenses.[17]

On 8 November President Bush announced he was doubling the number of U.S. forces in the Persian Gulf so as to give the growing coalition force an offensive capability. The extra punch would come in the form of additional Air Force and Marine troops and a reinforced U.S. VII Corps.[18]

The force that subsequently deployed to Saudi Arabia was a composite organization of units from the European-based V Corps (3d Armored Division), VII Corps (2d Armored Cavalry Regiment and 1st Armored Division), and the 1st Infantry Division (the "Big Red One" of

World War I and II fame) from Fort Riley, Kansas. With thirty-five heavy battalions (armor, mechanized infantry, and cavalry), seven attack helicopter battalions, and twenty-four self-propelled field artillery battalions, it was the most powerful mounted force ever deployed by the United States. General Schwarzkopf also assigned to it the British 1st Armoured Division in late December 1990. The 1st Cavalry Division was assigned to VII Corps from CENTCOM reserve to protect the Tapline Road and XVIII Corps moved west to tactical assembly areas (TAA). The 1st Cavalry Division returned to CENTCOM reserve at the beginning of the ground war on 24 February and was given back to VII Corps during the mid morning of 26 February.[19]

The first VII Corps unit in Saudi Arabia, the 2d Armored Cavalry Regiment, began arriving in early December. The corps' final elements, from the 3d Armored Division, would not arrive until just before the beginning of the ground campaign in February. While in the port areas of Al Jubayl and Ad Dammam, some units exchanged older equipment for the latest models: such as the M1 tank for the M1A1 or the M3 cavalry fighting vehicle for the M3A1. All units received a coat of specially formulated sand-colored camouflage paint to hide their woodland green European colors.[20]

Once the activities in the port were complete, VII Corps units moved initially to a TAA east of the Wadi al Batin. The XVIII Corps remained in its defensive posture along the coast. By 14 January almost sixty thousand soldiers were in the TAA, with the 2d Armored Cavalry Regiment and 1st Armored Division ready for combat.[21] While in the TAA units underwent a comprehensive training program to acclimate them to the desert environment and give them confidence in their equipment. Crews trained on all of their weapons systems, often firing service ammunition. The U.S. 1st Infantry and British 1st Armoured ("Desert Rats") Divisions, recently assigned to VII Corps, practiced for their upcoming breach of the Iraqi defenses and the subsequent passage of lines. Commanders and staffs at all levels conducted troop leading and communications exercises.[22]

Lieutenant Gen. John Yeosock, commander of the U.S. Third Army, directed his two corps (XVIII and VII) to move into their forward assembly areas (FAA) and be ready to fight by the third week in February.[23]

Yeosock ordered XVIII Corps to displace from its positions on the east coast and move west to new attack positions in a massive relocation by heavy equipment transporter (HET) and wheeled convoys mainly along the Tapline Road. It was a difficult yet disciplined movement that was masked from the Iraqi Army in the field.

VII Corps, by contrast, executed a dress rehearsal of the attack by

TWO-CORPS CONCEPT OF OPERATION

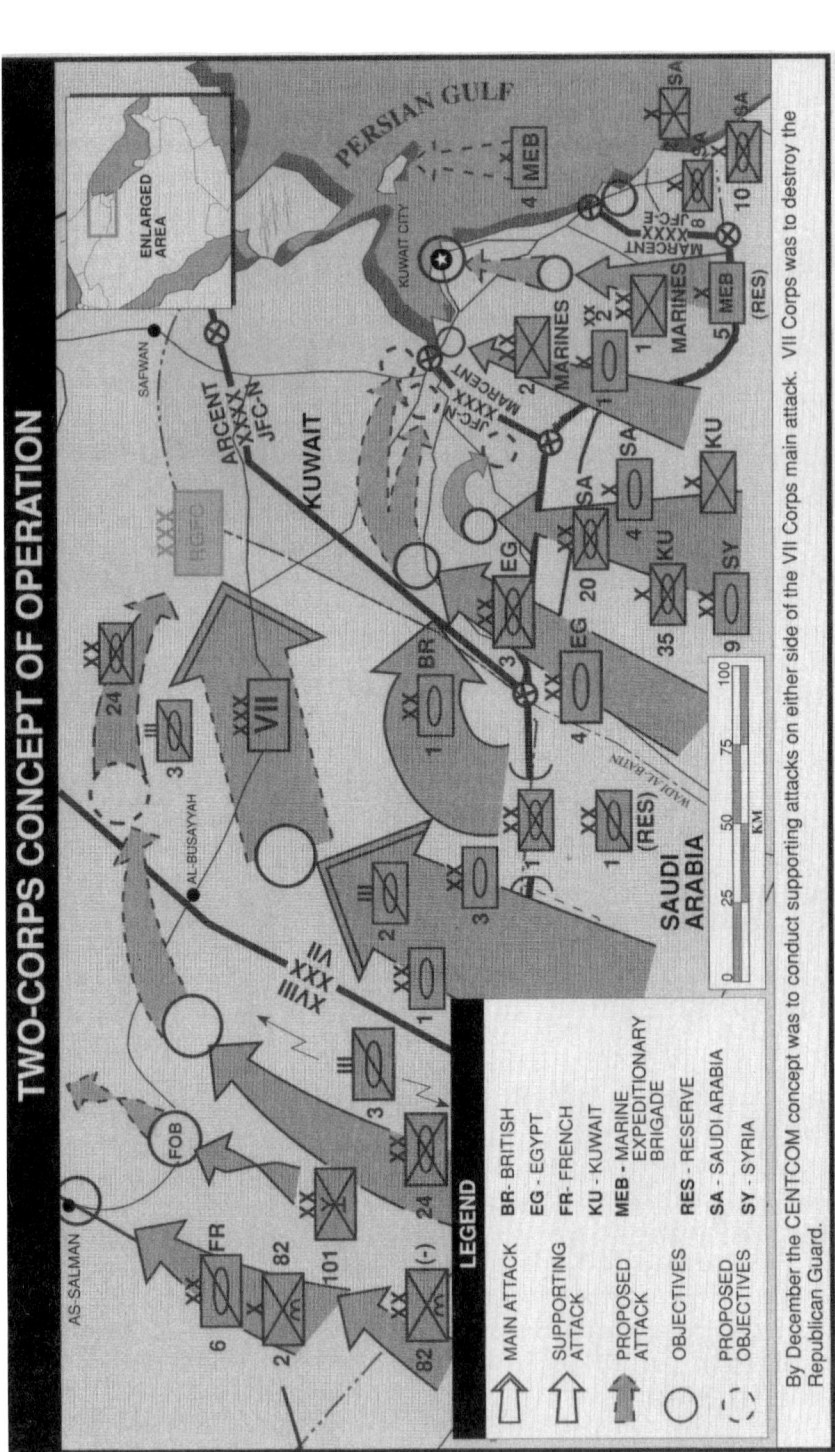

By December the CENTCOM concept was to conduct supporting attacks on either side of the VII Corps main attack. VII Corps was to destroy the Republican Guard.

maneuvering the 160-180 kilometers from the TAAs to its attack positions in the same tactical formations its units employed when the offensive began.[24]

When all his units were in position, General Yeosock had a force of almost two thousand tanks and more than fifteen hundred infantry fighting vehicles—more than twice the armored capability of General George Patton's Third Army on the eve of the 1944 Ardennes campaign.[25]

General Ayad Futayih al Rawaie's RGFC represented the best of the Iraqi Army and was critical to Saddam Hussein's hold on political power. The allied coalition's main effort thus focused on the destruction of this force. In the Kuwaiti theater of operations, the RGFC consisted of the three heavy divisions that spearheaded Iraq's invasion of Kuwait: the Medina and Hammurabi Armored Divisions and the Tawakalna Mechanized Division. These units had withdrawn from southern Kuwait and were, by 24 February, in a central position within the theater of operations. It was a formidable mechanized force with more than eight hundred T-72 tanks and six hundred BMP fighting vehicles.[26] The RGFC also contained several motorized infantry divisions. Each of these divisions contained three brigades of truck-mounted infantry and at least one tank battalion. These divisions (Al Faw, Nebuchadnezzar, and Adnan) protected the northern flank of the RGFC's heavy divisions.[27] The 10th and 12th Armored Divisions, part of the newly created Jihad Corps, were also under the RGFC commander's operational control. Although the 12th Armored Division was a new, poorly equipped unit, the 10th Armored Division was one of the best units in the Iraqi Army. The Iraqi high command placed these divisions to the east of the RGFC. Their mission was to act as a covering force for the RGFC heavy divisions, either leading an attack or blocking an enemy attack while the RGFC maneuvered before the main battle.[28]

Desert Storm, the code name for the effort to drive Iraq out of Kuwait, began on 16 January 1991. Allied coalition aircraft attacked Iraqi forces in one of the most concentrated bombing campaigns in history. American aircraft flew more than 112,000 individual sorties in less than forty days. However, the air campaign did not drive Iraq out of Kuwait. On 24 February the coalition's ground forces began the process of physically removing Saddam Hussein's troops from the emirate. When this short campaign ended, few had questions about the effectiveness of U.S. armored forces at end of the Cold War.

Third Army had devised a two corps attack to both interdict Highway 8 to prevent the RGFC's escape from the Kuwaiti theater and then to destroy the RGFC. VII Corps was assigned the main attack, aimed at

M2 Bradley crews make last-minute precombat checks before taking part in Operation Desert Storm. *ARMOR* magazine.

Generals Franks and Rhame. Fred Franks.

destroying the RGFC. Third Army published a contingency plan calling for both corps to execute a ninety-degree turn to the east to complete the RGFC's destruction if the Iraqis remained in their positions. VII Corps called its part of this contingency mission FRAGPLAN 7, one of seven branches Franks's headquarters developed from the original plan.[29]

VII Corps occupied the center of the coalition's line and had the mission of destroying the RGFC. First, the corps had to break through the Iraqi front lines held by the second-rate infantry divisions of the Iraqi VII Corps. Once beyond the front lines and in the open desert, the corps would conduct a movement to contact to find the Republican Guard. To break through the front lines as quickly and efficiently as possible, Lt. Gen. Frederick Franks Jr., the VII Corps commander, organized his initial attack into two components. One force, the 1st Infantry Division followed by the British 1st Armoured Division, was the breaching element. Initially the 1st Infantry Division would attack a weak point in the line, destroy the defending unit, and create a series of lanes through the Iraqi defenses. The British 1st Armoured Division would then pass through these lanes and complete the destruction of the Iraqi VII Corps and protect the U.S. VII Corps's right flank. Following the British 1st Armoured Division's passage, the 1st Infantry Division would later join Franks in the destruction of the Republican Guard.

While Maj. Gen. Thomas G. Rhame's Big Red One attacked into the breach, the second part of the U.S. VII Corps, the enveloping force, would launch its attack. It would pass by the exposed flank of the Iraqi frontline defenses and conduct a movement to contact toward the Republican Guard heavy divisions in the rear. Ultimately, Franks intended to smash into the RGFC with a "three division fist" and destroy it in detail.[30]

Lieutenant Gen. Franks, his commanders, and the logisticians were acutely aware of the massive logistics requirements—especially fuel—needed to sustain the attack without pause against the Republican Guard, located some 150 kilometers from the attack's start point along the border. To provide the needed support, Brig. Gen. Bob McFarlin, the VII Corps support command commander, ordered a logistics task force to follow the British 1st Armoured Division through the breach and establish Log Base Nelligen on the Iraqi side, thus lessening the travel distance required of the enveloping force's fuel tankers by as much as twelve hours.[31]

The 1st Infantry Division began its attack shortly after 5:30 A.M. on 24 February. Colonel Lon E. Maggart's 1st Brigade, reinforced by the 1st Squadron, 4th Cavalry, attacked on the left and Col. Anthony Moreno's 2d Brigade attacked on the right. Col. David Weisman's 3d Brigade remained in its attack position, prepared to exploit the breach. As the 1st and 2d Brigades moved toward the trench lines, the division's artillery support units began moving forward into the areas they cleared. Initially, they were to move close to the Iraqi main defenses and complete the attack the following morning. However, General Schwarzkopf decided to have Third Army, which was slated to attack on 25 February, continue the attack that morning. He ordered Yeosock to see how quickly he could get both his corps moving. VII Corps was ready at noon, but to remain coordinated with the Joint Forces Command North operation, the attack was to commence at at 3 P.M. Thirty minutes before the assault commenced, the 1st Infantry Division fired an intense artillery preparation on the Iraqi trench lines. During the next thirty minutes, more than twenty thousand Big Red One soldiers looked on as five artillery brigades performed an impressive firepower demonstration. More than eleven thousand 155mm and 175mm rounds and MLRS rockets slammed into the Iraqi forward positions. No Iraqi soldier dared to lift his head out of the trench during that incredible pounding.[32]

At 3 P.M. the 1st Infantry Division's two lead brigades plunged forward. To the defending Iraqis, they must have appeared as a solid wall of fire and iron. The lead battalions began moving with 120mm volleys from their tank platoons, and tanks and Bradleys poured machine-gun fire into any location from which the Iraqis might attempt to return fire. Mortars fired smoke rounds to mark assault sectors and to isolate Iraqi positions from each other. As the lead elements closed with the trenches, the Iraqis could see that many of the lead tanks were equipped with plows and mine rollers. Combat engineer vehicles with their short but lethal 165mm guns and plows headed for trenches and bunkers. The new armored combat earthmovers worked to plow lanes through the obstacle belt. Behind this six-thousand-meter-wide wall of iron came the remainder of the battalions, especially the Bradley infantry fighting vehicles, with soldiers ready to fire from the ports on the side and rear. In the turrets, the gunners prepared to engage with 7.62mm coaxial machine guns and 25mm Bushmaster cannon. Five hundred meters south of the trench line, the assaulting task forces dropped their plows from the carry position and continued north at fifteen kilometers per hour, firing their coaxial machine guns and 120mm cannon into the trenches. Trailing the four maneuver and two engineer task forces were the three exploitation battalions: the 1st Squadron, 4th Cavalry, and 1st Battalion, 34th Armor, from the 1st Brigade, and the 2d Brigade's 4th Battalion, 37th Armor.[33]

Finally, the attacking battalions reached the Iraqi positions. The

Iraqis probably anticipated dismounted American soldiers trying to fight for the bunkers and trench lines. The 1st Infantry Division, however, had no intention of playing by the Iraqi rules. Once across the trench, tanks on the flanks turned either right or left. Commanders called their ACEs forward to the fighting line and, with plows down, the ACEs moved along the back of the trenches, filling them in with sand. With the ACEs were Abrams tanks, Bradleys, and dismounted infantry, who captured and disarmed the Iraqi infantrymen pouring out of the trenches with hands held high in the air.[34]

Once the leading Big Red One troopers were through the first trench belt, the artillery shifted its fires farther north onto other positions. By 4:15, forty-five minutes after the attack began, the lead battalions had cleared sixteen lanes through the first trench line. Now, brigade commanders directed their exploitation forces to pass through the open lanes and attack the Iraqi defenders farther to the rear.[35]

By the end of the day, the 1st Division had taken about two thirds of the objective, including more than a thousand enemy prisoners of war. Friendly casualties were extremely light: only one dead and one wounded—much lower than the anticipated high casualty rates.[36]

As darkness approached on 24 February, General Franks faced one of those decisions that rests squarely on a corps Commander's shoulders: whether to continue the fight or stop and wait until dawn. Franks was focused on both the tactical situation and the operational objective and decisive point: the destruction of the RGFC. He knew he wanted maximum combat power available in the right combination in a rolling attack against the RGFC without pausing in front of their positions to refuel. He also knew that if the RGFC stayed and defended from its present locations, the attack to destroy it would be a two corps attack with XVIII Corps attacking west to east and VII Corps attacking north. At the tactical level, Franks was aware of the need to maintain momentum in the attack but at the same time he was also aware from talking to General Rhame that expanding the breach to accommodate three brigades abreast was a complex maneuver. Furthermore, trying to fit a third brigade in between the two leading brigades increased the risk of friendly fire casualties at night. Franks also wanted to get the British 1st Armoured Division through the breach as quickly as possible to defeat the Iraqi tactical reserve division and prevent it from interfering with the corps' enveloping operation. In addition, he was aware of the need to coordinate in time and space the availability of fuel and the third division for the corps' main attack against the RGFC some 150 kilometers distant. Finally, he considered the human factor of Soldier combat power for the main attack on the RGFC. He did not want to dissipate that power before the attacking units reached their objective. Taking all these things together, and considering the recommendations of his forward commanders, Rhame and Holder, Franks decided to order Rhame to continue his breach expansion at first light on 25 February.[37]

Although the unit symbols that displayed the leading edge of the advance on commander's map boards would not move for several hours, both sectors were far from quiet. Infantry and armor companies and cavalry troops searched the sectors to their front. Tanks, Bradleys, and field artillery forward observers attacked probable targets. Apaches attack helicopters hunted the forward sectors looking for Iraqi vehicles and weapons systems. Artillery harassment fires kept the Iraqis guessing as to the next blow. Service support crews refueled and rearmed the tanks and Bradleys and replenished artillery ammunition carriers. Meanwhile, Maj. Gen. Ruppert Smith's Desert Rats completed their movement into the forward staging area. The British were now in position to move through the breach and take on the remainder of the Iraqi VII Corps, still arrayed along the Saudi-Iraqi border.[38]

At 6 A.M. on 25 February, the 1st Infantry Division and supporting artillery pounded suspected Iraqi positions with a short but violent preparatory barrage. The Big Red One then resumed its attack. At 9:15 the 3d Brigade passed through the berm and through the 2d Brigade into the center of the enlarging semicircle. All three brigades then attacked together, rapidly overwhelming Iraqi resistance and moving toward their day's objective. Shortly after eleven, the 1st Infantry Division secured Phase Line New Jersey and began guiding the British 1st Armoured Division through the breach and into its forward positions. The 7th Armoured Brigade, leading the British unit, crossed Phase Line New Jersey, at 3:15 P.M. and attacked toward the Iraqi VII Corps. The 4th Brigade completed its passage at 7:30 and joined in the attack. Ultimately, the passage of the Desert Rat's seven thousand vehicles took until three the next morning.[39]

The 1st Infantry Division's deliberate attack was a competently planned, thoroughly rehearsed, and professionally executed operation. The mass of direct and indirect fires on the objective, followed by the division's violent attack against the Iraqi defenders, ensured the Big Red One's success. Meanwhile, as the 1st Infantry Division was completing its operations in the breach-head, the 2d Armored Cavalry Regiment searched for the Republican Guard in the enveloping sector.

Late on the afternoon of the twenty-fifth, Franks huddled with his chief of staff, G2, G3, and aviation brigade Commander at his forward tactical command post. It was apparent from intelligence information provided by the Third Army G2 that the RGFC would attempt to stand and fight. Seeing that the enemy was fixed, Franks ordered VII Corps to ninety degrees to the east into the flank and rear of the RGFC forces in accordance with FRAGPLAN 7. That the corps was able to execute this maneuver and also be able to remain in a posture to execute other branches of the Third Army contingency plan if necessary was a result of maneuvers the corps made during the first twenty-four hours. With both corps attacking abreast toward Basra and the Persian Gulf, a new boundary was established with XVIII Corps in the north and XVII Corps to the south.[40]

The 2d Armored Cavalry led VII Corps's march toward the Republican Guard heavy divisions. Several important characteristics of armored cavalry units set them apart from other ground combat units.[41] First, armored cavalry units have traditionally been combined arms organizations. Tank battalion scout platoons have, over time, included a scout section, tank section, infantry squad, and mortar section. Armored cavalry platoons were once organized along similar lines, and in later TO&Es troops were organized as combined arms teams. Armored cavalry regiments have been assigned three squadrons plus command and control, intelligence, and combat service support elements of various sizes. Considerable variants to the general scheme just described have been employed as the Army's equipment and operational needs changed, but the basic idea of combined arms from the lowest organizational level has generally prevailed in the cavalry. Second among armored cavalry's unique capabilities has been the presence of more robust and longer range communications capabilities from platoon to regiment. The ability to communicate reliably over longer distances has enabled armored cavalry units to operate along wider frontages and at greater depths than parent formations, reacting expeditiously to opportunities or crises in in extended battle space. Thirdly, armored cavalry units have always been equipped, organized, and trained to execute their varied missions. Long ago the U.S. Army decided that reconnaissance by stealth, while yet a necessary skill, should not be relied on as the primary method. Similarly, combat between a lighter force and a heavier enemy force would most likely not provide the security necessary for a friendly force to proceed on its mission unimpeded. In reconnaissance-in-force operations such as a movement to contact, where the mission is to find and define the enemy quickly and early in the battle, the ability to fight is essential. Finally, armored cavalry squadrons and regiments traditionally have been self-supporting organizations.[42]

Col. Don Holder's 2d Armored Cavalry Regiment represented the most sophisticated stage in armored cavalry's evolution. As soon as it began moving on the morning of 24 February it ran into Iraqi infantry. Throughout the day the 3d Squadron fought a series of small battles, mostly uneven firefights between Bradleys and dismounted infantry, destroying a few Iraqi tanks and armored personnel carriers. By 5 P.M. the regiment had advanced more than forty kilometers into Iraq, encountering only scattered resistance.[43]

At six the next morning the regiment, reinforced by the 210th Field Artillery Brigade, continued its attack, finding little initially but empty fighting positions. As it continued to move northeast the squadrons met armored and mechanized units from the 50th Armored Brigade of the Iraqi 12th Armored Division. This brigade, and the following 37th Mechanized Brigade, was trying to get into its positions to cover the Tawakalna Mechanized Division, which was deployed farther to the east. Troop O, flying AH-1 Cobra helicopters, engaged these Iraqis with TOW missiles and machine guns, while a supporting MLRS battery moved into firing positions behind the lead ground cavalry troop. With 3d Squadron in visual contact, the aviators passed to the rear and handed off the battle to the ground troops. Within seconds of their egress, volleys of American rockets landed on the hapless Iraqi battalion.[44]

At the end of the day, Colonel Holder ordered his regiment to assume hasty defensive positions for the night. Holder had good reasons for a pause. First, they were on the route that at least two brigades from the 12th Armored Division were using to move toward Al Busayyah. Holder held the terrain the Iraqis wanted and could inflict heavy casualties on them as they came into the regiment's sights. Second, the halt allowed the squadrons to reorient to the east so they could attack to gain contact with the main body of the Republican Guard. Most importantly, ammunition was running low. By noon most of the support squadron's ammunition trucks were stuck in sand that had, during the day's rain, turned to mud. An emergency resupply effort coordinated by the corps staff resulted in three CH-47 Chinook loads of ammunition being delivered to the regiment. However, bad weather canceled five additional resupply sorties. Other regimental wheeled vehicles moved back and forth between the stuck vehicles and the forward squadrons throughout the night.[45]

During the night of 25-26 February, the 2d Armored Cavalry Regiment formed up in a regimental laager on the road between Al Busayyah and the Kuwaiti border. Although he had stopped moving forward, Holder wanted to keep the pressure on the Iraqis to his east. Beginning at 10 P.M. on 25 February, Company M moved forward of the regimental position and cleared a zone for an MLRS battery. At 10:30 and again at midnight, the battery launched rockets into the center of the Iraqi position. After each engagement the tanks and MLRS launchers returned to the safety of the regimental defensive perimeter. Holder had also planned to use his aviation assets in these attacks but continued bad weather forced his helicopters to remain on the ground and allowed the Iraqi infantry to get close to the regiment's perimeter.

At 2 A.M. on 26 February a small Iraqi mechanized unit attacked the sector of the laager occupied by the 3d Squadron. Iraqi vehicles and crews mixed in among Bradleys and maintenance crews on the battle line. In the dark, Iraqi MT-LBs and the M113s used by the cavalry's maintenance crews look very much alike, even through thermal sights. Scouts reported that enemy vehicles had penetrated the sector and the danger level increased as Bradley and M1 turrets swiveled to the rear, searching for potential targets. By the time the shooting stopped the squadron had destroyed nine MT-LBs and an Iraqi tank and captured sixty-five Iraqi soldiers. The Iraqis damaged two Bradleys and destroyed two M113s, killing or wounding several U.S. Soldiers. American Soldiers probably shot some of their own friends during the confused encounter.

The regiment's three squadrons were on line by 6:20 A.M. and began moving slowly east through the two degraded brigades of the Iraqi 12th Armored Division. Terrible weather and visibility of less than a hundred meters kept the regiment's speed down. Gunners on tanks and Bradleys, straining to see enemy vehicles through their thermal sights, destroyed isolated T-55s[46] and MT-LBs in their zone. At 7:13 the 3d Squadron reported the first kill of a T-72 tank. All day, the regiment fought isolated detachments of Republican Guard vehicles along with the 12th Division's two brigades. At 8:45 the 3d Squadron received accurate artillery fire from the Iraqi defenders, damaging one M1 tank. It was one of the few times during the operation that Iraqi artillery had any noticeable effect.[47]

The regiment continued moving forward in a sandstorm, maintaining a slow but deadly efficient pace as the squadrons bounded toward the main defenses of the Tawakalna Division. This painstaking ritual consisted of one platoon halting and scanning the horizon for enemy positions while another platoon carefully moved forward to the next terrain feature or about a thousand meters forward of the stationary platoon. The maneuver platoon then halted and scanned the horizon while the other platoon moved forward. This technique ensured that the command did not stumble in mass into an Iraqi kill zone.[48]

In the southern portion of the regiment's sector, the 1st Squadron continued to advance in constant contact with Iraqi units.[49] The Iraqis never had a chance in these contests—even with the bad weather. Abrams tank and Bradley gunners could find the T-55s, MT-LBs, and other vehicles on the wide-open terrain and destroy them before coming into the Iraqis's view. By noon, the 1st Squadron alone had destroyed almost two dozen T-55 tanks and another dozen infantry carriers. Mixed in among the older kinds of equipment were several Republican Guard T-72 tanks and BMP fighting vehicles.[50]

In the northern part of the sector, Lt. Col. Mike Kobbe's 2d Squadron moved in a fairly narrow sector, with E and G Troops forward and F Troop and H Company in the rear. At 3:30 P.M. on 26 February the squadron broke into the Tawakalna Division's security zone and G Troop destroyed three tanks near the 68 Easting.[51] Ten minutes later the squadron received airburst artillery fire as it approached the Iraqi positions. Continuing to move through a howling sandstorm, E Troop ran right into an Iraqi battalion strong point. The Iraqis had built it around a small village at an intersection of several trails on the 68 Easting, just north of the 3d Squadron boundary.[52]

The E Troop Commander, Capt. H.R. McMaster, attacked the enemy with two tank platoons abreast and his scout platoons following the tanks and providing "scratching fires" to protect the tanks from dismounted infantry. The American tanks swept down on the enemy's defensive position with every weapon firing. Iraqi vehicles exploded as 120mm rounds found their mark. Enemy infantry attempted to fight back with RPGs and AK-47s, but were cut down by the fires of the following scouts. McMaster finally stopped his charge at the 73 Easting when he arrived at the rear of the Iraqi battalion strong point. Just as he was moving onto the only high ground in the area, the Iraqis launched a counterattack. Troop E stopped the attack in its tracks, destroying more than half of the attacking Iraqi battalion in just twenty-three minutes.[53]

Lt. Col. Scott Marcy's 3d Squadron moved just to the south of E Troop. Troop I, commanded by Capt. Dan Miller, moved onto the southern portion of the same strong point at about 3:30, hit the southern part of the same Iraqi battalion about twenty minutes after E Troop, and then destroyed whatever resistance remained. According to Captain Miller: "Enemy tank turrets were hurled skyward as 120mm SABOT rounds ripped through T-55s and T-72s. The fire balls that followed hurled debris one hundred feet into the air. Secondary explosions destroyed the vehicles beyond recognition. Resistance was sporadic. The unforgettable odor of burning diesel, melting metal and plastics, expended munitions and anything else that happened to be burning in bunkers, hung heavy in the air."[54]

At 4:45 the Iraqis launched a counterattack against I troop with a T-72 tank company. The Iraqi tanks opened fire on the Bradley scout vehicles at about twenty-five hundred yards, but the tank rounds struck the earth just short of their intended targets. The Iraqis were unable to get many more rounds off as the troop's M1 tanks bounded forward and, at about twenty-one hundred meters, destroyed most of the attackers.[55]

Colonel Holder's orders from General Franks were not to get decisively engaged. His troops had successfully destroyed one Iraqi battalion strong point, but there were still at least six or seven waiting for the regiment, which did not have the combat power to break through those defenses. Holder therefore ordered his squadrons to hold at their current positions and prepare to pass the 1st Infantry Division forward. The Iraqis, however, kept attacking.

Capt. Joseph Sartiano's G Troop, 2d Squadron, moved into positions along a small wadi near the 73 Easting at 4:15. An Iraqi 18th Mechanized Brigade battalion strong point lay on the other side of the wadi. Almost immediately the two sides began exchanging direct and indirect fire. Iraqi airburst artillery exploded overhead, forcing the M1 and Bradley crews to remain under cover. At about 5:45 an Iraqi round hit and destroyed a G Troop Bradley, killing the gunner. Fifteen minutes later the character of the battle changed as dismounted infantry and T-55 tanks and MT-LBs began a series of furious attacks on G Troop's positions.

Soon the troop was in a close-in firefight. Iraqi tanks and MT-LB personnel carriers raced toward the tank and Bradley platoons, firing their main guns and machine guns. Dismounted Iraqi infantry, believing that the darkness and poor visibility would protect them, charged the American positions firing their AK-47 assault rifles and RPG launchers. Troop G's defensive firepower stopped the Iraqi attacks cold. The Bradley's TOWs destroyed truck-mounted enemy soldiers before they could dismount and M1 tanks demolished T-55 and T-72 tanks long before they got within their own firing range. The troop's mortar section began firing airbursts at the dismounted Iraqi infantry, causing them to either retreat or dig-in. In several hours of combat, Sartiano's troopers knocked out at least two companies of Iraqi armor. Hundreds of Iraqi infantry and their lightly armored transporters lay scattered on the wadi floor. Low on TOW ammunition, Lieutenant Colonel Kobbe pulled G Troop off line to rearm and refuel and moved H Company into G Troop's sector.[56]

The 2d Armored Cavalry Regiment had done its job and found the left flank of the Tawakalna Mechanized Division. In the process it had destroyed the 50th Armored Brigade, a mechanized infantry battalion from the 18th Mechanized Brigade, and hundreds of other vehicles and infantry squads from a mixed bag of Iraqi units. It had led the VII Corps advance since crossing the border more than four days before. Now the tired troopers of the oldest mounted regiment in continuous active service in the U.S. Army passed the attack on to a unit with just as prestigious a lineage: the Big Red One, which Franks had ordered forward to take up position as the third division in VII Corps's rolling fist attack. The 1st Cavalry Division remained in CENTCOM reserve until it was too late to be used in this battle.[57]

That night, the 1st Infantry Division passed through the 2d Armored Cavalry and resumed the battle with the 9th Armored Brigade and the 18th and 37th Mechanized Brigades. In one of the most dramatic and violent battles of the war, the Big Red One's tankers demonstrated that they ruled the darkness. Throughout the night General Rhame's mounted troops moved relentlessly across their objective. Almost every target they shot at was hit. By dawn, three Iraqi brigades with more than two hundred tanks and hundreds of other combat and combat support vehicles were engulfed in a smoking cauldron.[58]

Simultaneously with the 2d Armored Cavalry Regiment's and 1st Infantry Division's battle, the 3d Armored Division, just to the north, was destroying the center of the Tawakalna's defenses.

As the corps turned to execute its FRAGPLAN 7 maneuver, the 3d Armored Division moved to the north of the 2d Armored Cavalry Regiment. Like the 2d Cavalry, the 3d Armored Division first fought with elements manning the enemy's security zone outposts. Brushing those and remnants of the 50th Armored Brigade aside, the division hit the center of the Iraqi line around 4:30 P.M. on 26 February.[59] The mainstays of the Iraqi defense were six battalions (two armored and four mechanized) from two Tawakalna brigades (the 29th Mechanized and 9th Armored). They faced the 3d Armored Division on a line approximately twenty kilometers wide. Additionally, at least one battalion from the 12th Armored Division's 46th Mechanized Brigade and at least one T-62 tank battalion, possibly from the 10th Armored Division, fought with the Tawakalna in this sector.[60] Approximately nine Iraqi heavy battalion groupings faced the attacking 3d Armored Division's ten heavy battalions. In terms of ground combat power, it was almost a one-to-one matchup.

Maj. Gen. Paul E. Funk deployed his division to maximize its flexibility in the face of the uncertain situation. The division's formation resembled an inverted V moving to the east. Col. Robert Higgins's 2d Brigade moved at the northern apex of the V while Col. William Nash's 1st Brigade moved to the south. Col. Leroy "Rob" Goff's 3d Brigade formed the base of the formation and was prepared to support the attacking brigades, assume the battle, or exploit their success. Behind each of the lead brigades were the direct support artillery battalions and batteries from the 42d Field Artillery Brigade. In addition, a company of Apaches from the 2d Battalion, 27th Attack Helicopter Regiment, flew in support of each brigade. To influence the battle, the division commander kept his MLRS batteries under his control. The 4th Squadron, 7th Cavalry, screened the division's southern boundary with the 2d Armored Cavalry Regiment.[61]

The 1st Brigade moved on a relatively narrow sector, less than five kilometers wide, in a brigade wedge. Task Force (TF) 3-5 Cavalry moved at the tip of the wedge with TF 4-32 Armor on the northern flank and TF 4-34 Armor in the south. Task Force 3-5 Cavalry's scouts screened the forward movement of the entire 1st Brigade. At 5:02 P.M. they ran into the northern battalion, reinforced, of the Iraqi 9th Armored Brigade. This battalion had sufficient time to prepare its positions and waited for the American attack. Rather than assault this complex hastily, the American battalion moved on line and used its superior fire-control systems and standoff range to pick apart the Iraqi position. Long-range tank and TOW fires and high explosive, dual-purpose improved conventional munitions (DPICM), and Copperhead artillery rounds ravaged the Iraqi complex. Nevertheless, for the next twelve hours the American battalion advanced no farther.[62]

In the dark, TF 4-32 Armor moved toward the Iraqi brigade's left flank. Around 7:20 P.M. the task force's scout platoon identified a T-72 covered with infantry heading toward it from the southeast, about five hundred meters away. In a short, confused fight, the scouts destroyed the tank and scattered its passenger infantry. A platoon of T-72s and other dismounted infantry joined the fight and stopped the Americans's forward progress. The battle in this sector ended by nine with the Iraqi defenders holding their own.[63]

Around 6 P.M., A Troop, 4th Squadron, 7th Cavalry, screening the division's southern flank, ran into part of a battalion strong point. Like the other Iraqi units, it had dug in and was waiting for a fight. After more than an hour of exchanging fire, the Americans began to pull back from the Iraqi position. In the confusion, a friendly tank, probably from the approaching TF 4-34 Armor, fired at one of A Troop's Bradleys, killing the gunner. Tanks from the 2d Armored Cavalry to the south also shot at and hit another vehicle. Before the squadron could back off and pass the fight on to TF 4-34 Armor, the Iraqis hit and damaged ten of thirteen M3 cavalry fighting vehicles besides the one hit by friendly fire. Troop A lost two soldiers killed and twelve more wounded.[64]

Task Force 4-34 Armor, attacking on the right flank, made no better progress than did the other two battalions. After closing with the Iraqis who had stopped the 4th Squadron, 7th Cavalry, the task force remained generally along the 71 Easting for the remainder of the night. The 1st Brigade, 3d Armored Division, pounded the Iraqi defenders for the next twelve hours. All night long A-10s, Apache helicopters, and artillery fire pounded identified and suspected enemy targets. Tanks and Bradleys fired at identified targets and, especially in TF 3-5 Cavalry's sector, the brigade had some success in clearing the defensive complex.[65]

Faced with a determined enemy defender, General Funk massed his forces in support of his main effort, Colonel Higgins's 2d Brigade in the northern portion of his sector. At 4:45 P.M. on 26 February the brigade attacked in a wedge formation with TF 4-8 Cavalry in the lead, TF 4-18 Infantry on the left, and TF 3-8 Cavalry on the right. It attacked on a relatively narrow front, less than five kilometers wide. Higgins had most of two artillery brigades and the 2d Battalion, 27th Attack Helicopter Regiment in support. Impatiently waiting less than ten kilometers behind the 2d Brigade was Colonel Goff's 3d Brigade, whose four battalions were eager to get into the fight at the first opportunity. With his forces thus massed, Funk's troops began to take the Iraqi defenders apart.[66]

The division's artillery pounded identified and suspected Iraqi positions within a nine square kilometer area. Then Funk launched his attack helicopter battalion across the forward line of troops against artillery and subsequent defensive positions in the rear.[67] Meanwhile, in the southern part of the division sector, the 1st Brigade continued its secondary attack against the reinforced battalion's defensive line. Finally, when he was convinced he had engaged the Iraqis on the flanks and throughout their sector, General Funk ordered his 2d Brigade forward.

At 10 P.M. Higgins's three battalions and supporting artillery provided the Iraqis a classic demonstration of a coordinated combined arms attack. For the next four hours disciplined tank and Bradley crews moved through the 29th Mechanized Brigade's defenses. Tank companies bounded forward by platoons, using their thermal sights and standoff range to engage Iraqi vehicles on their own terms. Out-ranged and fighting blind, the Republican Guard soldiers returned fire without any noticeable effect. Attack helicopters and MLRS launchers destroyed Iraqi artillery almost as soon as it fired. As the brigade line moved forward,

Iraqi infantrymen emerged from their hiding places and tried to engage American armor from close range. These Iraqi soldiers had little chance of success as a line of Bradley fighting vehicles, moving just behind the tanks, killed them with machine-gun fire.[68]

By 2 A.M. on 27 February, the 2d Brigade had fought through the 29th Brigade's first defensive echelon to the 73 Easting. Just beyond the forward line was the wreckage of the night's deep artillery and aviation battle. The situation was ripe for General Funk to order Colonel Goff's 3d Brigade forward. That morning Goff's battalions passed through the thin frontline troops and smashed through the Iraqi battalions that lay in their path. By early morning the 3d Armored Division had secured its objective, leaving the battlefield strewn with the wreckage of almost two brigades of Iraqi armor.[69]

In the northern portion of the Tawakalna's sector, one final Iraqi tank battalion stood in the way of the advancing 1st Armored Division. In a sharp fight waged at the same time the 1st Infantry and 3d Armored Divisions were attacking, the 1st Armored made short work of the Iraqi defenders. The VII Corps had defeated the Tawakalna Division by massing six brigades and an armored cavalry regiment against it and flanking it to the north and south with two more brigades. Attack helicopters and long-range artillery systems attacked the Iraqis behind the front lines throughout the battle. The Tawakalna's spirited defense demonstrated that the Republican Guard did not enter the battle already defeated. It was the application of the army's AirLand Battle doctrine—executed by well-trained and motivated soldiers—which ensured the Iraqi defeat. Fundamental to the attack's success was the superb performance of U.S. tank crews.

At dawn on 27 February the devastating effects of armored combat power stretched along forty kilometers of the 73 Easting. The 1st Armored Division was still on the move, and by late morning was approaching the Medina Armored Division, located southwest of the Al Rumaylah oil fields.

By noon on 27 February the Iraqi Army was trying to block the allied coalition's push toward Basra. In the VII Corps's sector, these defenses were based on the Medina Division, defending just west of the Al Rumaylah oil fields. Extending south for almost fifty kilometers, just west of the Kuwait border, was a line of mixed units from 10th, 12th, and 17th Armored Divisions and other surviving units. The Iraqi high command's goal was to stop the allied advance long enough to evacuate most of the remaining Iraqi troops from Kuwait.

Around 11:30 the soldiers of the Iraqi Medina Division's 2d Armored Brigade, began preparing lunch. Although they had been under constant attack from the air by artillery fire, they obviously believed there was no need to take special precautions. It was so rainy and overcast that air force A-10 aircraft were unable to find and attack them, and they failed to deploy a security zone five to ten kilometers forward of their sector. The brigade's alignment was designed as a rear slope defense behind a low ridge. Such a defensive plan is designed to draw the attacker's lead vehicles over a hill and into a prearranged kill zone while the following vehicles, still on the hill's forward slope, cannot see them. Defenders can improve the effectiveness of these kill zones by using mines and other obstacles, as well as prearranged artillery targets to destroy and disrupt the follow-on units.

However, the 2d Armored Brigade prepared its positions poorly. Vehicles were superficially dug in and there were very few obstacles. Nor had the Iraqi unit commanders verified their weapons' ranges to the top of the hill. They did not know that their battle line was too far back from the ridge to hit the attacking American armor when it first came over the crest. While these may have been some of Iraq's best troops, their combat training and discipline were well short of the standard expected of an elite force.[70]

Col. Montgomery C. Meigs IV's 2d Brigade began advancing west at almost the same time as the Medina Division's 2d Armored Brigade began eating lunch. For almost three days this brigade, and the rest of the 1st Armored Division, had driven through sporadic Iraqi defenses. Although tired, they were veterans in acquiring and destroying enemy vehicles. Each of the brigade's four battalions was organized as a combined arms task force. Visibility was terrible. Vehicle commanders could see no more than fifteen hundred meters without the aid of thermal sights, and Meigs had no idea that this large Iraqi formation was less than seven thousand meters away. At 12:17, Lt. Col. Roy S. Whitcomb's 4th Battalion, 70th Armor, moved over a small rise with his three tank companies abreast. As the battalion reached the top of the ridge, its thermal sights went wild, picking up the images of hundreds of Iraqi vehicles three thousand meters away. To the naked eye, or even with binoculars, the targets would have been missed. However, thanks to thermal imagery, an entire reinforced armored brigade lay arrayed before the attackers as though they were stationary targets on a tank range at Grafenwoehr or the NTC.[71]

Soon the U.S. vehicles opened fire, interrupting the Iraqis's lunch. Lacking thermal imaging devices and having an effective range of only two thousand meters, the Iraqi crews in their T-72 tanks could not see the American tanks on the ridge. Nevertheless, they fired at the flashes of the U.S. tank guns, then watched as their rounds landed harmlessly in the dirt in front of the attacking Americans. Whitcomb's tankers engaged the Iraqi tanks with disciplined, controlled, and deliberate fire, picking off their targets with impunity.[72]

Meanwhile, Lt. Col. Jerry Wiedenwitsh's 1st Battalion, 34th Armor, moved to the top of the ridge south of Whitcomb's battalion and began acquiring targets in the valley below. His D Company, the first on the hill, destroyed two BMPs about twenty-seven hundred meters away. Suddenly the valley came alive as Iraqi crews fired back. As Wiedenwitsh's companies each came on line, they replied to the ineffective Iraqi fire with platoon and company volleys of tank and TOW fire. Bradley TOW gunners registered hits out to thirty-three hundred meters—over two miles away. For almost forty minutes, the battalion fired away at the hapless Iraqi defenders. Meanwhile, Whitcomb's task force pulled on line in the northern portion of the brigade sector and began destroying an Iraqi mechanized battalion at ranges of twenty-six hundred to twenty-eight hundred meters.[73]

The Iraqis called for artillery fire, and the guns, all preregistered, overshot their targets, causing no initial damage. Lt. Col. James E. Unterheseher, Meigs's fire support coordinator and Commander of the 2d Battalion, 1st Field Artillery, a 155mm self-propelled unit, turned his guns loose on the Iraqi artillery, shooting at any Iraqi battery his target acquisition radar could identify. As in previous engagements, the Iraqis adjusted none of their artillery fires, so target identification was rather easy. Within a few minutes, the 1st Armored Division's counterbattery fires destroyed the Medina Division's firing batteries. Then Unterheseher shifted his artillery fires to help Meigs's committed battalions.[74]

Colonel Meigs positioned himself behind his center battalion, where he could observe as much of the battle as possible. Although he was unable to see it all, he had probably the best view of a battle of any brigade Commander in the war. While his tank crews and artillery pounded the Iraqi 2d Armored Brigade, he struggled to get Apache helicopters and air force ground attack aircraft into the battle. None showed up until after 1 P.M., too late to get in on the action. In less than an hour, Meigs's "Iron Brigade" had eliminated the Iraqi 2d Armored Brigade.[75]

Meanwhile, Col. James Riley's 1st Brigade, 1st Armored Division, encountered the Iraqi defensive line just south of Meigs's unit. In this sector many of the vehicles were still facing south, as though the attack was coming up the Wadi al Batin. Using the standoff capability of the M1A1 and Bradley, the "Phantom" Brigade shot first at almost maximum range, killing several T-72s and T-55s before they were able to rotate their turrets and return fire. Some T-72s, out in the open apparently receiving supplies, were quickly destroyed by the U.S. armor. The Iraqis fought back but were incapable of hitting the Americans from so far away.[76]

In the southern part of the 1st Armored Division sector, Col. Daniel Zanini's 3d Brigade struck the left of the Medina Division's line at about 1 P.M. Here, as in the 1st Brigade sector, the foe constituted a mixture of units caught generally out of position, and the earlier scenario repeated itself as U.S. tankers again methodically engaged and destroyed Iraqi tanks and fighting vehicles. This time Apache helicopters joined the firing line, sending Hellfire missiles over the heads of the American tankers. Air Force close support aircraft arrived on the scene and began attacking Iraqi forces beyond the range of the U.S. ground troops. In just a few hours of intense combat the Medina Armored Division and several smaller units ceased to exist, its demise marked by a valley littered with hundreds of burning tanks and armored personnel carriers.[77]

One of the key issues that the army worked on after the Vietnam War was the employment of the combined arms team, bringing all elements of a unit's combat power into play. The 1st Armored Division demonstrated how to do it. Direct support artillery battalions engaged Iraqi positions beyond main gun range. The MLRS and M110 howitzers fought and won the counterbattery battle. Apaches joined the fight from behind the tank firing line, shooting Hellfire ATGMs out to their maximum range, and the ceiling lifted enough for air force A-10 support aircraft to join the fight and wreak havoc in the rear areas along the Iraqi supply route. All the while tanks and Bradleys fired at all the hot spots they could identify with their thermal sights. General Franks flew forward and joined Maj. Gen. Ronald Griffith, the division Commander, behind the firing line at about 12:30. There, as lightning streaked through an overcast sky, they observed the cumulative results of twenty years' investment in soldiers, training, equipment, and doctrine.[78]

By 28 February the allied coalition had defeated the Iraqi Army. Iraqi units were in headlong flight toward the Euphrates Valley, with most concentrated around the city of Basra. Kuwait was free of organized Iraqi units. The U.S. VII Corps occupied all of northern Kuwait and the XVIII Corps's 24th Infantry Division (Mechanized), reinforced by the 3d Armored Cavalry Regiment, blocked escape routes south of the Euphrates River. The war ended at eight that morning when President Bush declared a unilateral suspension of hostilities and offered the Iraqi government a chance to negotiate a permanent cease-fire. Two of the Republican Guard's three heavy divisions were in shambles. The third division, the Hammurabi, was caught in the Basra Pocket.

The 24th Infantry Division (Mechanized), commanded by Maj. Gen. Barry McCaffrey, had provided the bulk of XVIII Corps's ground combat punch. From the beginning of the ground war until the cease-fire it had moved almost 250 miles into Iraq, to the banks of the Euphrates. After capturing the major airfields at Jalibah and Tallil, it continued to move southeast toward Basra along Highway 8.[79]

Although the cease-fire went into effect on 28 February, Iraqi units continued to shoot at the Americans in the hours that followed. In the early morning hours of 2 March a brigade or more from the Hammurabi Division tried to force its way out of the Al Rumaylah oil fields, through the 24th Infantry Division's positions on Highway 8. Apparently, they were trying to cross over a repaired portion of the causeway to assist in suppressing the growing Shia rebellion in southern Iraq.[80] Unfortunately for the Iraqis, they ran right into the 24th Infantry Division's 1st Brigade, commanded by Col. John LeMoyne. American attack helicopters, artillery, and long-range tank and TOW fires wreaked havoc among the Iraqi soldiers. For several hours the Hammurabi Division's T-72 tanks, BMPs, and support vehicles tried to move through the American lines without success.

Finally, at 10:45, LeMoyne unleashed Lt. Col. Bants J. Craddock's 4th Battalion, 64th Armor Battalion. Maneuvering from the south of the Iraqi forces, Craddock delivered the coup de grace in a scorching flank attack that cut right through the disorganized Iraqis and swept up to the edge of the causeway. Sweeping through the fire and devastation, the American tankers destroyed every item of equipment not burning, as thousands of panic-stricken Iraqi soldiers fled toward Basra. At the end of this fight, more than 185 armored vehicles, four hundred trucks, and three dozen artillery pieces lay scattered on the desert floor.[81]

By any measure, armor's performance in Operation Desert Storm vindicated the visions of mobility enthusiasts since the beginning of mechanization. Armor dominated the battlefield like no other system. Armor units punched through the front lines and raced hundreds of miles across the Iraqi desert. It was the armored combined arms team that annihilated seven out of nine Republican Guard heavy brigades and dozens of other Iraqi divisions. Ultimately it was the allied coalition's armor, smashing through the defenses in southern Kuwait, smashing into the Republican Guard, and driving toward objectives along the Euphrates River, that forced Saddam Hussein to retreat from his so-called Nineteenth Province.

Why was U.S. and British armor so effective in this conflict? Most importantly, armor units operated under the guidance of professional,

A catastrophic explosion in a Iraqi T-72 caused by a depleted uranium round from the Abrams 120mm main gun. Scales's *Certain Victory*.

competent, staffs that developed rational plans and provided the logistics support needed by frontline units. Armor leaders, from corps commander down to the individual tank commander, were technically and tactically proficient. They knew how to employ their weapons systems and motivate their soldiers to perform to the best of their ability. American armor equipment was superb: better fire control, heavier armor, faster cross-country speed, and thermal imagery for fighting at night and during inclement weather. Even more important, soldiers knew how to use this equipment and keep it running. Units and individual soldiers showed the effects of a prolonged, comprehensive, training program. Armor units performed within the context of AirLand Battle doctrine and its emphasis on fire and maneuver, the combined arms team, and initiative. Finally, the Iraqis were simply not as good as the OPFOR at the NTC, against whom most U.S. tankers and mechanized infantrymen had been tested. Iraqi armor units time and again were unable to move, shoot, and communicate at the same level as American armored units.[82]

Writing over 160 years ago, the oft-quoted German military theorist Carl von Clausewitz warned that: "Kind-hearted people might of

Bradley and M1A2. National Training Center. Greg Stewart.

course think there was some ingenious way to disarm or defeat an enemy without too much bloodshed. Pleasant as it sounds, it is a fallacy that must be exposed."[83] Clausewitz had it right; there are no simple, high-tech solutions to resolving conflicts between nations. After enduring economic embargoes and weeks of attacks by "smart" bombs, stealth fighters, cruise missiles, and special operations forces, Iraqi forces remained in Kuwait almost seven months after their arrival. It was armor, using its mobility, firepower, protection, and shock action that drove the Iraqi Army away. It is a lesson that should not be forgotten by those who shape America's future forces.

NOTES

1. U.S. heavy divisions in 1990 were the 1st, 3d, 4th, 5th, 8th and 24th Infantry Divisions (Mechanized), the 1st and 3d Armored Divisions, and the 1st Cavalry Division. In addition, the active army had three armored cavalry regiments (the 2d, 3d, and 11th) and three separate heavy brigades: the 177th and 194th Armored and 197th Infantry (Mechanized). Most CONUS-based divisions had only two brigades and depended on augmentation from one of the separate brigades or a National Guard brigade. There were also several units that were being inactivated. These included the 1st Brigade, 2d Armored Division, at Fort Hood, and brigades in Germany nominally assigned to the 1st Infantry Division and 2d Armored Division.

2. In some cases various combat units employed older or newer versions of these systems. For example, the 2d Brigade of the 1st Infantry Division still used the M1 with its 105mm gun.

3. Department of the Army, *FM 101-10-1/1, Staff Officer's Field Manual: Organizational, Technical, and Logistical Data* (Washington: GPO, 1987), pp. 1-166; Department of the Army, *FM 71-100, Division Operations* (Washington: GPO, 1990), pp. 2-2 to 2-3; and HQ, 1st Armored Division, Briefing Slides: "Desert Storm, 1990-1991," vol. 12A, Appendix B, "VII Corps After Action Report." The entire after-action report is housed at the CAC Historical Archives, Fort Leavenworth, Kans. The Swain and Scales Papers (both collections cited in this chapter) are also contained there.

4. *FM 101-10-1/1*, pp. 1-166; *FM 71-100*, pp. 2-2 to 2-3; and 1st Armored Division, Briefing Slides: "Desert Storm, 1990-1991." The 1st Infantry Division and 1st Cavalry Division had only one assigned MLRS battery.

5. *FM 71-100*, pp. 2-3 to 2-10; and 1st Armored Division, Briefing Slides: "Desert Storm, 1990-1991."

6. *FM 71-100*, pp. 2-12 to 2-17.

7. Robert H. Scales Jr., *Certain Victory: The U.S. Army in the Gulf War* (Washington: Brassey's, 1994; reprint, Fort Leavenworth, Kans.: U.S. Army CGSC Press, 1994), p. 239; and 1st Armored Division, Briefing Slides: "Desert Storm, 1990-1991."

8. *FM 100-5, Operations*, p. 185; and Gen. Frederick M. Franks Jr., U.S. Army (Ret.), to George F. Hofmann, 6 Mar. 1999.

9. Department of the Army, *Pamphlet 90-7, Winning in the Desert* (Fort Leavenworth, Kans.: Center for Army Lessons Learned, 1990), inside front cover.

10. John L. Romjue, Susan Canedy, and Anne W. Chapman, *Prepare the Army for War* (Fort Monroe, Va.: TRADOC, 1993), pp. 35-36; and Anne W. Chapman, *The Origins and Development of the National Training Center, 1976-1984* (Fort Monroe, Va.: TRADOC, 1992), pp. 81-109. A good narrative of what the NTC experience was like is found in James R. McDonough, *The Defense of Hill 781: An Allegory of Modern Mechanized Combat* (Novato, Calif.: Presidio Press, 1988).

11. Franks to Hofmann.

12. Crosbie E. Saint, "War Adds New Dimensions to Europe's Role," *ARMY* 41, no. 10 (1991): p. 97; Romjue, et al., *Prepare the Army for War*, p. 36; and Stephen P. Gehring, *From the Fulda Gap to Kuwait: U.S. Army Europe and the Gulf War* (Washington: CMH, 1998), p. 34.

13. The U.S. Government's official story of the Gulf War is found in U.S. Department of Defense, *Conduct of the Persian Gulf War: Final Report to Congress* (Washington: GPO, 1992). A good presentation of the political and international background is found in Lawrence Freedman and Efraim Karsh, *The Gulf Conflict 1990-1991: Diplomacy and War in the New World Order* (Princeton, N.J.: Princeton University Press, 1993). The U.S. Army's official story of its performance in the war is found in Scales, *Certain Victory*, and Frank N.

Schubert and Theresa L. Kraus, eds., *The Whirlwind War: The United States Army in Operations Desert Shield and Desert Storm* (Washington: CMH, 1995).

14. Friedman and Karsh, *The Gulf Conflict*, pp. 86-93.

15. Schubert and Kraus, eds., *The Whirlwind War*, p. 52; and Scales, *Certain Victory*, pp. 82-83.

16. Schubert and Kraus, eds., *The Whirlwind War*, pp. 73-77, and Scales, *Certain Victory*, pp. 82-83.

17. Nigel Pierce, *The Shield and the Sabre* (London: HMSO, 1992), pp. 23-25; and Charles H. Cureton, *With the First Marine Division in Desert Shield and Desert Storm* (Washington: Headquarters, U.S. Marine Corps, 1993), pp. 4-13.

18. Friedman and Karsh, *The Gulf Conflict*, pp. 208-10; Department of Defense, *Conduct of the Persian Gulf War*, p. 78.

19. U.S. Army Central Command (ARCENT), "Morning Briefing, 24 February 1991," Swain Papers, slide 11; and Franks to Hofmann, 6 Mar. 1999.

20. Most units from the United States, such as the 1st Infantry Division, were painted with this chemical agent resistant coating prior to their departure. See Stephen A. Bourque, "DESERT SABER: The VII Corps in the Gulf War" (Ph.D. diss., Georgia State University, 1996), pp. 98-103, 139-48.

21. 1st Infantry Division (Forward), Situation Report, 15 Jan. 1991, "Unit Status" chart, vol. 2B, "VII Corps After Action Report."

22. Frederick M. Franks Jr., taped interview with author, 12 Sept. 1994, TRADOC, Fort Monroe, Va.

23. Richard M. Swain, *"Lucky War:" Third Army in Desert Storm* (Fort Leavenworth, Kans.: U.S. Army CGSC Press, 1994), pp. 197-204; USARCENT, "Morning Briefing, 24 February 1991"; and Department of Defense, *Conduct of the Persian Gulf War*, pp. 114-24, 245-7.

24. Franks to Hofmann.

25. ARCENT, "Morning Briefing, 24 February 1991." In comparison, General Patton's Third Army possessed twenty-two medium tank battalions with 1,166 tanks and nine tank destroyer battalions with 324 tank destroyers or towed antitank guns. See Charles B. MacDonald, *A Time for Trumpets* (New York: William Morrow, 1985), pp. 629-41.

26. VII Corps, G2, "The 100-Hour Ground War," p. 80.

27. Department of the Army, S2, 177th Armored Brigade, *The Iraqi Army: Organization and Tactics*, NTC Handbook 100-91 (Fort Irwin, Calif.: NTC, 1991), p. 28; and USARCENT, "Desert Storm Intelligence Summary," briefing slide: "Iraqi Disposition 16 January 1991."

28. VII Corps, G2, "The 100-Hour Ground War," tab L. A weakness of the RGFC is that it had no organic covering forces, such as the U.S. 2d and 3d Armored Cavalry Regiments. This "corps" was an obvious attempt to address this problem.

29. Franks to Hofmann.

30. Frederick M. Franks Jr., interview by Richard Swain, 6 Mar. 1992,

Swain Papers; HQ, VII Corps, "OPLAN 1990-2 (DESERT SABER), vol. 4, "VII Corps After Action Report," pp. 5-9.

31. Franks to Hofmann.

32. Lon E. Maggart and Gregory Fontenot, "Breaching Operations: Implications for Battle Command and Battle Space," *Military Review* 74, no. 2 (1994): p. 32; and Franks to Hofmann.

33. Scales, *Certain Victory*, pp. 81, 221; Maggart and Fontenot, "Breaching Operations," 27; and Gregory Fontenot, "The 'Dreadnoughts' Rip the Saddam Line," *ARMY* 42, no. 1 (1992): pp. 34-36.

34. Jim Tice, "Coming Through: The Big Red One Raid," *Army Times*, 26 Aug. 1991, p. 20; and Scales, *Certain Victory*, p. 230.

35. VII Corps, "Tactical Command Post Daily Staff Journal," 24 Feb. 1991, entry 44. Unless otherwise noted, all VII Corps journals, fragmentary orders (FRAGO) and situation reports (SITREP) are found in vol. 2 of the "VII Corps After Action Report."

36. VII Corps, "Commander's SITREP #38," 24 Feb. 1991; and Tice, "Coming Through," p. 20.

37. Franks to Hofmann.

38. VII Corps, "Commander's SITREP #38," 24 February 1991.

39. VII Corps, "Commander's SITREP #39," 25 Feb. 1991; Pierce, *The Shield and the Sabre*, p. 99; and Gregory Fontenot to author, 13 Jan. 1998.

40. Franks to Hofmann.

41. Portions of this and the following sections previously appeared in the author's article, "Correcting Myths About the Persian Gulf War: The Last Stand of the Tawakalna," *Middle East Journal* 51 (Autumn 1997): pp. 566-83.

42. Donn Starry, "Operational Concept: Future Cavalry Scout Systems," Army Navy Club Forum, 18 Sept. 1997, Washington, D.C..

43. 2d Armored Cavalry Regiment, "Desert Storm, Operations Summary 23 Feb-1 Mar 91," vol. 16, "VII Corps After Action Report"; Steve Vogel, "A Swift Kick: 2nd ACR's taming of the Guard," *Army Times*, 5 Aug. 1991, p. 18; and VII Corps, "G3 Operations Daily Staff Journal," 24 Feb. 1991, entries 18, 41, 50.

44. VII Corps, "Commander's SITREP #39," 25 Feb. 1991; 2d Armored Cavalry Regiment, "Desert Storm, Operations Summary"; Peter S. Kindsvatter, "VII Corps in the Gulf War: Ground Offensive," *Military Review* 72, no. 2 (1992): p. 25; and Kevin Smith and Burton Wright III, *United States Army Aviation During Operations Desert Shield & Desert Storm: Selected Readings* (Fort Rucker, Ala.: U.S. Army Aviation Center, 1993), pp. 163-64.

45. VII Corps, "G3 Operations Daily Staff Journal," 25 Feb. 1991, entry 33; II Corps Support Command, "Second Corps Support Command Operation Desert Storm Battle Chronology," vol. 24, "VII Corps After Action Report"; and Leonard D. Holder, "Second Armored Cavalry Regiment: Operation Desert Storm, 1990-1991," vol. 16, "VII Corps After Action Report."

46. Iraqis also used Chinese built T-59 tanks that are almost identical. All tanks of this variety will be referred to as T-55 in this manuscript.

47. 2d Armored Cavalry Regiment, "Desert Storm, Operations Summary"; and Vogel, "A Swift Kick," p. 28; VII Corps, "G3 Operations Daily Staff Journal," 26 Feb. 1991, entries 4, 5, 6, 10.

48. *FM 17-95, Cavalry,* with Change 2 (Washington: GPO, 1981), pp. 4-11.

49. 2d Armored Cavalry Regiment, "Desert Storm, Operations Summary"; and Vogel, "A Swift Kick," p. 30.

50. Ibid.

51. Vince Crawley, "Ghost Troop's Battle at the 73 Easting," *ARMOR* 100 (May-June 1991): p. 8. An "easting" is the numbered line running from north to south on military maps that measure distance from east-west.

52. Michael D. Krause, *The Battle of 73 Easting, 26 February 1991: A Historical Introduction to a Simulation* (Washington: CMH, 1991), sketch map.

53. Krause, *The Battle of 73 Easting,* pp. 11, 25.

54. Ibid., p. 20.

55. Ibid., and Vogel, "A Swift Kick," p. 30.

56. Crawley, "Ghost Troop's Battle at the 73 Easting," pp. 9-10.

57. Franks to Hofmann.

58. Lon E. Maggart, "A Leap of Faith," *ARMOR* 101 (Jan.-Feb. 1992), pp. 27-29; and Gregory Fontenot, "Fright Night: Task Force 2/34 Armor," *Military Review* 73, no. 1 (1993): pp. 46-49.

59. VII Corps, "G3 Operations Daily Staff Journal," 26 Feb. 1991, entry 28; and 3d Armored Division, "Chronology of 3d Armored Division on Operation Desert Spear."

60. Vogel, "The Tip of the Spear," *Army Times,* 13 Jan. 1992, pp. 13, 16; and VII Corps, G2, "The 100-Hour Ground War;" 7th Engineer Brigade, "VII Corps Iraqi Material Denial Mission."

61. Paul E. Funk, interview with Richard Swain, 4 Apr. 1991, Swain Papers; and Vogel, "The Tip of the Spear," p. 13.

62. K. Webber and J. Aiello, "History of the Ready First Combat Team: First Brigade, 3rd Armored Division, November 1990-22 March 1991," vol. 14, "VII Corps After Action Report"; and Scales, *Certain Victory,* p. 273.

63. Webber and Aiello, "History of the Ready First Combat Team"; Scales, *Certain Victory,* pp. 273-74; and 3d Armored Division, "Chronology of 3rd Armored Division on Operation Desert Spear."

64. *U.S. News & World Report* Staff, *Triumph Without Victory: The Unreported History of the Persian Gulf Conflict* (New York: Times Books, 1992), pp. 351-56; and Vogel, "The Tip of the Spear," p. 13.

65. Webber and Aiello, "History of the Ready First Combat Team."

66. 2d Brigade, 3d Armored Division, "Operation Desert Shield, December 1990 through 27 February 1991," vol. 14, "VII Corps After Action Report"; and Scales, *Certain Victory,* pp. 273-74.

67. Smith and Wright, *United States Army Aviation During Operations*

Desert Shield & Desert Storm, pp. 55-67; and Scales, *Certain Victory,* p. 276.

68. Funk interview with Swain; 3d Armored Division, "Chronology of 3rd Armored Division on Operation Desert Spear"; and 2d Brigade, 3d Armored Division, "Operation Desert Shield."

69. VII Corps, "G3 Operations Daily Staff Journal," 27 Feb. 1991, entries 2, 15; Funk interview with Swain; and 3d Armored Division, "Chronology of 3rd Armored Division on Operation Desert Spear."

70. VII Corps G2, "100 Hour War," p. 132; and Scales, *Certain Victory,* p. 293.

71. Norm Johnson, group interview with 2d Brigade, 1st Armored Division leaders, 26 Mar. 1991, CMH, DSIT AS-091; Roy S. Whitcomb, "After Action Report, Operation Desert Viper," 5 June 1991; and Scales, *Certain Victory,* pp. 292-93.

72. Johnson group interview; and Whitcomb, "After Action Report."

73. Ibid.

74. Ibid.; Tom Carhart, *Iron Soldiers* (New York: Pocket Books, 1994), pp. 279-302; and Scales, *Certain Victory,* p. 293.

75. Johnson group interview; and Carhart, *Iron Soldiers,* pp. 279-302.

76. Steve Vogel, "Killer Brigade: 3rd Infantry Division 'Phantoms' hunt the enemy," *Army Times,* 11 Nov. 1991, p. 16; and 1st Armored Division, "The Fight."

77. Johnson group interview; and Vogel, "Metal Rain," *Army Times,* 16 Sep. 1991, p. 22.

78. Tom Clancy and Fred Franks Jr., *Into the Storm* (New York: G. P. Putnam's Sons, 1997), pp. 419-20; and Frederick M. Franks Jr., interview with author, 8 Sept. 1995.

79. Jason K. Kamiya, "A History of the 24th Mechanized Infantry Division Combat Team During Operation Desert Storm," Fort Stewart, Ga., 24th Infantry Division (Mechanized), 1992, p. 40. For more indepth historic detail see "24th Mechanized Infantry Division Combat Team Historical Reference Book, A Collection of Historical Letters, Briefings, Orders, and Other Miscellaneous Documents Pertaining to the Defense of Saudi Arabia and the Attack to Free Kuwait," Apr. 1991; and "24th Mechanized Infantry Division Combat Team, Operation Desert Storm, Action Plan," Feb. 1992. Both were compiled by the 24th Infantry Division (Mechanized), Fort Stewart, Ga.

80. Anthony H. Cordesman and Abraham R. Wagner, *The Lessons of Modern War,* vol. 4, *The Gulf War* (New York: Westview Press, 1996), pp. 648-49.

81. James Blackwell, "Georgia Punch: 24th Mech puts the squeeze on Iraq," *Army Times,* 2 Dec. 1991, p. 61; and Scales, *Certain Victory,* pp. 312-14.

82. A good analysis of American and Iraqi military capabilities is offered in Cordesman and Wagner, *The Lessons of Modern War,* vol. 4, *The Gulf War,* pp. 119-22, 145-49.

83. Carl von Clausewitz, *On War,* ed. and trans. Michael Howard and Peter Paret (Princeton, N.J.: Princeton University Press, 1976), p. 75.

16

Reflections

Donn A. Starry

Annually, April marks the anniversary of the 1917 arrival in France of the first elements of the AEF, the United States's contribution to the Allied defeat of Imperial Germany in the 1914-18 world war.

Subsequent deployments to the AEF included a fledgling group known as the Tank Corps. Tanks came to battle in that war as a means to counter the devastating effects of massed artillery and machine-gun fire on infantry. Some visionary tank persons of the day even foresaw a larger role for tanks—independent of mud, trenches, and massed infantry in collision along the static western front. At the operational level, tanks would strike deep, disrupting command and control, reserve forces, and support infrastructure, and turning forward-deployed enemy forces out of their fixed entrenchments. The battle would be won by encirclement, envelopment, and maneuver. It was a vision far beyond the capabilities of the machinery of the day. Indeed, tactical close action in support of infantry, despite some striking successes, was fraught with substantial mechanical challenges for the fragile machines of the time.

My father, Don Albert Starry, enlisted in the Tank Corps out of college. The Tank Corps and the Air Service were the premier branches of the time. Recruiters from both services worked college campuses of the nation, seeking to enlist the brightest and most active young men into these elite organizations rather than rely on conscript forces. They also sought—at least in places like rural Iowa, where my Dad went to college—young men from the farms, men who had at least some experience with engines and the running gear of machinery. Some tankers of the day enjoyed basic soldier training at Camp Colt, a site now buried in the town or on the battlefield at Gettysburg, Pennsylvania. Cpl. Starry's promotion to sergeant was signed by Capt. Dwight D. Eisenhower, the camp Commander. Since but a single tank was available for training at Camp Colt, Sergeant Starry and some of his buddies were trained with their tanks—French made Renaults—at the Tank Corps School at Langres, France. Lt. Col. George S. Patton Jr. was the Commandant there. Patton later accompanied his Renault tanks on foot through the wire and across the trenches in the Saint-Mihiel and Meuse-Argonne offensives in September 1918. He took a round through the leg and buttocks on the first day of the latter action—an event the tank was designed to prevent (for those inside), but of which he was, nonetheless, forever proud. Later, of course, Captain Eisenhower would become a General and then President, and the U.S. Army would name a couple of generations of tanks (the M46, M47, M48, and M60) after General Patton, who became the premier U.S. armor Commander of World War II.

World War I was soon finished for the AEF and its Tank Corps. Without a decent requiem for either, both just went away. It was called demobilization—the logical antithesis of mobilization. It was a process with considerable historical precedent in U.S. military affairs. Its political genesis was aggravated by a can-do willingness on the part of the military to simply do the best it could at whatever its civilian masters demanded of it.

If mobilization had seemed frenetic and helter-skelter, demobilization put its predecessor to shame. In 1918 demobilization took the U.S. Army quite by surprise. Numbers and time lines are instructive. For example, the Army's rolls on 1 April 1917 included fewer than 130,000 soldiers. In the succeeding nineteen months well over three million Soldiers enlisted or were conscripted. Then the war abruptly ended. Soon it was ruled that draftees and enlistees alike were eligible for immediate discharge. The "war to end all wars" was truly over. While a large staff section had been charged with mobilization planning, one lone Colonel, C.H. Conrad Jr., was charged with planning demobilization. Appointed to the task just a few weeks before 11 November 1918, he was sworn to secrecy lest word that demobilization was even being considered would be condemned as "peace propaganda." Eleven days after the Armistice, Colonel Conrad's demobilization recommendations were forwarded to Chief of Staff Peyton C. March by the War Plans Division of the General Staff. After several false starts, much confusion, and considerable meddling by the press and Congress, legislation governing demobilization was passed on 28 February 1919. By November, one year after the Armistice, 3,416,066 soldiers had been mustered out. Army strength on 30 June 1920 was reported at 209,901—only some 70,000 more than had been in uniform in April 1917. Clearly it was a case of get them all out and home—with back pay and a $60 bonus. The latter was the token reward of a grateful nation for helping make the world safe for democracy and for ending all wars.

So it was that Sergeant Starry and his young Tank Corps buddies—after rushing to the colors, undergoing partial training in the United States and France, and hustling off to combat—sojourned on leave at Monaco's casinos before returning to Camp Meade, Maryland. After a brief stint there, they then made their way back to the hinterlands, to the villages and farms from whence, scarcely a year earlier, they had sallied forth to war. Whatever other benefits of military experience he may have enjoyed, Sergeant Starry took with him back pay and a bonus totaling $81.53.

Many of them would go to war again. But in 1919 such a possibility was so remote as to be unthinkable to the men of C Company, 329th Battalion, Tank Corps, AEF.

While the mobilization and demobilization of personnel visibly occupied center stage during and after the conflict, the procurement of arms and equipment for an Army grown some thirteen times its prewar size in a matter of just over eighteen months was even more dysfunctional.

Traditionally arms, from rifles to artillery pieces, were the responsibility of the U.S. Army Ordnance Department, which designed, engineered, developed, and manufactured weapons and munitions within the bosom of the Army's arsenal system. It was a system developed in the eighteenth century, a time when there was virtually no industrial base in North America. Absent either a robust heavy machinery or arms industry, it was both necessary and expedient to create one internally. However, making rifles, pistols, and even artillery pieces was one thing; tanks were quite another matter. So it was that the AEF Tank Corps fought with French light tanks and British heavies while U.S. industry was consumed by start-up problems ranging from translating millimeters to inches to how to mount what cannon on which tractor-like chassis.

Conversely, tentage, uniforms, food, lumber, tar paper siding for barracks, and additional sinews of war flowed in fair order from a rapidly expanding civilian production base, albeit not without considerable difficulty. Demobilization struck all that preparation with a hurricane-like fury. Clearly most of it—from the design and development of arms to procurement contracts for more wrap leggings—was now a candidate for the dustbin. Whatever the status, it was all consumed by demobilization—conscripts and nonregular enlistees sent home, units disbanded, production lines shut down, procurement accounts closed out, everything back to normal. Furthermore, demobilization left the military with a research, development, and acquisition system quite out of tune with the demands of mechanization.

Despite some obvious opportunities demonstrated by tanks in the Great War just finished, the National Defense Act of 1920 set the policy azimuth for the next two decades. It relegated tank matters to the Chief of Infantry, who, with the other branch chiefs, enjoyed enhanced status in the Army bureaucracy under the new law. The 1923 *Field Service Regulations* proclaimed that: "The coordinating principle which underlies the employment of the combined arms is that the mission of the infantry is the general mission of the entire force."

The stage was set.

As the first post–World War I decade unfolded, many of that war's participants were moved to establish links with their military past, however brief it may have been. So it was that Sergeant Starry—formerly of C Company, 329th Battalion, Tank Corps—became First Lieutenant Starry, commander of Headquarters Company, 2d Battalion, 137th Infantry, 35th Division, Kansas National Guard. He had enjoyed military service. He liked being a soldier. Why he did not take advantage of opportunities to become a professional soldier at the time was never clear. In his civilian employment in Kansas City, Kansas, some of his business associates were National Guardsmen; it was likely they who prevailed on him to accept a commission.

Of course there was no longer a Tank Corps. While a troop of horse cavalry was part of the Kansas City garrison, for reasons now lost he elected to join the infantry. Shortly after his appointment to command Headquarters Company there arrived my own appointment as Brevet First Lieutenant, Kansas National Guard. It was to be the beginning of a long military career. The year was 1929, and I was four years and some months of age when Gov. Clyde M. Reed assigned me to my father for quarters, rations, discipline, and for such other duties as might be assigned by the company commander. Those included, as it turned out, periodic drills at the local armory—first located in an abandoned movie house and later a more substantial building—and attendance at all or part of an annual two-week summer camp at nearby Fort Riley.

In infantry battalions of the time, Headquarters Companies provided what is now called command and control. All battalion telephones, radios, and other electronic gear was assigned to Headquarters Company. Signal flags and other more primitive devices were more widely distributed. Headquarters Company operated the battalion message center and the Headquarters Company commander was the battalion adjutant—not the S1 but the adjutant, a sort of information staff officer on the model of staffs of earlier years.

It was a terribly lean Army. The National Guard was pretty much a mirror image of the Regular Army. Although Headquarters Company was authorized several high frequency radio sets, there was but one on hand. Radio operators and crews took turns operating this lone radio. Ammunition boxes salvaged from summer camp were painted to look like the real thing, complete with wooden knobs and dials and hand-painted scales. Operators and crews for whom there were not enough radios would go through the motions on their wooden mock-ups as the crew picked to operate the real radio practiced.

One older model Ford stake-and-platform truck was assigned to the Kansas City garrison. Companies assigned took turns using it for weekend field exercises. At night it was necessary to park the truck headed downhill for an easy, clutch-assisted start in the morning. It was easier than pushing.

Thus was the condition of the country's defense preparedness for whatever national security challenges might come next. It assured that there would be a reiteration of the mobilization frenzy of World War I, already noted, now further complicated by a host of new technology challenges either to be countered or taken advantage of.

Elsewhere there were new ideas—especially regarding mechanization. The story of Sir Ernest Swinton's invention, the tank—along with the history of the development of concepts for mechanization and mobile, all-arms warfare—began in World War I. Born independently in both the British and French Armies, tanks became the subject of considerable debate regarding design, development, and employment. In the United Kingdom a coterie of single-minded tank and mobility enthusiasts persisted in developing concepts for mobile, all-arms warfare built around tank-led striking forces. In France, Col. Jean Estienne, with the backing of industrialist Louis Renault, was able finally to convince the General Staff of the potential worth of light tanks employed in mass to break the trench-bound stalemate and restore maneuver to the battle-field. However, especially in the United Kingdom, these innovators struggled in the face of stubborn opposition by their less imaginative peers and, worse yet, superiors. They were forced to work around an organizational system dominated by foot infantry and horse cavalry, both of whose leaders abhorred change. Frustrated, many went public with their arguments, and by doing so incurred sufficient enmity from their superiors to bring on early retirement or relegation to remote and inconsequential postings.

Notwithstanding, field trials of a mechanized force were held on England's Salisbury Plain beginning in the late 1920s. These trials were designed to demonstrate new tentative tactics, equipment, and organization. Unable to reach post-trial agreement about what had been learned and what to do about it, the British did not much of anything. Thus it was that as war came to Europe in 1939 the British Army found itself absent agreed upon concepts for all-arms mechanized combat based on armored fighting vehicles. There were inadequate operational and tactical level concepts, structural and organizational alternatives, equipment requirement definitions, and training concepts to implement an idea of warfare they themselves had invented.

The Germans, meanwhile, took mechanization seriously. Armed with the writings of B.H. Liddell Hart and J.F.C. Fuller in the United Kingdom, and having studied reports of the Salisbury Plain trials, Heinz Guderian demonstrated what became the blitzkrieg concept to Adolf Hitler at the Kummersdorf test ground in 1934. With Hitler's approval, Guderian, in just eighteen short months, produced an all-arms panzer division. This division had a fairly well spelled out doctrinal framework. It included operational concepts for mobile warfare at the tactical and operational levels, force structure and organizational schemes, a preliminary array of the types of equipment that would be needed, and some carefully thought out ideas about how to train soldiers and, most importantly, units for mobile warfare. By 1939 the Wehrmacht had further developed Liddell Hart's operational concept of the "Expanding Torrent," foreseeing an all-arms mechanized force supported by tactical airpower capable of cutting deep into the enemy's rear. The mobile, mechanized, all-arms panzer divisions with which the Wehrmacht spearheaded its invasions of Poland, France, the Balkans, and the Soviet Union were spawned from this beginning. Despite the predominance of infantry divisions in the Wehrmacht structure, as well as the support of millions of horses for many transport tasks, it was the awesome power of those mechanized spearheads, supported by a fleet of Stuka dive-bombers, that made possible the Wehrmacht's most striking successes. This was especially so on the eastern front. Some participants in the operations there later reflected that, operationally and tactically, the skillful employment of those all-arms mechanized forces prolonged the war in the east by at least three years.

Meanwhile, scarcely anyone in the United States took mechanization seriously. Pioneers in thought and action were few in number, and the institution proved far more resistant to change than did even the British Army. The means for executing national strategy, if one existed, was restricted by America's abhorrence of large standing armies in peacetime. The chief of infantry clung stubbornly to a vision of the dismounted rifleman as the key actor in ground warfare. The Chief of Cavalry opposed mechanization for quite different reasons: he feared it would supersede his beloved horses. Basically it was internecine conflict over the extremely scarce resources provided by a pinchpenny Congress, further exacerbated by an abysmal lack of enlightened thinking by a pride of senior lions. While serving as Chief of Staff in the early 1930s, Gen. Douglas MacArthur testified before the Congress that the Army should not buy too many tanks because they were expensive and became quickly obsolete. Having issued that pronouncement, he decreed that each combat arms branch would pursue mechanization independently of the other branches. This single rationalization created, indeed invited, acrimonious and counterproductive branch contention that lasted well beyond World War II.

There were only two heroes in this drama: Lt. Gen. Daniel Van Voorhis and Maj. Gen. Adna R. Chaffee Jr. Without Chaffee the Army quite likely would have had no tank units at all in 1940. Using his key position on the Army staff in Washington to advantage, Chaffee squirreled away money in the procurement account to provide enough tanks for the equivalent of about three tank battalions. To keep them out of the grasp of the chief of infantry, they were designated GHQ battalions. Without Van Voorhis, who commanded the mechanized elements during most of the experimental armored force trials at Fort Knox in the 1930s, there would have been not even tentative operational or tactical level concepts for armored force employment. As it was, the Armored Force—first under General Chaffee and then, following his untimely death, under Maj. Gen. Jacob Devers—struggled to produce relevant doctrine. In the end, each armored division largely provided its own doctrine based on the study and experience of its leaders and Soldiers.

This was the bitter harvest of the long-standing branch impasse between infantry and cavalry over mechanization. The Chief of Infantry clung to his sacred charter under the 1920 National Defense Act, which had assigned tanks to the infantry and, by inference, to no one else. The Chief of Cavalry clung to his horse-mounted troops but in time accepted the "combat car" as the cavalryman's tank. It is also fair to charge that the search for an adequate tank for either branch was severely inhibited by the Ordnance Department's lack of resolution and constant head butting with maverick tank designer J. Walter Christie. An adequate tank for any employment was simply not available.

Some years ago, historian Edward Katzenbach laid the blame squarely on the horse cavalry in his fascinating paper, "The Horse

Cavalry in the Twentieth Century." He alleged that the Army of the most technically advanced nation on earth came to the threshold of World War II firmly wedded to strategy, operational concepts, and tactics deeply rooted in the nineteenth century. However, he erred on several counts. First, the United States on the eve of World War II was by no means the most technically advanced nation on earth. Second, by singling out the horse cavalry as the sole culprit in a bureaucracy whose most noteworthy characteristic was the intellectual inability to cope with both the need and opportunity for change, Katzenbach ignored what is likely the most persistent shortcoming in the history of American military thought: a blinding fixation on the infantryman as the centerpiece of all military action. It is a bias that ignores the truth that the mechanization of warfare was simply a means of providing more combat power with far fewer Soldiers. If Soldiers as human beings and as a national resource are precious national commodities, why then did America not wholeheartedly embrace the Soldier-saving capabilities of modern mechanized technology? One reason may be that there has been a persistent notion that manpower in conscript armies is a free resource. How many Soldiers, Sailors, Marines, and Airmen may have died unnecessarily as a result of this fixation is both hard to judge and frightening to contemplate.

Despite increasingly ominous developments in Europe in the second interwar decade, U.S. Army strength on 30 June 1939 stood at 187,893, with an additional 199,491 in the National Guard. It was a force somewhat—but only marginally—larger than it had been on 1 April 1917. The Regular Army alone was short nearly 100,000 Soldiers of the number authorized.

On 31 May 1940 President Roosevelt asked the Congress for authority to federalize the National Guard. Shortly after that a new draft law was passed and conscription began.

In armories everywhere the National Guard stood to arms. Headquarters Company, 2d Battalion, 137th Infantry, joined a host of others as it began readying itself for war. Fortunately for the war effort, Headquarters Company's Brevet First Lieutenant was deemed too young for mobilization. Without his services, the 35th Division moved to Camp Robinson, near Little Rock, Arkansas, for training. After some months equipment began to appear; industrial mobilization had begun. Extension of the law mobilizing the National Guard survived in the Congress by a single vote in 1941, just a few short months before the Japanese attacked Pearl Harbor. While some consideration was given to deploying divisions to the Philippines, the campaign and the force deployed there were lost before that could transpire. Meanwhile, mobilized National Guard units had sent off a number of officers and noncommissioned officers to serve as cadre in new divisions. Guard officers attended appropriate branch schools—infantry officers, for example, went to Fort Benning. Promotions flowed freely as the force expanded, and Capt. Don A. Starry soon left the 137th Infantry's Headquarters Company, 2d Battalion, to command L Company, 3d Battalion. Not long after that, he and other older Guard officers (he was forty-four) were replaced by younger officers and posted to other assignments. So, despite years of training and study and a desire to serve in the infantry, he fought as a headquarters warrior, finishing his war service in 1946 as a Colonel. Not at all new, the idea of using National Guardsmen as individual replacements would be repeated in subsequent wars. This led some thinking Soldiers to wonder why the Army should devote all the time, effort, and expense needed to build National Guard units in peacetime if all the system required was individual replacements in time of war.

Meanwhile, all facets of the Army's research, development, and acquisition process severely lagged behind demand, a demand for equipment for fighting units as well as for training units. The 1st Armored Division deployed to the United Kingdom in April 1942 equipped with obsolete M3 medium tanks, no antitank weapons, and short of other essentials as varied as binoculars, observation aircraft, and training ammunition.

By the summer of 1945 more than eight million Soldiers were under arms when demobilization of a sort began. The post–World War II demobilization did not elicit the same openmouthed astonishment as was the case after World War I. However, while poor Colonel Conrad's dilemma had been studied at length and considerable demobilization planning undertaken, things did not go well. So poorly did they go that the author of an Army study published in July 1952 concluded: "when future scholars evaluate the history of the United States in the first half of the twentieth century they will list the World War II demobilization as one of the cardinal mistakes."

However, with the war in Europe concluded, it was necessary to turn full attention to operations against Japan. Simultaneous establishment of a point system for releasing long-serving individual Soldiers and the need to beef up the Pacific-based force structure for the planned invasion of Japan's home islands were clearly at odds with one another. In the ensuing imbroglio between individual and unit deployments, unit effectiveness and cohesion were destroyed. Vocal dissent was rife, in some cases reaching near-riot proportions. The news media entered the debate to include, interestingly, the *Stars and Stripes*. Never able to refuse an opportunity to be heard, Congress joined the chorus. Fortuitously, nuclear weapons dropped on Hiroshima and Nagasaki not only induced Japan to surrender, they short-stopped what promised to be a most difficult national debate over the endgame in the Pacific.

Unlike Soldiers, tanks and guns are uncomplaining partners to whatever is undertaken. Divisions redeploying from Europe, in whatever combination of individual and unit deployment, simply marched their equipment into designated open areas, dismounted, and departed. Ammunition, weapons, and communications gear were removed; however, fuel, batteries, sighting and fire control equipment, and other impedimenta remained. As supplies and repair parts became scarce again in the postwar years, units stationed in Europe found these divisional parks to be welcome sources of everything from jeep windshields to tank power trains. Many of the removed components had been given over to a German government-run supply firm. Sold to whomever, it helped prime the postwar economic pump. Shortages of unit and individual equipment not available in the supply system were all too frequently made up at Soldier expense from these sources. Equipment readiness rates were sustained by units whose officers and Soldiers stood ready to spend their own scarce money buying back repair parts, major assemblies, and, in some cases, whole vehicles. Lean years had returned.

Back in the enclaves of visionary military thinking, revisions to doctrine, equipment, force structure and organization, and training for individuals and units were being drawn up. They were derived largely from the experience of U.S. forces, especially in the war in Europe, from studies of Soviet operations on the eastern front, and from considerable study of Wehrmacht operations. The changes that resulted struck deeply at some of the most dearly held underpinnings of traditional American military thought.

Possibly the foremost difficulty military forces face in the United States is a historic antipathy toward standing armies in peacetime. Born of America's pre-Revolution experience with the British Army and encouraged by the relative isolation of North America from the rest of the world, it led to a conviction that there would always be time to raise an Army, should the need arise. Politically, a two-part solution was adopted. The national military strategy was founded on strong naval forces. Deployed in far-flung battle lines, this "first line of defense" would buy time for raising, equipping, and training large ground forces. Secretary of War John C. Calhoun proposed the second part of the solution circa 1818: token ground forces in peacetime expanded by a flood of militiamen in wartime. Called the "expansible Army," the idea would last more than a century and a quarter, with Secretary Calhoun's militia reinforcements expanded to include conscripts.

These two fundamental precepts were reinforced by Napoleon's ideas on war, as reported by nineteenth century military theoretician Antoine Henri de Jomini: masses of men and fire (artillery) are necessary to win; one always wins by attacking. In addition, processes had been developed during the Industrial Revolution to turn masses of raw material into finished products quickly and efficiently. In sum, the expansible Army had to outnumber its foes to win, and to make the expansible Army possible one had only to invoke the processes of the Industrial Revolution. Later, modern technology in several forms would further enhance both the destructive power of armed forces and the productive power of factories—training factories and materiel factories.

By the end of World War II, one could detect signals that all this might be coming unraveled. First, in Europe especially, the Soviet Union's substantial postwar conventional military strength revealed the unpleasant truth that no longer could the United States be guaranteed numerical superiority, even with the aid of allies. Therefore, it would henceforth be prudently necessary to maintain substantial standing military forces in peacetime. Faced with a potentially overwhelming Soviet force—mostly mechanized—across the inter-German border, and with limited battle space in which to maneuver in defense of western Europe, a military decision would likely be reached between the deployed forces before mobilization and ensuing deployments could provide the means to salvage a difficult situation.

The expansible Army idea had further engendered the conviction that the U.S. Army could afford to lose the first few battles of its next war because mobilization would subsequently raise masses of Soldiers and materiel. Despite any early losses, the war would, in the long run, be won by sheer weight of arms and men. This arrangement clearly was no longer feasible.

Reinforcing this fracturing of sacred convictions was the advent of nuclear weapons—especially once it was understood that they could be employed at all levels of war: strategic, operational, and tactical. Nuclear weapons offered a relatively inexpensive way to invoke the gods of modern technology by substituting their mass killing power for that of masses of Soldiers and units, thus avoiding the expense and time consumed in mobilization. However, despite this perceived advantage, the politicization of nuclear weapons only aggravated the dilemma of how to fight and win the first and succeeding battles of the next war—with or without nuclear weapons—thereby avoiding early defeat of deployed forces.

The NATO dilemma is particularly instructive. Early on, General Eisenhower, as SACEUR, established a requirement for ninety-six divisions and nine thousand tactical fighter aircraft to defend NATO Europe against the Soviet mechanized threat from the east. However, the NATO member countries—including his own—turned thumbs down to his proposal. They were either unable or unwilling to provide the staggering resources required. Later, as president, Eisenhower settled for twenty-six divisions—twelve of them from the Bundeswehr (the West German Army)—fourteen hundred fighter aircraft, and fifteen thousand operational and tactical level nuclear weapons. Some seven thousand of the latter were deployed, eventually to be removed when the Warsaw Pact collapsed on the eve of Operation Desert Storm in 1991.

Then—interacting in complex ways with traditional concepts of war in U.S. military thinking—came North Korea's surprise invasion of South Korea on a fateful Sunday morning in June 1950. World War II demobilization had left overseas-deployed U.S. Army units with a mechanized capability only at division and regimental level. There was a tank battalion per division, a tank company per regiment, and an armored cavalry troop, all in various states of equipage and with serious personnel shortages and individual and unit training deficiencies. Two armored divisions remained—both at Fort Hood, Texas—considerably distant from the inter-German border where they were destined to fight. Fearful of some not quite understood connection between the North Korean attack and growing Soviet deployments to Europe, partial mobilization was ordered in the United States. Conscription was reinstated. Four divisions were called up, with two deployed to Europe and two to Korea. The Europe deployments included armored divisions. They became the basis for a two-corps mechanized force that included armored cavalry regiments at corps, and tank and armored infantry battalions, self-propelled artillery battalions, and armored cavalry squadrons in the divisions. A host of mechanized support units at both division and corps were envisioned, including separate tank and armored infantry battalions assigned to armor groups. The deployed force would initially be reinforced by active duty divisions deployed from the United States, and later reinforced by a host of National Guard divisions and U.S. Army Reserve support units to be mobilized. The deployment of heavy divisions would remain a complex and controversial transportation problem, as would the readiness of forces to be mobilized. Another political hot potato was the division of mobilization authority between the executive and legislative branches. Notwithstanding, U.S. mechanized force deployments and the designation of reserves to reinforce deployed forces in Europe continued for nearly forty years. Along with allied mechanized strength, it remained a strong and reasonably ready capability until the collapse of the Warsaw Pact nearly four decades later.

Mechanized or not, there can be no greater victory for a military force than to accomplish its mission without having to fight the campaigns and battles for which it so carefully prepared for so a long a time.

The Korean War, however, was a sort of field trial in a new era of military affairs. It exposed the bankruptcy of traditional U.S. military thought just described, a bankruptcy best reflected in the popular characterization of Korea as a "limited war." It suggested that all-out mobilization in order to win, after losing the first few battles of a war, was no longer a relevant concept. It brought again into stark focus the price of unpreparedness. Personnel turbulence and resulting unit unreadiness, as well as materiel shortfalls of all kinds, was reflected most tragically in the early losses of the nation's most precious resource: the Soldiers, Sailors, Marines, and Airmen called upon to fight those first battles.

American involvement in the Indochina War began as the Korean War ended. It began with U.S. naval forces deployed to assist with the evacuation of more than a million North Vietnamese into South Vietnam consistent with the 1954 Geneva Accords.

It is frequently said, yet as frequently ignored, that no two wars are ever the same. There is, however, a tragic sameness between Korea and Vietnam. It is a sameness that emphasizes the difficulty of defining a clear-cut connection between political goals and military actions ostensibly undertaken in pursuit thereof. With the absence of a crusade-like framework of total war, consensus over political goals becomes elusive, and on military action even more so—especially as field operations in progress are being beamed into the nation's homes with accompanying commentary that is sometimes relevant and too frequently not.

Mobilization for war in Vietnam was nominal at best. The largest single unit mobilized was an infantry brigade from the Kansas National Guard. The brigade never deployed. Some personnel deployed as individual replacements, reminiscent of earlier practice. Army Chief of Staff Harold K. Johnson publicly recalled making four or five trips to the White House to ask for mobilization, only to be turned away by a president paranoid about anything, including the Vietnam War, which militated against public and budget support for his Great Society. General Johnson's requests were based not so much on an urgent need for numbers of soldiers and units, but more on the likely effect of nonmobilization on the postwar Army. As we shall see, his fears were well founded. Thus, instead of traditional mobilization, structure was added to the Army by simply activating new units and filling them from the conscript replacement stream. Three additional divisions were added to the force structure in this fashion.

Despite considerable information about French and Vietnamese mechanized operations provided by the French Army and by U.S. observers of French operations in Indochina, the dominant conviction of U.S. force planners was that Vietnam was just another Pacific jungle. They saw it as a place where tanks and other armored vehicles could not function effectively and, if used at all, then only in support of dismounted infantry. Thus, U.S. infantry divisions were initially deployed to Vietnam minus their assigned tank and mechanized infantry battalions and armored cavalry squadrons. Once on the scene, deployed division Commanders normally sent for the mechanized units they had left behind, and advised the Commanders of units deploying after them to bring their mechanized units along. The largest single mechanized unit deployed was the 11th Armored Cavalry Regiment. As noted by Lewis Sorley in Chapter 10, it was not until a comprehensive examination of mounted operations by a Chief of Staff–chartered task force in 1966-67 that the battle utility of mechanized forces was set forth objectively. By that time, however, most important force structure decisions had been programmed, making substantial change very difficult. Nevertheless, that war's senior infantrymen clung to their antiarmor bias and their long-standing image of the individual rifleman as the principal weapon of combat to the bitter end.

New to battle in Vietnam was the helicopter. While frequently set aside in considering the mechanization process, helicopter-lifted units, especially so-called air cavalry units, must be included in considering mechanization at the tactical and operational levels of war. The airmobile concept was the brainchild of the 1960s Howze Board. Composed of a group of farseeing cavalrymen headed by Maj. Gen. Hamilton Howze, the board experimented with a concept called air cavalry. In it the helicopter was to replace ground vehicles as the means to move Soldiers and selected weapons quickly over the battlefield to bring force to bear at unexpected places and unanticipated times. A ready-made cavalry idea, it would no doubt have delighted General Howze's father, Robert L. Howze, who as a major led the 2d Squadron, 11th Cavalry, into Mexico with General Pershing's 1916 Punitive Expedition, in some of the most strenuous and demanding horse-mounted operations ever recorded. The helicopter was surely a more than adequate surrogate for Major Howze's exhausted horses! Full blown into a division-sized force, the air cavalry operational concept was both sophisticated and complex. Above all it demanded well-trained units and highly effective command and control.

Vietnam experience with one air cavalry division led to two conflicting observations. One says that helicopter lift has freed infantry from the tyranny of terrain. As a result, the rifleman can now be moved quickly and put down with supporting weapons in a new location in jolly quick time, rested and ready to fight. The counter view is that heliborne infantry suffers from serious limitations. Deploying onto unfamiliar ground and usually uncertain about the enemy, the airmobile force is less mobile, once landed, than its on-site opponent. The situation is aggravated by an individual replacement system that virtually guarantees lack of excellence in unit training. It is made even more difficult by a command-control lacuna from company level upward. Fearful of losing control either of assault forces or their support, most airmobile units prescribed that battalion-level and higher Commanders remain in command posts where communications were in place. This practice frequently inhibited higher level knowledge of the situation at lower levels, slowing decision cycles and reducing leader presence forward in the battle area—both of which increase risks to mission success.

Planning for deployment of U.S. forces from Vietnam began in late 1968. In April 1969 President Nixon issued National Security Study Memorandum 36, requiring, among other things, plans for the initial redeployment of twenty-five thousand of the nearly 549,000 U.S. troops in Vietnam and long-term plans for future redeployments. Limited mobilization, already noted, meant that most redeploying units would simply disappear from the force structure and their manpower from the end strength. Despite General Abrams's strong objections, redeployment of individuals rather than units was the solution chosen by Chief of Staff William Westmoreland. His decision, after all, had historic precedent. It was subsequently decided to rebalance the force in Vietnam by sending home long-serving Soldiers from all units, inactivating designated units, and transferring individuals with longer time remaining to other units. This practice increased personnel turbulence in the remaining units to unacceptable levels, adding to other factors already militating against unit effectiveness. Worse yet, large numbers of battle-weary Soldiers came home alone to a country where few, except their families, either knew or cared where they had been or what they had been doing for the last year or so, or gave any indication of being glad they had arrived home safely. That experience left a generation of young—and some not so young—Soldiers with scars that are yet unhealed.

Virtually hundreds of technical innovations were applied to existing equipment during Vietnam operations. Most were Soldier invented and applied. Many were as simple as sending home for spring-loaded clothespins to use as triggers for Claymore mines. Others were more complex. None were very expensive. The Army created a Limited War Laboratory to aid in developing modern technology solutions for battle-field problems. The Vietnam Mechanized and Armor Combat Opera-tions study group that met in 1966-67 found most laboratory contributions to be relatively ineffective, poorly designed, hard to use, fragile, or some combination of those factors. A case in point with mecha-nized units: mine/countermine operations, that is, equipment for the high-speed search, detection, location, identification, and clearing of antivehicular mines. It is a complex problem with no single technical solution. The search for technical solutions has been trickle funded for many years. As a result, the problem remains unsolved. But one example among many, this situation was further demonstration of the inability of the traditional research, development, and acquisition system to respond to demands for timely fielding of new and innovative technologies.

Battle damage was extensive. The Vietnam War ended in 1973—at least for the United States. Its wounds, however, fester to this day. The na-tional treasure in young lives sacrificed by death, and by physical as well as psychological trauma, has scarred generations. The revolt of the young against all authority so characteristic of most of the important war years has changed our society irrevocably in ways yet to be under-stood.

Post-Vietnam military attention turned back to the nation's com-mitment to NATO Europe. We discovered that the Soviets had been very busy while we were preoccupied with Vietnam. They had revised operational concepts at the tactical and operational levels, increased their fielded force structure, and introduced new equipment featuring one or more generations of new technology. With tactical and operational level nuclear forces deployed, the Soviets embraced the notion that they could fight and win at the operational level of war with or without nuclear weapons. Their preferred solution: without.

As always in Europe, numbers spoke loudly—so much so that several distinguished SACEURs in turn opined that while the conven-tional force might hold out for as long as ten days, when that threshold was reached there would likely be a compelling case for nuclear re-lease. Depending on the resulting status of the battle at the operational level, a decision would have to be made to use or not to use thermo-nuclear weapons delivered by intercontinental ballistic missiles, bomb-ers, or both.

While the Soviets had been building toward winning at the operational level, the disparity in theater force strength and the need for mobilization in many NATO countries, including the United States, inevitably tied the NATO operational level nuclear decision to an extremely high risk of an all-out thermonuclear weapons exchange. It was almost impossible to rationally project NATO strategy to any outcome except Armageddon.

As U.S. forces in Vietnam redeployed, military thinkers recognized the need for a new objective force for a new era. It was an era characterized by the expanded threat in Europe, a growing threat of conflict in the Third World (especially the Middle East), increasing worldwide economic interdependence, greater difficulty articulating political goals for the planners who design military activities to achieve them, and intrusive and abrasive media probing into all aspects of military operations.

One bright Pentagon morning in 1972, toward the end of Vietnam redeployment, force planners, of whom I was in charge, awakened to the reality that as force structure and end strength had declined, only that structure and strength specifically committed to NATO, either deployed or reinforcing, had been retained. There were, therefore, at that moment, no more than perhaps twelve divisions remaining in the structure, with a strength authorization of about 765,000. By comparison, the pre-Vietnam force structure had included sixteen divisions and nearly 986,000 Soldiers. Army Chief of Staff Abrams, General Westmoreland's successor, personally intervened with Secretary of Defense James Schlesinger and got the green light to retain sixteen divisions. However, additional personnel strength would have to be justified in follow-on budget negotiations with Congress. While the pre-Vietnam strength had admittedly included some residual holdings from the early 1960s partial mobilization in response to the Berlin Wall buildup, clearly the right answer was more than 765,000 but perhaps less than 986,000. Since increasing strength requires more money, only nominal strength increases were ever realized.

The U.S. Army was somewhere close to its nadir. Soldiers and units deployed to USAREUR saw themselves as minor speed bumps for Soviet forces en route to the Rhine River and beyond; they did not believe they could defend successfully, let alone win.

While there were many reasons for this, three stand out.

First, force modernization—that is, the promulgation of new doctrine, fielding of new equipment, organizational changes, and improvements in training and education for officers, noncommissioned officers, and Soldiers—had been at a standstill for nearly ten years.

Second, absent significant mobilization for the Vietnam War, it was necessary to use the entire Army—including the Army in Europe—as the rotation base for Vietnam. Combined with the one-year tour length in Vietnam, this increased personnel turbulence in units to a level well above that at which reasonable unit effectiveness could be achieved and sustained. The militant youth revolt against authority that characterized the 1960s in the United States had infested the country's armed forces. Military jails were full. The drug culture rampant in the country pervaded all but the best military units. Largely because of these factors, among others, conscription had been shut down in 1972—more than a year ahead of the draft law's expiration—without any sure knowledge that the Army could recruit sufficient numbers to fill its force structure.

Finally, Soldiers and leaders returned from a war in which they had won all the battles, only to find the nation had lost the war. There was a crisis in confidence; Soldier confidence in leaders, leader confidence in themselves as well as in the nation's political leadership, and Soldier and leader confidence in their Army and its units.

There was a widespread consensus that the Army needed substantial rebuilding. That rebuilding began with General Abrams's appointment as Chief of Staff in the fall of 1972. By the summer of 1973, reorganization of the Army's command structure was underway. That reorganization included, among other changes, dividing CONARC into FORSCOM and TRADOC. The former was responsible for all Army forces stationed in the United States, including the U.S. Army Reserve; the latter established doctrine, developed equipment and organizational requirements, and provided training and education for all ranks.

Born in the summer of 1973, TRADOC was commanded by Gen. William DePuy. At the same time I was posted to command the Armor Center and School at Fort Knox, armed with instructions from both Generals Abrams and DePuy that I was to define what the first and succeeding battles of the next war might require of the Army.

As that work began, seven thousand miles away the Arabs attacked Israel in October 1973. Within a few weeks General Abrams dispatched me and Brig. Gen. Bob Baer—the program manager of what would later become the Abrams tank—to Israel. There, Bob and I walked battlefields with the IDF Commanders who had fought on them, seeking answers to many questions about the future U.S. Army. The questions we posed focused on documenting requirements for the M1 tank, especially the need for a larger caliber gun than the planned 105mm, and most importantly, on critical operational lessons of the Yom Kippur War.

Answers to those questions framed the beginning of what grew into, some nine years later, the doctrine called AirLand Battle, a concept of war at the tactical and operational levels that U.S. and coalition commanders employed in Operation Desert Storm.

The armored battlefields of the Yom Kippur War yielded striking lessons about what to expect in first and succeeding battles of the next war.

First, we learned that the U.S. military should expect modern battlefields to be dense with large numbers of weapons systems whose lethality at extended ranges would surpass previous experience by nearly an order of magnitude. Direct-fire battle space would be expanded several orders of magnitude over that experienced in World War II and Korea.

Second, because of numbers and weapons lethality, the direct-fire battle will be intense, resulting in enormous equipment losses in a relatively short time. Significantly, we noted, combined tank losses in the first six critical days of the Yom Kippur War exceeded the total U.S. tank inventory deployed to NATO Europe—including both tanks in units and in war reserves.

Third, the air battle will be characterized by large numbers of highly lethal aerial platforms—both fixed- and rotary-wing—and by large numbers of highly lethal air defense weapons.

Fourth, the density-intensity-lethality equation will prevent domination of the battle by any single weapons system; to win, it will be necessary to employ all battlefield systems in closely coordinated all-arms action.

Fifth, the intensity of battle will make command and control at the tactical and operational levels ever more difficult. Effective command-control will be further degraded by the presence of large numbers of radio-electronic combat systems aimed at inhibiting effective command-control.

Sixth, at both the tactical and operational levels the complexity of modern battle demands clear thinking. Thinking takes time, and in battle there is no time to think. Therefore, to the extent possible, likely battle circumstances must be thought through in advance to reduce the chance of surprise and to ensure prompt, timely, and relevant decisions.

Finally, regardless of which side outnumbers the other, regardless of who attacks whom, the outcome of battle at the tactical and operational levels will be decided by factors other than numbers and other than who attacks and who defends. In the end, the side that somehow, at some time, somewhere during the battle seizes the initiative and holds it to the end is the side that wins. More often than not, the outcome of battle defies the traditional calculus employed to predict such outcomes. It is strikingly evident that battles will continue to be won by the courage of Soldiers, the character of leaders, and the combat excellence of well-trained units—beginning with crews and ending with corps and armies.

For those of us who crafted new doctrine to reflect the new environment, one single statement became the goal: The U.S. military must decide how to fight outnumbered and win the first and succeeding battles of the next war at the tactical and operational levels—without wasting Soldiers' lives, and without having to resort to the use of nuclear weapons to offset the military's likely numerical disadvantages or for any other purpose.

The first try at new doctrine focused on the tactical level of war. Several important facts drove that decision.

First there was the advent of long-range ATGMs in the early 1970s and their appearance and initial success early in the Yom Kippur War. Lethal at far greater ranges than any other infantry weapons, ATGMs promised to provide forward-deployed forces a new dimension in defense: the ability to maneuver to seize the initiative.

Second were General DePuy's convictions about inadequate tactics, ineffective training and inept leadership in small units in World War II, and most importantly the terrible cost in Soldier lives lost unnecessarily as a result. Drawn from his traumatic experience as a young officer in the 90th Infantry Division, recorded eloquently in his post-retirement oral history, and described in some detail in Richard Swain's discussion of the development of AirLand Battle doctrine in Chapter 11, those convictions burned deep in his soldier soul. While commanding the 1st Infantry Division in Vietnam in 1966-67, he was notorious for relieving Commanders, mostly at battalion level—a practice that was widely and bitterly criticized. In truth, however, his practice had serious purpose, best described in a statement he once made to me: "I'm just not willing to trust the lives of the soldiers to the command of officers in whom I don't have confidence."

Finally, in organizing TRADOC, General DePuy had assigned a disproportionate number of officers to write doctrine at the CGSC at Fort Leavenworth, where normally lay responsibility for writing operational level doctrine—the capstone ideas on which the writing of tactical level doctrine was based. The aim was to write the pacesetting doctrine at Leavenworth, while TRADOC's branch schools would write appropriate tactical doctrine for their respective branches. Early on it was quite clear that what was intended of Leavenworth would probably not be forthcoming, for reasons fairly objectively set out in Chapter 11. This created a situation in which doctrine, normally written at Leavenworth, was largely being put together elsewhere, primarily at Fort Knox, where work had begun early and in earnest to solve the Army's most important challenge: how to fight and win on the armored battlefields of NATO Europe. Time was of the essence. The DePuy watch at TRADOC could be expected to last four years; he wanted to finalize what had to be done within that window. In addition, General Abrams, while ill and before his untimely death in the fall of 1974, made imperative his approval of TRADOC's new direction.

Thus it was that in just a bit more than two years, a new capstone doctrine was drafted. Called "Active Defense," it appeared in a 1976 revision of *FM 100-5, Operations.* Since it had been written by a handful of officers at Fort Knox, and in General DePuy's closed sessions with his center Commanders at Fort A.P. Hill, the new doctrine attracted many critics—both inside and outside the Army. At root, most criticism stemmed from the fact that the doctrine had not been written at Fort Leavenworth. Some disdainfully called it "DePuy Tactics."

Active Defense was published in mid-1976. Earlier that year, armed with draft copies of the new manual and of the implementing armor force manuals written on my watch at Fort Knox, I took command of USAREUR's V Corps. Since I had authored critical parts of the new doctrine and was not completely happy with what had been written, I took as a first task to somehow test the ideas on the ground.

Issued in draft to all levels of command Active Defense became the basis for staff rides—terrain walks in which division, brigade, and battalion commanders met with me on-site to determine if the doctrine based on the Yom Kippur War's lessons was about right or all wrong. After more than six months of evaluating tactics, weapons, and organization using analytical models provided by BDM International, commanders at all levels had about convinced themselves that, with a little luck, the Soviet first echelon could be defeated well forward in the main battle area. It was quite clear to me at that juncture that, as I had feared earlier, the operational level problem had yet to be confronted, and it was the corps Commander's responsibility to do so.

In May 1977 I returned to Israel's battlefields to revisit action at the operational level and then translate that experience to Europe's environment. This led to a concept for extending the battlefield in time (the campaign) and distance (the theater of operations). Most importantly, it resulted in requirements for long-range surveillance and target acquisition systems and long-range weapons systems with which to find and attack Soviet style follow-on echelons. These systems would be used to disrupt and delay their advance into the main battle and to destroy as many as possible before they could reach assigned objectives deep in the main battle area. The primary goal was to involve the Army in the operational level of war business.

Two months later I succeeded Gen. Bill DePuy as TRADOC Commander, faced with the twin tasks of somehow defusing vocal criticism of Active Defense doctrine and expanding doctrine to integrate the tactical and operational levels of war. In addition, it would be necessary to build a broad consensus about the correctness of whatever ensued. Change is not possible without consensus. At least part of our problem with Active Defense was that it reflected the fact that the idea was developed quickly and in a fairly closed forum. Those who considered they had no voice in the matter tended to reject it out of hand. It was also obvious that somehow whatever was done had to be accomplished in the TRADOC schools, beginning with Leavenworth.

To defuse criticism and change the perception that new doctrine could only come from the big leather chair in the front office, I created in the TRADOC headquarters the position of Deputy Chief of Staff for Doctrine. There had been none before. I appointed Brig. Gen. Don Morelli, a very bright and persuasive officer, to that position. Morelli, assisted by a briefing team, did very little else for four years but expose the developing concept to staffs in the Congress and academia, even as the details were being written. Those who did not agree were invited to provide suggestions, with the assurance that their suggestions would, to the extent possible, be included or dealt with in the final product.

At Leavenworth, attrition had taken its toll of the reluctant phalanx of disappointed doctrine writers, and Lt. Gen. Bill Richardson, the new Commandant, gave the task of writing what became AirLand Battle to some very bright Lieutenant Colonels—Huba Wass de Czege, Don Holder, Richard Hart Sinnreich, and Richmond Henriques.

AirLand Battle doctrine, as first written in the 1982 edition of *FM 100-5* followed closely the concepts set forth in "Extending the Battlefield," a doctrinal essay taken from my briefings and published in *Military Review* in March 1981. The concept derived from my observations as corps commander, from visits to Israel's battlegrounds, and from a study of deep attack targeting for nuclear weapons done by Dr. Joseph V. Braddock of BDM International for the Defense Nuclear Agency. From the latter came convictions that advanced weapons delivery and fusing accuracies had reached the point at which we could now attack with conventional munitions what once required nuclear weapons to achieve desired target effects. A search for enabling surveillance technology found synthetic aperture radar. Funds were found to accelerate development, based on requirements which we could now write, for what became the JSTARS and ATACMS systems, designed to be the corps or joint task force Commander's surveillance and target acquisition and weapons systems, respectively. Thirteen years after we first set down requirements for those systems they performed with stunning success in Operation Desert Storm.

AirLand Battle, in my view at least, was an operational level concept; it combined the best tactical lessons of the Yom Kippur War with operational level schemes designed to defeat Soviet operational level concepts: mass, momentum, and continuous land combat—with or without nuclear weapons. Beginning with the Bolshevik Revolution in 1917, Soviet military thought had focused on operational level concepts, largely driven, some say, by the trauma of their defeat in World War I and subsequent civil war. In the West, as suggested earlier, while mechanization offered operational level opportunities to all nations, only the Germans took advantage of the offer.

From AirLand Battle doctrine came requirements for equipment. The long development time characteristic of the normal materiel acquisition cycle almost guarantees that either threat or technology, perhaps both, will change significantly during development, changing requirements and further prolonging development time, thus aggravating the inefficiency of an already severely dysfunctional system. At the outset, TRADOC inherited several ongoing programs. Principal among them were the armored reconnaissance scout vehicle (ARSV), the infantry fighting vehicle (IFV), the UH-60 Black Hawk troop carrying helicopter, and the AH-64 Apache attack helicopter. In addition, requirement documentation was in being for a main battle tank, the XM1, and was being developed for a division-level air defense system. Proponency for the ARSV, Apache, and the tank resided at Fort Knox.

Shortly after I arrived at Fort Knox in 1973, ARSV prototypes were delivered for testing. There were two candidates, one full tracked and one wheeled. One look was sufficient to suggest that both were wide of the requirement, by then some ten years old. Having just forced a decision to take the unsatisfactory M114 scout vehicle, and the equally deficient M551 Sheridan airborne assault/armored reconnaissance vehicle out of the inventory, I was extremely reluctant to buy into another uncertain program. We tested the candidates at Fort Knox and recommended the program's termination. In retrospect it might have been better if I had kept the program alive and tried to "fix" one of the can-

M114 Unsatisfactory Scout Vehicle. Patton Museum.

didates to meet the changed requirements, for despite considerable testing to develop relevant requirements, the Army still does not have a satisfactory ground scout vehicle.

Not long thereafter the MICV prototype was delivered for testing—again ten years after the requirements documentation was written. Just on observation it was worse than the ARSV. However, having just terminated ARSV, we feared cancellation of another major program would eliminate TRADOC as well. So it was decided to "fix" the MICV by redesigning the power train, adding armor, mounting the TOW ATGM system, finding a suitable cannon and fire control system, and including firing ports for mounted infantry. To help hold down rising costs we added the ARSV acquisition objective numbers to the IFV numbers to help reduce unit cost. Hence, the Army eventually fielded two versions of the Bradley fighting vehicle, one for infantry and one for cavalry scouts. However, neither model met the requirements.

Recognizing the Bradley's shortcomings for fighting the central battle alongside tanks, despite the serious upgrades just mentioned, the vehicle was inadequate for the task. Therefore a Heavy Infantry Fighting Vehicle Task Force was convened to draw up requirements for such a vehicle based on study of the Arab-Israeli wars, and IFV systems in other armies. We then considered revising the XM1 tank design to provide space inside for an infantry fire team, a concept similar to that of the Merkava tank then being developed for the IDF. Design change of that magnitude would have severely delayed the XM1 program, a risk we decided not to accept.

In 1974 an Advanced Scout Helicopter (ASH) Task Force was convened at Fort Knox to set forth requirements for a replacement for the Hughes OH-6 and Bell OH-58 scout helicopters used during the Vietnam War. The new scout aircraft would be a companion to the AH-64 Apache, then being developed. A highly successful ASH Task Force effort suffered rejection in some bitter bureaucratic infighting in the Pentagon. Nearly twenty-five years later, the Army scout helicopter requirement is being met by upgraded Bell OH-58s, with the Comanche intended as a future dual-purpose scout and attack helicopter.

Early in TRADOC's life the Air Defense Center at Fort Bliss, under Maj. Gen. C.J. LeVan, developed requirements for an air defense gun system based on observations of the Yom Kippur War. Twin Bofors 40mm cannons would be mounted on a refurbished M48 tank chassis, and an armored turret and F16 tracking radar added to make the Sergeant York Division Air Defense (DIVAD) gun system. The winning prototype candidate was defeated by last-minute changes to requirements—changes designed to make the cannon system less effective than required at extended ranges. Air defense missile enthusiasts who replaced the air defense gun enthusiasts were responsible for changing the criteria on which the DIVAD gun system was tested and subsequently failed.

Studying Yom Kippur War battles with IDF commanders led to a concept for an improved "over the hill" battle surveillance system, more recently styled as an unmanned aerial vehicle (UAV). Early requirements visualized a low-light television system mounted on a remotely controlled model airplane to demonstrate feasibility. The IDF did just that, flying the system in operations in southern Lebanon for several years and then employing two systems—Scout and Mastif—in the 1982 invasion of Lebanon, dubbed Operation Peace for Galilee. Meanwhile, in the United States, the artillery and intelligence communities started a fatal quarrel over a system called Aquila. In the end it was overloaded with sensors of all kinds, costs were considered exorbitant, and the vehicle displayed a penchant for crashing directly after launch, causing program termination. Only recently has the U.S. military developed a UAV capability—in a world where such machines have been fairly commonplace for nearly twenty years.

As Active Defense doctrine unfolded, General DePuy undertook a division-level test of doctrine and tentative organization at Fort Hood.

Called the Division Restructuring Study (DRS), instrumentation was provided by an early version of the Multiple Integrated Laser Exchange System (MILES). As is normally the case with large-scale tests, the attempt to gather definitive data about too many things with inadequate instrumentation produced less than adequate results. Returning to command TRADOC in 1977, I found the study had exceeded time and budget targets while producing suspect data outcomes. The study was terminated and TRADOC Center commanders met frequently with me over the next eighteen months to develop alternatives based on what had been learned in the DRS evaluation and in manual battle-lab exercises.

The resulting organization for heavy forces was called Division 86. Study and test outcomes strongly suggested a need for smaller combat units from platoon upward—with more leaders per soldiers led. However, the idea ran contrary to personnel management dictums since it increased the numbers of officers in the division. The question of numbers of tanks at platoon, company, and battalion was clouded by uncertain performance of the M1 tank power train in tanks then being conducted at White Sands, New Mexico. Resulting compromises produced an organization that was inconsistent with the organizational concepts in mind at the outset. We therefore styled Division 86 as an interim organization with opportunities for further improvement.

Training for all ranks underwent considerable change consistent with developing doctrine. Revised soldier training began in 1974 with one-station unit training for initial entry Soldiers at Fort Knox. A noncommissioned officer education system (NCOES) was implemented over several years. Tied to job performance, it became the basis for assignment and promotion, ending long-standing inequities in unit-level NCO academies. Meanwhile, officer branch advanced courses were halved in length—largely a budget decision.

The Combined Arms Staff and Services School (CAS[3]) was established at Fort Leavenworth for all officers in the grade of Captain in the 1970s. It was designed to make up for the foreshortened advanced courses and to accommodate the more than half the officer population that would not normally attend the year-long CGSC course at Leavenworth. The CGSC curriculum in turn was gradually changed to reflect the need for officer education at the operational level of war. A few select graduates of the year-long CGSC course were allowed to remain at Leavenworth for a second year in the School for Advanced Military Studies program, devoting themselves entirely to in-depth study of war at the operational level.

The resulting product, some seventeen years later, was a U.S. force that went to war as part of the coalition that executed Operation Desert Storm. During those seventeen years, almost everything about that force, especially the Army, had been remade from what existed in 1973 at the beginning of the post-Vietnam recovery.

It is well to remember that Desert Shield-Desert Storm was undertaken at a time when the United States was in the process of "downsizing" in the wake of the collapse of the long-standing Soviet threat in Europe. While the ultimate shape of U.S. post-Soviet national security arrangements was not at all clear, coalition operations to remove the Iraqi Army from Kuwait provided some useful calibrations.

Desert Shield-Desert Storm was a limited war. Its political goal was made clear at the outset and its accomplishment was contingent on joint operations by U.S. forces and coalition allies, and public support at home. When the announced political aim was accomplished, forces were redeployed. Despite some grumbling that more should have been done, the deployed military force did what was set forth at the beginning and then withdrew—for better or worse.

The military force fought outnumbered and won its first and succeeding battles, without invoking the specter of nuclear weapons, and virtually without casualties. The part of the force that brought the war to a successful termination was a corps-sized combined arms mechanized force employed with lightning-like speed and devastating lethality. The mechanized force was supported by deep surveillance and attack systems. Attack helicopter units were integrated into the tactical maneuver scheme, and fighter aircraft protected the battlefield from attack by enemy air forces and provided fire support—deep and close. Everything worked—battle tactics, Abrams tanks, Bradley fighting vehicles, Apache helicopters, the MLRS, JSTARS, and the ATACMS. New organizational schemes proved more effective than their predecessors. Soldier and leader training was more than adequate to the demand. In summary, the equipment, organization, and training designed to support AirLand Battle doctrine was an unqualified success—one might say in spite of and not because of all the compromises and shortfalls just recited.

There are hidden caveats, however. First, there was the mobilization of considerable numbers of small units or detachments whose numbers were nominal but whose skills were essential. Three National Guard mechanized brigades were called to active duty. Tested at the National Training Center at Fort Irwin, they were found not ready for combat.

The annoying and persistent truth about the Army's reserve components, especially National Guard combat and combat support units, is that no one knows what it takes to bring those units to a satisfactory level of readiness. Changing force requirements by the active Army have historically resulted in reorganization of National Guard units—to the end that they are forever unready. However, the reserve components have become so politicized that, despite an obvious need for dramatic systemic change, hardly any change is possible. In former years there was talk of "One Army." Conceptually it was a seamless structure including active, National Guard, and U.S. Army Reserve units—all ready, willing, and able to go to war quickly and effectively. However, concept and reality were poles apart. There was much rhetoric and little accomplishment. More recently "One Army" has given way to the "Total Army." Unfortunately, the conversation/reality gap is wider than ever. This is largely so because in the post-Soviet national security spectrum there are fewer deployed forces, hence increased requirements for ready forces and the means to deploy quickly to a more variegated selection of geography and contingency situations. Systemically, the Army Reserve Component System is a legacy of Secretary Calhoun's expansible Army concept and its eventual implementation (with conscripts instead of militia) and the mobilization system developed as a result of the reforms initiated by Secretary of War Elihu Root in the early twentieth century.

Second, unit performance, especially in Operation Desert Storm, strongly reflected several months of training in the region before the Hundred-Hour War began. Unit training, few if any individual replacements, plenty of ammunition, range area, and time, all enhanced unit cohesion and leadership. After some months, several unit commanders, satisfied with unit proficiency, observed that it was time for the war to begin. Once the Hundred-Hour War began, unexpectedly low casualty rates, especially in ground forces, spelled extremely low personnel turbulence rates and continued excellence in unit performance. Unit effectiveness is ever the victim of the individual replacement system, a fact commented on strongly and adversely after every war. The individual replacement system is a residual of the Army's Industrial Revolution heritage and of Secretary Calhoun's expansible Army idea. In the modern era it is clearly dysfunctional—an unacceptable inhibitor to unit effectiveness.

Third is logistics. Months of in theater train-up before the Hundred-Hour War enabled the establishment of a robust logistics infrastructure. Logistics support for the U.S. VII Corps and its supporting units in the great end run around Iraqi forces in Kuwait was surely one of the most impressive military logistics accomplishments ever. The period of preparation concealed both the complexity and difficulty of what was done. Since it all worked so well, hardly anyone is now looking closely for buried problems. There are some, however. Problems ranged from the enormous transportation effort to support such an operation over tremendous distances, to the vulnerability of large on-ground stores to enemy attack, especially ballistic missile attack. The U.S. Army logistics system is another significant part of the traditional mobilization system that, like the personnel replacement system, begs for change.

Fourth is the research, development, and acquisition process.

None of the marvelous new weapons used in Desert Storm appeared overnight. The Abrams tank had its beginnings in the MBT70 program. Conceived as a combined American-British-West German venture, it broke up in the late 1960s because of increasing costs and national differences over weapons and other integral systems. A unilateral U.S. version, the XM803, followed, but was unilaterally terminated by Chief of Staff Westmoreland in the wake of congressional termination of the Cheyenne helicopter program.

A user task force at Fort Knox in 1970-72 drew up the requirements for what became the Abrams. Eventually included was the exciting new British Chobham armor technology. Requirements were rigorously reviewed and amended based upon on the IDF's experience in the 1973 Yom Kippur War. Developed by competitive bid to industry, prototypes appeared in a relatively short five or six years; but the audit trail goes back to the late 1950s.

Requirements for what became the Bradley were first laid down in 1964 as a Mechanized Infantry Combat Vehicle (MICV). After substantive changes to armor, weapons, and power train, the Bradley finally appeared nearly twenty years later.

The acquisition system takes far too long, is far too encumbered by a complex maze of regulatory milestones, and is far too vulnerable to the funding whims of a Congress convinced that all weapons cost too much—as does the Defense Department establishment as a whole. These concerns—along with cost, competitive contracting, and ethical practices by defense contractors—head most congressional agendas. While the Goldwater-Nichols Defense Reform Act of 1986 sought to reform the system, the basic systemic flaws remain, despite the reform legislation.

The world has continued to change since Operation Desert Storm in 1991. Indeed, it changes even as we assemble this account of three-quarters of a century of mechanization in the U.S. armed forces. Nonetheless, some things do not change. Several of those have been cited above—for emphasis, more than once. There are others that have not been cited. However, I think it is necessary here to make two additional enduring observations.

The first is about technology. We like to think of ourselves as the most technically advanced nation on earth. Without debating that premise, permit me the following observation. An almost childlike faith in the marvels of modern technology is as much a part of U.S. history as is Manifest Destiny. Techno-utopianism has pervaded the last half of this century. This is true of society as a whole, but it is especially true of defense undertakings because of the size and visibility—and hence the vulnerability—of the defense budget. Techno-utopianism notwithstanding, we seem too frequently to deny out of hand some important achievement of modern technology that offers new capabilities and fresh opportunities. Again consider the tank, the centerpiece of modern mechanized warfare. The U.S. Army's Abrams tank has only one peer in world tank fleets, the Israeli designed and produced Merkava. The M1 Abrams and its variants embody lessons about tanks and tank warfare learned by tank Soldiers beginning in 1918, when the Army borrowed French light tanks so that Sergeant Starry and his Tank Corps comrades in C Company, 329th Tank Battalion, could cross the trenches and make it through the wire ahead of the infantry. Although considerable zeal has been displayed by those working to improve tanks technically, even more zeal has been shown by those trying to make the tank technically obsolete. The arrival of tactical nuclear weapons in the middle decades of this century was widely heralded as the death knell of the tank. However, with better understanding of nuclear weapons effects, and upon sober reflection, tanks were deemed probably the very best place to be on the nuclear battlefield. The appearance of ATGMs in the 1973 Yom Kippur War—along with considerable misinformation about their contribution to tank losses in that war—was widely heralded as marking the end of the tank in battle. After collecting and analyzing the facts, it still took considerable time and effort to persuade some, but not all, techno-zealots that ATGMs accounted for only about 8 percent of the IDF's total tank losses. Beginning with my time at Fort Knox, after visiting the Yom Kippur battlefields and examining closely the history of tank-on-tank engagements in that war, I created a brief lecture called "Tanks Forever." In it I set forth a concise statement of fact regarding the allegations of the ATGM enthusiasts. However compelling it may have been initially, it was subsequently necessary to revise and revive that briefing about every four years for the next fifteen years or more.

The second observation is about Soldiers. Modern battle is extremely intense and complex. The next modern battle will certainly be more intense than the last. It will witness increased range and lethality of a growing menu of weapons of all kinds, along with increased complexity of command and control amid a growing deluge of information. Added to this picture are ever more difficult strategic and operational level challenges in a yet dangerous world. As a result, Soldiers and their leaders will continue to constitute the deciding element. The systems people, the nuclear people, airpower advocates, maritime enthusiasts, and information technology wizards all too frequently misunderstand, ignore, or would forget that. Scientists will ever contend they have just discovered some new technology that makes courage, leadership, and Soldier and unit effectiveness obsolete. Let us hope that one day this may come to pass. History, however, instructs that it is unlikely to be so. The history of mechanization as recounted in these pages is a record of the search for the best doctrine, equipment design, force structure, organization, and training for soldiers and leaders. Yet battles will still be won by the courage of Soldiers, the character of leaders, and the excellence of unit training conducted before battle. Nothing in the superb performance of U.S. forces in Operation Desert Storm, indeed, nothing in the half-century-long standoff with the Soviets, goes against that eternal truth.

Taken altogether, we have come full circle with ideas set forth in the introduction to this book. In preparing for battle we must ever recognize the challenge of taking available technology and making with it the best force that can possibly be fielded. Clearly, as history has demonstrated time and again, the better tank is the one with the better crew—almost regardless of the level of technical sophistication. On the other hand, technological advances continue to provide new challenges and new opportunities. This relentless technical march can neither be avoided nor set aside. So it is that those who prepare for war must ever be mindful of their obligation to capture and harness technical advances to the doctrinal imperatives of battle—especially at the tactical and operational levels of war. That reality is why we began this anthology. We were impelled by the notion that: "we are . . . not so much in search of dominating technology as we are in search of the intellectual power to understand the possibilities and limitations of burgeoning technology, and the moral courage to step out in new directions. While there is no lack of new technology, the intellectual power and moral courage to use it properly seem ever wanting." That search for wisdom, courage, and leadership must never end.

Select
Bibliography

The following is a sampling of the books, journal articles, and occasional papers dealing with the history of armor and mechanization, as well as related supporting topics, that the editors and contributors used in crafting their histories. For reference to primary sources the authors used, see the endnotes accompanying each chapter. The notes direct the reader to a rich repository of material acquired from the Modern Military Records Division in the National Archives, the U.S. Army Military History Institute (USAMHI) and Army War College archives, the U.S. Army Command and General Staff College (CGSC), the Armor School Library at Fort Knox, the U.S. Marine Corps History and Museums Division, and the Command and General Staff College of the Marine Corps University. The Patton Museum of Cavalry and Armor at Fort Knox holds the huge Col. Robert J. Icks collection of books, photographs, and published and unpublished manuscripts on armor. Aberdeen Proving Ground has a collection of photographs and information on the test and evaluation of armored vehicles and is the home of the U.S. Army Ordnance Museum. In addition, there are listed in the chapter endnotes sources acquired from the U.S. Army TACOM and associated industrial functions, combat arms schools, congressional reports, master's theses, and doctoral dissertations that were used in crafting this armor history. The reader who wishes to further expand on a particular subject will find the chapter endnotes a very valuable resource.

For discussions on the evolution of ideas, technology, and doctrine regarding mechanization and armor, the reader will find a wealth of historical information on the maturation of concepts and debates in the numerous service journals cited, including *ARMY, Cavalry Journal, Armor Cavalry Journal, ARMOR, Field Artillery Journal, Infantry Journal, Army Ordnance, Ordnance, National Defense, Military Review, Parameters, Proceedings,* and the *Marine Corps Gazette. The Journal of Military History* (formerly *Military Affairs*) is a scholarly publication that has printed numerous articles pertaining to the history of armor and mechanization as well as doctrinal development. For an excellent history of U.S. armor technology, see the numerous works by Richard P. Hunnicutt. An excellent one-volume reference is Fred W. Crismon's *U.S. Military Tracked Vehicles* (1992). For an international reference for all fighting vehicles in the world, see the 18th edition of *Jane's Armour and Artillery*. A superb recent short history of U.S. tank development is Robert S. Cameron's three articles in *ARMOR,* "Armor Combat Development 1917-1945" (Sept.-Oct. 1997), "American Tank Development During the Cold War" (July-Aug. 1998), and "Pushing the Envelope of Battlefield Superiority" (Nov.-Dec. 1998).

Books and Booklets

Adan, Avraham (Bren). *On the Banks of the Suez.* San Rafael, Calif.: Presidio Press, 1980.

Albright, John, Cash, John A., and Sandstrum, Allan W. *Seven Firefights in Vietnam.* Washington: OCMH (OCMH), 1970.

Alexander, Arthur J. *Armor Development in the Soviet Union and the United States,* R-1860-NA. Santa Monica, Calif.: Rand Corporation, 1976.

Alexander, Joseph H. *Closing In: Marines in the Seizure of Iwo Jima.* Washington: History and Museums Division, Headquarters, USMC, 1994.

———. *Utmost Savagery: The Three Days of Tarawa.* Annapolis, Md.: Naval Institute Press, 1995.

———. *The Final Campaign: Marines in the Victory on Okinawa.* Washington: History and Museums Division, Headquarters, USMC, 1996.

———. *Storm Landings: Epic Amphibious Battles of the Central Pacific.* Annapolis, Md.: Naval Institute Press, 1997.

Alexander, Joseph H., and Bartlett, Merrill L. *Sea Soldiers in the Cold War: Amphibious Warfare, 1945-1991.* Washington: History and Museums Division, Headquarters, USMC, 1996.

Appleman, Roy E. *South to the Naktong, North to the Yalu: The U.S. Army in the Korean War.* Washington: OCMH, 1961.

Arnold, James R. *Armor.* New York: Bantam Books, 1987.

Bailey, Charles M. *Faint Praise: American Tanks and Tank Destroyers During World War II.* Hamden, Conn.: Archon Books, 1983.

Baldwin, Hanson W. *Tiger Jack.* Fort Collins, Colo.: Old Army Press, 1979.

Ball, Harry P. *Of Responsible Command: A History of the Army War College.* Carlisle Barracks, Pa.: Alumni Association of the U.S. Army War College, 1983.

Ballendorf, Dirk Anthony, and Merrill L. Bartlett. *Pete Ellis: An Amphibious Warfare Prophet, 1880-1923.* Annapolis, Md.: Naval Institute Press, 1997.

Barnett, Corelli. *The Desert Generals.* Bloomington: University of Indiana Press, 1983.

Bartlett, Merrill L. *Lejeune: A Marine's Life, 1867-1942*. Columbia: University of South Carolina Press, 1991.

Bittner, Donald F., ed. *Perspectives on Warfighting*. No. 3. Quantico, Va.: Marine Corps Association, 1994.

Blumenson, Martin. *Breakout and Pursuit. U.S. Army in World War II*. Washington: OCMH, 1961.

———. *The Patton Papers*. Vol. 1, *1885-1940*. Boston: Houghton Mifflin, 1972.

———. *The Patton Papers*. Vol. 2, *1940-1945*. Boston: Houghton Mifflin, 1974.

Bradley, Omar N. *A Soldier's Story*. New York: Henry Holt, 1951.

Bradley, Oman N., and Clay Blair. *A General's Life*. New York: Simon and Schuster, 1983.

Caidain, Martin. *The Tigers are Burning*. New York: Hawthorn Books, 1974.

Carhart, Thomas. *Iron Soldiers*. New York: Pocket Books, 1994.

Carver, R.M.P. *The Apostles of Mobility: The Theory and Practice of Armored Warfare*. New York: Holmes & Meier, 1979.

Chamberlain, Peter, and Chris Ellis. *British and American Tanks of World War II*. New York: Arco, 1969.

Chaney, Otto Preston. *Zhukov*. Norman: University of Oklahoma Press, 1996.

Chant, Christopher. *World Encyclopedia of the Tank: An international history of the armoured fighting machine*. London: Patrick Stephens, 1994.

Chapman, Anne W. *The Origins and Development of the National Training Center, 1976-1984*. Fort Monroe, Va.: U.S. Army Training and Doctrine Command (TRADOC), 1992.

Chynoweth, Bradford Grethen. *Bellamy Park: Memoirs By*. Hicksville, N.Y.: Exposition Press, 1975.

Clancy, Tom, with Fred Franks Jr. *Into the Storm: A Study in Command*. New York: G.P. Putnam's Sons, 1997.

Clausewitz, Carl von. *On War*. Edited and translated by Michael Howard and Peter Paret. Princeton, N.J. Princeton University Press, 1976.

Clifford, Kenneth J. *Progress and Purpose: A Developmental History of the United States Marine Corps, 1900-1970*. Washington: History and Museums Division, Headquarters, USMC, 1973.

Coffman, Edward M. *The War to End All Wars: The American Military Experience in World War I*. New York: Oxford University Press, 1968. Reprint, Lexington: University Press of Kentucky, 1998.

Cole, Hugh M. *The Ardennes: Battle of the Bulge. U.S. Army in World War II*. Washington: OCMH, 1965.

Collins, J. Lawton. *War in Peacetime*. Boston: Houghton Mifflin, 1969.

Cooper, Belton Y. *Death Traps: The Survival of an American Armored Division in World War II*. Novato, Calif.: Presidio Press, 1998.

Cordesman, Anthony H., and Abraham R. Wagner. *The Lessons of Modern War*. Vol. 4, *The Gulf War*. New York: Westview Press, 1996.

Corum, James S. *The Roots of Blitzkrieg: Hans von Seecht and German Military Reform*. Lawrence: University Press of Kansas, 1992.

Crismon, Fred W. *U.S. Military Tracked Vehicles.* Osceola, Wisc.: Motorbooks International, 1992.

Croizat, Victor J. *Across the Reef: the Amphibian Tracked Vehicle at War.* Quantico, Va.: Marine Corps Association, 1992 reprint edition.

Crow, Ducan, and Robert J. Icks. *Encyclopedia of Tanks.* Secaucus, N.J. Chartwell Books, 1975.

Crowell, Benedict, and Robert Forest Wilson. *The Armies of Industry: Our Nation's Manufacturing of Munitions for a World in Arms 1917-1918.* Vol. 1. New Haven, Conn.: Yale University Press, 1921.

Dastrup, Boyd L. *King of Battle: A Branch History of the U.S. Army's Field Artillery.* Fort Monroe, Va.: TRADOC, 1992.

Davidson, Phillip B. *Vietnam at War: The History, 1945-1975.* Novato, Calif.: Presidio Press, 1988.

Dayan, Moshe. *Diary of the Sinai Campaign.* New York: Harper & Row, 1966.

D'Este, Carlo. *Patton: A Genius for War.* New York: HarperCollins, 1995.

De Gaulle, Charles. *The Army of the Future.* London: Hutchinson, n.d.

Dietrich, Steve D., and Bruce R. Pirnie. *Developing the Armored Force.* Washington: OCMH, 1989.

Donnelly, Thomas M, and Sean Naylor, *Clash of Chariots: The Great Tank Battles.* Edited by Walter J. Boyne. New York: Berkley Books, 1996.

Doubler, Michael D. *Busting the Bocage: American Combined Arms Operations in France, 6 June-31 July 1944.* Fort Leavenworth, Kans.: U.S. Army CGSC, 1988.

———. *Closing With the Enemy: How GIs Fought the War in Europe, 1944-1945.* Lawrence: University Press of Kansas, 1994.

Doughty, Robert Allan. *The Seeds of Disaster: The Development of French Army Doctrine 1919-1939.* Hamden, Conn.: Archon Books, 1985.

Dunnigan, James F., and Austin Bay. *From Shield to Storm: High-Tech Weapons, Military Strategy, and Coalition Warfare in the Persian Gulf.* New York: William Morrow, 1992.

Dunstan, Simon. *Vietnam Tracks: Armor in Battle, 1945-1975.* Novato, Calif.: Presidio Press, 1982.

Eisenhower, Dwight D. *At Ease: Stories I Tell to Friends.* New York: Doubleday, 1967.

Fehrenbach, T.R. *This Kind of War.* New York: Macmillan, 1963. Reprint, Washington: Center of Military History (CMH), 1989.

Fletcher, David *The Great Tank Scandal: British Armour in the Second World War.* Part 1. London: HMSO, 1989.

———. *Mechanized Force: British tanks between the wars.* London: HMSO, 1991.

Forty, George. *United States Tanks of World War II in Action.* Poole, Dorset, U.K.: Blandford Press, 1983.

———. *Tank Commanders: Knights of the Modern Age.* Osceola, Wisc.: Motorbooks International, 1993.

———. *Tank Action: From the Great War to the Gulf.* London: Alan Sutton, 1995.

———. *Land Warfare.* London: Arms and Armour Press, 1997.

Foss, Christopher F. *Jane's World Armoured Fighting Vehicles.* New York: St. Martin's Press, 1976.

Frank, Benis M., and Henry I. Shaw Jr. *Victory and Occupation.* History of U.S. Marine Corps Operations in World War II. Washington: Headquarters, USMC, 1968.

Frank, Richard B. *Guadalcanal: the Definitive Account of the Landmark Battle.* New York: Random House, 1990.

Freedman, Lawrence, and Efraim Karsh. *The Gulf Conflict 1990-1991: Diplomacy and War in the New World Order.* Princeton, N.J.: Princeton University Press, 1993.

Friedman, Norman. *Desert Victory: The War for Kuwait.* Annapolis, Md.: Naval Institute Press, 1991.

Fuller, J.F.C. *Tanks in the Great War.* London: John Murray, 1920.

———. *The Reformation of War.* London: Hutchinson, 1923.

———. *Memoirs of an Unconventional Soldier.* London: Nicholson and Watson, 1936.

———. *Decisive Battles of the U.S.A.* New York: Thomas Yoseloff, 1942.

———. *Armored Warfare: An Annotated Edition of Lectures on F.S.R. III.* Harrisburg, Pa.: Military Service, 1943.

———. *Armaments and History: A Study of the Influence of Armaments on History from the Dawn of Classical Warfare to the Second World War.* New York: Charles Scribner's Sons, 1945.

Furgurson, Ernest B. *Westmoreland: The Inevitable General.* Boston: Little, Brown, 1968.

Gabel, Christopher R. *Seek, Strike, and Destroy: U.S. Army Tank Destroyer Doctrine in World War II.* Fort Leavenworth: Combat Studies Institute, Sept. 1985.

———. *The U.S. Army GHQ Maneuvers of 1941.* Washington: CMH, 1991.

Garand, George W., and Truman R. Strobridge. *Western Pacific Operations.* History of U.S. Marine Corps Operations in World War II. Washington: Headquarters, USMC, 1971.

Gehring, Stephen P. *From the Fulda Gap to Kuwait: U.S. Army Europe and the Gulf War.* Washington: CMH, 1998.

Germains, Victor Wallace. *The "Mechanization" of War.* London: Sifton Praed, 1927.

Gillie, Mildred Hanson. *Forging the Thunderbolt: A History of the Development of the Armored Force.* Harrisburg, Pa.: Military Service, 1947.

Glantz, David M. *Soviet Military Operational Art: In Pursuit of Deep Battle.* London: Frank Cass, 1991.

Gorman, Paul. *The Secret of Future Victories.* Fort Leavenworth, Kans.: U.S. Army CGSC Press, 1993.

Goulden, Joseph C. *Korea: The Untold Story of the War.* New York: Times Books, 1982.

Green, Constance McLaughlin, Harry C. Thomson, and Peter C. Roots. *The Ordnance Department: Planning Munitions for War.* U.S. Army in World War II. Washington: OCMH, 1955.

Greenfield, Kent Roberts, Robert R. Palmer, and Bell I. Wiley. *The Organization of Ground Combat Troops.* U.S. Army in World War II. Washington: Historical Division, U.S. Army, 1947.

Greer, Andrew. *The New Breed: The Story of the U.S. Marines in Korea.* New York: Harper & Brothers, 1952.

Greer, Thomas H. *The Development of Air Doctrine in the Army Air Corps, 1917-1940.* Washington: Office of Air Force History, 1955 reprint.

Guderian, Heinz. *Panzer Leader.* New York: E.P. Dutton, 1956.

———. *Achtung-Panzer!* Translated by Christopher Duffy. London: Arms and Armour Press, 1992 edition.

Gudgin, Peter. *Armoured Firepower: The Development of Tank Armament, 1939-45.* Gloucestershire, U.K.: Sutton, 1997.

Hammel, Eric. *Chosin: Heroic Ordeal of the Korean War.* New York: Vanguard, 1981. Reprint, Novato, Calif.: Presidio Press, 1990.

Hara, Tomio, and Akira Takeuchi. *Japanese Tanks and Armored Vehicles.* Tokyo: Shuppan Kyodo Sha, 1961.

Harmon, E.N., with Milton MacKaye and William Ross MacKaye. *Combat Commander: Autobiography of a Soldier.* Englewood Cliffs, N.J.: Prentice-Hall, 1970.

Harries, Meirion, and Susie. *Soldiers of the Sun.* New York: Random House, 1991.

Harris, J.P., and F.H. Toase, eds. *Armoured Warfare.* New York: St. Martin's Press, 1990.

Harrison, Gordon A. *Cross-Channel Attack.* U.S. Army in World War II. Washington: OCMH, 1951.

Hay, John H. Jr. *Tactical and Material Innovations.* Washington: Department of the Army, 1974.

Heinl, Robert Debs. *Soldiers of the Sea: The United States Marine Corps, 1775-1962.* Annapolis, Md.: Naval Institute Press, 1962.

Herbert, Paul H. *Deciding What Has to Be Done: General William E. DePuy and the 1976 Edition of FM 100-5, Operations.* Fort Leavenworth, Kans.: Combat Studies Institute, 1988.

Herr, John K., and Edward S. Wallace. *The Story of the U.S. Cavalry.* Boston: Little, Brown, 1953.

Herzog, Chaim. *The War of Atonement.* Boston: Little, Brown, 1975.

Hewes, James E. Jr. *From Root to McNamara: Army Organization and Administration, 1900-1963.* Washington: GPO, 1975.

Hilmes, Rolf. *Main Battle Tanks: Development in Design Since 1945.* Translated by Richard Simpkin. Washington: Brassey's, 1987.

Hogg, Ian. *Tank Killing: Anti-Tank Warfare by Men and Machines.* London: Sidgwick and Jackson, 1996.

Hofmann, George F. *The Super Sixth: History of the 6th Armored Division in World War II and Its Post-war Association.* Louisville, Ky.: Sixth Armored Division Association, 1975.

Holley, Irving. B. Jr. *Ideas and Weapons.* Washington: Office of Air Force History, 1953 reprint.

Holt, Daniel D., and James W. Leyerzapf. *Eisenhower: The Prewar Diaries and Selected Papers, 1905-1941.* Baltimore: Johns Hopkins University Press, 1998.

Hough, Frank O. Verle E. Ludwig, and Henry I. Shaw Jr. *Pearl Harbor to Guadalcanal.* History of U.S. Marine Corps Operations in World War II. Washington: GPO, 1958.

Houston, Donald E. *Hell on Wheels: The 2nd Armored Division.* Novato, Calif.: Presidio Press, 1977.

Howe, George F. *"Old Ironsides": The Battle History of the 1st Armored Division.* Washington: Combat Forces Press, 1954.

———. *Northwest Africa: Seizing the Initiative in the West.* U.S. Army in World War II. Washington: OCMH, 1957.

Howze, Hamilton H. *A Cavalryman's Story: Memoirs of a Twentieth-Century Army General.* Washington: Smithsonian Institution Press, 1996.

Hunnicutt, Richard P. *Pershing: A History of the Medium Tank T20 Series.* Berkeley, Calif.: Feist Publications, 1971.

———. *Sherman: A History of the American Medium Tank.* Belmont, Calif.: Taurus Enterprises, 1978.

———. *Patton: A History of the American Main Battle Tank.* Vol. 1. Novato, Calif.: Presidio Press, 1984.

———. *Firepower: A History of the American Heavy Tank.* Novato, Calif.: Presidio Press, 1988.

———. *Abrams: A History of the American Main Battle Tank.* Vol. 2. Novato, Calif.: Presidio Press, 1990.

———. *Stuart: A History of the American Light Tank.* Vol. 1. Novato, Calif.: Presidio Press, 1992.

———. *Sheridan: A History of the American Light Tank.* Vol. 2. Novato, Calif.: Presidio Press, 1995.

Icks, Robert J. *Famous Tank Battles: From World War I to Vietnam.* Garden City, N.Y.: Doubleday, 1972.

Isely, Jeter A., and Philip A. Crowl. *The U.S. Marines and Amphibious War: Its Theory and Its Practice in the Pacific.* Princeton, N.J.: Princeton University Press, 1951.

James, D. Clayton. *The Years of MacArthur, 1880-1941.* Boston: Houghton Mifflin, 1970.

Johnson, David E. *Fast Tanks and Heavy Bombers: Innovation in the U.S. Army, 1917-1945.* Ithaca, N.Y.: Cornell University Press, 1998.

Jones, Ralph E., George H. Rarey, and Robert J. Icks. *The Fighting Tanks From 1916 to 1933.* Old Greenwich, Conn.: WE, 1933.

Kahalini, Avigdor. *The Heights of Courage.* Westport, Conn.: Greenwood Press, 1984.

Kelly, Orr. *King of the Killing Zone: The Story of the M-1, America's Super Tank.* New York: Norton, 1989.

Krause, Michael D. *The Battle of 73 Easting, 26 Feb. 1991: A Historical Introduction to a Simulation.* Washington: CMH, 1991.

Kreidberg, Marvin A., and Henry G. Henry. *History of Military Mobilization in the United States Army 1775-1945.* Washington: Department of the Army, Nov. 1955.

Krulak, Victor, *First to Fight: An Inside View of the U.S. Marine Corps.* Annapolis, Md.: Naval Institute Press, 1984.

Larson, Robert H. *The British Army and the Theory of Armored Warfare, 1918-1940.* Newark: University of Delaware Press, 1984.

Lejeune, John A. *The Reminiscences of a Marine.* Philadelphia: Dorrance, 1930.

Lewin, Ronald. *Rommel as Military Commander.* London: B.T. Batsford, 1968.

Liddell-Hart, Basil H. *Strategy.* New York: Frederick A. Praeger, 1954.

———. *The Tanks.* 2 vols. New York: Frederick A. Praeger, 1959.

———. Ed. *The Rommel Papers.* New York: Harcourt Brace, 1953.

Lucas, James. *Panzer Army Afrika.* San Rafael, Calif.: Presidio Press, 1977.

Lung, Hoang Ngoc. *ARVN: The General Offensives of 1968-69.* Washington: OCMH, 1981.

MacArthur, Douglas. *Reminiscences.* New York: McGraw Hill, 1964.

MacDonald, Charles B. *A Time for Trumpets: The Untold Story of the Battle of the Bulge.* New York: Morrow, 1985.

Macksey, Kenneth. *Armoured Crusader: A Biography of Major General Sir Percy Hobart.* London: Hutchinson, 1967.

———. *Tank Warfare: A History of Tanks in Battle.* New York: Stein and Day, 1971.

———. *The Tanks: The History of the Royal Tank Regiment, 1945-1975.* London: Arms and Armour Press, 1979.

———. *The Tank Pioneers.* New York: Jane's, 1981.

Macksey, Kenneth, and John H. Batchelor. *Tank: A History of the Armoured Fighting Vehicle.* New York: Charles Scribner's Sons, 1970.

Mahan, Alfred T. *The Influence of Sea Power Upon History, 1660-1783.* Boston: Little, Brown, 1890.

Mahler, Michael D. *Ringed in Steel: Armored Cavalry in Vietnam, 1967-68.* Novato, Calif.: Presidio Press, 1986.

Manstein, Erich von. *Lost Victories.* Chicago: Henry Regnery, 1958.

Margiotta, Franklin D., ed. *Brassey's Encyclopedia of Land Forces and Warfare.* Washington: Brassey's, 1996.

Marshall, S.L.A. *Blitzkrieg.* New York: Morrow, 1940.

———. *The River and the Gauntlet.* New York: Morrow, 1953.

Martell, Giffard Le Q. *In the Wake of the Tank.* London: Sifton Praed, 1931.

———. *An Outspoken Soldier: His Views and Memoirs.* London: Sifton Praed, 1949.

McDonough, James R. *The Defense of Hill 781: An Allegory of Modern Mechanized Combat.* Novato, Calif.: Presidio Press, 1988.

McKercher, B.J.C., and Michael A. Hennessy, eds. *The Operational Art: Developments in the Theories of War.* Westport, Conn.: Praeger, 1996.

McRae, Vincent V., and Alvin D. Coox. *Tank-vs-Tank Combat in Korea.* Washington: Operations Research Office, Johns Hopkins University, 1955.

Meid, Pat, and James M. Yingling. *Operations in West Korea.* Vol. 5. U.S. Marine Operations in Korea, 1950-1953. Washington: Marine Corps Historical Center, 1972.

Mellenthin, F.W. von. *Panzer Battles.* Norman: University of Oklahoma Press, 1956.

Melson, Charles D. *Condition Red: Marine Defense Battalions in World War II.* Washington: History and Museums Division, Headquarters, USMC, 1996.

Mesko, Jim. *Armor in Vietnam.* Carrollton: Squadron/Signal Publications, 1982.

Messenger, Charles. *The Blitzkrieg Story.* New York: Charles Scribner's Sons, 1976.

Michaels, G.J. *Tip of the Spear: U.S. Marine Light Armor in the Gulf War.* Annapolis, Md.: Naval Institute Press, 1998.

Millett, Allan R. *Semper Fidelis: The History of the United States Marine Corps.* New York: Macmillan, 1980.

Millett, Allan R., and Williamson Murray, eds. *Military Effectiveness.* 3 vols. Boston: Allen and Unwin, 1987.

Millis, Walter, ed. *The Forrestal Diaries.* New York: Viking, 1951.

Mitchell, F. *Tank Warfare: The Story of the Tanks in the Great War.* London: Thomas Nelson and Sons, n.d.

Mitchell, Ralph M. *The 101st Airborne Division's Defense of Bastogne.* Fort Leavenworth, Kans.: U.S. Army CGSC, 1986.

Mitchell, William. *Winged Defense: The Development and Possibilities of Modern Air Power—Economic and Military.* New York: G.P. Putnam's Sons, 1925.

Montross, Lynn, and Nicholas A. Canzona. *The Pusan Perimeter.* Vol. 1. U.S. Marine Operations in Korea, 1950-1953. Washington: Marine Corps Historical Center, 1954.

———. *The Inchon Seoul Operations.* Vol. 2. U.S. Marine Operations in Korea, 1950-1953. Washington: Marine Corps Historical Center, 1955.

Montross, Lynn, Hubard D. Kuokka, and Norman W. Hicks. *The East-Central Front.* Vol. 4. U.S. Marine Operations in Korea 1950-1953. Washington: Marine Corps Historical Center, 1962.

Morelock. J.D. *Generals of the Ardennes: American Leadership in the Battle of the Bulge.* Washington: National Defense University Press, 1993.

Mossman, Billy C. *Ebb and Flow November 1950-July 1951.* The U.S. Army in the Korean War. Washington: CMH, 1990.

Murray, Williamson, and Allan R. Millett, eds. *Calculations: Net Assessment and the Coming of World War II.* New York: Free Press, 1992.

Newell, Clayton R., and Michael D. Krause, eds. *On Operational Art.* Washington: CMH, 1994.

Nolan, Keith William. *Into Cambodia.* Novato, Calif.: Presidio Press, 1990. Reprint, New York: Dell, 1991.

Nye, Roger H. *The Patton Mind: The Professional Development of an Extraordinary Leader.* Garden City, N.Y.: Avery Publishing Group Inc., 1993.

Odom, William O. *After the Trenches: The Transformation of U.S. Army Doctrine, 1918-1939.* College Station: Texas A&M University Press, 1999.

Ogorkiewicz, Richard M. *Armoured Forces: A History of Armoured Forces & Their Vehicles.* London: Arms and Armour Press, 1960.

———. *Technology of Tanks.* Vols. 1 and 2. London: Jane's, 1991.

Patton, George S. Jr. *War as I Knew It.* Boston: Houghton Mifflin, 1947.

Patton, Robert H. *The Pattons: A Personal History of an American Family.* New York: Crown, 1994.

Piekalkiewicz, Janusz. *Tank War, 1939-1945.* Poole, Dorset, U.K.: Blandford Press, 1986.

Pierce, Nigel. *The Shield and the Sabre.* London: HMSO, 1992.

Pitt, Barrie. *The Crucible of War: Western Desert, 1941.* London: Jonathan Cape, 1980.

———. *The Crucible of War : Year of Alamein, 1942.* London: Jonathan Cape, 1982.

Reid, Brian H. *J.F.C. Fuller: Military Thinker.* New York: St. Martin's Press, 1987.

Robinett, P. McD. *Armor Command.* Washington: McGregor & Werner, 1958.

Romjue, John L. *From Active Defense to AirLand Battle: The Development of Army Doctrine, 1973-1982.* Fort Monroe, Va.: TRADOC, 1984.

———. *The Army of Excellence: The Development of the 1980's Army.* Fort Monroe, Va.: TRADOC, 1993.

———, Susan Canedy, and Anne W. Chapman. *Prepare the Army for War.* Fort Monroe, Va.: TRADOC, 1993.

Ross, G. Macleod. *The Business of Tanks 1933 to 1945.* London: Arthur H. Stockwell, 1976.

Savkin, V. Ye. *The Basic Principles of Operational Art and Tactics (A Soviet view).* Washington: GPO, 1972.

Scales, Robert H. Jr. *Certain Victory: The U.S. Army in the Gulf War.* Washington: Brassey's, 1994.

Schmidt, Hans, *Maverick Marine: General Smedley D. Butler and the Contra-*

diction of American Military History. Lexington: University Press of Kentucky, 1987.

Schnabel, James F. *Policy and Direction: The First Year.* The U.S. Army in the Korean War. Washington: OCMH, 1972.

Schneider, James J. *The Structure of Strategic Revolution: Total War and the Roots of the Soviet Warfare State.* Novato, Calif.: Presidio Press, 1994.

Schreier, Konrad F. Jr. *The Classic Sherman.* Canoga Park, Calif.: Grenadier Books, 1969.

Schubert, Frank N., and Theresa L. Kraus, eds. *The Whirlwind War: The United States Army in Operations DESERT SHIELD and DESERT STORM.* Washington: CMH, 1995.

Semmes, Harry H. *Portrait of Patton.* New York: Appleton-Century-Croft, 1955.

Shaw, Henry I. Jr., Bernard C. Nalty, and Edwin T. Turnbladh. *Central Pacific Drive.* History and Museums Division, Headquarters, USMC, 1966.

Smith, E. Elberton. *The Army and Economic Mobilization.* U.S. Army in World War II. Washington: OCMH, 1959.

Smith, Holland M., with Percy Finch. *Coral and Brass.* New York: Charles Scribner's Sons, 1949.

Smith, Kevin, and Burton Wright III. *United States Army Aviation During Operations Desert Shield & Desert Storm: Selected Readings.* Fort Rucker, Ala.: U.S. Army Aviation Center, 1993.

Sorley, Lewis. *Thunderbolt: General Creighton Abrams and the Army of His Times.* New York: Simon & Schuster, 1992.

Spiller, Roger J., ed. *Dictionary of American Military Biography.* Westport, Conn.: Greenwood Press, 1984.

Stanton, Shelby L. *Vietnam Order of Battle.* Washington: U.S. News Books, 1981.

———. *World War II Order of Battle.* Novato, Calif.: Presidio Press, 1984.

Starry, Donn A. *Mounted Combat in Vietnam.* Washington: Department of the Army, 1978.

Stern, Albert G. *Tanks 1914-1918: The Log-Book of a Pioneer.* London: Hodder and Stoughton, 1919.

Stockwell, David B. *Tanks in the Wire.* Canton, Ohio: Daring Books, 1989.

Strahan, Jerry E. *Andrew Jackson Higgins and the Boats the Won World War II.* Baton Rouge: Louisiana State University Press, 1994.

Stubbs, Mary Lee, and Stanley Russell Connor. *Armor-Cavalry: Regular Army and Army Reserve.* Pt. 1. Army Lineage Series. Washington: OCMH, 1969.

Stubbs, Mary Lee, and Stanley Russell Connor, with Janice E. McKenney. *Armor-Cavalry: National Guard.* Pt. 2. Army Lineage Series. Washington: OCMH, 1972.

Summers, Harry G. Jr. *On Strategy: A Critical Analysis of the Vietnam War.* Novato, Calif.: Presidio Press, 1982.

———. *Historical Atlas of the Vietnam War.* Boston: Houghton Mifflin, 1995.

Swain, Richard., comp. *Selected Papers of General William E. DePuy.* Fort Leavenworth, Kans.: Combat Studies Institute, 1994.

————. *"Lucky War": Third Army in Desert Storm.* Fort Leavenworth: U.S. Army CGSC Press, 1994.

Swinton, Ernest D. *Eyewitness.* New York: Doubleday, Doran, 1933.

Thomson, Harry C., and Lida Mayo. *The Ordnance Department: Procurement and Supply.* U.S. Army in World War II. Washington: OCMH, 1960.

Toffler, Alvin, and Heidi Toffler. *The Third Wave.* New York: Bantam Press, 1980.

————. *War and Anti-War: Survival at the Dawn of the 21st Century.* New York: Little, Brown, 1993.

Triandafillov, V.K. *The Nature of the Operations of Modern Armies.* Edited by Jacob W. Kipp. Translated by William A. Burhans. London: Frank Cass, 1994.

Trythall, Anthony J. *'Boney' Fuller: The Intellectual General, 1878-1966.* London: Cassell, 1977.

Wallace, Brenton D. *Patton and His Third Army.* Harrisburg: Military Service, 1946.

Watson, Mark. *Chief of Staff: Prewar Plans and Preparations.* U.S. Army in World War II. Washington: OCMH, 1950.

Wedemeyer, Albert C. *Wedemeyer Reports!* New York: Henry Holt, 1958.

Weeks, John. *Men Against Tanks: A History of Anti-Tank Warfare.* New York: Mason/Charter, 1975.

Weigley, Russell F. *History of the United States Army.* New York: Macmillan, 1967. Revised and reprinted, 1984.

————. *The American Way of War: A History of United States Military Strategy and Policy.* New York: Macmillan, 1973.

————. *Eisenhower's Lieutenants: The Campaigns of France and Germany, 1944-1945.* Bloomington: Indiana University Press, 1981.

Williams-Ellis, Clough. *The Tank Corps.* New York: George H. Doran, 1919.

Wilson, Dale E. *Treat 'Em Rough!: The Birth of American Armor, 1917-20.* Novato, Calif.: Presidio Press, 1989.

Wilson, G. Murray. *Fighting Tanks: An Account of the Royal Tank Corps in Action, 1916-1919.* London: Seeley, Service, 1929.

Winton, Harold R. *To Change an Army: General Sir John Burnett-Stuart and British Armored Doctrine, 1927-1938.* Lawrence: University Press of Kansas, 1988.

Woodard, Larry L. *Before the First Wave: The 3rd Armored Amphibian Tractor Battalion—Peleliu and Okinawa.* Manhattan, Kans.: Sunflower University Press, 1994.

Woolcombe, Robert. *The First Tank Battle: Cambrai 1917.* London: Arthur Barker, 1967.

Ye Savkin, V. *The Basic Principles of Operational Art and Tactics (A Soviet View).* Washington: GPO, 1972.

Zaloga, Steven. *U.S. Amphibious Assault Vehicles.* London: Osprey, 1987.

Zumbro, Ralph. *Tank Sergeant.* Novato, Calif.: Presidio Press, 1986.

————. *Tank Aces.* New York: Pocket Books, 1997.

Journal Articles

Abrams, Creighton W. "Armor in the Team." *ARMOR* (July-Aug. 1948).

Alden, Herbert W. "Tanks." *Journal of the Society of Automobile Engineers* (May 1919).

Alexander, Joseph H. "Baptism by Fire: Sherman Tanks at Tarawa." *Leatherneck* (Nov. 1993).

Allmon, William B. "The Ontos." *Vietnam* (Aug. 1994).

Antal, John F. "Tanks at Chipyong-ni." *ARMY* (Mar. 1998).

Association of the U.S. Army. "Special Issue on Armor." *ARMY* (July 1987).

———. "Future Armor Systems: Special Issue." *ARMY* (May 1991).

Augustine, Norman R. "One Plane, One Tank, One Ship: Trend for the Future?" *Defense Management Journal* (Apr. 1975).

Bacevich, A.J., and Robert R. Ivany. "Deployable Armor Today." *Military Review* (Apr. 1987).

Bailey, Alfred D. "An LVT for the Battlefield." *Proceedings* (Nov. 1979).

———. "LVT-7A1 Status Report." *Proceedings* (June 1980).

Bale, E.L. Jr. "Ontos." *Marine Corps Gazette* (Oct. 1957).

Ballendorf, Dirk Anthony. "Earl Hancock Ellis: The Man and His Mission." *Proceedings* (Nov. 1983).

Bateman, Robert L. "Doctrine & Equipment." *ARMY* (Aug. 1997).

———. "Antipathy for the Military: Without Malice, Without Sympathy." *ARMY* (Jan. 1999).

Bellamy, Chris. "The Future of Land Warfare." *Parameters* (Dec. 1987).

Benson, C.C. "Armored Car Design." *Infantry Journal* (Apr. 1929).

———. "The New Christie, Model 1940." *Infantry Journal* (Sept.-Oct. 1929), also in *Army Ordnance* (Sept.-Oct. 1929).

———. "Tank Divisions." *Infantry Journal* (Jan. 1931).

Berry, F. Clifton Jr. "Doctrine Developed in a Vacuum." *Armed Forces Journal International* (Oct. 1976).

Betson, William R. "Sidi Bou Zid—A Case History of Failure." *ARMOR* (Nov.-Dec. 1982).

Boudinot, Burton S. "Sheridan Memoir: The Early Days." *ARMOR* (Jan.-Feb. 1997).

Bourque, Steven A. "Correcting Myths About the Persian Gulf War: The Last Stand of the Tawakalna." *Middle East Journal* (Autumn 1997).

———. "Incident at Safwan." *ARMOR* (Jan.-Feb. 1999).

Brett, Sereno E. "Tank Combat Principles." *Infantry Journal* (Feb. 1925).

———. "Tank Reorganization." *Infantry Journal* (Jan. 1930).

Cameron, Robert S. "Armor Combat Development, 1917-1945." *ARMOR* (Sept.-Oct. 1997).

———. "American Tank Development during the Cold War." *ARMOR* (July-Aug. 1998).

———. "Pushing the Envelope of Battlefield Superiority." *ARMOR* (Nov.-Dec. 1998).

Campbell, Levin H. Jr. "Motor Equipment of the Army." *Infantry Journal* (July 1928).

———. "Special Automotive Equipment of the Army." *Journal of the Society of Automotive Engineers* (Sept. 1928).

Carlisle, Norman. "The Military Inventor the U.S. Scorned." *True* (Aug. 1964).

Chaffee, Adna R. Jr. "James Harrison Wilson, Cavalryman." *Cavalry Journal* (July 1925).

———. "The Seventh Cavalry Brigade in the First Army Maneuvers." *Cavalry Journal* (Nov.-Dec. 1939).

Christmas, John K. "The Mechanization of Armies." *Military Engineer* (July-Aug. and Sept.-Oct. 1929).

———. "Is Mechanization Expensive? Cost Analysis Reveals Great Economy of Machine Warfare." *Army Ordnance* (July-Aug. 1930).

———. "Mechanization in Our Army Today." *Army Ordnance* (July-Aug. 1932).

———. "Tanks and Tactics." *Army Ordnance* (Jan.-Feb. 1937).

———. "The Manufacturing of High Speed Tanks." *Mechanical Engineering* (Jan. 1939).

———. "Tanks: The Ideal Combination of Fire Power, Mobility and Protection." *Army Ordnance* (Jan.-Feb. 1941).

Chynoweth, Bradford G. "Cavalry Tanks." *Cavalry Journal.* (July 1921).

Ciccarelli, John E. "America's Forgotten Tanker." *ARMOR* (July-Aug. 1965).

Clemens, Jon. "Waking Up from the Dream: The Crisis of Cavalry in the 1930s." *ARMOR* (May-June 1990).

Codd, Leo A. "The A.O.A. [American Ordnance Association] Story: the Early Years—1919-1924." *Ordnance* (Sept.-Oct. 1968).

———. "The A.O.A. Story: The Difficult Years—1925-1938." *Ordnance* (Nov.-Dec. 1968).

Connor Arthur W. Jr. "The Armor Debacle in Korea, 1950: Implications for Today." *Parameters* (Spring 1992).

Cossey, Gerald R. "Tank vs. Tank." *ARMOR* (Sept.-Oct. 1970).

Cranston, John. "1940 Louisiana Maneuvers Lead to Birth of Armored Force." *ARMOR* (May-June 1990).

Crawley, Vince. "Ghost Troop's Battle at the 73 Easting." *ARMOR* (May-June 1991).

Croizat, Victor J. "The Marines' Amphibians." *Marine Corps Gazette* (June 1953).

———. "Fifty years of Amphibian Tractors." *Marine Corps Gazette* (Mar. 1986).

Daley, John. "Patton Versus the 'Motor Maniacs': An Interwar Defense of Horse Cavalry." *ARMOR* (Mar.-Apr. 1997).

de Czege, Huba Wass. "How to Change an Army." *Military Review* (Nov. 1984).

DePuy, William E. "11 Men, 1 Mind." *ARMY* (Mar. 1958).

———. "FM 100-5 Revisited." *ARMY* (Nov. 1980).

———. "Toward a Balanced Doctrine." *ARMY* (Aug. 1984).

———. "Concepts of Operation: The Heart of Command, the Tool of Doctrine." *ARMY* (Aug. 1988).

DeVries, Paul T. "Maneuver and the Operational Level of War." *Military Review* (1983).

Dietrich, Steve D. "In-Theater Armored Force Modernization." *Military Review* (Oct. 1993).

Donovan, James A. "Saipan Tank Battles." *Marine Corps Gazette* (Oct. 1948).

Downing, Wayne A. "U.S. Operations Doctrine: A Challenge for the 1980s and Beyond." *Military Review* (Jan. 1981).

Eisenhower, Dwight D. "A Tank Discussion." *Infantry Journal* (Nov. 1920).

Erwin, Sandra I. "Tank Designers, Operators Profess 'Small is Beautiful.'" *National Defense* (Sept. 1998).

Estes, Kenneth W. "LAV—Quo Vadis." *Marine Corps Gazette* (Dec. 1981).

———. "Armor in the Amphibious Force." *Proceedings* (Apr. 1986).

———. "What is Heavy, What is Light?" *Marine Corps Gazette* (July 1987).

———. "Force of Choice." *Armed Forces Journal International* (Sept. 1995).

Field Artillery Board. "Horses, Tractors and Self-Propelled Mounts." *Field Artillery Journal* (Nov.-Dec. 1923).

Field Artillery School. "Implementing the Air Land Battle." *Field Artillery Journal* (Sept.-Oct. 1981).

Fleming, Thomas. "Tanks." *Invention & Technology* (Winter 1995).

Fogeleman, Ronald R. "Aerospace Doctrine: More than Just a Theory." *Airpower Journal* (Summer 1996).

Fontenot, Gregory. "The 'Dreadnoughts' Rip Saddam Line." *ARMY* (Jan. 1992).

———. "Fright Night: Task Force 2/34 Armor." *Military Review* (Jan. 1993).

Franks, Fred Jr. "Full-Dimensional Operations: A Doctrine for an Era of Change." *Military Review* (Dec. 1993).

Franz, W. "Maneuver: The Dynamic Element of Combat." *Military Review* (May 1983).

Fuller, J.F.C. "The Development of Sea Warfare on Land and Its Influence on Future Naval Operations." *Journal of the Royal United Service Institution* (May 1920).

———. "Tactics and Mechanization." *Infantry Journal* (May 1927).

Gabel, Christopher R. "Evolution of U.S. Armor Mobility." *Military Review* (Mar. 1984).

Garretson, Ralph B. Jr. "The Battle of Binh An." *ARMOR* (July-Aug. 1969).

Geier, Richard P. "Gain the Initiative with an Armored Raid." *Military Review* (June 1980).

Gill, Isaac Jr. "Value of Tanks in Action." *Infantry Journal* (Mar. 1921).

Glasgow, Wes, Christopher V. Cardine, and David Letson. "The M1A2 Current and Future Program Plans.: *ARMOR* (May-June 1996).

Goldstein, Jack. "The Army's Big Five: Irons in the Fire." *ARMY* (May 1973).

Grow, Robert W. "Mounted Combat: Lesson from the European Theater." *Cavalry Journal* (Nov.-Dec. 1945).

———. "An Epic of Brittany." *Armored Cavalry Journal* (Mar.-Apr. 1947).

———. "The Capture of Muhlhausen." *Armored Cavalry Journal* (Nov.-Dec. 1947).

———. "The Role of Armor." *ARMOR* (Sept.-Oct. 1961).

———. "The Ten Lean Years." Edited by Capt. Peter Mansoor. 4 Parts. *ARMOR* (Jan.-Feb., Mar.-Apr., May-June, and July-Aug. 1987).

Gunsburg, Jeffery A. "Samuel Dickerson Rockenbach: Father of the Tank Corps." *Virginia Cavalcade* (Summer 1976).

Hacker, Barton C. "Imaginations in Thrall: The Social Psychology of Military Mechanization 1919-1939." *Parameters* (Mar. 1982).

Hatcher, Michael J. "The Tank is Alive and Well." *Military Review* (Feb. 1978).

Heiberg, H.H.D. "Organize a Mechanized Force." *ARMOR* (Sept.-Oct. 1976).

Herr, John K. "Editorial Comment." *Cavalry Journal* (May-June 1946).

Hofmann, George F. "The Demise of the U.S. Tank Corps and Medium Tank Development Program." *Military Affairs* (Feb. 1973).

———. "Tactic vs. Technology." *ARMOR* (Sept.-Oct. 1973).

———. "The Tank that Was Ahead of Its Time." *ARMOR* (Jan.-Feb. 1976).

———. "A Yankee Inventor and the Military Establishment: The Christie Tank Controversy." *Military Affairs* (Feb. 1975).

———. "A Yankee Inventor and the Military Establishment." *ARMOR* (Mar.-Apr. 1976).

———. "A self-made automotive engineer finally convinced the military that an LVT existed in the 1920s." *Marine Corps Gazette* (Sept. 1977).

———. "The United States' Contribution to Soviet Tank Technology." *Journal of the Royal United Service Institute for Defense Studies* (Mar. 1980).

———. "The Troubled History of the Christie Tank." *ARMY* (May 1986).

———. "Robert J. Icks: The Man and a Brief Appraisal of His Military Works." *Military Affairs* (Oct. 1988).

———. "Christie's Last Hurrah: In 1941, the Army Reappraised the Christie Suspension for Use on Tank Destroyers." *ARMOR* (Nov.-Dec. 1991).

———. "Armor History and Operations in 1944: The 6th Armored Division Experience in the European Theater of Operations: A Study in Leadership Development and Execution." *ARMOR* (Sept.-Oct. 1994).

———. "Combatant Arms vs. Combined Arms: The U.S. Army's Quest for Deep Offensive Operations and an Operational Level of Warfare." *ARMOR* (Jan.-Feb. 1997).

———. "Doctrine, Tank Technology, and Execution: I.A. Khalepskii and the Red Army's Fulfillment of Deep Offensive Operations." *Journal of Slavic Military Studies* (June 1996).

———. "The Tactical and Strategic Use of Attaché Intelligence: The Spanish Civil War and the U.S. Army's Misguided Quest for a Modern Tank Doctrine." *Journal of Military History* (Jan. 1998).

Homan, Arthur Lee, and Keith Marvin, with Peter Helck and John M. Peckham. "Not Without Honor: An Account of the Life and Times of John Walter Christie." 2 Parts. *Antique Automobile* (May and July 1965).

Hooker, Richard D. Jr. "Redefining Maneuver Warfare." *Military Review* (Feb. 1992).

Icks, Robert J. "Four Decades of Mechanization: Our Record of Combat-Vehicle Development." *Army Ordnance* (May-June 1937).

Ireland, Merritte W. "MBT/XM803." *ARMOR* (July-Aug. 1970).

Irzyk, Albin F. "The Mystery of 'Tiger Jack.'" *ARMOR* (Jan.-Feb. 1990).

Jones, Philip Dwight. "U.S. Antitank Doctrine in World War II." *Military Review* (Mar. 1980).

Jones, Ralph E. "The Tactical Influence of Recent Tank Developments." *Infantry Journal* (May 1928).

———. "Our Tanks." 4 Parts. *Infantry Journal* (Oct. 1929, Nov. 1929, Dec. 1929, and Jan. 1930).

Jordan, Frank B. "Artillery and Tanks." *Field Artillery Journal* (1924).

Keegan, John. "The Specter of Conventional War." *Harpers* (May 1982).

Kindsvatter, Peter S. "VII Corps in the Gulf War: Deployment and Preparation for Desert Storm." *Military Review* (Jan. 1992).

———. "VII Corps in the Gulf War: Ground Offensive." *Military Review* (Feb. 1992).

———. "VII Corps in the Gulf War: Cease-Fire Operations." *Military Review* (June 1992).

King, James I., and Melvin A. Goers. "Modern Armored Cavalry Organization." *Armored Cavalry Journal* (July-Aug. 1948).

Kutz, C.R. "Break-Through Tanks: Will They Bring Freedom of Action to Armored Divisions?" *Army Ordnance* (Nov.-Dec. 1940).

Lane, Rufus H. "The Mission and Doctrine of the Marine Corps," *Marine Corps Gazette* (Mar. 1923).

Leonard, John W. "The Development of Tanks." *Infantry Journal* (Nov. 1925).

Lind, William S. "Some Doctrinal Questions for the United States Army." *Military Review* (Mar. 1977).

Longhouser, John E. "Converting Computing Power into Combat Power." *Army RD&A* (Mar.-Apr. 1996).

Ludvigsen, Eric C. "XM1 Face-Off to Have Strong Impact on Future Ground Forces." *ARMY* (Feb. 1976).

———. "The Failed Bluff of Task Force Smith: 'An Arrogant Display of Strength.'" *ARMY* (Feb. 1992).

Lyle, James M. "The Army's New Armor Battalion: Aggressive, Super Mobile, Lethal." *ARMY* (July 1983).

Maggart, Lon E. "A Leap of Faith." *ARMOR* (Jan.-Feb. 1992).

Maggart, Lon E., and Gregory Fontenot. "Breaching Operations: Implications for Battle Command and Battle Space." *Military Review* (Feb. 1994).

Mason, James D. "Tracked Landing Vehicles." *Ordnance* (Jan.-Feb. 1972).

McKenney, Janice E. "More Bang for the Buck in the Interwar Army: The 105mm Howitzer." *Military Affairs* (Apr. 1978).

McLaughlin, William P. "The Assault Amphibian Vehicle (AAV): Its Past, Present and Future." *ARMOR* (Apr.-May 1993).

Meadows, Sandra I. "Army Vehicle Programs Breed Information Age Warwagons." *National Defense* (Sept. 1996).

Melton, Stephen L. "The Future of Armor." *ARMOR* (May-June 1990).

Miller, H.W. "After the Tank, What? The Armored Vehicle May Have Reached Its Limit of Development." *Army Ordnance* (Jan.-Feb. 1944).

Miller, Martin J. Jr., and Konrad F. Schreier Jr. "Revolution in Tank Armament." *ARMY* (Mar. 1971).

Mills, Donna. "Thumbs Up for the M1A1!" *Soldiers* (Oct. 1991).

Mountcastle, John W. "Command and Control of Armor Units in Combat." *Military Review* (Nov. 1985).

Nenninger, Timothy K. "The Development of American Armor, 1917-1940: The World War I Experience." *ARMOR* (Jan-Feb. 1969).

———. "The Development of American Armor, 1917-1940: The Tank Corps Reorganized." *ARMOR* (Mar.-Apr. 1969).

———. "The Development of American Armor, 1917-1940: The Experimental Mechanized Forces." *ARMOR* (May-June 1969).

———. "The Development of American Armor, 1917-1940: A Revised Mechanization Policy." *ARMOR* (Sept.-Oct. 1969).

———. "Creating Officers: The Leavenworth Experience, 1920-1940." *Military Review* (Nov. 1989).

———. "Leavenworth and Its Critics: The U.S. Army Command and General Staff School, 1920-1940." *Journal of Military History* (Apr. 1994).

Nihart, Brooke. "Main Battle Tank Still in Trouble with Congress, OSD." *Armed Forces Journal* (21 June 1971).

———. "Armored Cars and the Marine Corps." *Fortitudine* (Fall 1991).

Ogorkiewicz, Richard M. "Evolution of the Amphibious Tank." *Marine Corps Gazette* (Aug. 1957).

———. "A Battle Tank for the 1970s: A U.S.-German Design with Innovations." *The Engineer* (2 Feb. 1968).

———. "Tanks in Tomorrow's Armies." *Military Review* (Feb. 1974).

———. "Composite Armour." *Composites* (Apr. 1976).

———. "LVTP7A1: Latest Tracked Landing Vehicle." Tracked Armoured Vehicles Series in *International Defense Review* (1982).

———. "Mechanized Infantry." *Military Review* (Aug. 1994).

Olinger, Mark A. "Too Late for the War: The U.S. Industrial Base and Tank Production 1950-1953." *ARMOR* (May-June 1997).

Owen, Stanley. "Marine Corps M1A1 Tanks in Operation Desert Storm." *Amphibious Warfare Review* (Summer-Fall 1992).

Patton, George S. Jr. "Tanks in Future Wars." *Infantry Journal* (May 1920).

———. "Comments on 'Cavalry Tanks.'" *Cavalry Journal* (July 1921).

Patton, George S. Jr., and C.C. Benson. "Mechanization and Cavalry." *Infantry Journal* (Jan.-June 1930).

Paulson, John C. "M1A2 Abrams Tank Trials in Southwest Asia." *ARMOR* (May-June 1993).

Pengilly, H.E. "Christie Motor Carriages." *Army Ordnance* (Mar.-Apr. 1922).

Polk, James H. "We Need a New Tank." *ARMY* (June 1972).

Reber, John J. "Pete Ellis: Amphibious Warfare Prophet." *Proceedings* (Nov. 1977).

Reilly, Henry J. "Proving Ground in Spain: Armament Trends as Revealed by the Spanish War." *Army Ordnance* (May-June 1939).

Richardson, William R. "FM 100-5: The AirLand Battle in 1986." *Military Review* (Aug. 1982).

Rockenbach, Samuel D. "Tanks and Their Cooperation with Other Arms." *Infantry Journal* (Jan. and Feb. 1920).

———. "Weight and Dimensions of Tanks." *Infantry Journal* (July 1922).

Saint, Crosbie E. "War Adds New Dimensions to Europe's Role." *ARMY* (Oct. 1991).

Sanders, Richard. "Mechanization Increases Industry's War-Time Task." *American Machinist* (19 Nov. 1931).

Sandrart, Hans-Henning von. "Forward Defense-Mobility and the Use of Barriers." *NATO's Sixteen Nations* (Feb.-Mar. 1985).

Schreier, Fred. "The XM1 Program-A Synopsis." *International Defense Review* (Mar. 1977).

Schreier, Konrad F. Jr. "U.S. Army Tank Development, 1925-1940." *ARMOR* (May-June 1990).

Scott, C.L. "The Armored Force: Newest Fast Moving Power of Our Army." *Army Ordnance* (May-June 1941).

Sendak, Theodore T. "The Airborne-Antiarmor Defense." *Military Review* (Sept. 1979).

Sheridan, Stan R. "Chariots of Fire: Building of Bradley Fighting Vehicle." *ARMOR* (Jan.-Feb. 1999).

Sinnreich, Richard Hart. "Tactical Doctrine of Dogma." *ARMY* (Sept. 1979).

Smith, Eugene Ferry. "Ideas for a Tank." *Infantry Journal* (Nov. 1930).

Smith, Gene. "The Seventeenth Largest Army." *American Heritage* (Dec. 1992).

Starry, Donn A. "A Tactical Evolution—FM 100-5." *Military Review* (Aug. 1978).

———. "Extending the Battlefield." *Military Review* (Mar. 1981).

———. "The Principles of War." *Military Review* (Sept. 1981).

———. "To Change an Army." *Military Review* (Mar. 1983).

———. "Combined Arms." *ARMOR* (May 1987).

———. "A Perspective on American Military Thought." *Military Review* (July 1989).

———. "The Profession of Arms in America." *Encyclopedia of the American Military* (1994).

Stevenson, F.A. "Mass Production of Combat Tanks: A Case History of Industrial Preparedness at Work." *Army Ordnance* (Mar.-Apr. 1941).

Stokes, Orville T. Jr. "After the Tank, Then What?" *Military Review* (Oct. 1980).

Sutton, Boyd D., John R. Landry, Malcolm B. Armstrong, Howell M. Estes, and Wesley K. Clark. "Deep Attack Concepts and the Defense of Central Europe." *Survival* (Mar.-Apr. 1984).

Tennison, Debbie C. "The Foundry Industry—Achilles' Heel of Defense?" *National Defense* (Mar.-Apr. 1976).

Thompson, Kris P. "Trends in Mounted Warfare: The Birth of Mounted Warfare in the United States Army." *ARMOR* (May-June 1998).

———. "Trends in Mounted Warfare: Blitzkrieg and the Operational Level of War." *ARMOR* (July-Aug. 1998).

———. "Trends in Mounted Warfare: Korea, Vietnam, and Desert Storm." *ARMOR* (Sept.-Oct. 1998).

Wagner, Robert E. "Active Defense and All That." *Military Review* (Aug. 1980).

———. ""The 'V' Maneuver Technique." *ARMOR* (Mar.-Apr. 1981).

Walker, Jim. "Vietnam: Tanker's War." *ARMOR* (May-June 1997).

Westervelt, William I. "A Challenge to American Engineers." *Army Ordnance*, 1 (1920).

Wik, Reynold M. "The American Farm Tractor as Father of the Military Tank." *Agricultural History* (Jan. 1980).

Williams, C.C. "Post-War Artillery: Developments in Large and Small Guns since Warring Days of 1918." *Scientific American* (Nov. 1922).

Woodmansee, John W. "Blitzkrieg and the AirLand Battle." *Military Review* (Aug. 1984).

USMC OCCASIONAL PAPERS

Bailey, Alfred Dunlop. *"Alligators, Buffaloes, and Bushmasters: The History of the Development of the LVT through World War II."* Washington: History and Museums Division, Headquarters, USMC, 1986.

Bartlett, Merrill M. *"John Archer Lejeune, 1869-1942: Register of His Personal Papers."* Washington: History and Museums Division, Headquarters, USMC, 1988.

Smith, Holland M. *"The Development of Amphibious Tactics in the U.S. Navy."* Washington: History and Museums Division, Headquarters, USMC, 1992.

ABOUT THE EDITORS

GEORGE F. HOFMANN served in the U.S Army as an instructor and cadre in the Special Training Regiment at the U.S. Armor Training Center, and is a Distinguished Member of the 13th Armored Regiment. He holds master's degrees from Xavier University and the University of Cincinnati, where he also earned a doctorate in history. In addition, he completed a postdoctoral program in military law and justice at the University of Cincinnati College of Law. He is a frequent contributor to *ARMOR* and the *Journal of Military History* (formerly *Military Affairs*). He also wrote articles for *ARMY*, the *Marine Corps Gazette*, the *Journal of the Royal United Services Institute for Defense Studies (RUSI)*, and the *Journal of Slavic Military Studies*. He is a contributing author to the *Dictionary of American Biography;* volume 3 of the Marine Corps University's *Perspective on Warfighting;* and author of *Super Six: A History of the 6th Armored Division* (1975) and *Cold War Casualty: The Court-Martial of Major General Robert W. Grow* (1993). Currently he is an adjunct associate professor of history at the University of Cincinnati, where he teaches courses in military history. He continues to serve by promoting local veteran activities, which included acquiring a M60A3 tank for the township's Veterans Park.

DONN A. STARRY, General, U.S. Army (Retired), enlisted in the army in World War II, won an appointment to and graduated from West Point, and subsequently commanded armor units from platoon through corps. He served two tours in Vietnam and commanded the famous 11th Armored Cavalry Regiment in the incursion into Cambodia in May 1970. In succession, he commanded the Armor Center and School at Fort Knox, the U.S. V Corps in USAREUR, TRADOC, and the U.S. Readiness Command, the predecessor of the U.S. Central Command of Desert Storm fame. He was the principal army staff designer of the post-Vietnam army force structure, then in the succession of commands noted, was primary architect of AirLand Battle—the army and joint force doctrine, equipment, organization, training, and education so dramatically successful in Desert Storm. He is the author of *Mounted Combat in Vietnam* (1977), and of more than a hundred articles for professional journals and encyclopedia. He is a graduate of the U.S. Army CGSC, the Armed Forces Staff College, and the Army War College. He holds a master's degree in International Affairs from George Washington University. He is chairman of the board of the U.S. Cavalry Memorial Foundation and Cavalry Association, and continues to serve as honorary colonel of the 11th Armored Cavalry Regiment, which is currently serving as the Opposing Force at the National Training Center.

ABOUT THE CONTRIBUTORS

JOSEPH H. ALEXANDER, Colonel, USMC (Retired), served twenty-eight years as an assault amphibian officer in the Marine Corps. He holds an undergraduate degree in history from North Carolina and master's degrees from Jacksonville and Georgetown universities. He is also a distinguished graduate of the Naval War College. He commanded a company in Vietnam and a battalion in Okinawa. As colonel, he was chief of staff of the 3d Marine Division in the western Pacific and director of the Research and Development Center. He is the author of *Utmost Savagery: Three Days of Tarawa* (1995), *Storm Landings: Epic Amphibious Battles in the Central Pacific* (1997), and *A Fellowship of Valor: The Battle History of the United States Marines* (1997). As chief historian for Lou Reda Productions, he has created fifteen television documentaries for the Arts and Entertainment Network and the History Channel.

PHILIP L. BOLTÉ, Brigadier General, U.S. Army (Retired), graduated from the U.S. Military Academy in 1950. He is also a graduate of the Canadian Army Staff College and the U.S. Army War College, and holds a master's degree in Electrical Engineering from the Georgia Institute of Technology. During his thirty years of army service he commanded armor units from platoon to brigade. He served in combat and was wounded in both Korea and Vietnam. Other assignments include service on the army staff and in the army secretariat; as assistant PM for tank main armament in the Abrams tank program; and as PM, Bradley Fighting Vehicle Systems. He is a frequent contributor to *ARMOR* magazine. He is the president of the United States Cavalry Association and serves on the board of the Patton Museum foundation.

STEPHEN A. BOURQUE is a graduate of the U.S. Army CGSC and completed his master's degree at Ball State University and his doctorate at Georgia State University. From 1975-91 he served in various command and staff assignments in cavalry and armor units, as a CGSC tactics instructor, and as assistant operations officer of the 1st Infantry Division during the 1991 Persian Gulf War, where he coordinated deployment and redeployment operations and managed the tactical operations center. His recent publication "Correcting Myths About the Persian Gulf War: The Last Stand of the Tawakalna" appeared in the *Middle East Journal*, Autumn 1997. He recently completed a forthcoming book, *JAYHAWK: The VII Corps in Operation Desert Storm*, for the U.S. Army Center of Military History. He currently teaches history at California State University, Northridge.

OSCAR C. DECKER, Major General, U.S. Army (Retired), is a graduate of the U.S. Army CGSC, the Armed Forces Staff College, and the Navy War College. He holds a master's degree in International Affairs from George Washington University. During World War II he served as an enlisted man in a tank battalion. After being commissioned in 1951, he served in a tank battalion and a reconnaissance battalion before reverting to his basic branch, Ordnance. Subsequent assignments enabled him to stay closely aligned with armor systems and activities during his military career, which ended with command of the U.S. Army TACOM. He holds the Distinguished Service Medal, the Legion of Merit with three oak leaf clusters, and was inducted into the Ordnance Hall of Fame for his work with tanks.

KENNETH W. ESTES, Lieutenant Colonel, USMC (Retired), is a 1969 Naval Academy graduate who earned a master's degree in history from Duke University and a doctorate in Modern European History from the University of Maryland. He has taught at Duke University, the Naval Academy, and the Center for International Studies in Madrid, Spain. He held a variety of command and staff assignments in the Marine Corps, commanding tank, headquarters, and service companies; and served as both an operations and logistics staff officer in the Marine Corps's 2d Tank Battalion from 1970 to 1981. During that period, he qualified crews in both the M48A3 and M60A1 as tank commander, and trained in the M103 and M67A2 tanks. He is the editor of the *Marine Officer's Guide* and the *Handbook for Marine NCOs*, and has written extensively in military journals. In 1992 he was made an honorary legionnaire in the Spanish Legion.

CHRISTOPHER R. GABEL is a professor of history on the faculty of the Combat Studies Institute, U.S. Army CGSC, where he teaches survey courses, conducts a seminar on interwar armies, and leads staff rides to Civil War battlefields. He earned his master's degree and doctorate at Ohio State University. His publications include *The U.S. Army GHQ Maneuvers of 1941* (1991), and *Seek, Strike, and Destroy: U.S. Army Tank Destroyer Doctrine in World War II* (1985), plus articles and book reviews in *Military Affairs*, *Military Review*, and *Reviews in American History*.

TIMOTHY K. NENNINGER completed his master's and doctoral degrees in American military history at the University of Wisconsin-Madison. Excerpts from his master's thesis, "The Development of American Armor, 1917-1940," were published in *ARMOR* in 1969. In addition, he has published *The Leavenworth Schools and the Old Army, 1888-1918* (1978), and is the author of "Leavenworth and Its Critics: The U.S. Army Command and General Staff School, 1920-1940," which appeared in the *Journal of Military History* (April 1994). Since 1970 he has been an archivist at the National Archives, where he is currently the chief of the Modern Military Records Division. In 1987-88 he was the John F. Morrison Professor of Military History at the U.S. Army CGSC. In 1996 he was elected vice president of the Society for Military History.

LEWIS SORLEY, a former U.S. Army armor officer and civilian official at the CIA, is a graduate of the U.S. Military Academy. He holds a master's degree from the University of Pennsylvania and a doctorate from Johns Hopkins University. His army assignments included leadership of tank and armored cavalry units in Germany, Vietnam, and the United States; service on the faculties at West Point and the Army War College; and staff positions in the offices of the Secretary of Defense and the Army Chief of Staff. He is the author of *Thunderbolt: General Creighton Abrams and the Army of His Times* (1992) and *Honorable Warrior: General Harold K. Johnson and the Ethics of Command* (1998).

ROBERT J. SUNELL, Major General, U.S. Army (Retired), is a graduate of the U.S. Marine Corps Command and Staff College and the Army War College. He holds a master's degree in communications from Shippensburg State College. He was the deputy director, Cavalry Scout Study; director, Armored Family of Vehicles Task Force; deputy PM, XM1 Tank; commander of the 11th Armored Cavalry Regiment; commanding general of the Army Training Support Center during the development of the National Training Center; and PM, U.S. Army Tank Systems. He has been on various steering committees, including the Center for Strategic and International Studies, the Defense Industrial Base Study, and the future of Armor/Anti-Armor Warfare. Recently he was appointed as a consultant to the Army Science Board's Tank Modernization Study and to a three-year term as a member of the Association of the United States Army's Advisory Board of Directors and assigned to the Land Warfare Committee.

RICHARD M. SWAIN, Colonel, U.S. Army (Retired), is a graduate of the U.S. Military Academy and U.S. Army CGSC. He completed his senior service college education as an Advanced Operational Studies Fellow at the SAMS at Fort Leavenworth and holds master's and doctoral degrees in history from Duke University. He has served on the History Department faculty at West Point, as a seminar leader at SAMS, and as director of the Combat Studies Institute at the U.S. Army CGSC. In 1991 he was appointed theater army historian for Third Army during Operation Desert Storm. He wrote an account of the war titled *"Lucky War": Third Army in Desert Storm* (1994). At present he is director of Fellows at the SAMS.

DIANE L. URBINA served as a Department of the Army TACOM civilian employee in Warren, Michigan. She spent more than twelve years specializing in U.S. Army weapons system acquisition and program management, where she gained considerable knowledge on the development of the Bradley Fighting Vehicle. She has undergraduate and graduate degrees from Texas A&M University. Currently she is employed as a consultant with Radian, Incorporated, in Warren, Michigan.

DALE E. WILSON, Major, U.S. Army (Retired), is an independent scholar and freelance editor and teaches graduate courses for American Military University. He was an assistant professor of military history at the U.S. Military Academy after earning his master's and doctoral degrees at Temple University. He is the author of *Treat 'Em Rough!: The Birth of American Armor, 1917-20* (1989), and was awarded the Society for Military History's Moncado Prize for his July 1992 article "Recipe for Failure: Major General Edward M. Almond and Preparation of the U.S. 92d Infantry Division for Combat in World War II," published in the *Journal of Military History.* An enlisted infantryman and combat correspondent in Vietnam, he was commissioned in armor after completing Officer Candidate School in 1979. His tank company was awarded the Draper Armor Leadership Award at Fort Carson, Colorado, in FY 1985, and he subsequently received the Silver Medallion of the Order of Saint George from the United States Armor Association.

Index

Note: f *following page number indicates figure;* t *indicates table.*